OXFORD REFEREN

# A Concise Dictionary of Business

# A Concise Dictionary of Business

OXFORD NEW YORK
OXFORD UNIVERSITY PRESS

*Oxford University Press, Walton Street, Oxford* OX2 6DP
*Oxford New York Toronto*
*Delhi Bombay Calcutta Madras Karachi*
*Petaling Jaya Singapore Hong Kong Tokyo*
*Nairobi Dar es Salaam Cape Town*
*Melbourne Auckland*
*and associated companies in*
*Berlin Ibadan*

*Oxford is a trade mark of Oxford University Press*



*First published 1990 as an Oxford University Press paperback*
*and simultaneously in a hardback edition*
*Paperback reprinted 1991*



*British Library Cataloguing in Publication Data*
*A concise dictionary of business*
*1. Economics*
*I. Oxford Reference*
*330*
*ISBN 0–19–211667–3*
*ISBN 0–19–285231–0 pbk*

*Text prepared by*
*Market House Books Ltd., Aylesbury*
*Printed in Great Britain by*
*Clays Ltd.*
*Bungay, Suffolk*

# Preface

*A Concise Dictionary of Business* is intended for students of all kinds of business courses, ranging from GCSE Business Studies to the various degree and postgraduate courses in business and allied subjects. The dictionary has also been prepared on the assumption that it will be of use to businessmen and their professional advisers (lawyers, bankers, accountants, advertising agents, insurers, etc.)

At this range of levels the dictionary provides an extensive coverage of the terms commonly used in commerce; dealings in stocks, shares, commodities, and currencies; taxation and accountancy; marketing and advertising; shipping; insurance; business management and policy; personnel management and industrial relations; and banking and international finance. It also has a more restricted coverage of the law and of economics as they apply to businessmen, the terms a businessman might be expected to know in the field of computers, and a smattering of the vocabulary of printers.

In compiling this book the contributors and editors have attempted to make the entries as concise and comprehensible as possible; they have made a particular point of avoiding jargon in the definitions, although most of the terms used in businessmen's jargon are themselves entries. Indeed a feature of the book is its wide coverage of the new City jargon.

Another feature of the book is its extensive network of cross references. An asterisk (*) placed before a word in an entry indicates that this word can be looked up in the dictionary and will provide further explanation or clarification of the entry being read. However, not every word that appears in the dictionary has an asterisk placed before it. Some entries simply refer the readers to another entry, indicating either that they are synonyms or that they are most conveniently explained in one of the dictionary's longer articles. Synonyms and abbreviations are usually placed within brackets immediately after the headword.

A.I. 1990

## Editors

(Market House Books Ltd.)

Alan Isaacs BSc, PhD    Elizabeth Martin MA
Hazel Egerton BA    Kate Smith BA
Edmund Wright MA, DPhil

## Contributors

Graham Betts
Barry Brindley BA
S. L. Williams MA, Lecturer in Law, University of Buckingham
Leslie de Chernatony PhD, City University Business School
Joan Gallagher BA, Personnel Consultant
Peter Lafferty MSc
Clive Longhurst ACII
Edward Philips LLB, BCL, Lecturer in Law, University of Buckingham
Anne Stibbs BA
Stefan Szymanski BA, MSc, PhD, London Business School
R. M. Walters MA, FCA, Chartered Accountant
Matthew Wright MA, MPhil

**A1** A description of property or a person that is in the best condition. In marine insurance, before a vessel can be insured, it has to be inspected to check its condition. If it is "maintained in good and efficient condition" it will be shown in *Lloyd's Register of Shipping as 'A' and if the anchor moorings are in the same condition the number '1' is added. This description is also used in life assurance, in which premiums are largely based on the person's health. After a medical examination a person in perfect health is described as "an A1 life".

**AA** Abbreviation for *Advertising Association.

**abandonment** The act of giving up the ownership of something covered by an insurance policy and treating it as if it has been completely lost or destroyed. If the insurers agree to abandonment, they will pay a total-loss claim (*see* actual total loss; constructive total loss). This often occurs in marine insurance if a vessel has run aground in hazardous waters and the cost of recovering it would be higher than its total value and the value of its cargo. It also occurs during wartime when a vessel is captured by the enemy. If the owner wishes to declare a vessel and its cargo a total loss, he gives the insurer a **notice of abandonment**; if, subsequently, the vessel or its cargo are recovered, they become the property of the insurer.

**ABC** Abbreviation for *Audit Bureau of Circulation.

**ABI** Abbreviation for *Association of British Insurers.

**ability-to-pay taxation** A form of taxation in which taxes are levied on the basis of the taxpayers' ability to pay. This form of taxation leads to the view that as income or wealth increases, its marginal utility (its value to its owner) decreases so that progressive rates of tax can be levied on the higher slices. Typical taxes of this sort in the UK are *income tax and *inheritance tax. *Compare* benefit taxation.

**above par** *See* par value.

**above-the-line** **1.** Denoting entries above the horizontal line on a company's *profit and loss account that separates the entries that establish the profit (or loss) from the entries that show how the profit is distributed. **2.** Denoting advertising expenditure on mass media advertising, including press, television, radio, and posters. It is traditionally regarded as all advertising expenditure on which a commission is payable to an *advertising agency. **3.** Denoting transactions concerned with revenue, as opposed to capital, in national accounts. *Compare* below-the-line.

**ABP** Abbreviation for *Associated British Ports.

**absenteeism** Absence from work for which there is no legitimate reason; it is often self-certified sick leave lasting for one day at a time. Most prevalent in large organizations, it can be a major problem. In order to combat it some organizations have introduced flexible working hours, increased annual leave, introduced personal days leave in addition to normal holiday entitlement, and devised incentive schemes for full attendance.

**absolute advantage** The relative efficiency of an individual or group of individuals in an economic activity, compared to another individual or group. Adam Smith (1723–90) proposed that free trade would be beneficial if countries specialized in activities in which they possessed an absolute advantage. While this may be true, it was superseded by the theory of *comparative advantage of David Ricardo (1772–1823), which showed that even if an individual or group (a country, say) possesses an absolute advantage in *all* activities, there could still be gains from trade.

**absorption costing** The process of costing products or activities by taking into account the total costs incurred in producing the product or service, however remote. This method of costing ensures that full costs are recovered provided that goods or services can always be sold at the price implied by full-cost pricing. However, if sales are lost due to the high sales price opportunities may be lost of making some contribution to overheads. *Compare* marginal costing.

**abstract of title** A document used in conveyancing land that is not registered to show how the vendor derived good title. It consists of a summary of certain documents, such as conveyances of the land, and recitals of certain events, such as marriages and deaths of previous owners. The purchaser will check the abstract against title deeds, grants of probate, etc. This document is not needed when registered land is being conveyed, as the land certificate shows good title.

**ACA** Abbreviation for Associate of the *Institute of Chartered Accountants.

**ACAS** Abbreviation for *Advisory Conciliation and Arbitration Service.

**ACC** Abbreviation for *Agricultural Credit Corporation Ltd.

**ACCA** Abbreviation for Associate of the Chartered Association of Certified Accountants. *See* certified accountant; chartered accountant.

**accelerated depreciation** A basis for a form of tax relief available in the UK until 1984 that enabled certain assets of a business organization to be deemed to waste away at a rate faster than that applicable to their normal useful lives. This gave tax relief earlier than would otherwise have been the case and therefore gave some incentive to businesses to invest in assets for which accelerated depreciation was a permitted basis upon which to calculate tax liability.

**accelerator theory** The theory that the level of *investment is proportional to the output of an economy. Based on the observation that investment tends to fluctuate more than the rate of growth of the economy, the theory played a major role in early Keynesian models of the economy. However, it is not clear that there is a fixed capital-output ratio for an economy; investment decisions are based on a variety of factors, such as *interest rates, business confidence, and profitability. The accelerator theory fails to provide an adequate account of these factors.

**acceptance** **1.** The signature on a *bill of exchange indicating that the person on whom it is drawn accepts the conditions of the bill. Acceptance is usually written: "Accepted, payable at . . . (name and address of bank): (*Signature*)". *See also* non-acceptance. **2.** A bill of exchange that has been so accepted. **3.** Agreement to accept the terms of an *offer; for example, the agreement of an insurance company to provide a specified insurance cover or of a trader to accept a specified parcel of goods at the offer price.

**acceptance credit** A means of financing the sale of goods, particularly in international trade. It involves a commercial bank or merchant bank extending credit to a foreign importer, whom it deems creditworthy. An acceptance credit is opened against which the exporter can draw a *bill of exchange. Once accepted by the bank, the bill can be discounted on the *money market or allowed to run to maturity.

**acceptance supra protest (acceptance for honour)** The acceptance or payment of a *bill of exchange, after it has been dishonoured, by a person wishing to save the honour of the drawer or an endorser of the bill.

**accepting house** An institution specializing in accepting or guaranteeing *bills of exchange. A service fee is charged for guaranteeing payment, enabling the bill to be discounted at preferential rates on the *money market. The decline in the use of bills of exchange has forced the accepting houses to widen their financial activities, many of whom have returned to their original function of *merchant banking.

**Accepting Houses Committee** A committee representing the 16 *accepting houses in the City of London. Members of the committee are eligible for finer discounts on bills bought by the Bank of England, although this privilege has been extended to other banks.

**acceptor** The drawee of a *bill of exchange after he has accepted the bill, i.e. has accepted liability by signing the face of the bill.

**access** To obtain data from, or to place data into, a computer memory. *See also* access time.

**access time** The time taken to obtain information from a computer *memory, or to write information to the memory. Access times can vary from a thousand millionth of a second (a nanosecond) in a fast electronic memory to a second or longer with magnetic-tape memories.

**accident insurance** An insurance policy that pays a specified amount of money to the policyholder in the event of the loss of one or more eyes or limbs in any type of accident. It also pays a sum to the dependants of the policyholder in the event of his or her death. These policies first appeared in the early days of railway travel, when passengers felt a train journey was hazardous and they needed some protection for their dependants if they were to be killed or injured.

**accommodation bill** A *bill of exchange signed by a person (the accommodation party) who acts as a guarantor. He is liable for the bill should the *acceptor fail to pay at maturity. Accommodation bills are sometimes known as **windbills** or **windmills**. *See also* kite.

**accord and satisfaction** A device enabling one party to a contract to avoid an obligation that arises under the contract, provided that the other party agrees. The accord is the agreement by which the contractual obligation is *discharged and the satisfaction is the *consideration making the agreement legally operative. Such an agreement only discharges the contractual obligation if it is accompanied by consideration. For example, under a contract of sale the seller of goods may discharge his contractual obligation by delivering goods of different quality to that specified in the contract, provided there is agreement with the buyer (the accord) and he offers a reduction in the contract price (the satisfaction). The seller has therefore 'purchased' his release from the obligation. Accord and satisfaction refer to the discharge of an obligation arising under the law of tort.

**account 1.** A statement of indebtedness from one person to another; an invoice. A provider of professional services or of goods may render an account to his client or customer, and a solicitor selling a house on a person's behalf will render him an account of the sale, which may show that the solicitor owes the seller the proceeds of the sale, less expenses. **2.** A named segment of a ledger recording transactions relevant to the person or the matter named (*see* double-entry book-keeping). Accounts consist of two sides; increases are recorded on one side and decreases on the other. Accounts may be kept in a written form in a ledger, they may be on loose cards, or they may be on magnetic tape in a computer. **3.** An account maintained by a *bank or a *building society in which a depositor's money is kept. *See* cheque account; current account; deposit account. **4.** A period during which dealings on the London Stock Exchange are made without immediate cash settlement. There are 24 accounts in the year, most of two weeks duration, but a few last for three weeks to accommodate public holidays. Up to the end of each account transactions are recorded but no money changes hands, which enables speculators to operate with the minimum capital outlay (for shares bought and sold within an account only differences have to be settled). Settlement of all transactions made within an account is made ten days after the account ends (*see* account day). **5.** In an advertising, marketing, or public-relations agency, a client of the agency from whom a commission or fee is derived, in return for the services. **6.** *See* annual accounts.

**accountant** A person trained to keep books of account, which record all the financial transactions of a business or other organization, and to prepare periodic accounts. The accounts normally consist of a *balance sheet; a *profit and loss account, or in the case of a non-trading organization, an *income and expenditure account; and sometimes a statement of sources and application of funds. Other roles of accountants are to *audit the accounts of organizations and to give advice on taxation and other financial matters. Qualified accountants are normally members of one of several professional bodies to which they gain admission through a period of work experience and examinations (*see* chartered accountant; certified accountant). *See also* cost accountant; financial accountant; management accountant.

**account day (settlement day)** The day on which all transactions made during the previous *account on the London Stock Exchange must be settled. It falls on a Monday, ten days after the account ends.

**account executive** The person in an advertising, marketing, or public-relations agency responsible for implementing a client's business. This involves carrying out the programme agreed between the agency and client, coordinating the activities, and liaising with the client.

**accounting concepts** The basic theoretical ideas devised to support the activity of accounting. As accounting developed largely from a practical base, it has been argued that it lacks a theoretical framework. Accountants have therefore tried to develop such a framework; although various concepts have been suggested, few have found universal agreement. However, four are deemed to be important (*see* statements of standard accounting practice): the first, the **on-going concern concept**, assumes that the business is a going concern until there is evidence to the contrary, so that assets are not stated at their break-up value; the second, the ***accruals concept**, involves recording income and expenses as they accrue, as distinct from when they are received or paid; the third, the **consistency concept**, demands that accounts be prepared on a consistent basis from one period to another; and the fourth, the **prudence concept**, calls for accounts to be prepared on a conservative basis, not taking credit for profits or income before they are realized but making provision for losses when they are foreseen. Other accounting concepts might be *depreciation and *deferred taxation, which are concepts relating to accounting, but which are not often considered when reference is made to accounting concepts.

**accounting cost** *See* cost.

**accounting package** *See* business software package.

**accounting period 1.** The period for which an organization makes up its accounts, usually an annual period. *Profit and loss accounts (or *income and expenditure accounts) are drawn up to cover that period and *balance sheets are prepared for the closing date. It is often a legal requirement that these accounts should be published. However, an organization might make use of shorter accounting periods for internal management purposes, e.g. monthly, quarterly, or six-monthly periods. **2.** An accounting period for tax purposes, as defined by the taxation Acts. For unincorporated businesses it is the period for which the businesses prepare their accounts. For companies it is broadly the same except that if an accounting period exceeds twelve months it is divided into two or more accounting periods of twelve months, each excluding the last.

**accounting rate of return (ARR)** The net profit to be expected from an investment, calculated as a percentage of the book value of the assets invested. *Compare* net present value (NPV).

**accounting technician** A person qualified by membership of an appropriate body (such as the Association of Accounting Technicians) to undertake tasks in the accountancy field without being a fully qualified *accountant.

**account management group** A group within an advertising, marketing, or public-relations agency responsible for planning, supervising, and coordinating all the work done on behalf of a client. In large agencies handling large accounts the group might consist of an account director, account manager, account or media planner, and *account executive.

**account of profits** A legal remedy available as an alternative to *damages in certain circumstances, especially in breach of *copyright cases. The person whose copyright has been breached sues the person who breached it for a sum of money equal to the gain he has made as a result of the breach.

**account payee** The words used in crossing a *cheque to ensure that the cheque is paid into the bank account of the payee only. However, it does not affect the negotiability of the cheque.

**account rendered** An unpaid balance appearing in a *statement of account, details of which have been given in a previous statement.

**accounts 1.** The *profit and loss account and the *balance sheet of a company. **2.** *See* account; books of account.

**account sale** A statement giving details of a sale made on behalf of another person or firm, often as an *agent. The account sale shows the proceeds of the sale less any agreed expenses, commission, etc.

**accounts payable** The amounts due to suppliers of goods and services to an organization. Originally a US term, it is gaining popularity in the UK (*see* trade creditor).

**accounts receivable** The amounts owed to an organization for goods and services that it has supplied. Originally a US term, it is gaining popularity in the UK.

**accrual (accrued charge)** An amount incurred as a charge in a given accounting period but not paid by the end of that period, e.g. the last quarter's electricity charge. *See* accruals concept.

**accruals concept** One of the four principal *accounting concepts. Merely to record cash received or paid would not give a fair view of an organization's profit or loss, since it would not take account of goods sold but not yet paid for nor of expenses incurred but not yet paid. Accordingly, it is considered good accounting practice to prepare accounts taking note of such accruals. This is also akin to the **matching concept**, which suggests that costs should, as far as possible, be matched with the income to which they give rise.

**accrued benefits** The benefits that have accrued to a person in respect of his pension, for the service he has given up to a given date, whether or not he continues in office.

**accrued charge** *See* accrual.

**accumulated depreciation** The total amount written off the value of an asset. It is the sum of the yearly instalments of depreciation since the asset was acquired.

**accumulated dividend** A dividend that has not been paid to a company's preference shareholders. It is, therefore, shown as a liability in its accounts.

**accumulated profits** The amount showing in the *appropriation of profits account that can be carried forward to the next year's accounts, i.e. after paying dividends, taxes, and putting some to reserve.

**accumulating shares** Ordinary shares issued to holders of ordinary shares in a company, instead of a dividend. Accumulating shares are a way of replacing annual income with capital growth; they avoid income tax but not capital-gains tax. Usually tax is deducted by the company from the declared dividend, in the usual way, and the net dividend is then used to buy additional ordinary shares for the shareholder.

**accumulation unit** A unit in an *investment trust in which dividends are ploughed back into the trust, after deducting income tax, enabling the value of the unit to increase. It is usually linked to a life-assurance policy.

**acid-test ratio** *See* liquid ratio.

**ACII** Abbreviation for Associate of the *Chartered Insurance Institute.

***ACORN*** Abbreviation for *A Classification of Residential Neighbourhoods*. This directory classifies 39 different types of neighbourhoods in the UK, assuming that people living in a particular neighbourhood will have similar behaviour patterns, disposable incomes, etc. It is used by companies to provide target areas for selling particular products or services (e.g. swimming pools, double glazing, etc.) or alternatively to exclude areas (particularly finance and insurance-related) from a sales drive. It is also used extensively for selecting representative samples for questionnaire surveys.

**acoustic coupler** A device used to connect a computer to an ordinary telephone set so that the computer can send information along the telephone line and receive information sent along the line by other computers. To send information, the telephone handset is placed in a cradle on the coupler, after the receiving number has been dialled. The output of the computer is then converted by a *modem in the coupler to a sequence of different tones that can be sent over the telephone line. When information is being received, the modem converts the tones from the telephone line into electronic signals that the computer can understand.

**acquisition accounting** The accounting procedures adopted when one company is taken over by another. This often involves controversial issues as to the way in which *goodwill is to be treated.

**across the network** Denoting a TV advertisement or programme series that is broadcast across all ITV regions simultaneously.

**ACT** Abbreviation for *advance corporation tax.

**active partner** A partner who has contributed to the business capital of a partnership and who participates in its management. All partners are deemed to be active partners unless otherwise agreed. *Compare* sleeping partner.

**active stocks** The stocks and shares that have been actively traded, as recorded in the Official List of the London Stock Exchange.

**act of God** A natural event that is not caused by any human action and cannot be predicted. It is untrue (as is sometimes thought) that insurance policies covering homes and businesses exclude acts of God. In fact, both cover such natural events as storms, lightning, and floods. However, some contracts exclude liability for damage arising from acts of God (*see* force majeure).

**act of war** Anything that causes loss or damage as a result of hostilities or conflict. Such risks are excluded from all insurance policies (except life assurances). In marine and aviation insurance only, any extra premium may be paid to include war risks.

**actuals (physicals)** Commodities that can be purchased and used, rather than goods traded on a *futures contract, which are represented by documents. *See also* spot goods.

**actual total loss** The complete destruction or loss of an insured item or one that has suffered an amount of damage that makes it cease to be the thing it originally was. For example, a motor car would be an actual total loss if it was destroyed, stolen and not recovered, or damaged so badly that the repair cost exceeded its insured value. *See also* constructive total loss.

**actuary** A person trained in the mathematics of statistics and probability theory. Some are employed by insurance companies to calculate probable lengths of life and advise insurers on the amounts that should be put aside to pay claims and the amount of premium to be charged for each type of risk. Actuaries also advise on the administration of pension funds; the *government actuary is responsible for advising the government on National Insurance and other state pension schemes. *See also* Institute of Actuaries.

**Ada** A computer language first developed in the USA for controlling military equipment. It is now becoming more widely used for civilian uses. It is a *high-level language based on *Pascal. An advantage of the language is that programs written in Ada are portable (i.e. can be run on different computers). The language is named after Ada Lovelace, who worked with Charles Babbage (1791–1871), one of the important pioneers of computing.

**ADC** Abbreviation for *advice of duration and charge.

**additional personal allowance** An income-tax allowance available, in addition to the *personal allowance, to a single person who has a child under 16 living with him or her or to a married man with such a child and a totally incapacitated wife.

**address** The label, name, or number that identifies the location of a particular piece of data in a computer memory. There are many different types of addresses, with such names as 'absolute address', 'indirect address', and 'base address'. Sometimes the term may be used synonymously with *access.

**adjudication** **1.** The judgment or decision of a court, especially in bankruptcy proceedings. **2.** An assessment by the Commissioners of Inland Revenue of the amount of stamp duty due on a document. A document sent for adjudication will either be stamped as having no duty to pay or the taxpayer will be advised how much is due. An appeal may be made to the High Court if the taxpayer disagrees with the adjudication.

**adjuster** *See* loss adjuster.

**administration order 1.** An order made in a county court for the administration of the estate of a judgment debtor (*see* judgment creditor). The order normally requires the debtor to pay his debts by instalments; so long as he does so, the creditors referred to in the order cannot enforce their individual claims by other methods without the leave of the court. Administration orders are issued when the debtor has multiple debts but it is thought that his *bankruptcy can be avoided. **2.** An order of the court under the Insolvency Act (1986) made in relation to a company in financial difficulties with a view to securing its survival as a going concern or, failing that, to achieving a more favourable realization of its assets than would be possible on a *liquidation. While the order is in force, the affairs of the company are managed by an **administrator**.

**administrative receiver** *See* receiver.

**administrator 1.** Any person appointed by the courts, or by private arrangement, to manage the property of another. **2.** Any person appointed by the courts to take charge of the affairs of a deceased person, who died without making a will. This includes collection of assets, payment of debts, and distribution of the surplus to those persons entitled to inherit, according to the laws of *intestacy. The administrator must be in possession of *letters of administration as proof of his authority.

**ADR** Abbreviation for *American Depository Receipt.

**ad referendum** (Latin: to be further considered) Denoting a contract that has been signed although minor points remain to be decided.

**ADST** Abbreviation for *approved deferred share trust.

**ad valorem** (Latin: according to value) Denoting a tax or commission that is calculated as a percentage of the total invoice value of goods rather than the number of items. For example, *VAT is an ad valorem tax, calculated by adding a fixed percentage to an invoice value.

**advance corporation tax (ACT)** A feature of the *imputation system of taxation that has applied in the UK since 1972. When dividends (or other distributions) are made by UK companies to their shareholders, the companies must account to the Inland Revenue for advance corporation tax at a rate that would equal the basic rate of income tax on a figure consisting of the distribution plus the ACT. Thus, if the basic rate of income tax is 30%, the rate of ACT would be 3/7, so that a dividend of £700 with its ACT of 3/7 × £700 = £300 would total £1000, on which the tax at 30% would be £300 (the amount of the ACT). The ACT thus paid serves two purposes: (1) it is a payment on account of the individual shareholder's personal income tax on the dividend, and (2) for the paying company it constitutes a payment on account of that company's corporation tax for the period in which the dividend is paid. There are limits to the amount of ACT that may be set against corporation tax liabilities for any given period. Unrelieved ACT may also be carried backwards or forwards or surrendered to other companies.

**adverse balance** A deficit on an account, especially a *balance of payments account.

**adverse selection** *See* moral hazard.

**advertising** A communication intended both to inform and persuade. The media that carry advertising range from the press, television, cinema, radio, and posters to company logos on apparel. Advertising creates awareness of a product, extensive advertising creates confidence in the product, and good advertising creates a desire to buy the product. This series of emotions is known mnemonically as AIDA: attention, interest, desire, action. Only half the money spent on advertising in the UK is accounted for by producers of goods and services; the remainder is spent by individuals (mostly on *classified advertising), the government, charities, and marketing intermediaries (e.g. banks, institutions, and retailers). *See* above-the-line; below-the-line; consumer advertising; trade advertising.

**Advertising Association (AA)** An organization representing the interests of advertisers, *advertising agencies, and the media. Founded in 1926, it collects and assesses statistics on advertising expenditure as well as running an annual programme of seminars and training courses for people working in advertising, marketing, and sales promotion.

**advertising agency** A business organization specializing in planning and handling *advertising on behalf of clients. A full-service agency provides a range of services to clients, including booking advertising space, designing and producing advertisements, devising media schedules, commissioning research, providing sales promotion advice, and acting as a marketing consultant. The departments within an agency include research, planning, creative design, media bookings, production, and accounts. Most advertising agents work on the basis of a commission on the total sums spent by the client.

**advertising allowance** A price concession given by a manufacturer of a product to a retailer to allow him to pay for local advertising. It is an effective way of advertising both the product and the retail outlet.

**advertising brief** An agreement between an *advertising agency and a client on the objectives of an advertising campaign. It is important that the client knows exactly what the objectives are, helps to plan the overall strategy, and sets the budgets. Once the brief has been agreed the agency can prepare and evaluate the advertisements themselves and develop the media plan.

**advertising rates** The basic charges made by the advertising media for use of their services or facilities.

**Advertising Standards Authority (ASA)** An independent body set up and paid for by the advertising industry to ensure that its system of self-regulation works in the public interest. The ASA must have an independent chairman, who appoints individuals to serve on the council, two-thirds of which must be unconnected with the advertising industry. The ASA maintains close links with central and local government, consumer organizations, and trade associations. All advertising, apart from television and radio commercials, which are dealt with by the *Independent Broadcasting Authority (IBA), must be legal, decent, honest, and truthful; it must adhere to the **British Code of Advertising Practice (BCAP)**, which provides the rules for all non-broadcast advertising. This applies not only to what it said in an advertisement, but also what is shown. If it is claimed that one bar of chocolate contains ½ pint of milk, then the chocolate bar must contain that amount of milk. If the advertisement makes no claim, but shows a ½ pint bottle or carton of milk, then the chocolate must still contain this amount of milk. The ASA controls the contents of advertisements by continuous monitoring of publications and by dealing with complaints from members of the public.

**advice note** A note sent to a customer by a supplier of goods to advise him that an order has been fulfilled. The advice note may either accompany the goods or be sent separately, thus preceding the *invoice and any *delivery note. The advice note refers to a particular batch of goods, denoting them by their marks and numbers (if more than one package); it also details the date and method of dispatch.

**advice of duration and charge (ADC)** The length of a telephone call and its cost. In the UK a caller who wishes to know the cost of a call after he has made it, dials the operator and asks for an ADC call to a particular number. After the call the operator will ring back with the information.

**advise fate** A request by a collecting bank wishing to know, as soon as possible, whether a cheque will be paid on its receipt by the paying bank. The cheque is sent direct and not through the Bankers' Clearing House, asking that its fate should be advised immediately.

**Advisory Conciliation and Arbitration Service (ACAS)** A UK government body set up in 1975 to mediate in industrial disputes in both the public and private sectors. Its findings are not binding on either side, but carry considerable weight with the government. It consists of a panel of ten members, three each appointed by the TUC and the CBI, who elect three academics and an independent chairman. It does not, itself, carry out arbitrations but may recommend an arbitration to be held by other bodies.

**AFBD** Abbreviation for *Association of Futures Brokers and Dealers.

**affidavit** A sworn written statement by a person (the deponent), who signs it in the presence of a *commissioner for oaths. It sets out facts known to the deponent. In certain cases, particularly proceedings in the Chancery division of the High Court, evidence may be taken by affidavit rather than by the witness appearing in person.

**affirmation of contract** Treating a contract as being valid, rather than exercising a right to rescind it for a good reason. Affirmation can only occur if it takes place with full knowledge of the facts. It may take the form of a declaration of intention, be inferred from such conduct as selling goods purchased under the contract, or allowing time to pass without seeking a remedy.

**afloat** Denoting goods, especially commodities, that are on a ship from their port of *origin to a specified port of destination; for example, "afloat Rotterdam" means the goods are on their way to Rotterdam. The price of such goods will usually be between the price of spot goods and goods for immediate shipment from origin.

**after date** The words used in a *bill of exchange to indicate that the period of the bill should commence from the date inserted on the bill, e.g. "... 30 days after date, we promise to pay ...". *Compare* after sight; at sight.

**after-hours deals** Transactions made on the *London Stock Exchange after its official close at 3.30 pm. Such deals are recorded as part of the following day's trading and are therefore also known as **early bargains**.

**after-sales service** Maintenance of a product by its manufacturer or his agent after it has been purchased. This often takes the form of a guarantee (*see* warranty), which is effective for a stated period during which the service is free in respect of both parts and labour, followed by a maintenance contract for which the buyer of the product has to pay. Efficient and effective after-sales service is an essential component of good marketing policy, especially for such consumer durables as cars and computers; in the case of exported goods it is of overriding importance.

**after sight** The words used in a *bill of exchange to indicate that the period of the bill should commence from the date on which the drawee is presented with it for acceptance, i.e. has sight of it. *Compare* after date; at sight.

**AG** Abbreviation for *Aktiengesellschaft*. It appears after the name of a West German, Austrian, or Swiss company, being equivalent to the British abbreviation plc (i.e. denoting a public limited company). *Compare* GmbH.

**agency 1.** The business carried on by a commercial (or mercantaile) *agent. **2.** The relationship between an agent and his principal.

**agenda** The list of items to be discussed at a business meeting. For the *annual general meeting or an *extraordinary general meeting of a company, the agenda is usually sent to shareholders in advance. *See also* order of business.

**agent** A person appointed by another (the **principal**) to act on his behalf, often to negotiate a contract between the principal and a third party. If an agent discloses his principal's name (or at least the existence of a principal) to the third party with whom he is dealing, the agent himself is not normally liable on the contract. An **undisclosed principal** is one whose existence is not revealed by the agent to a third party; he may still be liable on the contract, but in such cases the agent is also liable. However, an undisclosed principal may not be entitled to the benefit of a contract if the agency is inconsistent with the terms of the contract or if the third party shows that he wished to contract with the agent personally.

Agents are either **general agents** or **special agents**. A general agent is one who has authority to act for his principal in all business of a particular kind, or who acts for the principal in the course of his (the agent's) usual business or profession. A special agent is authorized to act only for a special purpose that is not in the ordinary course of the agent's business or profession. The principal of a general agent is bound by acts of the agent that are incidental to the ordinary conduct of the agent's business or the effective performance of his duties, even if the principal has imposed limitations on the agent's authority. But in the case of a special agent, the principal is not bound by acts that are not within the authority conferred. In either case, the principal may ratify an unauthorized contract. A commercial (or mercantile) agent selling goods on behalf of his principal sometimes agrees to protect his principal against the risk of the buyer's insolvency. Such an agent is called a *del credere agent.

***agent de change*** A stockbroker on the Paris Bourse (*see* bourse).

**age of consent** An age that depends upon the legal circumstances to which it refers. For commercial purposes, it is set at 18 years by the Family Law Reform Act (1969). A *contract entered into by a minor (i.e. someone below the age of consent) is not always capable of being enforced.

**age relief** An additional *personal allowance set against income for income-tax purposes in the UK for both single people and married couples over 65. At the age of 75 both the personal allowance and the married couple's allowance are increased.

**aggregate demand** The sum of demands for all the goods and services in an economy at any particular time. Made a central concept in *macroeconomics by Keynes, it is usually defined as the sum of consumers' expenditure (*see* consumption), *investment, government expenditure, and imports less exports. Keynesian theory (*see* Keynesianism) proposes that the free market will not always maintain a sufficient level of aggregate demand to ensure *full employment and that at such times the government should seek to stimulate aggregate demand (*see* pump priming). However, monetarists and new classical macroeconomists have questioned the feasibility of such policies and this remains a critical issue in macroeconomics. *See also* aggregate supply.

**aggregate supply** The total supply of all the goods and services in an economy. Keynes made *aggregate demand the focus of macroeconomics; however, since the 1970s many economists have questioned the importance of aggregate demand in determining the health of an economy, suggesting instead that governments should concentrate on establishing conditions to encourage the supply of goods and services. This could entail *deregulation, encouraging *competition, and removing restrictive practices in the labour market.

**AGM** Abbreviation for *annual general meeting.

**Agricultural Bank (Land Bank)** A credit bank specifically established to assist agricultural development, particularly by granting loans for longer periods than is usual with commercial banks.

**Agricultural Credit Corporation Ltd (ACC)** A corporation established in 1964 to extend the availability of medium-term bank credit for buildings, equipment, livestock, and working capital to farmers, growers, and co-operatives. The ACC offers a guarantee to the farmer's bank for such loans and promises to repay the bank should the farmer fail to do so. In return for this service the farmer pays a percentage charge to the ACC.

**Agricultural Mortgage Corporation Ltd (AMC)** A corporation established to grant loans to farmers against mortgages on their land by the Agricultural Credits Act (1928). The AMC offers loans for periods of 5 to 30 years. The capital of the corporation is supplied by the Bank of England, the joint-stock banks, and by the issue of state-guaranteed debentures. The corporation's loans are irrevocable except in cases of default and are usually made through the local branches of the commercial banking system.

**AIDA (attention, interest, desire, action)** A 19th-century mnemonic for the progressive steps of customer reaction in the process of making a sale. *See* advertising.

**aids to trade** The formal study of commerce recognizes four aids to trade: advertising, banking, insurance, and transport.

**aid trade provision (ATP)** A major component of the British aid programme, which seeks to combine aid to developing countries with creating business for UK companies. Subsidized loans and credits are offered to developing countries on condition that goods and services are purchased from UK-based enterprises.

**air consignment note** *See* air waybill.

**air date** The date of first transmission of a commercial or an advertising campaign on television.

**air freight 1.** The transport of goods by aircraft, either in a scheduled airliner or chartered airliner carrying passengers (an all-traffic service) or in a freight plane (an all-freight service). Air cargo usually consists of goods that have a high value compared to their weight. **2.** The cost of transporting goods by aircraft. It is usually quoted on the basis of a price per kilogram.

**airspace 1.** The space that lies above a state's land and sea territory and is subject to its exclusive jurisdiction. **2.**

The space above a piece of land. The owner of the land is entitled to the ownership and possession of the airspace above his land. This is not exclusive, however, and is limited by the reasonable and necessary use of that airspace by his neighbours as well as for the flight of aircraft.

**airtime** **1.** The amount of time allocated to an advertisement on radio or television. **2.** The time of transmission of an advertisement on radio or television.

**air waybill (air consignment note)** A document made out by a consignor of goods by *air freight to facilitate swift delivery of the goods to the consignee. It gives the name of the consignor and the loading airport, the consignee and the airport of destination, a description of the goods, the value of the goods, and the marks, number, and dimensions of the packages.

**Aktb** Abbreviation for *Aktiebolaget*. It appears after the name of a Swedish joint-stock company.

**Algol** Acronym for ALGOrithmic Language. This computer programming language is designed for mathematical and scientific use. It is a *high-level language that allows easy translation of algebraic formulae into program instructions. Although never as popular as its contemporary language, *FORTRAN, it has influenced the design of all subsequent programming languages.

**algorithm** A set of well-defined rules for solving a problem in a finite number of steps. Algorithms are extensively used in computer science. The steps in the algorithm are translated into a series of instructions that the computer can understand. These instructions form the computer program.

**allonge** An attachment to a *bill of exchange to provide space for further endorsements when the back of the bill itself has been fully used. With the decline in the use of bills of exchange it is now rarely needed.

**allotment** A method of distributing previously unissued shares in a limited company in exchange for a contribution of capital. An application for such shares will often be made after the issue of a *prospectus on the *flotation of a public company or on the privatization of a state-owned industry. The company accepts the application by dispatching a **letter of allotment** to the applicant stating how many shares he has been allotted; he then has an unconditional right to be entered in the *register of members in respect of those shares. If the number of shares applied for exceeds the number available (oversubscription), allotment is made by a random draw or by a proportional allocation. If an applicant has been allotted fewer shares than he has applied for, he receives a cheque for the unallotted balance (an application must be accompanied by a cheque for the full value of the shares applied for). *See also* multiple application.

**allowance** **1.** A tax-free amount that may be deducted before a particular tax is calculated. Typical allowances in assessing a person's liability to income tax are the personal allowances, e.g. the single-person's allowance, married-man's allowance, and wife's-earned-income relief. It also includes tax-free annual amounts, such as the annual allowances available in respect of capital-gains tax or inheritance tax. **2.** Money paid to an employee for expenses he has incurred in the course of his business. **3.** A deduction from an invoice for a specified purpose, such as substandard quality of goods, late delivery, etc.

**all-risks policy** An insurance policy covering personal possessions against many risks but not, of course, all risks. A policy of this kind does not list the risks covered; instead it lists only the exclusions. Such wide cover often merits very high *premiums and items covered on this basis often include jewellery, photographic or electronic equipment, and other valuables.

**alphanumeric** Consisting of the letters of the alphabet, numbers, and special characters used by computers. The typewriter or computer keyboard is called an alphanumeric keyboard since it has keys for all these letters, numbers, and characters. A computer may have an alphanumeric display, which shows only alphanumeric characters, as opposed to a display with a full graphics capability, which can show such drawings as bar charts.

**alpha stocks** The most actively traded *securities on the *SEAQ trading system of the *London Stock Exchange. Since October 1986 (*see* Big Bang) the approximately 4000 securities have been divided into four categories, alpha, beta, gamma, and delta stocks, according to the frequency with which they are traded. This classification defines the degree of commitment to trading that must be provided by the *market makers. For alpha stocks, currently the 100 or so largest quoted companies on SEAQ, at least 10 market makers must continuously display firm buying and selling prices on the *TOPIC screens installed throughout the City; furthermore, all transactions must be immediately published on TOPIC. Other criteria for alpha stocks relate to the quarterly turnover and the market capitalization of the company concerned. **Beta stocks** are the second rank of quoted companies (of which there are approximately 500) and are less actively traded. For these, continuous prices must also be displayed but immediate publication of transactions is not required. For **gamma stocks** and **delta stocks**, the relatively small companies in which trade is much more infrequent, prices displayed on the screens are treated as indicative but market makers are not necessarily required to buy and sell at these prices. The prices of delta stocks need not be displayed at all.

**alteration of share capital** An increase, reduction, or any other change in the *authorized capital of a company (*see* share capital). If permitted by the *articles of association, a limited company can increase its authorized capital as appropriate. It can also rearrange its existing authorized capital (e.g. by consolidating 100 shares of £1 into 25 shares of £4 or by subdividing 100 shares of £1 into 200 of 50p) and cancel unissued shares. These are reserved powers, passed – unless the articles of association provide otherwise – by an ordinary resolution.

**alternate director** A person who can act temporarily in place of a named director of a company in his absence. An alternate director can only be present at a meeting of the board of directors if the *articles of association provide for this eventuality and if the other directors agree that the person chosen is acceptable to undertake this role.

**amalgamation** The joining together of two or more businesses. *See* merger.

**ambulance stocks** High performance stocks recommended by a broker to a client whose portfolio has not fulfilled his expectations. They either refresh the portfolio – and the relationship between broker and client – or they confirm the client's worst fears.

**AMC** Abbreviation for *Agricultural Mortgage Corporation Ltd.

**American Depository Receipt (ADR)** A certificate issued by a US bank containing a statement that a specific number of shares in a foreign company has been deposited with them. The certificates are denominated in US dollars and can be traded as a security in US markets.

**American option** *See* option.

**amongst matter** The position of an advertisement within a newspaper or magazine so that it is amongst editorial material.

**amortization** The process of treating as an expense the annual amount deemed to waste away from a fixed asset. The concept is particularly applied to leases, which are acquired for a given sum for a specified term at the end of which the lease will have no value. It is customary to divide the cost of the lease by the number of years of its term and treat the result as an annual charge against profit. While this method does not necessarily reflect the value of the lease at any given time, it is an equitable way of allocating the original cost between periods. *Compare* depreciation.

Goodwill may also be amortized. The *statements of standard accounting practice recommend as its preferred method the writing-off in the year of purchase of all purchased goodwill. The charge should be to the reserves and not to the *profit and loss account. However the standard also permits the writing-off of goodwill to the profit and loss account in regular instalments over the period of its economic life. Home-grown goodwill, if it is in the balance sheet at all, should be similarly dealt with by one of the two methods above.

**amounts differ** The words stamped or written on a cheque or bill of exchange by a banker who returns it unpaid because the amount in words differs from that in figures.

**AMSO** Abbreviation for *Association of Market Survey Organizations.

**analog computer** A machine that performs arithmetical calculations on numbers that are represented by physical quantities. For example, in mechanical analog computers, the numbers are represented by the rotations of gear wheels. In electrical analog computers, voltages are used to represent numbers. The essential characteristic of an analog computer is that the quantities representing the numerical data vary continuously with time. Analog computers, therefore, differ from the more common digital computers, which deal only with digits, or quantities that vary in steps.

Analog computers are essentially mechanical or electrical devices that can perform the operations of addition, subtraction, multiplication, and division. The output may be in the form of a graph plotted by a pen, or a trace on an oscilloscope, or an electrical signal used to control the operation of a machine or process. As they are able to respond immediately to changes in input data, they are ideal for the automatic control of industrial processes. They are also used for scientific research, particularly where cheap electrical or mechanical devices can mimic the situation being studied.

**ancient lights** Light enjoyed for 20 years or more through a defined aperture (e.g. a window) in a building. Under the Prescription Act (1832) the owner of the building has a right to such light, which may not thereafter be obstructed. Before the passing of this Act it was very difficult to obtain rights to any light, as the common law recognizes no natural right to light.

**ancillary credit business** A business involved in credit brokerage, debt adjusting, debt counselling, debt collecting, or the operation of a credit-reference agency (*see* commercial agency). **Credit brokerage** includes the effecting of introductions of individuals wishing to obtain credit to persons carrying on a consumer-credit business. **Debt adjusting** is the process by which a third party negotiates terms for the discharge of a debt due under consumer-credit agreements or consumer-hire agreements with the creditor or owner on behalf of the debtor or hirer. The latter may also pay a third party to take over his obligation to discharge a debt or to undertake any similar acitivity concerned with its liquidation. **Debt counselling** is the giving of advice (other than by the original creditor and certain others) to debtors or hirers about the liquidation of debts due under consumer-credit agreements or consumer-hire agreements. A **credit-reference agency** collects information concerning the financial standing of individuals and supplies this information to those seeking it. The Consumer Credit Act (1974) provides for the licensing of ancillary credit businesses and regulates their activities.

***Annual Abstract of Statistics*** An annual publication of the *Central Statistical Office giving UK industrial, vital, legal, and social statistics. *Compare Monthly Digest of Statistics.*

**annual accounts** An organization's financial statements published annually, usually to comply with a statutory obligation to do so. They include a *balance sheet, *profit and loss account (or *income and expenditure account), and possibly a statement of sources and application of funds. *See also* modified accounts. The most common are those produced by companies and filed at Companies House, in accordance with the provisions of the Companies Acts. However, other bodies are regulated by different statutes, for example Friendly Societies report to the Registrar of Friendly Societies. Sole traders and partnerships have no statutory obligation to produce annual accounts, although accounts are required of them in order to agree assessments raised by the Inland Revenue for taxation purposes.

**annual general meeting (AGM)** An annual meeting of the shareholders of a company, which must be held every year; the meetings may not be more than 15 months apart. Shareholders must be given 21 days' notice of the meeting. The usual business transacted at an AGM is the presentation of the audited accounts, the appointment of directors and auditors, the fixing of their remuneration, and recommendations for the payment of dividends. Other business may be transacted if notice of it has been given to the shareholders. *See also* agenda; order of business.

**annual percentage rate (APR)** The annual equivalent *rate of return on a loan or investment in which the rate of interest specified is chargeable or payable more frequently than annually. Most investment institutions are now required by law to specify the APR when the interest intervals are more frequent than annual. Similarly those charge cards that advertise monthly rates of interest (say, 2%) must state the equivalent APR (24% in this case).

**annual report** The *annual accounts and directors' report of a company, issued to shareholders and filed at Companies House in accordance with the provisions of the Companies Acts.

**annual return** A return made annually to the Registrar of Companies in accordance with the Companies Acts. It records details of the share capital and assets subject to charges of directors, the company secretary, and shareholders. The return is made up to the date 14 days after the date of the company's annual general meeting.

**annual value (net annual rentable value; rateable value)** *See* rates.

**annuitant** A person who receives an *annuity.

**annuity 1.** A contract in which a person pays a premium to an insurance company, usually in one lump sum, and in return receives periodic payments for an agreed period or for the rest of his life. An annuity has been described as the opposite of a life assurance as the policyholder pays the lump sum and the insurer makes the regular payments. Annuities are often purchased at a time of prosperity to convert capital into an income during old age. *See also* annuity certain; deferred annuity. **2.** A payment made on such a contract.

**annuity certain** An *annuity in which payments continue for a specified period irrespective of the life or death of the person covered. In general, annuities cease on the death of the policyholder unless they are annuities certain.

**Ansoff Matrix** *See* product-market strategy.

**ante-date** To date a document before the date on which it is drawn up. This is not necessarily illegal or improper. For instance, an ante-dated cheque is not in law invalid. *Compare* post-date.

**anti-trust laws** Laws passed in the USA, from 1890 onwards, making it illegal to do anything in *restraint of trade, set up monopolies, or otherwise interfere with free trade and competition.

**Anton Piller order** A court injunction ordering the defendant to allow the plaintiff to enter named premises to search for and take copies of specified articles and documents. These orders are obtained by the plaintiff 'ex parte' (without the other party being present in court) to allow him to preserve evidence in cases in which he has grounds to think it will be destroyed. It is especially useful in 'pirating' cases. The order is not a search warrant, so entry cannot be forced, but the defendant will be in contempt of court if entry is refused. A solicitor must serve the order. It is named after an order made in the High Court in 1976 against Anton Piller KG.

**APACS** Abbreviation for *Association for Payment Clearing Services.

**APEX** Abbreviation for Advance-Purchase Excursion. It refers to a form of international return airline ticket offered at a discount to the standard fare, provided that bookings both ways are made 21 days in advance for international flights and 7 – 14 days for European flights, with no facilities for stopovers or cancellations.

**APL** Acronym for A Programming Language, a language used to program computers. In its simplest form, APL performs the functions of an intelligent calculator. It allows the user to perform easily such tasks as solving a set of linear equations, or finding the inverse of a matrix. APL is a powerful tool for the scientist or engineer and is sometimes used for business programs.

**appellant** A person or organization that appeals against the decision of a court. The party resisting the appeal is called the respondent.

**application form** A form, issued by a newly floated company with its *prospectus, on which members of the public apply for shares in the company. *See also* allotment; multiple application; pink form.

**application for quotation** An application by a *public limited company for a quotation on the *London Stock Exchange. The company is scrutinized by the Quotations Committee to see if it complies with the regulations and if its directors have a high reputation. If the application is accepted the company is given a quotation on one of the Stock Exchange's markets.

**applications software** Computer programs that are designed for a particular purpose or application. For example, accounts programs, games programs, and educational programs are all applications software. *Compare* system software.

**appreciation 1.** An increase in the value of an asset, usually as a result of inflation. This usually occurs with land and buildings; the directors of a company have an obligation to adjust the nominal value of land and buildings and other assets in balance sheets to take account of appreciation. *See* asset stripping. **2.** An increase in the value of a currency with a *floating exchange rate relative to another currency. *Compare* depreciation; devaluation.

**apprentice** A young employee who signs a contract (an **indenture** or **articles of apprenticeship**) agreeing to be trained in a particular skill for a set amount of time by a specific employer. During this time the wages will be relatively low but on completion of the apprenticeship they increase to reflect the increased status of the employee and to recognize the skills acquired.

**appropriation 1.** An allocation of the net profit of organizations in its accounts. Some payments may be treated as expenses and deducted before arriving at net profit; other payments are deemed to be appropriations of profit, once that profit has been ascertained. Examples of the former are such normal trade expenses as wages and salaries of employees, motor running expenses, light and heat, and most interest payments on external finance. Appropriations of the net profit include payments of income tax or corporation tax, dividends to shareholders, transfers to reserves, and, in the case of partnerships, salaries and interest on capital paid to the partners. *See also* accumulated profits. **2.** The allocation of payments to a particular debt out of several owed by a debtor to one creditor. The right to make the appropriation belongs first to the debtor but if he fails to make the appropriation the creditor has the right to do so. **3.** A document identifying a particular batch of goods to be supplied in fulfilment of a forward contract for a commodity. In some cases, for example, a forward contract may call for goods to be shipped in six months' time. At the time the contract is made the goods may not be identifiable. As the period for shipment approaches the supplier will notify the customer exactly which parcel of goods he is going to ship against the contract by identifying them (e.g. by the marks and numbers on the packages) in an appropriation.

**approved deferred share trust (ADST)** A trust fund set up by a British company, and approved by the Inland Revenue, that purchases shares in that company for the benefit of its employees. Tax on dividends is deferred until the shares are sold and is then paid at a reduced rate.

**APR** Abbreviation for *annual percentage rate.

**arbitrage** The non-speculative transfer of funds from one market to another to take advantage of differences in interest rates, exchange rates, or commodity prices between the two markets. It is non-speculative because an arbitrageur will only switch from one market to another if he knows exactly what the rates or prices are in both markets and he will only make the switch if the profit to be gained outweighs the costs of the operation. Thus, a large stock of a commodity in a user country may force its price below that in a producing country; if the difference is greater than the cost of shipping the goods back to the producing country, this could provide a profitable opportunity for arbitrage. Similar opportunities arise with *bills of exchange and foreign currencies.

**arbitration** The determination of a dispute by an arbitrator or arbitrators rather than by a court of law. Any civil (i.e. noncriminal) matter may be settled in this way; commercial contracts often contain **arbitration clauses** providing for this to be done in a specified way. If each side appoints its own arbitrator, as is usual, and the arbitrators fail to agree, the arbitrators are often empowered to appoint an **umpire**, whose decision is final. Arbitration is made binding on the parties by the Arbitration Acts (1950 and 1975). Various industries and *chambers of commerce set up tribunals for dealing with disputes in their particular trade or business. *See* award.

**Architects Registration Council of the UK** A council established under the Architects (Registration) Act (1931) to maintain a register of persons entitled to practise as architects; to recognize the qualifying examinations for registration; to provide scholarships and maintenance grants for students of architecture; and to act as a disciplinary body for the profession.

**archive** A store for documents and magnetic disks or tapes containing records that are seldom used. Most computer users maintain an archive holding copies of disks or tapes containing vital information. If the original disk or tape becomes damaged, the archive copy is used to reinstate the information lost from the damaged master disk or tape.

**arithmetic mean** An average obtained by adding together the individual numbers concerned and dividing the total by their number. For example, the arithmetic mean of 7, 20, 107, and 350 is 484/4 = 121. This value, however, gives no idea of the spread of numbers. *Compare* geometric mean.

**ARR** Abbreviation for *accounting rate of return.

**arrangement 1.** A method of enabling a debtor to enter into an agreement with his creditors (either privately or through the courts) to discharge his debts by partial payment, as an alternative to bankruptcy. This is generally achieved by a **scheme of arrangement**, which involves applying the assets and income of the debtor in proportionate payment to his creditors. For instance, a scheme of arrangement may stipulate that the creditors will receive 20 pence for every pound that is owed to them. This is sometimes also known as **composition**. Once a scheme of arrangement has been agreed a **deed of arrangement** is drawn up, which must be registered with the Department of Trade and Industry within seven days. **2.** *See* voluntary arrangement.

**articled clerk** A trainee solicitor. The Law Society lays down provisions regulating the training of solicitors. All trainees are now graduates and will have taken professional examinations. They are then required to be articled to (i.e. to sign an agreement to learn from) a qualified solicitor for two years before being admitted as solicitors themselves.

**articles of apprenticeship** *See* apprentice.

**articles of association** The document that governs the running of a company. It sets out voting rights of shareholders, conduct of shareholders' and directors' meetings, powers of the management, etc. Either the articles are submitted with the *memorandum of association when application is made for incorporation or the relevant model articles contained in the Companies Regulations (Tables A to F) are adopted. Table A contains the model articles for companies limited by shares. The articles constitute a contract between the company and its members but this applies only to the rights of shareholders in their capacity as members. Therefore directors or company solicitors (for example) cannot use the articles to enforce their rights. The articles may be altered by a special resolution of the members in a general meeting.

**artificial intelligence** The ability of a computer to perform tasks normally associated with human intelligence, such as reasoning and learning from experience. There has been considerable progress in the field recently, particularly in those applications that make use of the computer's calculating power. Chess-playing computers that can beat most human players seem to be intelligent, yet their skill relies only on their ability to calculate better than their human opponents. A more important development, the *expert system, makes use of the computer's ability to store, organize, and retrieve large volumes of information. Also called intelligent knowledge-based systems, these store the knowledge and experience of an expert in a particular field. The system can be questioned by a non-expert and will give the answer that the expert would give. These systems are used for a wide range of tasks, such as analysis of company results, review of loan applications, buying stocks and shares, medical diagnosis, identifying poisons, and prospecting for oil.

**artificial person** A person whose identity is recognized by the law but who is not an individual. For example, a company is a person in the sense that it can sue and be sued, hold property, etc. in its own name. It is not, however, an individual or real person.

**ASA** Abbreviation for *Advertising Standards Authority.

**ASCII** Acronym for American Standard Code for Information Interchange. This is a standard code adopted by many computer manufacturers to simplify the transfer of information between computers. The code represents the numbers, letters, and symbols used in computing by a standard set of numbers. For example, the capital letter A is represented by the number 65, B is represented by the number 66, and so on. Many computers can convert their output to ASCII code, in which form it can be transferred to, and recognized by, other computers.

**ASEAN** Abbreviation for *Association of South East Asian Nations.

**A shares** Ordinary shares in a company that usually do not carry voting rights. **Non-voting shares** are issued by a company when it wishes to raise additional capital without committing itself to a fixed dividend and without diluting control of the company. They are, however, unpopular with institutional investors (who like to have a measure of control with their investments) and are therefore now rarely issued.

**as per advice** Words written on a *bill of exchange to indicate that the drawee has been informed that the bill is being drawn on him.

**assay** A chemical test to determine the purity of a sample of metal or to determine the content of an alloy. In the UK assays are carried out by Official Assay Offices, the marks of which appear in the *hallmarks of silver and gold articles.

**assembler** A computer program that takes instructions prepared by the computer user in a kind of shorthand (called *assembly language) and converts them into a form that the computer can understand.

**assembly language** A type of *low-level language used to program computers. Each instruction is a short mnemonic, or 'memory-jogger', that describes one operation to be performed by the machine. For instance, for a particular machine the assembly-language instruction ADD B adds a number to the total already in the computer memory. A special program, called an **assembler**, is needed to convert the mnemonics into a form, called *machine code, that the computer can understand. In practice, most programming is done using *high-level languages, such as BASIC or PASCAL, that use abstract constructs, which have no one-for-one correspondence with machine-code instructions. In this case the translation into machine code is done by a program called an *interpreter, or a *compiler.

**assented stock** A security, usually an ordinary share, the owner of which has agreed to the terms of a *takeover bid. During the takeover negotiations, different prices may be quoted for assented and **non-assented stock**.

**assessment** The method by which a tax authority raises a bill for a particular tax and sends it to the taxpayer or his agent. The assessment may be based on figures already agreed between the authority and the taxpayer or it may be an estimate by the tax authorities. The taxpayer normally has a right of appeal against an assessment within a specified time limit. *See also* deferred asset.

**assessor** *See* loss assessor.

**asset** Any object, tangible or intangible, that is of value to its possessor. In most cases it either is cash or can be turned into cash; exceptions include prepayments, which may represent payments made for rent, rates, or motor licences, in cases in which the time paid for has not yet expired. Tangible assets include land and buildings, plant and machinery, fixtures and fittings, trading stock, investments, debtors, and cash; intangible assets include goodwill, patents, copyrights, and trademarks. *See also* deferred asset.

**asset-backed fund** A fund in which the money is invested in tangible or corporate assets, such as property or shares, rather than being treated as savings loaned to a bank or other institution. Asset-backed funds can be expected to grow with inflation in a way that bank savings cannot. *See also* equity-linked policy; unit-linked policy.

**asset stripping** The acquisition or takeover of a company whose shares are valued below their *asset value, and the subsequent sale of the company's most valuable assets. Asset stripping was a practice that occurred primarily in the decade after World War II, during which property values were rising sharply. Having identified a suitable company, an entrepreneur would buy up its shares on the stock exchange until he had a controlling interest; after revaluing the properties held, he would sell them for cash, which would be distributed to shareholders, including himself. He would then either revitalize the management of the company and later sell off his shareholding at a profit or, in some cases, close the business down. Because the asset stripper is totally heedless of the welfare of the other shareholders, the employees, the suppliers, or creditors of the stripped company, the practice is now highly deprecated.

**asset value (per share)** The total value of the assets of a company less its liabilities, divided by the number of ordinary shares in issue. This represents in theory, although probably not in practice, the amount attributable to each share if the company was wound up. The asset value may not necessarily be the total of the values shown by a company's balance sheet, since it is not the function of balance sheets to value assets. It may, therefore, be necessary to substitute the best estimate that can be made of the market values of the assets (including goodwill) for the values shown in the balance sheet. If there is more than one class of share, it may be necessary to deduct amounts due to shareholders with a priority on winding up before arriving at the amounts attributable to shareholders with a lower priority.

**assignment** The act of transferring, or a document (a **deed of assignment**) transferring, property to some other person. Examples of assignment include the transfer of rights under a contract or benefits under a trust to another person. *See also* assignment of lease.

**assignment of copyright** *See* copyright.

**assignment of insurable interest** Assigning to another party the rights and obligations of the *insurable interest in an item of property, life, or a legal liability to be insured. This enables the person to whom the interest is assigned to arrange insurance cover, which would not otherwise be legally permitted.

**assignment of lease** The transfer of a lease by the tenant (assignor) to some other person (assignee). Leases are freely transferable at common law although it is common practice to restrict assignment by conditions (covenants) in the lease. An assignment that takes place in breach of such a covenant is valid but it may entitle the landlord to put an end to the lease and re-enter the premises. An assignment of a legal lease must be by deed. An assignment puts the assignee into the shoes of the assignor, so that there is 'privity of estate' between the landlord and the new tenant. This is important with regard to the enforceability of covenants in the lease (*see* covenant). An assignment transfers the assignor's whole estate to the assignee, unlike a sub-lease (*see* head-lease).

**assignment of life policies** Transfer of the legal right under a life-assurance policy to collect the proceeds. Assignment is only valid if the life insurer is advised and agrees; life assurance is the only form of insurance in which the assignee need not possess an *insurable interest. In recent years policy auctions have become a popular alternative to surrendering *endowment assurances. In these auctions, a policy is sold to the highest bidder and then assigned to him or her by the original policyholder.

**Associated British Ports (ABP)** A statutory corporation set up by the Transport Act (1981) to administer the 19 ports previously controlled by the British Transport Docks Board. ABP now administers 21 ports and is controlled by Associated British Ports Holding plc.

**Association for Payment Clearing Services (APACS)** An association set up by the UK banks in 1985 to manage payment clearing and overseas money transmission in the UK. The three operating companies under its aegis are: BACS Ltd, which provides an automated service for interbank clearing in the UK; Cheque and Credit Clearing Co. Ltd, which operates a bulk clearing system for interbank cheques and paper credits; and CHAPS and Town Clearing, which provides same-day clearing for high value cheques and electronic funds transfer. In addition EftPos UK Ltd is a company set up to develop electronic funds transfer at the point of sale. APACS also oversees London Dollar Clearing, the London Currency Settlement Scheme, and the cheque card and *eurocheque schemes in the UK.

**Association of British Insurers (ABI)** A trade association representing over 440 insurance companies offering any class of insurance business, whose members transact over 90% of the business of the British insurance market. It was formed in 1985 by a merger the British Insurance Association, the Accident Offices Association, the Fire Offices Committee, the Life Offices Association, and the Industrial Life Offices Association.

**Association of Futures Brokers and Dealers (AFBD)** An association, established in 1984 by the major London futures exchanges, as a *Self-Regulatory Organization to protect investors by providing a regulatory framework for the supervision of futures brokers and dealers.

**Association of Market Survey Organizations (AMSO)** An association of 27 of the largest UK survey research organizations. Member companics adhere to a strict code of conduct to ensure the highest standards of market research.

**Association of South East Asian Nations (ASEAN)** A political and economic grouping of the capitalist nations of South East Asia, formed in 1967 and comprising: Thailand, Malaysia, Singapore, Philippines, Indonesia, and Brunei. The countries are very diverse. For example the per capita income of Singapore in 1986 was some 12 times that of Indonesia; interests often diverge accordingly. While committed to strengthening economic ties, progress has been limited. There has also been political cooperation, for example over policy towards Indochina. There are regular consultations between ASEAN and the major industrialized countries.

**assurance** *Insurance against an eventuality (especially death) that must occur. *See* life assurance.

**assured** The person named in a life-assurance policy to receive the proceeds in the event of maturity or the death of the *life assured. As a result of the policy, the person's financial future is 'assured'.

**assured tenancy** *See* regulated tenancy.

**at and from** Denoting a marine hull insurance cover that begins when the vessel is in dock before a voyage, continues during the voyage, and ends 24 hours after it has reached its port of destination.

**at best** An instruction to a broker to buy or sell shares, stocks, commodities, currencies, etc., as specified, at the best possible price. It must be executed immediately irrespective of market movements. *Compare* at limit.

**at call** Denoting money that has been lent on a short-term basis and must be repaid on demand. *Discount houses are the main borrowers of money at call.

**at limit** An instruction to a broker to buy or sell shares, stocks, commodities, currencies, etc., as specified, at a stated limiting price (i.e. not above a stated price if buying or not below a stated price if selling). When issuing such an instruction the principal should also state for how long the instruction stands, e.g. for a day, a week, etc. *Compare* at best.

**ATP** Abbreviation for *aid trade provision.

**at par** *See* par value.

**at sight** The words used on a *bill of exchange to indicate that payment is due on presentation. *Compare* after date; after sight.

**attachment** The procedure enabling a creditor, who has obtained judgment in the courts (the judgment creditor), to secure payment of the amount due from his debtor. The judgment creditor obtains a further court order (the garnishee order) to the effect that money or property due from a third party (the garnishee) to the debtor must be frozen and paid instead to the judgment creditor to satisfy the amount due to him. For instance, a judgment creditor may, through a garnishee order, attach the salary due to the debtor from the debtor's employer (the garnishee).

**attention, interest, desire, action** *See* AIDA.

**attest** To bear witness to an act or event. The law requires that some documents are only valid and binding if the signatures on them have been attested to by a third party. This also requires the third party's signature on the document. For instance, the signature of the purchaser of land under a contract must be attested to by a witness.

**at-the-money option** *See* intrinsic value.

**attitude research** An investigation into the attitudes of people towards an organization or its products. Attitude research is important to advertising specialists in planning *campaigns. For example, it might reveal a 'Buy British' attitude among respondents, which would present a marketing and advertising problem to a US manufacturer.

**at warehouse** Delivery terms for goods that are available for immediate delivery, in which the buyer pays for delivery of the goods, including the cost of loading them onto the road or rail transport. *Compare* ex warehouse.

**auction** A method of sale in which goods are sold in public to the highest bidder. Auctions are used for any property for which there are likely to be a number of competing buyers, such as houses, second-hand and antique furniture, works of art, etc., as well as for certain commodities, such as tea, bristles, wool, furs, etc., which must be sold as individual lots, rather than on the basis of a standard sample or grading procedure. In most auctions the goods to be sold are available for viewing before the sale and it is usual for the seller to put a *reserve price on the articles offered, i.e. the articles are withdrawn from sale unless more than a specified price is bid. The auctioneer acts as agent for the seller in most cases and receives a commission on the sale price. An auctioneer who declares himself to be an agent of the seller promises that he has authority to sell and that he knows of no defect in the seller's title to the goods; he does not, however, promise that a buyer will receive good title for a specific object. An advertisement that an auction will be held does not bind the auctioneer to hold it. It is illegal for a dealer (a person who buys at auction for subsequent resale) to offer a person a reward not to bid at an auction. *See also* Dutch auction.

**audience research** Research to establish readership, audience, and circulation data, which is vital information in *advertising. Research into television audiences is undertaken by the **Broadcasters Audience Research Board (BARB)** and into commercial radio audiences by the **Joint Industry Committee for Radio Audience Research (JICRAR)**. The BBC undertakes its own research through the **BBC Audience Research Unit.**

**audioconferencing** A national and international telephone service enabling several users in different places to conduct a business meeting over the telephone. Connections can be made between callers in several different countries. Details are available from British Telecom. *See also* videoconferencing.

**audio response device** A output device, connected to a computer, that produces human speech. These devices are used, for example, in computerized telephone enquiry systems. They work by storing words, syllables, or phrases in the computer and linking them together to form more-or-less recognizable speech.

**audit** The inspection of an organization's *annual accounts. An **external audit** is carried out by a qualified *accountant, in order to obtain an opinion as to the veracity of the accounts. Under the Companies Acts, companies are required to appoint an *auditor to express an opinion as to whether the annual accounts give a true and fair view of the company's affairs and whether they comply with the provisions of the Companies Acts.

To give such an opinion, the auditor needs to examine the company's internal accounting systems, inspect its assets, make tests of accounting transactions, etc. Some bodies have different requirements for audit; for example, accountants report to the Law Society on solicitors' accounts under the Accountants' Report Rules. Many companies and organizations now appoint internal auditors to carry out an **internal audit**, with the object of reporting to management on the efficacy and security of internal systems. In some cases these systems may not even be financial; they may include, for example, an audit of health and safety in the workplace or an audit of compliance with equal opportunities legislation. *See also* auditors' report.

**Audit Bureau of Circulation (ABC)** An organization to which most newspaper, magazine, and periodical publishers belong. Its function is to collect and audit sales figures from publishers regularly and to publish monthly circulation figures in its quarterly *Circulation Review*, a publication of great value to advertisers. A newspaper, magazine, or periodical must have been publishing for a minimum of six months before joining the ABC.

**auditor** A person who carries out an *audit. **External auditors** are normally members of a body of *accountants authorized by the Companies Acts, such as the *Institute of Chartered Accountants or the Chartered Association of Certified Accountants. The principal requirement for an external audit is that the auditor should be independent of the organization audited and professionally qualified. The same is not true of **internal auditors,** who may be members of such professional bodies as the Institute of Internal Auditors or any of the accountancy bodies, but who may also be appropriately trained employees of the organizations being audited. Every company must have properly qualified auditors. If none have been appointed, the Secretary of State for Trade and Industry must be informed within one week. Failure to do so may result in the company and its officers being liable to a fine. Auditors are usually appointed at the *annual general meeting of the company; the appointment of a new auditor or the removal of an auditor requires special notice. An auditor's notice of resignation must contain a statement of any matters that should be brought to the notice of the shareholders or creditors or a statement that there are no such matters. An auditor's duties must be carried out with due care and skill; an auditor may be liable to negligence to the company, the members, and third parties relying on the audited accounts. The **auditors' remuneration** has to be approved at a general meeting of the company and should be distinguished in the accounts from the cost of other accounting work. *See also* auditors' report.

**auditors' remuneration** *See* auditor.

**auditors' report** A report by the auditors appointed to *audit the accounts of a company or other organization. Auditors' reports may take many forms depending on who has appointed the auditors and for what purposes. Some auditors are engaged in an internal audit while others are appointed for various statutory purposes. The auditors of a limited company are required to form an opinion as to whether the annual accounts of the company give a true and fair view of its profit or loss for the period under review and of its state of affairs at the end of the period; they are also required to certify that the accounts are prepared in accordance with the requirements of the Companies Act (1985). The auditors' report is technically a report to the members of the company and it must be filed together with the accounts with the Registrar of Companies under the Companies Act (1985). Under this Act, the auditors' report must also include an audit of the directors' report. *See also* qualified report.

**autarky** (from Greek: self-sufficiency) An economic policy of self-sufficiency that aims to prevent a country from engaging in international trade. It has been condemned by economists since Adam Smith (1723–90), although some countries have chosen autarky for political reasons, notably the Soviet Union before World War II, on the ground that an underdeveloped country will be better able to build up its industries if it is free from international competition.

**authorized share capital** *See* share capital.

**automatic debit transfer** *See* giro.

**automatic identification** A means of identifying a product mechanically and entering the data obtained automatically into a computer. The most widely used method involves *bar codes. Other methods include *optical character recognition (OCR), magnetic ink character recognition (MICR), magnetic stripes, and voice systems.

**automatic stabilizers** Adjustments to *fiscal policy that occur automatically during *business cycles and smooth the path of economic growth. For example, in a *recession the government will pump money into the economy by paying more in unemployment benefit without a change in policy. Automatic stabilizers counterbalance the effect of the *multiplier on economic activity, although in practice their effectiveness is limited.

**auxiliary storage** *See* backing store.

**available earnings** *See* earnings per share.

**average 1.** A single number used to represent a set of numbers; mean. *See* arithmetic mean; geometric mean; median. **2.** A partial loss in *marine insurance (from French: *avarie*, damage). In **general average (GA)**, a loss resulting from a deliberate act of the master of the ship (such as throwing overboard all or part of the cargo to save the ship) is shared by all the parties involved, i.e. by the shipowners and all the cargo owners. *See* average adjuster; average bond. In a **particular average (PA)**, an accidental loss is borne by the owners of the particular thing lost or damaged, e.g. the ship, an individual cargo, etc. Cargo can be insured either **free of particular average (FPA)** or **with average (WA)**. An FPA policy covers the cargo against loss by perils of the sea, fire, or collision and includes cover for any contribution payable in the event of a general average. A WA policy gives better cover as it also includes damage by heavy seas and sea-water damage. In addition, marine cargo can be covered by an *all-risks policy. *See also* free of all averages; Institute cargo clauses. **3.** A method of sharing losses in property insurance to combat underinsurance. This is usually applied in an **average clause** in a fire insurance policy, in which it is stated that the sum payable in the event of a claim shall not be more than the proportion that the insured value of an item bears to its actual value.

**average adjuster** A person who handles marine insurance claims on behalf of the insurers. If there is a claim that involves general *average and contributions have to be made by all the parties involved it is the average adjuster who is responsible for apportioning payment.

**average bond** A promise to pay general *average contributions, if required. If a general-average loss occurs during a marine voyage the carrier has a right to take part of the cargo as payment of the cargo owners' contribution to the loss. As an alternative to the possibility of losing part of his cargo, the cargo owner may take out an average bond with insurers, who agree to pay any losses arising in this way.

**average clause** *See* average.

**average cost (average unit cost)** The average cost of a unit of production. A unit of production is whatever is being produced by a manufacturer at any given time; it might be a complete car or it might be a car component, such as a carburettor. Costs of production will vary from time to time and for budgeting purposes it may be convenient to take an average.

**average fixed cost** An average taken over a specified period of the fixed costs incurred by a *cost centre. It may seem to be a contradiction that if costs are fixed they remain the same and there can therefore be no average. However, for some purposes costs, such as rents, are taken as fixed, although they may vary from time to time; in such cases, for budgeting purposes, it is reasonable to take an average. A fixed cost in this sense is not necessarily a cost that will never change from one accounting period to the next, although it is one that will not change solely with the level of production. *Compare* average variable cost.

**average stock** A method of accounting for stock movements that assumes goods are taken out of stock at the average cost of the goods in stock. *See* base stock method; FIFO; LIFO; stock.

**average variable cost** An average taken over a specified period of the variable cost of producing units of production (*see* average fixed cost; average cost). The variable costs (such as the cost of raw materials, direct labour, machine time, etc.) of producing a unit are those that vary directly with the number of units produced. As they are likely to change from time to time it may be convenient for budgeting purposes to take an average.

**averaging** Adding to a holding of particular securities or commodities when the price falls, in order to reduce the average cost of the whole holding. **Averaging in** consists of buying at various price levels in order to build up a substantial holding of securities or commodities over a period. **Averaging out** is the opposite process, of selling a large holding at various price levels over a long period.

**aviation broker** A broker who arranges chartering of aircraft, airfreight bookings, insurance of air cargo and aircraft, etc.

**aviation insurance** The insurance of aircraft, including accident or damage to aircraft, insurance of air cargo, loss of life or injury while flying, and loss or damage to baggage.

**award** A judgment delivered as a result of an *arbitration. This is binding on the parties concerned unless an appeal is lodged. On a question of law an arbitrator's award can be the subject of an appeal to the High Court under the provisions of the Arbitration Act (1979).

# B

**BAA** Abbreviation for *British Airports Authority plc.

**backdate 1.** To put an earlier date on a document than that on which it was compiled, in order to make it effective from that earlier date. **2.** To agree that salary increases, especially those settled in a pay award, should apply from a specified date in the recent past.

**back door** One of the methods by which the *Bank of England injects cash into the *money market. The bank purchases Treasury bills at the market rate rather than by lending money directly to the discount houses (the front door method) when it acts as *lender of last resort.

**back freight** The cost of shipping goods back to the port of destination after they have been overcarried. If the master overcarried the goods for reasons beyond his control, the shipowner may be responsible for paying the back freight. If delivery was not accepted in reasonable time at the port of destination and the master took the goods on or sent them back to the port of shipment, the back freight would be the responsibility of the cargo owner.

**background** Low-priority work performed when computer facilities are not otherwise required. Background work may not be less important than other work: it is more likely to be a large or time-consuming program whose operation is interrupted for a series of small tasks that require immediate attention.

**backing store** A computer memory that supplements the computer's internal memory by holding information not in immediate use, also called **auxiliary storage**, **external storage**, or **secondary storage**. The main types are *floppy disks, *hard disks, and *magnetic tape. The information stored in the backing store is not lost when the power is turned off. The capacity of a backing store is usually much greater than the computer's internal memory.

**back-to-back credit (countervailing credit)** A method used to conceal the identity of the seller from the buyer in a credit arrangement. When the credit is arranged by a British finance house, the foreign seller provides the relevant documentation. The finance house, acting as an intermediary, issues its own documents to the buyer, omitting the seller's name and so concealing the seller's identity.

**back-up copy** A copy of information held in a computer taken in case the original is lost or destroyed. If the original information is on disk, the back-up copy should be on a completely different disk, or tape, and stored in a separate location from the original. Any sensible business will have back-up copies of all information held on its computer. How frequently the copies are made will depend upon how rapidly the information changes, its difficulty of replacement, and its importance.

**backwardation** **1.** The difference between the spot price of a commodity, including rent, insurance, and interest accrued, and the forward price. **2.** A situation that occasionally occurs on the London Stock Exchange when a market maker quotes a buying price for a share that is lower than the selling price quoted by another market maker. However, with prices now displayed on screens, this situation does not now last long.

**backward integration** *See* integration.

**bad debt** An amount owing from a *debtor that is very unlikely to be paid. Such an amount can be treated as a loss and written off in the *profit and loss account. Doubtful debts may appear in the accounts of an organization as a *provision for bad debts.

**bailment** A delivery of good from the **bailor** (the owner of the goods) to the **bailee** (the recipient of the goods), on the condition that the goods will ultimately be returned to the bailor. The goods may thus be hired, lent, pledged, or deposited for safe custody. A delivery of this nature is usually also the subject of a contract; for example, a contract with a bank for the deposit of valuables for safekeeping. Nonetheless, in English law a bailment retains its distinguishing characteristic of a business relationship that arises outside the law of contract and is therefore not governed by it.

**balanced-budget multiplier** The effect on national income caused by a change in government expenditure that has been offset by an equal change in taxation. For example, an increase in government spending will inject more demand into the economy than the equal increase in taxation takes out, since some of the income absorbed by the tax would have been used for savings and therefore did not contribute to the *aggregate demand. In effect, individuals have reduced savings, they feel worse off, and therefore work harder to build up their savings again. The balanced-budget multiplier is not usually pursued explicitly as an instrument of fiscal policy as taxation is generally unpopular.

**balance of payments** The accounts setting out a country's transactions with the outside world. They are divided into various sub-accounts, notably the **current account** and the **capital account**. The former includes the trade account, which records the balance of imports and exports (*see* balance of trade). Overall, the accounts must always be in balance. A deficit or surplus on the balance of payments refers to an imbalance on a sub-account, usually the amount by which the foreign-exchange reserves of the government have been depleted or increased. The conventions used for presenting balance-of-payments statistics are those recommended by the *International Monetary Fund.

**balance of trade** The accounts setting out the results of a country's trading position. It is a component of the *balance of payments, forming part of the *current account. It includes both the *visibles (i.e. imports and exports in physical merchandise) and the *invisible balance (receipts and expenditure on such services as insurance, finance, freight, and tourism).

**balance sheet** One of the principal statements comprising a set of *accounts, showing the financial state of affairs of an organization on a given date, usually the last day of an *accounting period. A balance sheet has three main headings: assets, liabilities, and capital. The assets must always equal the sum of the liabilities and the capital, as the statement can be looked at in either of two ways: (1) as a statement of the organization's wealth, in which case the assets less the liabilities equal the capital (the amount of wealth attributable to the proprietors); or (2) as a statement of how the assets have been funded, i.e. partly by borrowing (the liabilities) and partly by the proprietors (the capital). Although a balance sheet balances, i.e. its two sides are equal, it is actually so named since it comprises balances from the accounts in the ledgers.

**balloon** A large sum repaid as an irregular instalment of a loan repayment. **Balloon loans** are those in which repayments are not made in a regular manner, but are made, as funds become available, in balloons.

**ballot** A random selection of applications for an oversubscribed *new issue of shares (*see also* flotation). The successful applicants may be granted the full number of shares for which they have applied or a specified proportion of their applications. Applicants not selected in the ballot have their applications and cheques returned.

**Baltic Exchange** A commodity and freight-chartering exchange in the City of London. It takes its name from the trade in grain with Baltic ports that was the mainstay of its business in the 18th century. International commodity trading continues in grain, potatoes, and meat under the auspices of the Grain and Free Trade Association (GAFTA). The exchange is also a centre for chartering freight for goods shipped by sea and for chartering aircraft for both goods and passengers. Dealings in forward freight are supervised by the **Baltic International Freight Futures Exchange (BIFFEX)**.

**Bancogiro** *See* giro.

**banded pack** A special offer of goods in which two or more related (or sometimes unrelated) items are bound together to form a single pack. This pack is offered at a lower price than the combined price of the individual items. Banded packs are sometimes offered with the slogan "buy one, get one free".

**bank** A commercial institution involved in a variety of financial activities. Banks are concerned mainly with making and receiving payments on behalf of their customers, accepting deposits, and making short-term loans to private individuals, companies, and other organizations. In the UK, the banking system comprises the *Bank of England (the central bank), the *commercial banks, *merchant banks, branches of foreign and Commonwealth banks, the *T.S.B. Group, the *National Savings Bank, and the National Girobank (*see* giro). *See also* building society.

**bank account** *See* account; cheque account; current account; deposit account; savings account.

**bank advance** *See* bank loan.

**bank bill** A bill of exchange issued or guaranteed (accepted) by a bank. It is more acceptable than a trade bill as there is less risk of non-payment and hence it can be discounted at a more favourable rate.

**bank certificate** A certificate, signed by a bank manager, stating the balance held to a company's credit on a specified date. It may be asked for by a company's auditors if there is any reason for them to doubt the figures given by the company.

**bank charges** The amount charged to a customer's account by a bank as a handling charge or a charge for any special services. Charges may be made for clearing cheques, making deposits, or settling standing orders and direct debits. However, modern practice is to provide 'free banking' by waiving such charges on current accounts as long as they remain in credit for a specified period (normally 3 months). The commercial banks have largely been forced to take this step as a result of the free banking services offered by *building societies.

**bank draft (banker's cheque; banker's draft)** A cheque drawn by a bank on itself or its agent. A person who owes money to another buys the draft from a bank for cash and hands it to the creditor who need have no fear that it might be dishonoured. A bank draft is used if the creditor is unwilling to accept an ordinary cheque.

**banker's cheque** *See* bank draft.

**banker's order** *See* standing order.

**banker's reference** A reference provided by a bank with regard to the creditworthiness of a customer and his suitability for being given trade credit. Increasingly this information is being provided by private credit-reference agencies (*see* commercial agency), who collate information regarding debtors, ratepayers, bankruptcies, and court judgments in order to build a financial picture of each customer.

**Bank for International Settlements (BIS)** An international bank originally established in 1930 as a financial institution to coordinate the payment of war reparations between European central banks. It was hoped that the BIS, with headquarters in

Basle, would develop into a European central bank but many of its functions were taken over by the *International Monetary Fund (IMF) after World War II. Since then the BIS has fulfilled several roles including acting as a trustee and agent for various international groups, such as the OECD, European Monetary Agreement, etc. The frequent meetings of the BIS directors have been a useful means of cooperation between central banks, especially in combatting short-term speculative monetary movements. Since 1986 the BIS has acted as a clearing house for inter-bank transactions in the form of *European Currency Units.

The original members were France, Belgium, West Germany, Italy, and the UK but now most European central banks are represented as well as the USA, Canada, and Japan. The London agent is the Bank of England, whose governor is a member of the board of directors of the BIS.

**Bank Giro** *See* giro.

**bank guarantee** An undertaking given by a bank to settle a debt should the debtor fail to do so. A bank guarantee can be used as a security for a loan but the banks themselves will require good cover before they issue a guarantee.

**Bank Holidays** Public holidays in the UK, when the banks are closed. They are New Year's Day, Easter Monday, May Day (the first Monday in May), Spring Bank Holiday (the last Monday in May), August Bank Holiday (last Monday in August), and Boxing Day. In Scotland, Easter Monday is replaced by 2 January and the August Bank Holiday is on the first Monday in August. In Northern Ireland St Patrick's Day (17 March) is added. In the Channel Islands Liberation Day (9 May) is included. Bank Holidays have a similar status to Sundays in that *bills of exchange falling due on a Bank Holiday are postponed until the following day and also they do not count in working out *days of grace. Good Friday and Christmas Day are also public holidays, but payments falling due (including *bills of exchange) on these days are payable on the preceding day. When Bank Holidays fall on a Sunday, the following day becomes the Bank Holiday.

**Banking Ombudsman** *See* Financial Ombudsman.

**bank loan (bank advance)** A specified sum of money lent by a bank to a customer, usually for a specified time, at a specified rate of interest. In most cases banks require some form of security for loans, especially if the loan is to a commercial enterprise. *See also* loan account; overdraft; personal loan.

**banknote** An item of paper currency issued by a central bank. Banknotes developed in England from the receipts issued by London goldsmiths in the 17th century for gold deposited with them for safe-keeping. These receipts came to be used as money and their popularity as a *medium of exchange encouraged the goldsmiths to issue their own banknotes, largely to increase their involvement in banking and, particularly, moneylending. Now only the Bank of England and the Scottish and Irish banks in the UK have the right to issue notes. Originally all banknotes were fully backed by gold and could be exchanged on demand for gold; however, since 1931 the promise on a note to "pay the bearer on demand" simply indicates that the note is legal tender. *See also* promissory note.

**Bank of England** The central bank of the UK. It was established in 1694 as a private bank by London merchants in order to lend money to the state and to deal with the national debt. It came under public ownership in 1946 with the passing of the Bank of England Act. The Bank of England acts as the government's bank, providing loans through ways and means advances and arranging borrowing through the issue of gilt-edged securities. The bank helps to implement the government's financial and monetary policy as directed by the Treasury. It also has wide statutory powers to supervise the banking system, including the commercial banks to which, through the discount market, it acts as *lender of last resort.

The Bank Charter Act (1844) divided the bank into an issue department and a banking department. The issue department is responsible for the issue of banknotes and coins as supplied by the *Royal Mint. The banking department provides banking services (including accounts) to commercial banks, foreign banks, other central banks, and government departments. The bank manages the national debt, acting as registrar of government stocks. It also administers *exchange control, when in force, and manages the *exchange equalization account. The bank is controlled by a governor, deputy-governor and a court (board) of 16 directors, appointed by the Crown for periods of 4–5 years.

**bank rate** An obsolete term for the rate of interest at which the central bank (Bank of England) lends to the banking system, which in practice meant the rate charged for either rediscounting 'eligible paper' or making loans to the *discount market. This was replaced between 1972 and 1981 by the *minimum lending rate (MLR). *See* base rate.

**bank reconciliation statement** A statement, often in a firm's cash book, showing how the entries in the cash book can be reconciled with the firm's bank statement. Differences will arise owing to the record in the cash book of cheques drawn that have not yet passed through the banking system, cheques paid in but not credited, and bank charges, interest payments, and standing orders not yet recorded.

**bankruptcy** The state of an individual who is unable to pay his debts and against whom a **bankruptcy order** has been made by a court. The order deprives the bankrupt of his property, which is then used to pay his debts. Bankruptcy proceedings are started by a petition, which may be presented to the court by (1) a creditor or creditors; (2) a person affected by a voluntary arrangement to pay debts set up by the debtor under the Insolvency Act (1986); (3) the Director of Public Prosecutions; or (4) the debtor himself. The grounds for a creditors' petition are that the debtor appears to be unable to pay his debts or to have reasonable prospects of doing so, i.e. that the debtor has failed to make arrangements to pay a debt for which a statutory demand has been made or that a judgment debt has not been satisifed. The debts must amount to at least £750. The grounds for a petition by a person bound by a voluntary arrangement are that the debtor has not complied with the terms of the arrangement or has withheld material information. The Director of Public Prosecutions may present a petition in the public interest under the Powers of Criminal Courts Act (1973). The debtor himself may present a petition on the ground that he is unable to pay his debts.

Once a petition has been presented, the debtor may not dispose of any of his property. The court may halt any other legal proceedings against the debtor. An interim receiver may be appointed. This will usually be the *official receiver, who will take any necessary action to protect the debtor's estate. A *special manager may be appointed if the nature of the debtor's business requires it.

The court may make a bankruptcy order at its discretion. Once this has happened, the debtor is an undischarged bankrupt. He is deprived of the ownership of all his property and must assist the official receiver in listing it, recovering it, protecting it, etc. The official receiver becomes manager and receiver of the estate until the appointment of a **trustee in bankruptcy**. The bankrupt must prepare a statement of his affairs for the official receiver within 21 days of the bankruptcy order. A **public examination** of the bankrupt may be ordered on the application of the official receiver or the creditors, in which the bankrupt will be required to answer questions about his affairs in court.

Within 12 weeks the official receiver must decide whether to call a **meeting of creditors** to appoint a trustee in bankruptcy. The trustee's duties are to collect, realize, and distribute the bankrupt's estate. He may be appointed by the creditors, the court, or the Secretary of State; he must be a qualified insolvency practitioner, or the official receiver. All the property of the bankrupt is available to pay the creditors, except for the following: equipment necessary to him in his employment or business, necessary domestic equipment; and income required for the reasonable domestic needs of the bankrupt and his family. The court has discretion whether to order sale of a house in which a spouse or children are living. All creditors must prove their claims to the trustees. Only unsecured claims can be proved in bankruptcy. When all expenses have been paid, the trustee will divide the estate. The Insolvency Act (1986) sets out the order in which creditors will be paid (*see* preferential creditor). The bankruptcy may end automatically after two or three years, but in some cases a court order is required. The bankrupt is discharged and receives a certificate of discharge from the court.

**bankruptcy order** *See* bankruptcy.

**bank statement** A regular record, issued by a bank or building society, showing the credit and debit entries in a customer's cheque account, together with the current balance. The frequency of issue will vary with the customer's needs and the volume of transactions going through the account. Modern cash dispensers enable the customer to ask for a statement whenever he needs one.

**banques d'affaires** The French term for an *investment bank.

**bar** *Slang* One million.

**bar code** A code, consisting of an array of parallel rectangular bars and spaces, printed on a package for sale in a retail shop. When a laser scanner is passed over the bar code at the retailer's checkout till (*see* electronic point of sale), the price and description of the goods are displayed on the till screen and the computer-controlled stock record is simultaneously reduced.

**bareboat charter (demise charter)** A form of *chartering a ship in which the expenses incurred during the period of the charter are paid by the hirer, including the hiring of the master and crew, the provision of fuel and stores, etc. All the hirer gets is the ship.

**bargain** A transaction on the London Stock Exchange. The bargains made during the day are included in the Daily Official List.

**bargaining theory** The theory that analyses economic outcomes between individuals possessing market power. Under conditions of *perfect competition no individual has any market power; in most situations of practical interest, however, individuals have some influence over the prices at which they buy and sell. Bargaining theory can be applied to the determination of wages (*see* bilateral monopoly), competition between oligopolists, or even the relationship between the government and the electorate. Modern bargaining theory mainly centres on *game theory and focuses on some of the main problems in contemporary economies. *See also* Nash equilibrium; predatory pricing; prisoners' dilemma; strategic behaviour.

**barometer stock** A security whose price is regarded as an indicator of the state of the market. It will be a widely held blue chip with a stable price record.

**barratry** Any act committed wilfully by the master or crew of a ship to the detriment of its owner or charterer. Examples include scuttling the ship and embezzling the cargo. Illegal activities (e.g. carrying prohibited persons) leading to the forfeiture of the ship also constitute barratry. Barratry is one of the risks covered by *marine insurance policies.

**barrel 1.** A unit of capacity used in the oil industry equal to 42 US gallons (35 Imperial gallons). **2.** A unit of capacity used in the brewing industry equal to 36 Imperial gallons.

**barriers to entry** Factors that prevent competitors from entering a particular market. These factors may be innocent, e.g. an absolute cost advantage on the part of the firm that dominates the market, or deliberate, such as high spending on advertising to make it very expensive for new firms to enter the market and establish themselves. Barriers to entry reduce the level of competition in a market, i.e. they make it less contestable, thereby by enabling incumbents to charge higher-than-competitive prices. *See also* contestable markets theory; limit price.

**barter** A method of trading in which goods or services are exchanged without the use of money. It is a cumbersome system, which severely limits the scope for trade. Means of exchange, such as money, enable individuals to trade with each other at much greater distance and through whole chains of intermediaries, which are inconceivable in a barter system.

**base date** *See* base year.

**base metals** The metals copper, lead, zinc, and tin. *Compare* precious metals. *See also* London Metal Exchange.

**base rate 1.** The rate of interest used by the commercial banks as a basis for the rates they charge their customers. **2.** An informal name for the rate at which the Bank of England lends to the *discount houses, which effectively controls the lending rate throughout the banking system. The abolition of the *minimum lending rate in 1981 heralded a loosening of government control over the banking system, but the need to increase interest rates in the late 1980s (to control inflation and the *balance of payments deficit) led to the use of this term in this sense.

**base stock method** A method of accounting for stock movements that assumes a given amount of the stock never moves and therefore retains its original cost (*see* average stock; FIFO; LIFO; stock).

**base year (base date)** The first of a series of years in an index. It is often denoted by the number 100, enabling percentage rises (or falls) to be seen at a glance. For example, if a price index indicates that the current value is 120, this will only be meaningful if it is compared to an earlier figure. This may be written: 120 (base year 1985 = 100), making it clear that there has been a 20% increase in prices since 1985.

**BASIC** Acronym for Beginner's All-purpose Symbolic Instruction Code. BASIC is a popular language used especially to program small computers. It is an easy-to-learn easy-to-use language, particularly suitable for the novice, hobbyist, and small businessman. Schools have found it suitable as a first introduction to computer programming. BASIC is a *high-level language, which uses a small number of ordinary English words (called keywords) to give instructions to the computer. For instance, some words used are PRINT, LET, NEXT, and READ. A computer program in BASIC consists of a numbered list of instructions formulated using these words. The computer executes the program by working sequentially through the list of instructions. All the usual arithmetic operations can be handled easily in BASIC, which also enables the automatic calculation of frequently used mathematical and trigonometric functions.

A disadvantage of BASIC is that there are many versions, known as dialects, of the language. Almost every microcomputer uses a slightly different version of the language. Each version has a slightly different collection of commands and each is said, by its manufacturer, to be more 'powerful' than its rivals. This means that BASIC programs are not portable and can only be used on the machine they were written for.

**basket of currencies** A group of selected currencies used to establish a value for some other unit of currency. The *European Currency Unit's value is determined by taking a weighted average of a basket of European currencies.

**batch costing** The method of attributing costs to batches of items being produced rather than individual items. This method is mainly used where it is not appropriate to attribute cost to a single unit of production.

**batch processing** A method of processing data, using a computer, in which the programs to be executed are collected together into groups, or batches, for processing. All the information needed to execute the programs is loaded into the computer at the start so that it can work without further intervention. This contrasts with interactive processing, in which information is fed into the computer during processing. Batch processing is used for such tasks as preparing payrolls, maintaining inventory records, and producing reports.

**baud** A unit that measures the rate at which information is transmitted along a communications link. In normal computer usage, it is equivalent to *bits per second. Thus, a 300-band communications link sends 300 bits of information per second.

**Bay Street 1.** The street in Toronto in which the Toronto Stock Exchange is situated. **2.** The Toronto Stock Exchange itself. **3.** The financial institutions of Toronto collectively.

**BCAP** Abbreviation for British Code of Advertising Practice. *See* Advertising Standards Authority.

**BCC** Abbreviation for *British Coal Corporation (formerly the National Coal Board).

**bear** A dealer on a stock exchange, currency market, or commodity market who expects prices to fall. A **bear market** is one in which a dealer is more likely to sell securities, currency, or goods than to buy them. He may even sell securities, currency, or goods that he does not have. This is known as selling short or establishing a **bear position**. The bear hopes to close (or cover) his short position by buying in at a lower price the securities, currency, or goods that he has contracted to deliver. The difference between the purchase price and the original sale price represents the successful bear's profit. A concerted attempt to force prices down by one or more bears by sustained selling is called a **bear raid**. In a **bear squeeze**, sellers force prices up against someone known to have a bear position that he has to cover. *Compare* bull.

**bearer** A person who presents for payment a cheque or *bill of exchange marked "pay bearer". As a bearer cheque or bill does not require endorsement it is considered a high-risk form of transfer.

**bearer security (bearer bond)** A security for which proof of ownership is possession of the security certificate; this enables such bonds to be transferred from one person to another without registration. This is unusual as most securities are registered, so that proof of ownership is the presence of the owner's name on the security register. *Eurobonds are bearer securities, enabling their owners to preserve their anonymity, which can have taxation advantages. Bearer bonds are usually kept under lock and key, often deposited in a bank. Dividends are usually claimed by submitting coupons attached to the certificate.

**bear hug** An approach to the board of a company by another company indicating that an offer is about to be made for their shares. If the target company indicates that it is not against the merger, but wants a higher price, this is known as a **teddy bear hug**.

**bear note** *See* bull note.

**bed and breakfast** An operation on the London Stock Exchange in which a shareholder sells a holding one evening and makes an agreement with the broker to buy the same holding back again when the market opens the next morning. The object is to establish a loss, which can be set against other profits for calculating capital-gains tax. In the event of an unexpected change in the market, the deal is scrapped.

**below par** *See* par value.

**below-the-line 1.** Denoting entries below the horizontal line on a company's *profit and loss account that separates the entries that establish the profit (or loss) from the entries that show how the profit is distributed or where the funds to finance the loss have come from. **2.** Denoting advertising expenditure in which no commission is payable to an advertising agency. For example, direct mail, exhibitions, point-of-sale material, and free samples are regarded as below-the-line advertising. **3.** Denoting transactions concerned with capital, as opposed to revenue, in national accounts. *Compare* above-the-line.

**benchmark** A standard set of computer programs used to measure what a computer can do and how fast it can do it. The benchmark programs are designed to accomplish simple tasks, such as performing a large number of additions or retrieving data from memory.

**beneficial interest** The right to the use and enjoyment of property, rather than to its bare legal ownership. For example, if property is held in trust, the trustee has the legal title but the beneficiaries have the beneficial interest in equity. The beneficiaries, not the trustee, are entitled to any income from the property.

**beneficial owner** The owner of a *beneficial interest in property.

**beneficiary 1.** A person for whose benefit a *trust exists. **2.** A person who benefits under a will.

**benefit segmentation** *See* market segmentation.

**benefits in kind (income in kind)** Remuneration other than wages and salaries; they include company cars, free lunches, health insurance, cheap loans, etc. In an economy in which the main form of taxation is an income tax, benefits in kind become common because, without special provisions, they avoid tax. There is usually therefore a substantial body of legislation to enable the authorities to tax these benefits.

**benefit taxation** A form of taxation in which taxpayers pay tax according to the amounts of benefit that they receive from the system. Such a system of taxation is, in practice, very difficult to apply unless specific charges are made for specific services, such as metered electricity charges. *Compare* ability-to-pay taxation.

**Benelux** An association of countries in western Europe, consisting of Belgium, the Netherlands, and Luxembourg. Apart from geographical proximity these countries have particularly close economic interests, recognized in their 1947 *customs union. In 1958 the Benelux countries joined the *European Economic Community.

**Berne Union** The informal name for the International Union of Credit and Investment Insurers, an association of credit insurers from the main industrial countries, except Japan. Its main function is to facilitate an exchange of information, especially over credit terms. The *Export Credits Guarantee Department of the UK government is a member.

**beta stocks** *See* alpha stocks.

**BEXA** Abbreviation for *British Exporters Association.

**bid 1.** The price at which a buyer is willing to close a deal. If the seller has made an *offer, which the buyer considers too high, he may make a bid at a lower price (or on more advantageous terms). Having received a bid, the seller may accept it, withdraw, or make a *counteroffer. Once the buyer has made a bid the original offer no longer stands. **2.** *See* bid price. **3.** *See* takeover bid.

**bid price** The lower of the two prices quoted by a *market maker or institution marketing shares or units in a unit trust, i.e. it is the price at which the market maker or institution will buy shares or units from an investor. *Compare* offer price.

**BIFFEX** Abbreviation for Baltic International Freight Futures Exchange. *See* Baltic Exchange.

**Big Bang** The upheaval on the *London Stock Exchange (LSE) when major changes in operation were introduced on 27 October, 1986. The major changes enacted on that date were: (a) the abolition of LSE rules enforcing a dual-capacity system; (b) the abolition of fixed commission rates charged by *stockbrokers to their clients. The measures were introduced by the LSE in return for an undertaking by the government (given in 1983) that they would not prosecute the LSE under the Restrictive Practices Act. Since 1986 the Big Bang has also been associated with the *globalization and modernization of the London securities market.

**Big Eight** The eight largest firms of accountants in the world, i.e. Arthur Andersen, Coopers and Lybrand, Deloitte Haskins and Sells, Ernst and Whinney, Peat Marwick Mitchell, Price Waterhouse, Touche Ross, and Arthur Young.

**Big Four** The four largest UK commercial banks, i.e. Barclays, Lloyds, Midland, and National Westminster.

**bilateral monopoly** A situation in which a *monopoly seller bargains with a monopoly buyer (*see* monopsony). The classic application of bilateral monopoly is to the negotiation of wages between a union and a firm. The most famous solution (1950) is that stated by John Nash (*see* Nash equilibrium), which suggests that both sides will settle for the wage that maximizes the benefits from cooperation.

**bilateral trade agreement** An agreement on trade policy between two countries, usually concerning a reduction in tariffs or other protective barriers. Although bilateral deals tend to fragment world trade, governments are attracted by their relative simplicity. The USA, in particular, has used such deals in free-trade agreements with Mexico and Canada and to hasten the resolution of problems with particular trading partners, notably Japan and other Asian countries.

**bill broker (discount broker)** A broker who buys *bills of exchange from traders and sells them to banks and *discount houses or holds them to maturity. Many now deal exclusively in *Treasury bills.

**billion** Formerly, one thousand million ($10^9$) in the USA and one million million ($10^{12}$) in the UK; now it is almost universally taken to be one thousand million.

**bill of entry** A detailed statement of the nature and value of a consignment of goods prepared by the shipper of the consignment for *customs entry.

**bill of exchange** An unconditional order in writing, addressed by one person (the drawer) to another (the drawee) and signed by the person giving it, requiring the drawee to pay on demand or at a fixed or determinable future time a specified sum of money to or to the order of a specified person (the payee) or to the bearer. If the bill is payable at a future time the drawee signifies his *acceptance, which makes him the party primarily liable upon the bill; the drawer and endorsers may also be liable upon a bill. The use of bills of exchange enables one person to transfer to another an enforceable right to a sum of money. A bill of exchange is not only transferable but also negotiable, since if a person without an enforceable right to the money transfers a bill to a *holder in due course, the latter obtains a good title to it. Much of the law on bills of exchange is codified by the Bills of Exchange Act (1882). *See* accommodation bill; bills in a set; dishonour.

**bill of lading** A document acknowledging the shipment of a consignor's goods for carriage by sea. It is used primarily when the ship is carrying goods belonging to a number of consignors (a general ship). In this case, each consignor receives a bill issued (normally by the master of the ship) on behalf of either the shipowner or a charterer under a charterparty. The bill serves three functions: it is a receipt for the goods; it summarizes the terms of the contract of carriage; and it acts as a document of title to the goods. A bill of lading is also issued by a shipowner to a charterer who is using the ship for the carriage of his own goods. In this case, the terms of the contract of carriage are in the charterparty and the bill serves only as a receipt and a document of title. During transit, ownership of the goods may be transferred by delivering the bill to another if it is drawn to bearer or by endorsing it if it is drawn to order. It is not, however, a negotiable instrument. The bill gives details of the goods; if the packages are in good order a **clean bill** is issued; if they are not, the bill will say so (*see* dirty bill of lading). *See also* containerization.

**bill of quantities** A document drawn up by a quantity surveyor showing in detail the materials and parts required to build a structure (e.g. factory, house, office block), together with the price of each component and the labour costs. The bill of quantities is one of the tender documents that goes out to contractors who wish to quote for carrying out the work.

**bill of sale 1.** A document by which a person transfers the ownership of goods to another. Commonly the goods are transferred conditionally, as security for a debt, and a **conditional bill of sale** is thus a mortgage of goods. The mortgagor has a right to redeem the goods on repayment of the debt and usually remains in possession of them; he may thus obtain false credit by appearing to own them. An **absolute bill of sale** transfers ownership of the goods absolutely. The Bills of Sale Acts (1878 and 1882) regulate the registration and form of bills of sale. **2.** A document recording the change of ownership when a ship is sold; it is regarded internationally as legal proof of ownership.

**bill of sight** A document that an importer, who is unable fully to describe a cargo he is importing, gives to the Customs and Excise authorities to authorize them to inspect the goods on landing. After the goods have been landed and the importer supplies the missing information the entry is completed and the importer is said to have **perfected the sight**.

**bill of sufferance** A document issued by the Customs and Excise enabling goods to be landed in the absence of detailed documentation, subject to the Customs being able to examine the goods at any time.

**bill rate (discount rate)** The rate on the *discount market at which *bills of exchange are discounted (i.e. purchased for less than they are worth when they mature). The rate will depend on the quality of the bill and the risk the purchaser takes. First-class bills, i.e. those backed by banks or well-respected finance houses, will be discounted at a lower rate than bills involving greater risk.

**bills in a set** One of two, or more usually three, copies of a foreign bill of exchange. Payment is made on any one of the three, the others becoming invalid on the payment of any one of them. All are made out in the same way, except that each refers to the others. The first copy is called the **first of exchange**, the next is the **second of exchange**, and so on. The duplication or triplication is to reduce the risk of loss in transit.

**bills payable** An item that may appear in a firm's accounts under current liabilities, summarizing the *bills of exchange being held, which will have to be paid when they mature.

**bills receivable** An item that may appear in a firm's accounts under current assets, summarizing the *bills of exchange being held until the funds become available when they mature.

**BIM** Abbreviation for *British Institute of Management.

**binary digit (bit)** Either of the digits 0 or 1. *See also* binary notation.

**binary notation** A way of writing numbers using two symbols only. The symbols are usually written as 0 and 1, and called **binary digits** or **bits**. These bits are usually grouped in eights, called *bytes. Numbers written in binary notation are generally much longer than their decimal equivalents. For example, the binary number 1111 is equivalent to the decimal number 15. In a binary number the columns from the right represent 'units', 'twos', 'fours', 'eights', and so on. Binary notation is used in computing because digital computers represent numbers in terms of the presence (1) or absence (0) of an electrical pulse.

**BIS** Abbreviation for *Bank for International Settlements.

**bit** Abbreviation for binary digit. *See* binary notation.

**black economy** Economic activity that is undisclosed, as to disclose it would render the earnings involved liable to taxation or even cause those engaged to be imprisoned (if they are claiming state benefits and have lied about their earnings). In economies in which production and exchange are regulated by the state (*see* command economies), the black economy is generally considered to be very large, although by its nature it is unmeasurable. Even in *mixed economies, however, the black economy is thought to be significant, largely due to the benefits of evading tax. Earnings made in the black economy do not appear in national statistics.

**black knight** A person or firm that makes an unwelcome *takeover bid for a company. *Compare* grey knight; white knight.

**blackleg 1.** An employee who refuses to join a trade union. **2.** An employee who refuses to stop work when the rest of his co-unionists have declared a strike.

**black market** An illegal market for a particular good or service. It can occur when regulations control a particular trade (as in arms dealing) or a particular period (as in wartime). *Compare* grey market.

**Black Monday** Either of the two Mondays on which the two largest stock market crashes occurred in this century. The original Wall Street crash occurred on Monday, 28 October, 1929 when the *Dow Jones Industrial Average fell by 13%. On Monday, 19 October, 1987 the Dow Jones Average lost 23%. In both cases Black Monday in the USA triggered heavy stock market falls around the world.

**blank bill** A *bill of exchange in which the name of the payee is left blank.

**blank cheque** *See* cheque.

**blank endorsed** *See* endorsement.

**blanket policy** An insurance policy that covers a number of items but has only one total sum insured and no insured sums for individual items. The policy can be of any type, e.g. covering a fleet of vehicles or a group of buildings. A **blanket motor-insurance certificate** covers any vehicle owned by the policyholder without specifying the registration number of individual vehicles.

**blank transfer** A share transfer form in which the name of the transferee and the transfer date are left blank. The form is signed by the registered holder of the shares so that the holder of the blank transfer has only to fill in the missing details to become the registered owner of the shares. Blank transfers can be deposited with a bank, when shares are being used as a security for a loan. A blank transfer can also be used when shares are held by *nominees, the beneficial owner holding the blank transfer.

**blind testing** A market-research technique in which unidentified products are tested by consumers in order to determine their preference. It is often used to test a new product against an established product, especially one with a strong market position.

**block** A group of numbers, characters, or words that is transferred as a unit between the parts of a computer system, for example between the computer terminal and a disk drive. Also, when information is stored on disks or tape, it is divided into blocks, stored on separate physical areas of the disk or tape. The blocks are separated from each other by inter-block gaps.

**blocked account 1.** A bank account from which money cannot be withdrawn, for any of a number of reasons, of which the most likely is that the affairs of the holder of the account are in the hands of a receiver owing to his *bankruptcy, or *liquidation in the case of a company. **2.** A bank account held by an exporter of goods in another country into which the proceeds of the sale of his goods have been paid but from which they cannot be transferred to the exporter's own country. This is usually a result of a government order, when that government is so short of foreign currency that it has to block all accounts that require the use of a foreign currency.

**blocked currency** A currency that cannot be removed from a country as a result of *exchange controls.

***Blue Book*** *See UK National Accounts.*

**blue button** A trainee stockbroker on the London Stock Exchange, who can be distinguished by the small blue button badge worn in the lapel.

**blue chip** An ordinary share in a substantial company with a well-known name, a good growth record, and large assets. Blue chips are not precisely definable, but the main part of an institution's equity portfolio will consist of blue chips.

**blue-collar worker** A manual worker, normally one working on the shop floor, as opposed to an office worker, who is known as a *white-collar worker. The blue collar refers to the blue overalls often worn in factories and the white collar to the normal office attire of a white shirt and a tie.

**Board of Customs and Excise** The government department responsible for collecting and administering customs and excise duties and *VAT. The Commissioners of Customs were first appointed in 1671 by Charles II; the Excise department, formerly part of the Inland Revenue Department, was merged with the Customs in 1909. The Customs and Excise have an investigation division responsible for preventing and detecting evasions of revenue laws and for enforcing restrictions on the importation of certain goods (e.g. arms, drugs, etc.). Their statistical office compiles overseas trade statistics from customs import and export documents.

**board of directors** *See* director.

**Board of Inland Revenue** A small number of higher civil servants, known individually as Commissioners of Inland Revenue, responsible to the Treasury for the administration and collection of the principal direct taxes in the UK, but not the indirect *VAT and *excise duties. They are responsible for income tax, capital-gains tax, corporation tax, inheritance tax, petroleum revenue tax, and stamp duties. They also advise on new legislation and prepare statistical information.

**boilerplate** A copy intended for use in making other copies. It is sometimes used to describe a group of instructions that is incorporated in different places in a computer program or the detailed standard form of words used in a contract, guarantee, etc.

**boldface** Any style of typeface in which the characters are printed in heavy black ink, such as those in the first word of this entry.

**bona fide** In good faith, honestly, without collusion or fraud. A bona fide purchaser for value without notice is a person who has bought property in good faith, without being aware of prior claims to it (for example, that it is subject to a trust). He will not be bound by those claims, unless (if the property is land) they were registered.

**bona vacantia** (Latin: empty goods) Goods without an owner. By royal prerogative, the Crown is entitled to any personal property that has no other owner, such as the goods of a person who dies intestate with no living relatives. The prerogative does not apply to real property, but the Crown may be entitled to that as well under the doctrine of escheat (the return of ownerless land to the superior landowner). *See* fee simple.

**bond** An IOU issued by a borrower to a lender. Bonds usually take the form of *fixed-interest securities issued by governments, local authorities, or companies. However, bonds come in many forms: with fixed or variable rates of interest, redeemable or irredeemable, short- or long-term, secured or unsecured, and marketable or unmarketable. *See also* debenture; deposit bonds; in bond; income bond, premium bonds.

**bonded goods** Imported goods on which neither customs duty nor excise have been paid although the goods are dutiable. They are stored in a *bonded warehouse until the duty has been paid or the goods re-exported.

**bonded warehouse** A warehouse, usually close to a sea port or airport, in which goods that attract customs duty or excise are stored after being imported, pending payment of the duty or the re-export of the goods. The owners of the warehouse are held responsible for ensuring that the goods remain in bond until the duty is paid; they may only be released in the presence of a customs officer.

**bond note** A document, signed by an officer of the Customs and Excise, that enables goods to be released from a bonded warehouse, usually for re-export.

**bond washing** *See* dividend stripping.

**Bonn Index FAZ** *See* Commerzbank Index.

**bonus 1.** An extra payment made to employees by management, usually as a reward for good work, to compensate for something (e.g. dangerous work) or to share out the profits of a good year's trading. **2.** An extra amount of money additional to the proceeds, which is distributed to a policyholder by an insurer who has made a profit on the investment of a life-assurance fund. Only holders of *with-profits policies are entitled to a share in these profits and the payment of this bonus is conditional on the life assurer having surplus funds after claims, costs, and expenses have been paid in a particular year. **3.** Any extra or unexpected payment. *See also* no-claim bonus; reversionary bonus; terminal bonus.

**bonus issue** *See* scrip issue.

**book-keeping** The keeping of the *books of account of a business. The records kept enable a *profit and loss account and the *balance sheet to be compiled. Most firms now use *business software packages of programs to enable the books to be kept by computer.

**book of prime entry** A book or record in which certain types of transaction are recorded before becoming part of the *double-entry book-keeping system. The most common book of prime entry are the *day books, the *cash book, and the *journal.

**books of account** The books in which a business records its transactions. *See* book of prime entry; double-entry book-keeping system.

**book value** The value of an asset as recorded in the *books of account of an organization. This is normally the historical cost of the asset reduced by amounts written off for *depreciation. If the asset has ever been revalued, the book value will be the amount of the revaluation less amounts subsequently written off for depreciation. Except at the time of purchase of the asset, the book value will rarely be the same as the market value of the asset.

**boom** The part of the *business cycle that follows a recovery, in which the economy is working at full capacity. Demand, prices, and wages rise, while unemployment falls. If government control of the economy is not sufficiently tight a boom can lead to a recession and ultimately to a slump.

**bootstrap 1.** A cash offer for a controlling interest in a company. This is followed by an offer to acquire the rest of the company's shares at a lower price. The purpose is twofold: to establish control of the company and to reduce the cost of the purchase. **2.** A technique enabling a computer to load a program of instructions. Before computer hardware can function, a program must be loaded into it. However, as a program is needed in the computer to enable it to load a program, preliminary instructions are stored permanently in the computer making it possible for longer programs to be accepted. These preliminary instructions are called a **bootstrap loader**. To **boot** a program is to load the program using the bootstrap loader.

**BOTB** Abbreviation for *British Overseas Trade Board.

**bottom** An old mercantile name for a ship, which still occurs in references to shipments of cargo being "made in one bottom" (i.e. in the same ship) and in bottomry bonds (*see* hypothecation).

**bottomry bond** *See* hypothecation.

**bought deal** A method of raising capital for acquisitions or other purposes, used by quoted companies as an alternative to a *rights issue or *placing. The company invites *market makers or banks to bid for new shares, selling them to the highest bidder, who then sells them to the rest of the market in the expectation of making a profit. Bought deals originated in the USA and are becoming increasingly popular in the UK, although they remain controversial as they violate the principle of *pre-emption rights. *See also* vendor placing.

**bought note** *See* contract note.

**bounded rationality** A form of *rationality in which it is assumed that some economic agents will not go to the lengths of calculation required by full rationality in order to attain either *utility or *profit maximization but will instead follow various empirical rules determined by the complexity of the real situations they encounter. The concept was formulated by Herbert Simon (1916– ) and introduced into his analysis of managerial decision-making in the 1950s. Bounded rationality produces different predictions about economic behaviour from strict rationality, although it has proved extremely difficult to specify how far the bounds of rationality extend.

**bourse** A French *stock exchange (from the French *bourse*, purse). 'The Bourse' usually refers to the stock exchange in Paris, but other continental stock markets are also known by this name. Members of the Paris Bourse, who have to buy their membership for a large sum, are known as *agents de change*.

**boutique** An office, usually with a shop front and located in a shopping parade, that offers financial advice to investors, often on a walk-in basis.

**bracket indexation** A change in the upper and lower limits of any particular *taxation bracket in line with an index of inflation. This is needed in times of inflation to avoid fiscal drag (the tax system collecting unduly large amounts of tax).

***BRAD*** Abbreviation for **British Rates and Data.*

**brainstorming** A group discussion in order to invoke ideas and solve business problems. No idea is rejected, no matter how irrelevant it appears, until it has been thoroughly discussed and evaluated. A major rule of brainstorming is that the discussion of ideas should not be inhibited.

**brand (brand name)** A tradename used to identify a specific product, manufacturer, or distributor. The sale of most branded products began in the UK at the turn of the century (some, such as Bovril (Trademark) and Horlicks (Trademark), were mid-Victorian) when manufacturers wanted to distinguish their goods from those of their competitors. As consumers became more sophisticated, manufacturers placed more emphasis upon promoting their brands directly to consumers (rather than to distributors), spending considerable sums on advertising the high quality of their products. Manufacturers believe that if they invest in the quality of their brands, consumers will respond by asking for their goods by their brand names and by being willing to pay a premium for them; manufacturers also believe they will be less susceptible to demands from distributors for extra discounts to stock their brands. For some products (e.g. perfumes and alcoholic drinks), considerable effort has gone into promoting brands to reflect the personality of their likely purchasers; market research has indeed shown that for these products consumers can be persuaded to buy brands that enhance the image they have of themselves. *See* generic; own brand.

**brand loyalty** Support by consumers for a particular *brand or product. Brand loyalty is usually the result of continued satisfaction with a product or its price and is reinforced by effective and heavy advertising. Strong brand loyalty, which is often subjective or subconscious, reduces the impact of competitive brand promotions and discourages switching between brands, unless it is for an improved product.

**brand manager (product manager)** The executive responsible for the overall marketing and promotion of a particular branded product. His responsibilities range from setting objectives for the product to managing and coordinating its sale. Brand managers emerged in the 1930s at Proctor and Gamble but are now to be found employed by most producers of consumer goods.

**breach of contract** A failure by a party to a contract to perform his obligations under that contract or an indication of his intention not to do so. An indication that a contract will be breached in the future is called **repudiation** or an **anticipatory breach**; it may be either expressed in words or implied from conduct. Such an implication arises when the only reasonable inference from a person's acts is that he does not intend to fulfil his part of the bargain. For example, an anticipatory breach occurs if a person contracts to sell his car to A but sells and delivers it to B before the delivery date agreed with A. The repudiation of a contract entitles the injured party to treat the contract as discharged and to sue immediately for *damages for the loss sustained. The same procedure only applies to an actual breach if it constitutes a **fundamental breach**, i.e. a breach of a major term (*see* condition) of the contract. In either an anticipatory or an actual breach, the injured party may, however, decide to affirm the contract instead (*see* affirmation of contract). When an actual breach relates only to a minor term of the contract (a warranty) the injured party may sue for damages but has no right to treat the contract as discharged. The process of treating a contract as discharged by reason of repudiation or actual breach is sometimes referred to as *rescission. Other remedies available under certain circumstances for breach of contract are an *injunction and *specific performance.

**breach of warranty** *See* warranty.

**break-even point** The sales quantity of a product required to product an income to cover costs of overheads and production without profit or loss. Once the break-even point is passed the producer will be making a profit.

**break-forward** A contract on the money market that combines the features of a *forward-exchange contract and a currency option. The forward contract can be undone at a previously agreed rate of exchange, enabling the consumer to be free if the market moves in his favour. There is no premium on the option; the cost is built into the fixed rate of exchange.

**breaking a leg** *See* straddle.

**break-up value** **1.** The value of an asset on the assumption that an organization will not continue in business. On this assumption the assets are likely to be sold piecemeal and probably in haste. **2.** The *asset value per share of a company.

**bribery and corruption** Offences relating to the improper influencing of people in positions of trust. The offences commonly grouped under this expression are now statutory. Under the Public Bodies Corrupt Practices Act (1889), amended by the Prevention of Corruption Act (1916), it is an offence corruptly to offer to a member, officer, or servant of a public body any reward or advantage to do anything in relation to any matter with which that body is concerned; it is also an offence for a public servant or officer to corruptly receive or solicit such a reward. The Prevention of Corruption Act (1906) amended by the 1916 Act is wider in scope. Under this Act it is an offence corruptly to give or offer any valuable consideration to an agent to do any act or show any favour in relation to his principal's affairs.

**bridging loan** A loan taken on a short-term basis to bridge the gap between the purchase of one asset and the sale of another. It is particularly common in the property and housing market.

**Britannia coins** A range of four British *gold coins (£100, £50, £25, and £10 denominations). They were introduced in October 1987 for investment purposes, in competition with the *Krugerrand. Although all sales of gold coins attract VAT, Britannia coins are widely dealt in as bullion coins.

**British Airports Authority plc (BAA)** A public limited company floated on the London Stock Exchange in 1987 and formed from the former British Airports Authority (founded in 1966). It owns and operates London airports (Heathrow, Gatwick, and Stansted) as well as Aberdeen, Edinburgh, Prestwick, and Glasgow airports. It is responsible for the construction and maintenance of buildings, fire and security services, passenger services, and terminal management.

**British Bankers Association** An organization founded in 1920 to represent the views of the recognized UK banks (approximately 300). Membership is open to all banks established under British law with head offices in the UK. The Association is a member of the European Community Banking Federation.

**British Coal Corporation (BCC)** The nationalized corporation, formerly called the **National Coal Board (NCB)**, which owns and runs all British coal mines. The NCB itself took control of the mines in 1947.

**British Code of Advertising Practice (BCAP)** *See* Advertising Standards Authority.

**British Exporters Association (BEXA)** An association, formerly the British Export Houses Association (BEHA), that puts UK suppliers in touch with Association members, who trade and finance trade throughout the world.

**British Institute of Management (BIM)** An institution set up in 1974 by the then Board of Trade to promote professionalism in management practice and to provide information for its members. It promotes courses in management and those with a diploma in management can become associate members (AMBIM); there are also fellows (FBIM) and members (MBIM) of the institute.

**British Insurance and Investment Brokers Association** A trade association for insurance brokers registered with the *Insurance Brokers Registration Council and investment brokers registered under the Financial Services Act (1986). Formed in 1977 as the British Insurance Brokers Association by the amalgamation of a number of insurance broking associations, it changed to its current name in 1988 to widen its membership to include investment advisors. It provides public relations, free advice, representation in parliament, and a conciliation service for consumers.

**British Overseas Trade Board (BOTB)** An organization set up by the Department of Trade and Industry, whose members are drawn mainly from industry and commerce. The Board was formed in 1972 to advise on overseas trade and the official export-promotion programme. It liaises between government and private industry to expand overseas trade, advises new exporters on foreign tariffs and regulations, assists exporters in displaying products in trade fairs, etc.

***British Rates and Data* (*BRAD*)** A monthly publication listing addresses, cover price, circulation, frequency, rate cards, copy and cancellation requirements, and advertising representatives for national and provincial newspapers, consumer, trade, technical, and professional publications, and for television and radio stations.

**British Standards Institution (BSI)** An institution founded in 1901, which received a royal charter in 1929 and took its present name in 1931. Its function is to formulate standards for building, engineering, chemical, textile, and electrical products, ensuring that they maintain a specified quality. Products so standardized make use of the **Kite mark** logo as a symbol of quality. Manufacturers who use the Kite mark do so under licence from the BSI on condition that products are subject to regular inspection. Apart from maintaining quality standards in this way, the BSI attempts to ensure that the design of goods is restricted to a sensible number of patterns and sizes for one purpose, to avoid unnecessary variety. The BSI, which collaborates closely with the *International Standards Organization, is also actively concerned in metrology, providing information on units of measurement and issuing glossaries defining technical words.

**British Technology Group (BTG)** A government-appointed organization formed in 1981 by the merger of the **National Enterprise Board** (**NEB**) and the **National Research and Development Corporation** (**NRDC**). Its purpose is to encourage technological development by providing finance for new scientific and engineering products and processes discovered through research at UK universities, polytechnics, research councils, and government research establishments.

**British Telecom** The British Telecommunications Corporation was formed in 1981 as a public corporation to control the UK telephone and telecommunications system, which had previously been the responsibility of the Post Office. In 1984 this corporation became British Telecommunications plc, when 51% of the shares were sold to the public. British Telecom is now licensed to run telecommunications throughout the UK.

**British Textile Confederation** An organization formed in 1972 to represent the interests of textile manufacturers, on issues of trading and economic policies, to the UK government and within the European Community.

**British Waterways Board** A board set up under the Transport Act (1962) to provide services and facilities for UK inland waterways. The Board's responsibilities extend over approximately 2000 miles of waterways and 90 reservoirs.

**broad money** An informal name for M3 (*see* money supply). *Compare* narrow money.

**broker** An agent who brings two parties together, enabling them to enter into a contract to which he is not a principal. His remuneration consists of a **brokerage**, which is usually calculated as a percentage of the sum involved in the contract but may be fixed according to a tariff. Brokers are used because they have specialized knowledge of certain markets or to conceal the identity of a principal, in addition to introducing buyers to sellers. *See* aviation broker; bill broker; commodity broker; insurance broker; shipbroker; stockbroker.

**brokerage** The commission earned by a *broker.

**broker/dealer** A member of the *London Stock Exchange who, since the *Big Bang, has functioned both as a stockbroker and a jobber, but only in a single capacity in one particular transaction.

**brown goods** Televisions, hi-fi equipment, etc., which are usually housed in wood or imitation wood cabinets. *Compare* white goods.

**BSI** Abbreviation for *British Standards Institution.

**BTEC** Abbreviation for Business and Technical Education Council. *See* vocational training.

**BTG** Abbreviation for *British Technology Group.

**bucket shop** *Slang*. A derogatory term for a firm of brokers, dealers, agents, etc., of questionable standing and frail resources, that is unlikely to be a member of an established trade organization.

**budget 1.** A financial plan setting targets for the revenues, expenditures, etc., of an organization for a specified period. **2. (the Budget)** The UK financial plan for the coming year, presented to Parliament, usually in March or April, by the Chancellor of the Exchequer. It reveals his predictions for the economy and his taxation changes for the coming year.

**budgetary control** A means of controlling the progress of a business in which budgets submitted by each department are agreed by senior management and subsequently compared with actual results during the course of the ensuing accounting period. The use of computers enables any discrepancies to be rapidly noted so that the appropriate action can be taken, or the budget altered, without waiting for the accounts to be prepared at the end of the accounting period. The budgets will normally include output, sales revenue, expenses, efficiency, cash flow, etc., and it is regarded as a principle of budgetary control that departmental managers should only be held responsible for those revenues or expenditures that they can themselves control.

**budget constraint** The different combinations of goods that can be bought with a given income at prevailing prices; this can be represented as a curve on a graph. In *consumer theory the point on a curve of an individual's budget constraint that touches his highest *indifference curve represents the point of *utility maximization. It is also possible to think of governments or firms as facing a budget constraint when choosing their level of expenditure.

**budget deficit** The excess of government expenditure over government income, which must be financed either by borrowing or by printing money. Keynesians have advocated that governments should run budget deficits during *recessions in order to stimulate *aggregate demand (*see* pump priming). Monetarists and new classical macroeconomists, however, argue that budget deficits simply stimulate *inflation and crowd out private investment. Most economists now argue that, at least on average, governments should seek a balanced budget and that persistent deficits should be eliminated, either by reducing expenditure or increasing taxation. In some cases a **budget surplus** can be used during a *boom to collect more revenue than is being spent.

**budget surplus** *See* budget deficit.

**buffer** A computer store or memory used when information is transmitted from one unit to another working at different speeds. For example, a slow printer has a built-in buffer to allow it to be connected to a high-speed computer. The computer sends information to the buffer far more rapidly than the printer can handle it. The information is temporarily stored in the buffer and printed out at the natural speed of the printer. When the buffer empties it sends a message to the computer that it is ready to receive more information. The process continues until all the information has been printed.

**buffer stock** A stock of a commodity owned by a government or trade organization and used to stabilize the price of the commodity. Usually the manager of the buffer stock is authorized to buy the commodity in question if its price falls below a certain level, which is itself reviewed periodically, to enable producers to find a ready market for their goods at a profitable level. If the price rises above another fixed level, the buffer stock manager is authorized to sell the commodity on the open market. Thus, producers are encouraged to keep up a steady supply of the commodity and users are reassured that its price has a ceiling. This arrangement is often effective, but may collapse during a boom or slump.

**bug** An error in a computer program, or a malfunction in a computer system. To **debug** a program is to find and correct all bugs.

**building society** A financial institution that accepts deposits, upon which it pays interest, and makes loans for house purchase or house improvement secured by *mortgages. They developed from the *Friendly Society movement in the late 17th century and are non-profit making. They are regulated by the Building Society Act (1986). The societies accept deposits into a variety of accounts, which offer different interest rates and different withdrawal terms, or into 'shares', which often require longer notice of withdrawal. Interest on all building-society accounts is paid net of income tax, the society paying the tax direct to the Inland Revenue. The societies attract both large and small savers, with average holdings being about £5000.

Loans made to persons wishing to purchase property are usually repaid by regular monthly instalments of capital and interest over a number of years. Another method, which is growing in popularity, is an endowment mortgage in which the capital remains unpaid until the maturity of an assurance policy taken out on the borrower's life; in these arrangements only the interest and the premiums on the assurance policy are paid during the period of the loan.

Since the 1986 Act, building societies have been able to widen the range of services they offer; this has enabled them to compete with the high-street banks in many areas. They offer cheque accounts, which pay interest on all credit balances, cash cards, credit cards, loans, money transmission, foreign exchange, personal financial planning services (shares, insurance, pensions, etc.), estate agency, and valuation and conveyancing services. The distinction between banks and building societies is fast disappearing, indeed some building societies have obtained the sanction of their members to become *public limited companies. These changes have led to the merger of many building societies to provide a national network that can compete with the *Big Four banks. Competition is well illustrated in the close relationship of interest rates between banks and building societies as they both compete for the market's funds. Moreover, the competition provided by the building societies has forced the banks into offering free banking services, paying interest on current accounts, and Saturday opening.

**Building Societies Ombudsman** *See* Financial Ombudsman.

**built-in obsolescence** *See* obsolescence.

**built-in software** Computer programs that are built into a computer and available automatically when the computer is switched on. Examples include the *wordprocessors, *spreadsheets, and graphics programs that are built into some desktop terminals used by business executives.

**bulk carrier** A ship in which the cargo is carried in bulk, rather than in bags, containers, etc. Bulk cargoes are usually homogeneous and capable of being loaded by gravity; examples include ores, coal, wheat, etc.

**bull** A dealer on a stock exchange, currency market, or commodity market who expects prices to rise. A **bull market** is one in which a dealer is more likely to be a buyer than a seller, even to the extent of buying for his own account and establishing a **bull position**. A bull with a long position hopes to sell his purchases at a higher price after the market has risen. *Compare* bear.

**bulldog bond** A fixed-interest *bond issued in the UK by a foreign borrower.

**bullet 1.** A security offering a fixed interest and maturing on a fixed date. **2.** The final repayment of a loan, which consists of the whole of sum borrowed. In a **bullet loan**, interim repayments are interest-only repayments, the principal sum being repaid in the final bullet.

**bullion** Gold, silver, or some other precious metal used in bulk, i.e. in the form of bars or ingots rather than in coin. Central banks use gold bullion in the settlement of international debts. In the **London bullion market**, bullion brokers act as agents for both buyers and sellers and also trade as principals.

**bull note** A bond whose redemption value is linked to a price index (e.g. FTSE 100 Index; *see* Financial Times Share Indexes) or a commodity price (e.g. the price of gold). Thus, a holder of a bull note will receive on redemption an amount greater than the principal of the bond if the relevant index or price has risen (but less if it has fallen). With a **bear note** the reverse happens. Bull and bear notes are therefore akin to an ordinary bond plus an *option, providing opportunities for hedging and speculating.

**bundling** The marketing ploy of giving away a relatively cheap product with a relatively expensive one to attract customers; for example, giving a number of free audio cassettes with each purchase of a music centre.

**Bureau of Customs** The US equivalent of the British Board of Customs and Excise. It is a part of the Department of the Treasury.

**burn-out turnaround** The process of restructuring a company that is in trouble by producing new finance to save it from liquidation, at the cost of diluting the shareholding of existing investors.

**burn rate** The rate at which a new company uses up its venture capital to fund fixed overheads before cash begins to come in from its trading activities.

**Business and Technical Education Council (BTEC)** *See* vocational training.

**business cycle (trade cycle)** The process by which investment, output, and employment in an economy tend to fluctuate up and down in a regular pattern causing *booms and *depressions, with *recession and recovery as intermediate stages. The business cycle is one of the major unsolved mysteries in economics. Both *market

clearing (*see also* new classical macroeconomics) and Keynesian theories have failed to account convincingly for the facts, and *fine-tuning policies aimed at smoothing over business cycles have rarely been successful.
One explanation of business cycles is that they operate as a result of political activity. The earliest suggestion of the political aspect came from Karl Marx and was developed by Michal Kelecki (1899–1970). He suggested that in a boom period workers become economically powerful and demand higher wages – the subsequent recession is engineered by capitalists in order to create unemployment and reduce the power of the workers. More recent political theories ascribe the cause to the desire of political parties to achieve re-election; it is easier for a party to impose hard decisions shortly after an election and then stimulate demand in the run-up to the next election, thus creating a business cycle.

**business-interruption policy (consequential-loss policy; loss-of-profits policy)** An insurance policy that pays claims for financial losses occurring if a business has to stop or reduce its activities as a result of a fire or any other insurable risk. Claims can be made for lost profit, rent, rates, and other unavoidable overhead costs that continue even when trading has temporarily ceased.

**business judgment rule** The rule that the courts will not generally interfere in the conduct of a business. For example, the courts will not substitute their judgment for that of the directors of a company unless the directors are acting improperly. The rule is often invoked when directors are accused of acting out of self-interest in *takeover bids.

**business marketing research** *See* industrial marketing research.

***Business Monitor*** A government publication produced by the *Business Statistics Office giving information and data on production, services and distribution, civil aviation, car registrations, and cinema audiences.

**business name** The name under which a business trades. According to the Business Names Act (1985), if a business is conducted under a name other than that of the proprietor, it must display at its place of business the business name, its activity, and the name of the proprietor. All business stationery must carry both the business name and the names of the owners. The name of the business must be registered with the *Registrar of Companies.

**business plan** A detailed plan setting out the objectives of a business over a stated period, often three, five, or ten years. A business plan is drawn up by many businesses, especially if the business has passed through a bad period or if it has had a major change of policy. For new businesses it is an essential document for raising capital or loans. The plan should quantify as many of the objectives as possible, providing monthly *cash flows and production figures for at least the first two years, with diminishing detail in subsequent years; it must also outline its strategy and the tactics it intends to use in achieving its objectives. Anticipated *profit and loss accounts should form part of the business plan on a quarterly basis for at least two years, and an annual basis thereafter. For a group of companies the business plan is often called a **corporate plan**.

**business reply service** A service offered by the Post Office enabling a company to supply its customers with a prepaid business reply card, envelope, or label (either first- or second-class postage) so that they can reply to direct-mail shots, ask for follow-up literature, pay bills promptly, etc., free of postal charges.

**business software package** One of a wide range of software programs sold in packages to enable computers to be used for a variety of business uses. They range in complexity and expense from those needed to operate a PC to the suite of programs required by a mainframe. A typical package would include one or more of: book-keeping programs, which provide facilities for keeping sales, purchase, and nominal ledgers; accounting packages, enabling balance sheets, budgetary control, and sale and purchase analysis to be undertaken automatically; payroll packages, dealing with wages, salaries, PAYE, National Insurance, pensions, etc.; *database management systems to maintain company records; communications software to allow two or more computers to work together; and *wordprocessors. The programs comprising the package are designed to work together and use each other's data; sometimes a single program provides one or more of these functions.

**Business Statistics Office** Until August, 1989, a department of the Department of Trade and Industry, since then a department of the *Central Statistical Office. It collects statistics of British businesses and publishes **Business Monitors*.

**butterfly** A strategy used by dealers in traded *options. It involves simultaneously purchasing and selling call options (the right to buy) at different *exercise prices or different *expiry dates. A butterfly is most profitable when the price of the underlying security fluctuates within narrow limits. *Compare* straddle.

**buy-back** The buying back by a company of its shares from an investor, who put venture capital up for the formation of the company. The shares are bought back at a price that satisfies the investor, which has to be the price the company is willing to pay for its independence. The buy-back may occur if the company is publicly floated or is taken over.

**buy earnings** To invest in a company that has a low *yield but whose earnings are increasing, so that a substantial capital gain can be expected.

**buyers' market** A market in which the supply exceeds the demand, so that buyers can force prices down. At some point, however, sellers will withdraw from the market if prices fall too low; when this happens the supply will fall off and prices will begin to rise again. *Compare* sellers' market.

**buyers over** A market in securities, commodities, etc., in which the sellers have sold all they wish to sell but there are still buyers. This is clearly a strong market, with an inclination for prices to rise. *Compare* sellers over.

**buygrid** The three classes of industrial buying: rebuying, which usually entails reordering from existing suppliers; modified rebuying, which involves reassessing existing policy, often seeking improved quality; and new-task buying, in which an organization buys something it has not bought before.

**buying in** The buying of securities, commodities, etc., by a broker because the original seller has failed to deliver. This invariably happens after a rise in a market price (the seller would be able to buy in himself if the market had fallen). The broker buys at the best price available and the original seller is responsible for any difference between the buying-in price and the original buying price, plus the cost of buying in. *Compare* selling out.

**buyout** **1.** An option, open to a member of an *occupational pension scheme on leaving, of transferring the benefits already purchased to an insurance company of his or her own choice. **2.** *See* leveraged buyout. **3.** *See* management buyout.

**byte** A unit of information in a computer, consisting of a group of *binary digits, that represents a number or character. Most small computers use a byte with 8 binary digits; larger computers use a byte with 16 digits. Byte and character are often used synonymously. For example, a *floppy disk holding 180 000 bytes will store 180 000 characters. (However, slightly less than 180 000 characters are available to the user because some disk space is needed by the computer to index and lay out the data on the disk.)

# C

**C** A language used to program computers. It was developed in the mid-1970s by the Bell laboratories in the USA and is now much used in university research. There are also many business programs written in C. It is a flexible *high-level language that produces efficient programs. Programs written in C run much faster than equivalent programs written in some other languages, such as *BASIC.

**CA 1.** Abbreviation for *chartered accountant. **2.** Abbreviation for *Consumers' Association.

**CAA** Abbreviation for *Civil Aviation Authority.

**cable** An international telegram. *See also* telegraphic address. To a large extent international telegrams and cables have now been replaced by *Telex and *Fax.

**CAD 1.** Abbreviation for *cash against documents. **2.** Abbreviation for *computer-aided design.

**call** A demand from a company to its shareholders to pay a specified sum on a specified date in respect of their *partly paid shares (*see also* share capital).

**call bird** A low-priced article used in the retail trade to encourage members of the public to come into a shop, in the hope that purchases for higher-priced goods will follow.

**call-cost indicator** A device installed by a telephone company on business premises to monitor calls as they are being made or over a prescribed period. A printout provides information on the duration and cost of each call (which is identified by date, time, and the number called).

**called-up capital** *See* share capital.

**call money 1.** Money put into the money market that can be called at short notice (*see also* money at call and short notice). **2.** *See* option money.

**call-of-more option** *See* option to double.

**call option** *See* option.

**callover** A meeting of commodity brokers and dealers at fixed times during the day in order to form a market in that commodity. The callover is usually used for trading in futures, in fixed quantities on a standard contract, payments usually being settled by differences through a *clearing house. Because traders usually form a ring around the person calling out the prices, this form of market is often called **ring trading**. This method of trading is also called **open outcry**, as bids and offers are shouted out during the course of the callover.

**CAM** Abbreviation for *computer-aided manufacturing.

**camera-ready copy (CRC)** Text and illustrations printed in such a form that they are ready for photographing to make the film from which printing plates are made. The CRC may be produced by a phototypesetter for high-class work or for less high quality it may be produced by a laser printer or daisywheel printer driven by a word-processor.

**campaign** An organized course of action, carefully planned to achieve set objectives, especially in advertising, sales, public relations, or marketing.

***Campaign*** A weekly magazine that gives details of new advertising *campaigns, job vacancies, personnel moves in *advertising agencies, and general advertising gossip.

**c & f** Abbreviation for cost and freight. It is the basis of an export contract in which the seller pays the cost of shipping the goods to the port of destination but not the cost of insuring the goods once they have been loaded onto a ship or aircraft. It is, in other respects, similar to a *c.i.f. contract. *Compare* FOB.

**c & m** Abbreviation for *care and maintenance.

**cannibalization** A market situation in which increased sales of one brand results in decreased sales of another brand within the same *product line, usually because there is little differentiation between them. For instance, two drinks marketed by the same manufacturer and packaged in almost the same colour could cause increased sales of one to be achieved at the expense of decreased sales of the other.

**cap** *See* collar.

**CAP** Abbreviation for *Common Agricultural Policy.

**capital 1.** The total value of the assets of a person less liabilities. **2.** The amount of the proprietors' interests in the assets of an organization, less its liabilities. **3.** The money contributed by the proprietors to an organization to enable it to function; thus **share capital** is the amount provided by way of shares and the **loan capital** is the amount provided by way of loans. However, the capital of the proprietors of companies not only consists of the share and loan capital, it also includes retained profit, which accrues to the holders of the ordinary shares. *See also* reserve capital. **4.** In economic theory, a factor of production, usually either machinery and plant (**physical capital**) or money (**financial capital**). However, the concept can be applied to a variety of other assets (*see* human capital). Capital is generally used to enhance the productivity of other factors of production (e.g. combine harvesters enhance the productivity of *land; tools enhance the value of *labour) and its return is the reward following from this enhancement. In general, the rate of return on capital is called *profit.

**capital account 1.** An *account recording capital expenditure on such items as land and buildings, plant and machinery, etc. **2.** A budgeted amount that can only be spent on major items, especially in public-sector budgeting. *Compare* revenue account. **3.** An *account showing the interest of a sole trader in the net assets of his business. **4.** A series of *accounts recording the interests of the partners in the net assets of a partnership. Capital accounts can embrace both the amounts originally contributed and the *current accounts; it may also refer, more narrowly, to the amounts originally contributed, adjusted where necessary by agreement between the partners.

**capital allowances** Income tax reliefs given against business and some other profits to reflect the depreciation of certain types of asset owned by the business. Because the Inland Revenue cannot control the amount of depreciation of fixed assets that traders charge in their accounts against profits, it is customary for the trader's own depreciation charge to be disallowed for tax purposes and the Revenue's own charges (the capital allowances) substituted. These may be targeted to some types of asset (e.g. plant and machinery) and not others (office buildings); where the authorities want to create incentives for traders to invest, the allowances may be accelerated (i.e. allowed at a higher rate than would be expected if normal depreciation rates were applied), even to the extent of allowing 100% capital allowances in the year of purchase.

**capital asset (fixed asset)** An asset that is expected to be used for a considerable time in a trade or business (*compare* current asset). Examples of capital assets in most businesses are land and buildings, plant and machinery, investments in subsidiary companies, goodwill, and motor vehicles, although in the hands of dealers these assets would become current assets. The costs of these assets are normally written off against profits over their expected useful life spans by deducting an item for *depreciation from their book value each year.

**capital budget** The sums allocated by an organization for future *capital expenditure. The capital budget may well encompass a longer period than the next accounting period.

**capital commitments** Firm plans, usually approved by the board of directors in the case of companies, to spend sums of money on *capital assets. Capital commitments must by law be shown by way of a note, or otherwise, on company balance sheets.

**capital consumption** The total depreciation in the value of the capital goods in an economy during a specified period. It is difficult to calculate this figure, but it is needed as it has to be deducted from the *gross national product (GNP) and the *gross domestic product (GDP) to obtain the net figures.

**capital-conversion plan** An *annuity that converts capital into income. Capital-conversion policies are often used to provide an income later in life for a person who might be liable to capital-gains tax if his capital is not reinvested in some way.

**capital duty** A stamp duty payable in the UK on the formation of companies, the issue of shares, or the immigration to Great Britain of companies other than from within the EEC.

**capital expenditure** Expenditure on *capital assets. Capital expenditure is not deducted from profits, as an asset is acquired rather than a loss being made. However, as a capital asset loses value by *depreciation, the amount of the depreciation is charged against profit.

**capital gain** A gain on an asset not bought and sold in the normal course of trade by the person making the gain. These gains are taxed in the UK and in many other countries by means of a *capital-gains tax.

**capital-gains tax (CGT)** A tax on *capital gains. Most countries have a form of income tax under which they tax the profits from trading and a different tax to tax substantial disposals of assets either by traders for whom the assets are not trading stock (e.g. a trader's factory) or by individuals who do not trade (e.g. sales of shares by an investor). The latter type of tax is a capital-gains tax. Short-term gains taxes are taxes sometimes applied to an asset that has only been held for a limited time. In these cases the rates tend to be higher than for the normal capital-gains tax. In the UK, capital-gains tax applies to the net gains (after deducting losses) accruing to an individual in any tax year, with an exemption to liability if the individual's gains do not exceed a specified figure (£5000 in 1990–91); this exemption applies separately to husbands and wives after April 1990. Other exemptions include gains on private cars, government securities and savings certificates, loan stocks, options, gambling, life-assurance and deferred-annuity contracts, main dwelling house, and works of art. The rate of tax is the taxpayer's *marginal tax rate. An indexation allowance is available when calculating a chargeable gain or allowable loss based on the *Retail Price Index between March 1982 (or, if later, the month of purchase) and the date of sale.

**capital gearing** The ratio of the amount of fixed interest loan stock and preference shares in a company to its ordinary share capital. A company with a preponderance of ordinary share capital is **low-geared** while one in which fixed-interest capital dominates is **high-geared**. With high gearing, when profits are rising, the amounts available to ordinary shareholders rise, in percentage terms, faster than the percentage rise in profits. However, when profits are falling shareholders in high-geared companies suffer a larger percentage drop in their dividends than the percentage fall in profits. In the USA capital gearing is known as **leverage**. *See also* degearing.

**capital good (producer good)** A *good that is used to produce other goods rather than being bought by consumers (*see* consumer goods). Capital goods include plant and machinery, industrial buildings, and raw materials.

**capital growth** An increase in the value of invested capital. Investment in *fixed-interest securities or bonds provides income but limited capital growth (which may be improved in *index-linked gilts). To have a chance of making substantial capital growth it is necessary to invest in equities (*see* ordinary share), the value of which should increase with *inflation. Investing in equities is thus said to be a hedge against inflation. *See* growth stocks.

**capital-intensive** *See* labour-intensive.

**capital investment** *See* investment.

**capitalism** An economic and political system in which individual owners of *capital are free to dispose of it as they please; in particular, for their own profit. Most of the developed economies outside the Eastern bloc are capitalistic, even though governments do interfere significantly with the movements of capital. Neoclassical economics is usually associated with capitalism because of its traditional advocacy of free trade. However, neoclassical economists have little to say about redistribution of capital. Marxist economists see capitalism as a stage in the development of society – a stage to be superseded by socialism, in which private capital will be abolished.

**capitalization 1.** The act of providing *capital for a company or other organization. **2.** The structure of the capital of a company or other organization, i.e. the extent to which its capital is divided into share or loan capital and the extent to which share capital is divided into ordinary and preference shares. *See also* thin capitalization. **3.** The conversion of the reserves of a company into capital by means of a *scrip issue.

**capitalization issue** *See* scrip issue.

**capitalized value 1.** The value at which an asset has been recorded in the balance sheet of a company or other organization, usually before the deduction of *depreciation. **2.** The capital equivalent of an asset that yields a regular income, calculated at the prevailing rate of interest. For example, a piece of land bringing in an annual income of £1000, when the prevailing interest rate is 10%, would have a notational capitalized value of £10,000 (i.e. £1000/0.1). This may not reflect its true value.

**capital loss** A loss arising from the disposal, loss, or destruction of a *capital asset or from a long-term liability. For tax purposes a capital loss can be set off against a capital profit. *See* capital-gains tax.

**capital maintenance** Maintaining the value of the share capital of an organization or of its share capital and reserves. More vaguely, it could refer to maintaining the operating capacity of an organization, regardless of the composition of its assets.

**capital market** A market in which long-term *capital is raised by industry and commerce, the government, and local authorities. The money comes from private investors, insurance companies, pension funds, and banks and is usually arranged by *issuing houses and *merchant banks. *Stock exchanges are also part of the capital market in that they provide a market for the shares and loan stocks that represent the capital once it has been raised. It is the presence and sophistication of their capital markets that distinguishes the industrial countries from the *developing countries, in that this facility for raising industrial and commercial capital is either absent or rudimentary in the latter.

**capital movement** The transfer of capital between countries, either by companies or individuals. Restrictions on *exchange controls and capital transfers between countries have been greatly reduced in recent years. Capital movements seeking long-term gains are usually those made by companies investing abroad, for example to set up a factory. Capital movements seeking short-term gains are often more speculative, such as those taking advantage of temporarily high interest rates in another country or an expected change in the exchange rate.

**capital profit** A profit arising from the disposal of a *capital asset. Capital profits, if taxable at all, are normally subject to *capital-gains tax, which applies both to individuals and to companies. The Companies Act (1980) allows capital profits made by limited companies to be distributed as revenue profits, provided that they are real, realized, have not been previously capitalized, and take into account any accumulated realized capital losses.

**capital-redemption reserve** A reserve fund required to be created by the UK Companies Act (1985) when shares are redeemed out of *retained profits and not out of a new issue of share capital. The reserve is created by making a transfer out of the *profit and loss account to a specially designated account, the **capital-redemption reserve account**. Amounts held in this account cannot be distributed to shareholders by way of dividend, although they may be used to make bonus issues of share capital. The purpose of the reserve is to ensure that the company's capital is not diluted by the redemption of some of the shares.

**capital reserves** Undistributed profits of a company that for various reasons are not regarded as distributable to shareholders as dividends. These include certain profits on the revaluation of capital assets and any sums received from share issues in excess of the nominal value of the shares, which are shown in a *share premium account. Capital reserves are now known as **undistributable reserves** under the Companies Act (1985). *See also* reserve.

**capital stock** The aggregate of an organization's *capital assets. *Compare* current assets; stock-in-trade.

**capital structure** The elements, such as shares, loan stock, etc., from which the *capital of a company or other organization is composed.

**capital-transfer tax** A tax levied when capital is transferred from one person's estate usually into that of another, as by lifetime gifts or inheritances. There was such a tax in the UK from 1974 to 1986. It continued in a truncated form from 1986, when it became known as *inheritance tax.

**capital turnover** The ratio of sales of a company or other organization to its capital employed (i.e. its assets less liabilities). It is presumed that the higher this ratio, the better the use that is being made of the assets in generating sales.

**captive audience** An audience that is unlikely to be able to escape being exposed to an advertising message in toto. Examples include cinema audiences and conference audiences.

**captive insurance company** An insurance company that is totally owned by another organization and insures only, or mostly, the parent company's risks. In this way the parent organization is able to obtain insurance cover (particularly those classes that are compulsory by law) without having to pay premiums to an organization outside its trading group.

**captive market** A group of purchasers who are obliged to buy a particular product as a result of some special circumstance, such as the absence of an alternative supplier or product.

**CAR** Abbreviation for *compound annual return.

**carat** **1.** A measure of the purity (fineness) of gold. Pure gold is defined as 24 carat; 14-carat gold contains 14/24ths gold, the remainder usually being copper. **2.** A unit for measuring the weight of a diamond or other gemstone, equal to 0.2 gram.

**care and maintenance (c & m)** Denoting the status of a building, machinery, ship, etc., that is not currently in active use, usually as a result of a fall in demand, but which is being kept in a good state of repair so that it can be brought back into use quickly, if needed.

**cargo insurance** An insurance covering cargoes carried by ships, aircraft, or other forms of transport. On an *FOB contract the responsibility for insuring the goods for the voyage rests with the buyer. The seller's responsibility ends once the goods have been loaded onto the ship or aircraft (or train in the case of FOR contracts). On a *c.i.f. contract, insurance is the responsibility of the seller up to the port of destination. On a *c & f contract, the seller arranges the shipment and pays the freight but the buyer is responsible for the insurance during the voyage (as in an FOB contract). *See also* average; floating policy; open cover.

**Caribbean Community and Common Market (CARICOM)** An association of Caribbean states established in 1973 to further economic cooperation through the Caribbean Common Market, coordinate foreign policy among member states, and to provide common services in health, education and culture, communications, and industrial relations. The organs are the Conference of Heads of Government and the Common Market Council of Ministers (usually trade ministers). The members are Antigua and Barbuda, the Bahamas, Barbados, Belize, Dominica, Grenada, Guyana, Jamaica, Montserrat, St Christopher and Nevis, St Lucia, St Vincent and the Grenadines, and Trinidad and Tobago.

**CARICOM** Abbreviation for *Caribbean Community and Common Market.

**carriage cost** The cost of delivering goods within the UK. **Carriage forward** means that the cost of delivery has to be paid by the buyer. **Carriage paid** or **carriage free** means that they are paid by the seller.

**carrier** A person or firm that carries goods or people from place to place, usually under a contract and for a fee. A **common carrier**, who has to have an A licence in the UK, provides a public service and must carry any goods or people on his regular routes; he must charge a reasonable rate; and he is liable for all loss or damage to goods in transit. A **limited carrier**, who requires a B licence, carries only certain types of goods (e.g. liquid chemicals in tankers), and may refuse to carry anything else. A **private carrier**, with a C licence, carries only his own goods.

**carry-over** **1.** The quantity of a *commodity that is carried over from one crop to the following one. The price of some commodities, such as grain, coffee, cocoa, and jute, which grow in annual or biannual crops, is determined by the supply and the demand. The supply consists of the quantity produced by the current crop added to the quantity in the hands of producers and traders that is carried over from the previous crop. Thus, in some circumstances the carry-over can strongly influence the market price. **2.** *See* contango.

**car tax** An excise duty levied on motor cars at the wholesale stage. It applies to both home-produced and imported cars and is additional to VAT.

**cartel** An association of independent companies formed to regulate the price and sales conditions of the products or services they offer. A cartel may be national or international, although some countries, including the UK and the USA, have legislation forbidding cartels to be formed on the grounds that they are *monopolies that function against the public interest. The International Air Transport Association is an example of an international price-fixing cartel that is regarded as acceptable because of fears that price-cutting of fares between airlines would jeopardize safety.

**car telephone** *See* cellular network.

**case of need** An endorsement written on a *bill of exchange giving the name of someone to whom the holder may apply if the bill is not honoured at maturity.

**cash** *Legal tender in the form of banknotes and coins that are readily acceptable for the settlement of debts.

**cash against documents (CAD)** Payment terms for exported goods in which the shipping documents are sent to a bank, agent, etc., in the country to which the goods are being shipped, and the buyer then obtains the documents by paying the invoice amount in cash to the bank, agent, etc. Having the shipping documents enables the buyer to take possession of the goods when they arrive at their port of destination; this is known as **documents against presentation**. *Compare* documents against acceptance.

**cash and carry** **1.** A wholesaler, especially of groceries, who sells to retailers and others with businesses at discounted prices on condition that they pay in cash, collect the goods themselves, and buy in bulk. **2.** An operation that is sometimes possible on the *futures market (*see* futures contract), especially the *London Metal Exchange. In some circumstances the spot price of a metal, including the cost of insurance, warehousing, and interest for three months, is less than the futures-market price for delivery in three months. Under these conditions it is possible to buy the spot metal, simultaneously sell the forward goods, and make a profit in excess of the yield the capital would have earned on the money market.

**cash and new** An arrangement enabling speculators on the London Stock Exchange to carry forward a transaction from one *account day to a subsequent account day. It replaces the former *contango arrangements but is now discouraged. It involves selling the stock at the end of one account and buying it back at the start of the next. Commissions have to be paid but settlement of the transaction can be delayed.

**cash book** The *book of prime entry in which are recorded receipts into and payments out of the organization's bank account (*compare* petty cash). The cash book, unlike most other books of prime entry, is also an *account, as its balance shows the amount due to or from the bank.

**cash budget** *See* cash flow.

**cash card (cash-point card)** *See* cash dispenser.

**cash cow** A product, with a well-known brand name, that commands a high market share in old-established markets and therefore produces a steady flow of cash.

**cash crop** *See* subsistence crop.

**cash deal** A transaction (either buying or selling) on the *London Stock Exchange in which settlement is made immediately, i.e. usually on the following day.

**cash discount 1.** A reduction in the price of goods offered by a seller to a buyer who pays promptly (usually within seven days). **2.** A reduction in the price of goods bought from a wholesaler and paid for in cash.

**cash dispenser** An automatic machine for supplying cash to cardholders, often installed outside banks and building societies to provide a 24-hour service. The customer is supplied with an embossed plastic card (**cash card** or **cash-point card**) bearing account details. When it is inserted into the dispenser with the appropriate *personal identification number, it will allow cash to be withdrawn direct from the customer's account. Other services may also be requested, such as the provision of a statement of account, a cheque book, or account balance. Cheques and cash may also be paid in through a cash dispenser.

**cash flow** The amount of cash being received and expended by a business, which is often analysed into its various components. A **cash-flow projection** (or **cash budget**) sets out all the expected payments and receipts in a given period. This is different from the projected profit and loss account and, in times of cash shortage, may be more important. It is on the basis of the cash-flow projection that managers arrange for employees and creditors to be paid at appropriate times. *See also* discounted cash flow.

**cash management account** A bank account in which deposits are invested by the bank, usually on the money market; it is, however, a cheque account and the client is able to obtain loans if required.

**cash on delivery (COD)** Terms of trade in which a supplier will post goods to a customer, provided the customer pays the postman or delivery man the full invoice amount when they are delivered. It was extensively used in mail order (*see* mail-order house), but the use of telephone ordering using credit cards has reduced the amount of COD business.

**cash price** The price at which a seller is prepared to sell goods provided that he is paid immediately in cash, i.e. he does not have to give credit or give a commission to a credit-card company. This is invariably below the price that includes a *hire-purchase agreement.

**cash ratio (liquidity ratio)** The ratio of the cash reserve that a bank keeps in coin, banknotes, etc., to its total liabilities to its customers, i.e. the amount deposited with it in current accounts and deposit accounts. Because cash reserves earn no interest, bankers try to keep them to a minimum, consistent with being able to meet customers' demands. The usual figure for the cash ratio is 8%.

**cash settlements (cash deals)** Deals on the London Stock Exchange in which gilt-edged securities or new issues are bought by investors. These have to be paid for immediately (normally by the next business day) rather than on the next *account day.

**casting vote** A second or deciding vote. It is common practice to give the chairman of a meeting a second vote to be used to resolve a deadlock. In the case of a company meeting, the chairman is generally given this right by the articles of association, but he has no common-law right to a casting vote.

**cast-off** An estimate of the number of characters, words, or pages that text for printing contains. If the cast-off is given in pages, this will include any illustrations.

**CAT** Abbreviation for *computer-assisted trading.

**catalogue store** A store that combines the techniques of selling by catalogue with those of a *discount house. The stores themselves are not designed for customer comfort but by keeping the premises simple they achieve competitive prices.

**catastrophe risk** A risk in which the potential loss is of the greatest size, such as the explosion of a nuclear power station or a major earthquake.

**catch crop** A quick-growing crop planted on land that is available for a short period, as between main crops, or sometimes planted between the rows of a main crop. *Compare* subsistence crop.

**catching bargain (unconscionable bargain)** An unfair contract, often one in which one party has been taken advantage of by the other. Such a contract may be set aside or modified by a court.

**caveat** A proviso or qualification that limits liability by putting another party on notice. For example, a retailer may sell goods subject to the caveat that he does not guarantee their suitability for a particular purpose. The purchaser is thereby put on notice that he has no remedy against the retailer should the goods in fact turn out to be unsuitable for that particular purpose.

**caveat emptor** (Latin: let the buyer beware) A maxim implying that the purchaser of goods must take care to ensure that they are free from defects of quality, fitness, or title, i.e. that the risk is borne by the purchaser and not by the seller. If the goods turn out to be defective, the purchaser has no remedy against the seller. The rule does not apply if the purchaser is unable to examine the goods, if the defects are not evident from a reasonable examination, or if the seller has behaved fraudulently. Some measure of protection for the unwary purchaser is afforded by a number of statutes, including the Unfair Contract Terms Act (1977), the Sale of Goods Act (1979), and the Consumer Protection Act (1987).

**CBI** Abbreviation for *Confederation of British Industry.

**CBT** Abbreviation for *computer-based training.

**CCA** Abbreviation for *current-cost accounting.

**CD 1.** Abbreviation for *certificate of deposit. **2.** Abbreviation for corps diplomatique, especially as used on the diplomatic plates on the cars of members of the diplomatic corps.

**Ceefax** *See* Teletext.

**cellular network** A radiotelephone network for mobile subscribers that connects them to the main telephone system. Operated in the UK jointly by British Telecom and Securicor (as Cellnet), it consists of a number of adjacent cells, each containing a transmit/receive station connected to the main telephone network and reached by the mobile subscriber using a battery-operated portable radiotelephone. As subscribers move from one geographical cell to another (e.g. by car), they are automatically switched to receive signals from the new area. Car telephones, operating on this system, have been a great boon to business people who can remain in contact with their offices while travelling.

**census** An official count of a population for demographic, social, or economic purposes. In the UK, population censuses have been held every ten years since 1801 and since 1966 supplementary censuses have been held halfway through the ten-year period. A **census of distribution**, quantifying the wholesale and retail distribution, has been carried out approximately every five years since 1950. A **census of production**, recording the output of all manufacturing, mining, quarrying, and building industries, together with that of all public services, has been held every year since 1968.

**Central Arbitration Committee** A committee set up by the Employment Protection Act (1975) to arbitrate on matters voluntarily submitted to it through *ACAS by the parties to a trade dispute. It has powers to enforce the disclosure of certain bargaining information in these disputes and to arbitrate when a statutory joint industrial council is deadlocked. It does not charge for its services and cannot award costs.

**central bank** A bank that provides financial and banking services for the government of a country and its commercial banking system as well as implementing the government's monetary policy. The main functions of a central bank are: to manage the government's accounts; to accept deposits and grant loans to the commercial banks; to control the issue of banknotes; to manage the public debt; to help manage the exchange rate when necessary; to influence the interest rate structure and control the money supply; to hold the country's reserves of gold and foreign currency; to manage dealings with other central banks; and to act as lender of last resort to the banking system. Examples of major central banks include the *Bank of England in the UK, the Federal Reserve Bank of the USA (*see* Federal Reserve System), the Deutsche Bundesbank in Germany, and France's Banque de France.

**Central Government Borrowing Requirement (CGBR)** In the UK, the *Public Sector Borrowing Requirement (PSBR) less any borrowings by local authorities and public corporations from the private sector. Since local authorities and public corporations both have a measure of freedom in deciding how much they borrow, the government does not have the complete control over the PSBR as it does over the CGBR.

**central processing unit (CPU)** The main part of a computer that controls all other parts of the system, especially in medium to large computers, in which the CPU is a physically separate unit. In small microcomputers, the functions of the CPU are carried out by a *microprocessor and other silicon chips inside the computer and the CPU is therefore not a distinct unit. The CPU consists of an arithmetic and logic unit, a control unit, and a memory unit. The arithmetic and logic unit carries out the calculations required by the program; the memory unit holds the program instructions while they are being executed and any intermediate results generated by the calculations; and the control unit coordinates the activities of the various parts of the CPU and the computer system as a whole. *See also* computer.

**Central Statistical Office** The UK government's statistical unit. Among its publications are the **Monthly Digest of Statistics*, **Financial Statistics* (monthly), **Economic Trends* (monthly), and **UK National Accounts* (the *Blue Book*; annual), *UK*

*Balance of Payments* (the *Pink Book*; annual), and the **Annual Abstract of Statistics*. It is also responsible for the *Business Statistics Office.

**certificate of damage** A certificate issued by a dock or wharfage company when it takes in damaged goods. The certificate, signed for or on behalf of the dock surveyor, states the nature of the damage and the cause, if known.

**certificate of deposit (CD)** A negotiable certificate issued by a bank in return for a term deposit of up to five years. They originated in the USA in the 1960s. From 1968, a sterling CD was issued by UK banks. They were intended to enable the *merchant banks to attract funds away from the *clearing banks with the offer of competitive interest rates. However, in 1971 the clearing banks also began to issue CDs as their negotiability and higher average yield had made them increasingly popular with the larger investors.

A secondary market in CDs has developed, made up of the *discount houses and the banks in the interbank market. They are issued in various amounts between £10,000 and £50,000, although they may be subdivided into units of the lower figure to facilitate negotiation of part holdings.

**certificate of incorporation** The certificate that brings a company into existence; it is issued to the shareholders of a company by the Registrar of Companies. It is issued when the *memorandum and *articles of association have been submitted to the Registrar of Companies, together with other documents that disclose the proposed registered address of the company, details of the proposed directors and company secretary, the nominal and issued share capital, and the capital duty. The statutory registration fee must also be submitted. Until the certificate is issued, the company has no legal existence.

**certificate of insurance** A certificate giving abbreviated details of the cover provided by an insurance policy. In a *motor-insurance policy or an *employers'-liability policy, the information that must be shown on the certificate of insurance is laid down by law and in both cases the policy cover does not come into force until the certificate has been delivered to the policyholder.

**certificate of origin** A document that states the country from which a particular parcel of goods originated. In international trade it is one of the shipping documents and will often determine whether or not an import duty has to be paid on the goods and, if it has, on what tariff. Such certificates are usually issued by a chamber of commerce in the country of origin.

**certified accountant** A member of the Chartered Association of Certified Accountants. Its members are trained in industry, in the public service, and in the offices of practising accountants. They often attend sandwich courses in technical colleges while still working and take the Association's exams. Members are recognized by the UK Department of Trade and Industry as qualified to audit the accounts of companies. They may be associates (ACCA) or fellows (FCCA) of the Association and although they are not *chartered accountants, they fulfil much the same role.

In the USA the equivalent is a **certified public accountant (CPA)**, who is a member of the **Institute of Certified Public Accountants**.

**certified check** *See* marked cheque.

**certified stock (certificated stock)** Stocks of a commodity that have been examined and passed as acceptable for delivery in fulfilment of contracts on a futures market (*see* futures contract)

**cesser clause** A clause in a charter-party (*see* chartering), inserted when the charterer intends to transfer to a shipper his right to have goods carried. It provides that the shipowner is to have a lien over the shipper's goods for the freight payable under the charterparty and that the charterer's liability for freight ceases on shipment of a full cargo.

**CET** Abbreviation for *Common External Tariff.

**ceteris paribus** (Latin: all other things being equal) The widely used assumption in economics that the variables under consideration have no other effects on the economy than those specified and are not themselves affected by any other variables. Usually hopelessly unrealistic, the ceteris paribus assumption is the basis of *partial equilibrium analysis from which most economic theories are derived. On the other hand, *general-equilibrium analysis excludes the ceteris paribus assumption but has difficulty in reaching any worthwhile conclusions.

**CFC** Abbreviation for Common Fund for Commodities. *See* United Nations Conference on Trade and Development (UNCTAD).

**CFTC** Abbreviation for *Commodity Futures Trading Commission.

**CGBR** Abbreviation for *Central Government Borrowing Requirement.

**CGT** Abbreviation for *capital-gains tax.

**chain stores** *See* multiple shops.

**chairman** The most senior officer in a company, who presides at the *annual general meeting of the company and usually also at meetings of the board of directors. He may combine the roles of chairman and *managing director, especially in a small company of which he is the majority shareholder, or he may be a figurehead, without executive participation in the day-to-day running of the company. He is often a retired managing director. In the USA the person who performs this function is often called the president. If this office is filled by a woman, she is known as a **chairwoman** or **chairperson**. To avoid this complication the officer is now sometimes referred to as the **chair**.

**chairman's report** A report, often included in the *annual report of a company, giving a summary of the year's activities and a brief survey of what can be expected in the coming year. It is signed by the chairman, who usually reads it at the *annual general meeting.

**chamber of commerce** In the UK, a voluntary organization, existing in most towns, of commercial, industrial, and trading businessmen who represent their joint interests to local and central government. The London Chamber of Commerce is the largest such organization in the UK; it also fulfils an educational role, running several commercial courses, for which it also sets examinations. Most UK chambers of commerce are affiliated to the Association of British Chambers of Commerce. *Compare* chamber of trade.

**chamber of trade** An organization of local retailers set up to protect their interests in local matters. They are a much narrower organization than a *chamber of commerce and most in the UK are affiliated to the *National Chamber of Trade (NTC).

**Chambre Agent General Index** An arithmetically weighted index of 430 shares on the Paris Bourse.

**channel captain** The most powerful member in the distribution channel of goods. The channel consists of the manufacturer, wholesaler, and retailer, the channel captain usually being the manufacturer. There are, however, exceptions, most notably when the retailer is a major department store or chain, whose buying power enables it to become the channel captain. The channel captain controls the distribution channel and can insist that products should be made to their own specifications. *See also* channel conflict.

**channel conflict** A disagreement between members of a distribution channel (*see* channel captain). Horizontal channel conflict can occur among retailers, if one retailer feels another is competing too vigorously through pricing or is invading his sales territory by means of advertising. Vertical conflict occurs between retailers and their suppliers when either side feels dominated by the other.

**Channel Tunnel (Chunnel; Eurotunnel)** The Anglo-French railway tunnel being built beneath the English Channel. It is due to open in 1993, when it will greatly facilitate the transport of goods between the UK and mainland Europe.

**CHAPS** Abbreviation for Clearing House Automatic Payments System. *See* Association for Payment Clearing Services.

**character** One of the numbers, letters, or symbols (such as punctuation marks) that a computer can print or display on a screen. *See also* byte.

**character printer** A device that prints one *character at a time. An ordinary typewriter is a type of character printer. *Compare* line printer.

**character set** The collection of *characters, such as letters, numbers, and symbols, that can be produced by a device, such a printer or typewriter.

**charge 1.** A legal or equitable interest in land, securing the payment of money. It gives the creditor in whose favour the charge is created (the **chargee**) the right to payment from the income or proceeds of sale of the land charged, in priority to claims against the debtor by unsecured creditors. **2.** An interest in company property created in favour of a creditor (e.g. as a *debenture holder) to secure the amount owing. Most charges must be registered by the *Registrar of Companies (*see also* register of charges). A *fixed charge (or specific charge) is attached to a specific item of property (e.g. land); a *floating charge is created in respect of circulating assets (e.g. cash, stock in trade), to which it will not attach until **crystallization**, i.e. until some event (e.g. winding-up) causes it to become fixed. Before crystallization, unsecured debts can be paid out of the assets charged. After, the charge is treated as a fixed charge and therefore unsecured debts (except those given preference under the Companies Acts) rank after those secured by the charge (*see also* fraudulent preference). A charge can also be created upon shares. For example, the articles of association usually give the company a *lien in respect of unpaid *calls, and company members may, in order to secure a debt owed to a third party, charge their shares, either by a full transfer of shares coupled with an agreement to retransfer upon repayment of the debt or by a deposit of the share certificate.

**chargeable event** Any transaction or event that gives rise to a liability to income tax or to capital-gains tax.

**charge account (credit account)** An account held by a customer at a retail shop that allows him to pay for any goods purchased at the end of a stated period (usually one month). While the large stores usually offered this facility without charging interest, it is now usual for interest to be charged on any amounts unpaid after the stated period. The customer identifies himself with a plastic **charge card**. If this is lost or stolen it is the customer's responsibility to notify the store immediately.

**charge card** *See* charge account.

**charges forward** An instruction to the effect that all carriage charges on a consignment of goods will be paid by the consignee after he receives them.

**charges register 1.** *See* land registration. **2.** *See* register of charges.

**charitable trust** A trust set up for a charitable purpose that is registered with the **Charity Commissioners**, a body responsible to parliament. Charitable trusts do not have to pay income tax if they comply with the regulations of the Charity Commissioners.

**chartered accountant** A qualified member of the *Institute of Chartered Accountants in England and Wales, the Institute of Chartered Accountants of Scotland, or the Institute of Chartered Accountants in Ireland. These were the original bodies to be granted royal charters. Other bodies of accountants now have charters (the Chartered Association of Certified Accountants, the Chartered Institute of Management Accountants, and the Chartered Institute of Public Finance and Accountancy) but their members are not known as chartered accountants. Most firms of chartered accountants are engaged in public practice concerned with auditing, taxation, and other financial advice; however, many trained chartered accountants fulfil management roles in industry.

**Chartered Association of Certified Accountants** *See* certified accountant; chartered accountant.

**chartered company** *See* company.

**Chartered Institute of Management Accountants** *See* chartered accountant; management accountant.

**Chartered Institute of Public Finance and Accountancy** *See* chartered accountant.

**Chartered Insurance Institute (CII)** An association of insurers and brokers in the insurance industry. Its origins date back to 1873; its first Royal Charter was granted in 1912. It provides training by post and at its own college, examinations leading to its associateship diploma (ACII) and fellowship diploma (FCII), and sets high standards of ethical behaviour in the industry.

**chartered secretary** *See* Institute of Chartered Secretaries and Administrators.

**chartered surveyor** *See* Royal Institution of Chartered Surveyors.

**chartering** Hiring the whole of a ship or aircraft. The hirer is called the **charterer** and the document setting out the terms and conditions of the contract is called the **charterparty**. In a *time charter the ship or aircraft is hired for a specified period, usually with the owner operating it and the charterer paying fuel and stores costs. However, in a *bareboat charter (or demise charter), which is a form of time charter, the charterer pays for all crew expenses. In a *voyage charter, the ship or aircraft is hired for a single voyage between stated ports to carry a stated cargo and the owner pays all the expenses of running the ship. In an **open charter** the charterer may use the ship or aircraft to carry any cargo to any port.

**charterparty** *See* chartering.

**chartist** An investment analyst who uses charts and graphs to record past movements of the share prices, P/E ratios, turnover, etc., of individual companies to anticipate the future share movements of these companies. Claiming that history repeats itself and that the movements of share prices conform to a small number of repetitive patterns, chartists have been popular, especially in the USA, in the past. It is now more usual for analysts to use broader techniques in addition to those used by chartists.

**chattels** All property of whatever kind, excluding freehold land and anything permanently affixed to freehold land. Interests in land (e.g. leaseholds) are **chattels real**. **Chattels**

**personal** are all movable and tangible articles of property. Chattels include timber growing on land (whether freehold or leasehold) and articles of personal use.

**cheap money (easy money)** A monetary policy of keeping *interest rates at a low level. This is normally done to encourage an expansion in the level of economic activity by reducing the costs of borrowing and investment. It was used in the 1930s to help recovery after the depression and during World War II to reduce the cost of government borrowing. *Compare* dear money.

**check** The US spelling of *cheque.

**cheque** A preprinted form on which instructions are given to an account holder (a bank or building society) to pay a stated sum to a named recipient. It is the most common form of payment of debts of all kinds (*see also* cheque account; current account). In a **crossed cheque** two parallel lines across the face of the cheque indicate that it must be paid into a bank account and not cashed over the counter (a **general crossing**). A **special crossing** may be used in order to further restrict the negotiability of the cheque, for example by adding the name of the payee's bank. An **open cheque** is an uncrossed cheque that can be cashed at the bank of origin. An **order cheque** is one made payable to a named recipient "or order", enabling the payee to either deposit it in an account or endorse it to a third party, i.e. transfer the rights to the cheque by signing it on the reverse. In a **blank cheque** the amount is not stated; it is often used if the exact debt is not known and the payee is left to complete it. However, the drawer may impose a maximum by writing "under £. . . " on the cheque. *See also* marked cheque; returned cheque; stale cheque.

**cheque account** An account with a *bank or *building society on which cheques can be drawn. In general, building societies pay interest on the daily credit balances in a cheque account but banks traditionally did not; they are now often doing so to meet competition from building societies. *See also* current account.

**cheque card** A card issued by a UK bank or building society to an approved customer, guaranteeing any cheque drawn by that customer up to an amount stated on the cheque card (usually either £50 or £100). This enables the customer to use cheques for purchases (up to the value of the cheque card) from establishments in which they are unknown. The cheque card can also be used by a bank customer to draw cash from any branch of his bank.

**chief executive 1.** The person with responsibility for ensuring that an organization functions efficiently if it is non-profitmaking and makes a profit acceptable to the shareholders if it is profitmaking. Although the term was formerly more common in North America than the UK, it is now sometimes used in place of *managing director in the UK. **2.** The President of the USA.

**child benefit** A payment made direct to a parent in the UK by the DSS through a post office in respect of each child under the age of 16 (or 18 if they are in full-time education). This payment replaces the former family allowance set against the parents' income tax.

**children's assurance** An assurance policy arranged on the life of a child or a minor (anyone aged below 18). The main use of life assurance is to replace income lost following the death of a breadwinner. As a child does not usually earn any income, the law insists that children's-assurance policies only pay the sum assured when the policy matures or death occurs after the child reaches the age of 18. If the child dies before reaching this age no payment is made but the premiums are refunded in full. This restriction was applied to policies of this kind to prevent unscrupulous parents from murdering their children to claim the benefit.

**Chinese wall** A notional information barrier between the parts of a business, especially between the market-making part of a stockbroking firm and the broking part. It would clearly not be in investors' interests for brokers to persuade their clients to buy investments from them for no other reason than that the market makers in the firm, expecting a fall in price, were anxious to sell them.

**chip** *See* integrated circuit.

**CHIPS** Abbreviation for *Clearing House Inter-Bank Payments System.

**chose in action** A right of proceeding in a court of law to obtain a sum of money or to recover damages. Examples include rights under an insurance policy, a debt, and rights under a contract. A chose in action is a form of property and can be assigned, sold, held in trust, etc. *See also* chose in possession.

**chose in possession** Moveable chattels that a person has in his possession. Examples include goods, merchandise, etc. All assets are either choses in possession or *choses in action.

**Chunnel** *See* Channel Tunnel.

**churning 1.** The practice by a broker of encouraging an investor to change his investments frequently in order to make him pay excessive commissions. **2.** The practice by a bank, building society, insurance broker, etc., of encouraging a householder with an endowment *mortgage to surrender the policy and to take out a new one when seeking to increase a mortgage or to raise extra funds, instead of topping up the existing mortgage. The purpose is to increase charges and commissions at the expense of the policyholder. **3.** A government policy of paying a benefit to a wide category of persons and taxing it so that those paying little or no taxes receive it while the well-off return it through the tax system. There have been suggestions that a higher child benefit is suitable for churning.

**c.i.f.** Abbreviation for cost, insurance, and freight. It is the basis of an export contract in which the seller pays the cost of shipping the goods to the port of destination and of insuring the goods up to this point. On a c.i.f. contract, the seller sends the documents giving title to the goods to the buyer, usually through a bank. These documents include the *bill of lading, the *insurance policy, the commercial invoice, and sometimes such additional documents as a *certificate of origin, quality certificate, or export licence. In order to obtain the goods the buyer is obliged to pay for the documents when they are presented to him. It is said that with a c.i.f. contract the buyer is paying for documents rather than goods, because provided the documents are in order, he is obliged to pay for them even if the goods themselves have been lost at sea during the voyage. In these circumstances the buyer would hold the insurance policy entitling him to recompense for the lost goods. *Compare* c & f; c.i.f.c.i.; FOB.

**c.i.f.c.i.** Abbreviation for cost, insurance, and freight plus commission and interest. A price quoted c.i.f.c.i. is a *c.i.f. price, which also includes an agreed commission payable by the seller to the buyer and the interest charged by the seller's bank for negotiating the documents.

**CII** Abbreviation for *Chartered Insurance Institute.

**circuity of action** The return of a *bill of exchange, prior to maturity, to the person who first signed it. Under these circumstances it may be renegotiated, but the person forfeits any right of action against those who put their names to it in the intervening period.

**circular letter of credit** *See* letter of credit.

**circular flow of income** The process by which money and goods pass between different groups in the economy. A concept first developed by the French economist François Quesnay (1694–1774), it is used as the basis for studying *macroeconomic relationships. In its simplest forms, it postulates that households provide labour to firms in exchange for money, which the households use to buy the goods produced by firms. Household *savings represent a leakage from the economy, as money saved is removed from the circular flow, but this is partially compensated for by *investment, which is an injection into the circular flow. In the real world, circular flow is complicated by such factors as taxation (leakage) and government spending (injection), exports (leakage) and imports (injection). *National income accounts are based on the concept of the circular flow.

**circulating capital (working capital)** The part of the *capital of a company or other organization that is used in the activities of trading, as distinct from its fixed capital (*see* capital assets). The circulation of this capital occurs thus: suppliers provide stock; the stock is sold to customers who become debtors, eventually paying cash; the cash is used to pay suppliers, who provide more stock, etc. The concept was introduced by the economist Adam Smith (1723–90) and is popular with judges in tax cases.

**City** The financial district of London in which are situated the head offices of the banks, the money markets, the foreign exchange markets, the commodity and metal exchanges, the insurance market (including *Lloyd's), the *London Stock Exchange, and the offices of the representatives of foreign financial institutions. Occupying the square mile on the north side of the River Thames between Waterloo Bridge and Tower Bridge, it has been an international merchanting centre since medieval times.

**City Call** A financial information service provided by British Telecom over the telephone in the UK. It gives nine bulletins of updated information each day.

**City Code on Takeovers and Mergers** A code first laid down in 1968, and subsequently modified, giving the practices to be observed in company takeovers (*see* takeover bid) and *mergers. Encouraged by the Bank of England, the code was compiled by a panel including representatives from the London Stock Exchange Association, the Issuing Houses Association, the London Clearing Bankers, and others. The code does not have the force of law but the panel can admonish offenders and refer them to their own professional bodies for disciplinary action.

The code attempts to ensure that all shareholders, including minority shareholders, are treated equally, are kept advised of the terms of all bids and counterbids, and are advised fairly by the directors of the company receiving the bid on the likely outcome if the bid succeeds. Its many other recommendations are aimed at preventing directors from acting in their own interests rather than those of their shareholders, ensuring that the negotiations are conducted openly and honestly, and preventing a spurious market arising in the shares of either side.

**Civil Aviation Authority (CAA)** An independent UK body set up in 1971 to be responsible for: the development and economic regulation of the UK civil air-transport industry; safety regulations by certification of airlines and aircraft; and the provision of air-traffic control and telecommunications.

**civil law 1.** The law applied by the civil courts in the UK, as opposed to ecclesiastical, criminal, or military law. It is, thus, the law that regulates dealings between private citizens that are not subject to interference by the state. Its chief divisions include the law of contract, torts, and trusts. **2.** Roman law. **3.** The law generally in force on the Continent, which has its basis in Roman law.

**Classical school** The school of economics founded by Adam Smith (1723–90). Smith was primarily concerned with explaining the origins of wealth creation and with advocating the benefits of free trade. He achieved this by analysing the economic relationships between the classes: workers, who earn their living by wage labour; capitalists, who derive income from profits; and landlords, whose income derives from rent. Supply and demand in each class determined prices. David Ricardo (1772–1823) extended this analysis, in particular elucidating the concept of value, which in the Classical school is seen as a product of labour. The *labour theory of value was used by Marx (1818–83) as a basis for his analysis of the capitalist economy and Marxists have remained firmly wedded to the Classical school. The *Marginalists of the late 19th century overturned this thinking by defining value in relation to scarcity alone; this remains the basis of the *neoclassical school.

**classical system of corporation tax** A system of taxing companies in which the company is treated as a taxable entity separate from its own shareholders. The profits of companies under this system are therefore taxed twice, first when made by the company and again when distributed to the shareholders as dividends. *Compare* imputation system of taxation.

**classified advertising** The form of *advertising that consists of small typeset or semi-display advertisements grouped together in such categories as cars for sale, furniture for sale, flats to let, etc., usually in a paper or magazine. They are usually inserted by direct contact between the advertiser and the advertising department of the publication.

**Classified Directories** UK trade directories (known colloquially as Yellow Pages because they are printed on yellow paper) issued by British Telecom for each business area. Businesses are listed by trade.

**claused bill of lading** *See* dirty bill of lading.

**clawback** Money that a government takes back from members of the public by taxation, especially by *higher-rate tax, having given the money away in benefits, such as increased retirement pensions. Thus the money is clawed back from those who have no need of the extra benefit (because they are paying higher-rate taxes).

**clean bill of lading** *See* bill of lading.

**clean floating** A government policy allowing a country's currency to fluctuate without direct intervention in the foreign-exchange markets. In practice, clean floating is rare as governments are frequently tempted to manage exchange rates by direct intervention by means of the official reserves, a policy sometimes called **managed floating** (*see also* managed currency). However, clean floating does not necessarily mean that there is no control of exchange rates, as they can still be influenced by the government's monetary policy.

**clean price** The price of a *gilt-edged security excluding the accrued interest since the previous dividend payment. Interest on gilt-edged stocks accrues continuously although dividends are paid at fixed intervals (usually six months). Prices quoted in newspapers are usually clean prices, although a buyer will normally pay for and receive the accrued income as well as the stock itself.

**clear days** The full days referred to in a contract, i.e. not including the days on which the contract period starts or finishes.

**clearing bank** A bank that is a member of the London Bankers' Clearing House. It is often used as a synonym for a *commercial bank or a joint-stock bank. *See also* Association for Payment Clearing Services.

**clearing house** A centralized and computerized system for settling indebtedness between members. The best-known in the UK is the *Association for Payment Clearing Services (APACS), which enables the member banks to offset claims against one another for cheques and orders paid into banks other than those upon which they were drawn. Similar arrangements exist in some commodity exchanges, in which sales and purchases are registered with the clearing house for settlement at the end of the accounting period. *See* International Commodities Clearing House.

**Clearing House Inter-Bank Payments System (CHIPS)** The US electronic *clearing house, situated in New York, that clears most dollar cheques between US banks.

**client account** A bank or building society account operated by a professional person (e.g. a solicitor, stockbroker, agent, etc.) on behalf of a client. A client account is legally required for any company handling investments on a client's behalf; it protects the client's money in case the company becomes insolvent and makes dishonest appropriation of the client's funds more difficult. For this reason money in client accounts should be quite separate from the business transactions of the company or the professional person.

**cliometrics** The application of *econometrics to the study of history, a technique developed in the 1960s and 1970s. For example, cliometrics has been used to study the impact of railways on the development of the US economy in the 19th century (and found that impact to be very small). Cliometrics has been described as a method "born of the marriage contracted between historical problems and advanced statistical analysis, with economic theory as bridesmaid and the computer as best man."

**close company** A company controlled by its directors or by a small number of shareholders. Since these companies can be easily manipulated by a small group of persons for their own advantage, the Inland Revenue have powers to counteract these manipulations that they do not have in respect of larger companies. In practice the extra powers are now comparatively few, being largely limited to the ability to apportion investment income to the shareholders so that they can be taxed by means of income tax. In the USA it is known as a **closed company**.

**closed economy** A theoretical model of an economy that neither imports nor exports and is therefore independent of economic factors in the outside world. No such economy exists, although the foreign sector of the US economy is relatively small and most economic decisions are taken by the US government independently of the rest of the world.

**closed indent** *See* open indent.

**closed shop** Any organization in which there is an agreement between the employers and a trade union that only members of that union will be taken on as employees. Under the Employment Protection (Consolidation) Act (1978), as amended by the Employment Acts (1980, 1982, and

1988), all employees are free to join a trade union or not, as they wish. If an employer takes action, short of dismissal, against an employee to enforce membership of a union, the employee can complain to an *industrial tribunal, which can order the employer to pay him compensation. Dismissal for failure to belong to a trade union is automatically unfair (*see* unfair dismissal). In this case there are special minimum rates of compensation payable. If, as a result of trade-union pressure, an employer dismisses an employee for failing to belong to a union, the employer can join the union as a party to the dismissal proceedings and pass the liability to pay compensation on to the union. Under the Employment Act (1988), a union that attempts to enforce a closed shop by industrial action loses the immunity from legal action that it would otherwise have if the action was in furtherance of a trade dispute. The effect of these provisions is that, while closed-shop agreements are not in themselves illegal, they are unenforceable by either employers or unions. In an **open-shop organization** the employer is free to take on employees of any trade union or employees who belong to no trade union.

**close price** The price of a share or commodity when the margin between the bid and offer prices is narrow.

**closing deal** A transaction on a commodity market or stock exchange that closes a long or short position or terminates the liability of an *option holder.

**closing prices** The buying and selling prices recorded at the end of a day's trading on a commodity market or stock exchange. *See* after-hours deals.

**cluster sampling** A type of sampling used in market research, in which the respondents are selected in groups, often chosen according to their geographical areas in order to reduce travelling costs.

**CMEA** Abbreviation for *Council for Mutual Economic Assistance.

**CNAR** Abbreviation for compound net annual rate. *See* compound annual return.

**coaster** A small cargo ship that makes short trips around the coast of the UK from port to port.

**COBOL** Acronym for COmmon Business-Oriented Language. This language is normally used to program computers for business applications, such as invoice and payroll production. It is an easy-to-learn *high-level language that uses a limited number of English words. Programs written in COBOL can be run on different computers, provided that the standard version has been used. Because it was designed to make program writing easy, COBOL occupies considerable amounts of computer memory, making it more popular on larger systems than on microcomputers. However, microcomputer versions of COBOL are available.

**COD** Abbreviation for *cash on delivery.

**coding notice** A document issued by the Inland Revenue as part of the PAYE system showing the amount of *income-tax allowances to be set against an individual's taxable pay. The total of allowances is reduced to a code by dropping the last digit and adding a letter depending on the taxpayer's circumstances (e.g. H for higher personal allowance, L for lower single person's allowance, etc.).

**coemption** The act of buying up the whole stock of a commodity. *See* corner.

**cognitive dissonance** A state of mental conflict caused by a difference between a consumer's expectations of a product and its actual performance. As expectations are largely formed by advertising, dissonance may be reduced by making only realistic and consistent claims about a product's performance.

**coinsurance** The sharing of an insurance risk between several insurers. An insurer may find a particular risk too large to accept because the potential losses may be out of proportion to their claims funds. Rather than turning the insurance away, the insurer can offer to split the risk with a number of other insurers, each of whom would be asked to cover a percentage of the risk in return for the same percentage of the premium. The policyholder deals only with the first or *leading insurer, who issues all the documents, collects all the premiums, and distributes shares to the others involved. A coinsurance policy includes a schedule of all the insurers involved and shows the percentage of the risk each one is accepting.

A policyholder can also become involved in coinsurance. In this case, a reduction of the premium by an agreed percentage is given, in return for an acceptance by the policyholder that all payments of claims are reduced by the same proportion.

**cold calling** A method of selling a product or service in which a sales representative makes calls, door-to-door, by post, or by telephone, to people who have not previously shown any interest in that product or service. In view of the high cost of maintaining a sales force, many companies direct their sales representatives to those potential customers who are favourably disposed to the supplier (either as an existing customer or by replying to a press advertisement). When companies wish to extend their market to new customers, the sales representative must be willing to call on a large number of people, many of whom will show no interest in the product. The primary objective in cold calling is to establish a favourable relationship quickly, enabling the representative to explain the benefits of their product or service before being dismissed.

**collar** Two interest-rate *options combined to protect an investor against wide fluctuations in interest rates. One, the **cap**, covers the investor if the interest rate rises against him; the other, the **floor**, covers him if the rate of interest falls too far.

**collateral** A form of *security, especially an impersonal form of security, such as life-assurance policies or shares, used to secure a bank loan. In some senses such impersonal securities are referred to as a secondary collateral, rather than a primary security, such as a guarantee.

**collective bargaining** Bargaining between employers and employees over wages, terms of employment, etc., when the employees are represented by a trade union or some other collective body.

**collectivism** An economic system in which much of the planning is carried out by a central government and the means of production owned by the community. This is the system practised in several eastern-bloc countries.

**Collector of Taxes** A civil servant responsible for the collection of taxes for which assessments have been raised by *Inspectors of Taxes and for the collection of tax under *PAYE.

**collusion 1.** An agreement between two or more parties in order to prejudice a third party, or for any improper purpose. Collusion to carry out an illegal, not merely improper, purpose is punishable as a conspiracy. **2.** In legal proceedings, a secret agreement between two parties as a result of which one of them agrees to bring an action against the other in order to obtain a judicial decision for an improper purpose. **3.** A secret agreement between the parties to a legal action to do or to refrain from doing something in order to influence the judicial decision. For instance, an agreement between the plaintiff and the defendant to supress certain evidence would amount to collusion. Any judgment obtained by collusion is a nullity and may be set aside.

**collusive duopoly** A form of *duopoly in which producers collude with each other in a price-fixing agreement, thus forming a virtual *monopoly, and negotiate a profit-sharing arrangement. *See* Monopolies and Mergers Commission.

**colophon** A publisher's identifying emblem on a book or other published material. It was formerly an inscription at the end of the book giving the title, date of publication, and the printer.

**column inches (column centimetres)** A measurement of area of typeset matter equal to the width of a column of type in a newspaper or magazine multiplied by its depth. Column centimetres are now replacing column inches in practice, although most *public relations consultancies still refer to column inches when measuring the coverage their activities have achieved.

**COM** Abbreviation for *computer output microfilm.

**combined-transport bill of lading** *See* containerization.

**COMECON** Abbreviation for *Council for Mutual Economic Assistance.

**COMEX** Abbreviation for the Commodity Exchange in New York. It deals in commodities, metal, financial futures, etc.

**command economy (planned economy)** An economy in which the activities of firms and the allocation of productive resources is determined by government direction rather than market forces. The USSR and China are typical examples. Western economists have always pointed out that the command economy is likely to lead to inefficiency because the bureaucrats who run them will not have sufficient information to allocate resources in a way that will satisfy consumer demands. Soviet economists argue that the command system prevents the exploitation of workers by capitalists.

**commercial** An advertisement on television or radio.

**commercial agency (credit-reference agency)** An organization that gives a credit reference on businesses or persons. They specialize in collating information regarding debtors, bankruptcies, and court judgments in order to provide a widely based service to their clients. *See also* ancillary credit business; banker's reference.

**commercial bank** A privately owned bank that provides a wide range of financial services both to the general public and to firms. The principal activities are operating cheque current accounts, receiving deposits, taking in and paying out notes and coin, and making loans. Additional services include trustee and executor facilities, the supply of foreign currency, the purchase and sale of securities, insurance, a credit-card system, and personal pensions. They also compete with the *finance houses and *merchant banks by providing venture capital and with *building societies by providing mortgages.

The number of commercial banks has gradually reduced following a series of mergers. The main banks with national networks of branches are the *Big Four (National Westminster, Barclays, Lloyds and the Midland), the Royal Bank of Scotland, the Bank of Scotland, the Ulster Bank, and the T.S.B. Group plc. They are also known as *joint-stock banks. *See also* clearing bank.

**commercial code** Any of a number of codes used to reduce the cost of sending cables. Single five-letter code words can represent long phrases; however, the international use of Telex and Fax has reduced the need for cables and writing in commercial codes.

**commercial invoice** *See* invoice.

**commercial law** *See* mercantile law.

**commercial paper** A short-term (less than 270 days) bond, bill of exchange, etc., used to make payments, especially in the USA.

**Commerzbank Index** An arithmetically weighted index of 60 West German shares representing 75% of the Bonn market. It has largely replaced the **Bonn Index FAZ**, which is based on 100 industrial shares.

**commission** A payment made to an intermediary, such as an agent, salesman, broker (*see also* brokerage), etc., usually calculated as a percentage of the value of the goods sold. Sometimes the whole of the commission is paid by the seller (e.g. an estate agent's commission in the UK) but in other cases (e.g. some commodity markets) it is shared equally between buyer and seller. In *advertising, the commission is the discount (usually between 10% and 15%) allowed to an *advertising agency by owners of the advertising medium for the space or time purchased on behalf of their clients. A **commission agent** is an agent specializing in buying or selling goods for a principal in another country for a commission.

**Commissioner for Local Administration** A local ombudsman. Established separately in England, Wales, and Scotland, they are responsible for investigating complaints from the public against local authorities, but they exclude personnel and commercial matters. *See also* Parliamentary Commissioner for Administration.

**Commissioners of Customs and Excise** *See* Board of Customs and Excise.

**Commissioners of Inland Revenue** *See* Board of Inland Revenue; General Commissioners; Special Commissioners.

**Committee of Marketing Organizations (COMO)** A body formed to encourage the development of marketing in all sectors of UK business. COMO represents all the main organizations concerned with marketing, including the Advertising Authority (AA), the Association of Market Survey Organizations (AMSO), the Incorporated Advertising Management Association (IAMA), the *Incorporated Society of British Advertisers (ISBA), the Industrial Marketing Research Association (IMRA), the *Institute of Marketing (IM), the *Institute of Practitioners in Advertising (IPA), the *Institute of Public Relations (IPR), the *Institute of Sales Promotion (ISP), the *Market Research Society (MRS), and the Public Relations Consultants Society (PRCA).

**commodity** **1.** A raw material, such as grain, coffee, cocoa, wool, cotton, jute, or rubber (sometimes known as **soft commodities**), that is traded on a *commodity market. In some contexts these raw materials are referred to as **produce**. **2.** A good regarded in economics as the basis of production and exchange. A commodity in this sense is characterized by its physical attributes and where and when it is available. *See also* service.

**commodity broker** A *broker who deals in *commodities, especially one who trades on behalf of his principals in a *commodity market (*see also* futures contract). The rules governing the procedure adopted in each market vary from commodity to commodity and the function of brokers may also vary. In some markets brokers pass on the names of their principals, in others they do not, and in yet others they are permitted to act as principals. Commodity brokers, other than those dealing in metals, are often called **produce brokers**. *See* Association of Futures Brokers and Dealers Ltd; London FOX; London Metal Exchange.

**commodity exchange** *See* commodity market.

**Commodity Futures Trading Commission (CFTC)** A US government body set up in 1975 in Washington to control trading in commodity futures (*see* futures contract) and *options.

**commodity market** A market in which *commodities are traded. The main *terminal markets in commodities are in London and New York, but in some commodities there are markets in the country of origin. Some commodities are dealt with at auctions (e.g. tea), each lot being sold having been examined by dealers, but most dealers deal with goods that have been classified according to established quality standards (*see* certified stock). In these commodities both *actuals and futures (*see* futures contract) are traded on **commodity exchanges**, often with daily *callovers, in which dealers are represented by *commodity brokers. Many commodity exchanges offer *option dealing in futures, and settlement of differences on futures through a *clearing house. As commodity prices fluctuate widely, commodity exchanges provide users and producers with *hedging facilities with outside speculators and investors helping to make an active market, although amateurs are advised not to gamble on commodity exchanges.

The fluctuations in commodity prices have caused considerable problems in developing countries, from which many commodities originate, as they are often important sources of foreign currency, upon which the economic welfare of the country depends. Various measures have been used to restrict price fluctuations but none have been completely successful. *See also* International Commodities Clearing House; London FOX; London Metal Exchange; United Nations Conference on Trade and Development.

**Common Agricultural Policy (CAP)** A policy set up by the *European Economic Community to support free trade within the Common Market and to protect farmers in the member states. The European Commission fixes a **threshold price**, below which cereals may not be imported into the European Community (EC), and also buys surplus cereals at an agreed **intervention price** in order to help farmers achieve a reasonable average price, called the **target price**. Prices are also agreed for meats, poultry, eggs, fruit, and vegetables, with arrangements similar to those for cereals. The *European Commission is also empowered by the CAP to subsidize the modernization of farms within the community. The common policy for exporting agricultural products to non-member countries is laid down by the CAP. In the UK, the Intervention Board for Agricultural Produce is responsible for the implementation of EC regulations regarding the CAP. *See also* Common Budget.

**Common Budget** The fund, administered by the *European Commission, into which all levies and customs duties on goods entering the European Community (EC) are paid and from which all subsidies due under the *Common Agricultural Policy are taken.

**common carrier** *See* carrier.

**Common External Tariff (CET)** The tariff of import duties payable on certain goods entering any country in the European Community from non-member countries. Income from these duties is paid into the *Common Budget.

**Common Fisheries Policy** A 20-year fishing policy agreed between members of the European Community

(EC) in 1983. It lays down annual catch limits for major species of fish, a 12-mile exclusive fishing zone for each state, and an equal-access zone of 200 nautical miles from its coast, within which any member state is allowed to fish. There are also some exceptions to these regulations.

**common law 1.** The law common to the whole of the UK, as opposed to local law. **2.** Case law as opposed to legislation, i.e. the law that has evolved by judicial precedent rather than by statute. For instance, the rules relating to the formation of contracts is a product of the common law, and is not contained in any Act of parliament. **3.** The law of the UK as opposed to foreign law. In this sense, the common law would include case law as well as legislation.

**Common Market** *See* European Community; European Economic Community.

**common stock** The US name for *ordinary shares.

**Commonwealth Development Corporation** A public corporation set up in 1948 to assist commercial and industrial development in any Commonwealth or other developing country. It is authorized to borrow up to £750M from the Treasury and the minimum investment permitted is £1M. Investments are in the form of low-interest loans.

**Commonwealth preference** *See* preferential duty.

**community charge** An annual lump-sum tax to be paid by every adult in a community; it was introduced by the Conservative government in the late 1980s as an alternative to the *rates for local-government finance. There are some allowances for personal circumstances. The charge is often referred to as a *poll tax.

**Communication, Advertising and Marketing Education Foundation** An educational charity funded and supported by the Advertising Association and the Institute of Public Relations. It provides education and training in communications, public relations, advertising, and marketing, operating diploma courses in these subjects.

**commutation** The right to receive an immediate cash sum in return for accepting smaller annual payments at some time in the future. This is usually associated with a pension in which certain life-assurance policyholders can, on retirement, elect to take a cash sum from the pension fund immediately and a reduced annual pension.

**COMO** Abbreviation for *Committee of Marketing Organizations.

**Companies House (Companies Registration Office)** The office of the *Registrar of Companies, formerly in London but now in Cardiff. It contains a register of all UK private and public companies, their directors, shareholders, and balance sheets. All this information has to be provided by companies by law and is available to any member of the public for a small charge.

**company** A corporate enterprise that has a legal identity separate from that of its members; it operates as one single unit, in the success of which all the members participate. An **incorporated company** is a legal person in its own right, able to own property and to sue and be sued in its own name. A company may have limited liability (a *limited company), so that the liability of the members for the company's debts is limited. An **unlimited company** is one in which the liability of the members is not limited in any way. There are various different types of company: a **chartered company** is one formed under Royal Charter. This was the earliest type of company to exist and they were influential in the development of both foreign trade and colonization; an example is the

Hudson's Bay Company. Chartered companies, however, are now rare, unless a charter is required for prestige purposes, as it might be for a new university. A **joint-stock company** is a company in which the members pool their stock, trading on the basis of their joint stock. This differs from the earlier **merchant corporations** or **regulated companies** of the 14th century, in which each member traded with his own stock but agreed to obey the rules of the company.

A **registered company**, one registered under the Companies Acts, is the most common type of company. A company may be registered either as a public limited company or a private company. A **public limited company** must have a name ending with the initials 'plc' and have an authorized share capital of at least £50,000, of which at least £12,500 must be paid up. The company's memorandum must comply with the format in Table F of the Companies Regulations (1985). It may offer shares and securities to the public. The regulation of such companies is stricter than that of private companies. Most public companies are converted from private companies, under the re-registration procedure laid down in the Companies Act. A **private company** is any registered company that is not a public company. The shares of a private company may not be offered to the public for sale. The legal requirements for such a company are less strict; for example, there is no minimum issued or paid-up share capital requirement and small and medium-sized private companies need not file full accounts. A **statutory company** is a company formed by special Act of parliament. These are generally public utilities that were either not nationalized (for example, certain water authorities) or that have been privatized (such as British Gas and British Telecom). Their powers and privileges depend upon the Act under which they were formed.

**company doctor 1.** A businessman or accountant with wide commercial experience, who specializes in analysing and rectifying the problems of ailing companies. He may either act as a consultant or may be given executive powers to implement the policies he recommends. **2.** A medical doctor employed by a company, either full-time or part-time, to look after its staff, especially its senior executives, and to advise on medical and public-health matters.

**company formation** The procedure to be adopted for forming a company in the UK. The *subscribers to the company must send to the *Registrar of Companies a statement giving details of the registered address of the new company together with the names and addresses of the first directors and secretary, with their written consent to act in these capacities. They must also give a declaration (**declaration of compliance**) that the provisions of the Companies Acts have been complied with and provide the *memorandum of association and the *articles of association. Provided all these documents are in order the Registrar will issue a *certificate of incorporation and a certificate enabling it to start business. In the case of a *public limited company additional information is required.

**company law** The laws governing the formation, conduct, and control of companies. These are largely statute-based law; the principal Acts are the Companies Act (1985), the Insolvency Act (1986), the Company Directors' Disqualification Act (1986), and the Financial Services Act (1986).

**company seal** The common seal with the company's name engraved on it in legible characters. It is used to authenticate share certificates and other important documents issued by the company. The articles of association set out how and when the seal is to be affixed to contracts. Unless it is affixed to any contract required by English law to be made under seal, the company will not be bound by that contract.

**company secretary** An *officer of a company. The appointment is usually made by the directors. The secretary's duties are mainly administrative, including preparation of the agenda for directors' meetings. However, the modern company secretary has an increasingly important role; he may manage the office and enter into contracts on behalf of the company. Duties imposed by law include the submission of the annual return and the keeping of minutes. The secretary of a public company is required to have certain qualifications, set out in the Companies Act (1985). *See* Institute of Chartered Secretaries and Administrators.

**comparative advantage** The relative efficiency in a particular economic activity of an individual or group of individuals over another economic activity, compared to another individual or group. One of the fundamental propositions of economics is that if individuals or groups specialize in activities in which their comparative advantage lies, then there are gains from trade. This proposition, first outlined by David Ricardo (1772–1823), is one of the main arguments for free trade and against such restrictions as tariffs and quotas. It still holds even if one group holds an *absolute advantage in all economic activities over another group. For example, even if Japan is better at producing both cars and ships than the UK, there will still be gains from trade if the UK specializes in the production of the goods in which it holds a comparative advantage.

**compatibility** The ability of two or more different types of computer to use the same programs and data. Two computers are said to be compatible if the same *machine code can run on both without alteration. Computers from different manufacturers are rarely compatible, unless this is a deliberate feature of the design. For example, many microcomputers are designed to be compatible with the IBM PC. Manufacturers are increasingly making their small machines compatible with their larger ones. This is called **upward compatibility**: programs written for the smaller computer will run on the larger, but not vice versa. This ensures that customers can easily upgrade their machines when necessary. The term is also used of parts of a computer system, either hardware (e.g. terminals) or software (e.g. spreadsheets), meaning that two or more specified brands or versions can be substituted for each other.
The term **plug compatible** describes *peripheral devices that can be joined to a computer by a standard interface, or plug.

**compensated demand function** *See* Hicksian demand function.

**compensation for loss of office** A payment, often tax-free, made by a company to a director, senior executive, or consultant who is forced to retire before the expiry of his service contract, as a result of a merger, takeover, or any other reason. This form of **severance pay** (*see also* redundancy) may be additional to a retirement pension or in place of it; it must also be shown separately in the company's accounts. Because these payments can be very large, they are known as **golden handshakes**. *See also* golden parachute.

**compensation fund** A fund set up by the *London Stock Exchange, to which member firms contribute. It provides compensation to investors who suffer loss as a result of a member firm failing to meet its financial obligations.

**competition** Rivalry between suppliers providing goods or services for a market. The consensus of most economic theory is that competition is beneficial, in the sense of achieving *Pareto optimality. Governments usually pursue policies aimed at increasing competition in markets (*see* regulation; deregulation), although there may often be a conflict between policies that increase competition and those that promote purely national interest. *See also* imperfect competition; monopoly; perfect competition.

**competition and credit control** The subject of an important paper issued in 1971 by the *Bank of England. It outlined a number of changes affecting the banking system and the means of controlling credit. From October 1971 a new system of reserve requirements was implemented, the banks agreed to abandon collusion on setting interest rates, and the Bank of England changed its operations in the gilt-edged securities market. The main aim of these changes was to stimulate more active competition between the banks and to move towards greater reliance upon interest rates as a means of credit control. Further changes were made in 1981 with the abolition of the *minimum lending rate and the reserve asset ratio.

**competitive advantage** The factors that give a company an advantage over its rivals. For companies marketing similar products, one may achieve a competitive advantage by creative and memorable advertising, innovative package design, or superior distribution methods.

**competitiveness** The ability of an economy to supply increasing *aggregate demand and maintain exports. A loss of competitiveness is usually signalled by increasing imports and falling exports. Competitiveness is often measured in a narrower sense by comparing relative inflation rates. For instance, if the sterling-dollar exchange rate remains constant, but prices rise faster in the UK than in the US, UK goods will become relatively more expensive, reflecting a loss in competitiveness; this in turn may lead to a falling demand for exports.

**compiler** A computer program that translates a program written in a *high-level language into the detailed instructions (called *machine code) that the computer can execute. The program must be translated in its entirety before it can be executed; however, it can then be executed any number of times (*compare* interpreter).

**complement** A good for which the demand changes in the same direction as the demand for some other good whose price has changed. For example, a rise in the price of bread may lead to a fall in demand for bread; if the demand for butter fell at the same time, butter would be called a complement of bread. Another way of saying this is that the price *elasticity of the demand for butter with respect to bread is less than zero. *Compare* substitute.

**completion** The conveyance of land in fulfilment of a contract of sale. The purchaser will have obtained an equitable interest in the land at the date of the *exchange of contracts but will not become the full legal owner until completion. If the seller refuses to complete, the court may grant a decree of *specific performance. The date of completion is stated in the contract of sale.

**compliments slip (comp slip)** A slip of paper with the words "With the compliments of . . ." printed on it, followed by the name and address of the firm sending it. It is enclosed with material sent to another person or firm to identify the sender, when a letter is not necessary.

**composite rate tax** A special rate of tax (introduced in the UK in 1951) that building societies and banks must deduct from interest paid to investors (resident in the UK). The current rate of tax (until April 1991) is about 3% below the basic rate of income tax, but this tax cannot be claimed back under any circumstances. The result of the tax is that basic-rate taxpayers with such deposits gain some 3% and non-taxpayers (e.g. some children, married women, pensioners, etc.) lose 22%. The composite rate tax is abolished from April 1991. Thereafter taxpayers will pay the full basic rate (which will still be deducted at source) and non-taxpayers can opt to be paid gross interest or will be able to reclaim the tax.

**composition** *See* arrangement.

**compositor** A person who sets type, especially metal type, which has largely been replaced by phototypesetting.

**compound annual return (CAR)** The total return available from an investment or deposit in which the interest is used to augment the investment. The more frequently the interest is credited, the higher the CAR. The CAR is usually quoted on a gross basis. The return, taking into account the deduction of tax at the basic rate on the interest, is known as the **compound net annual rate (CNAR)**.

**compound interest** *See* interest.

**compound net annual rate (CNAR)** *See* compound annual return.

**comprehensive income tax** An income tax for which the tax base consists not only of income but also of capital gains as well as other accretions of wealth, such as legacies. Although this is not a tax currently levied in the UK, tax theorists find it attractive since sometimes clear distinctions between income, capital gains, etc., are difficult to sustain.

**comprehensive insurance** *See* motor insurance.

**comptroller** The title of the financial director in some companies or chief financial officer of a group of companies. The title is more widely used in the USA than in the UK.

**compulsory liquidation (compulsory winding-up)** The winding-up of a company by a court. A petition must be presented both at the court and the registered office of the company. Those by whom it may be presented include: the company, the directors, a creditor, an official receiver, and the Secretary of State for Trade and Industry. The grounds on which a company may be wound up by the court include: a special resolution of the company that it be wound up by the court; that the company is unable to pay its debts; that the number of members is reduced below two; or that the court is of the opinion that it would be just and equitable for the company to be wound up. The court may appoint a *provisional liquidator after the winding-up petition has been presented; it may also appoint a *special manager to manage the company's property. On the grant of the order for winding-up, the official receiver becomes the *liquidator and continues in office until some other person is appointed, either by the creditors or the members. *Compare* members' voluntary liquidation.

**compulsory purchase** The compulsory acquisition of land by the state when it is required for some purpose under the Town and Country Planning legislation. Compensation is paid on the basis of market value. *Compare* requisitioning.

**compulsory purchase annuity** An annuity that must be purchased with the fund built up from certain types of pension arrangements. When retirement age is reached, a person who has been paying premiums into this type of pension fund is obliged to use the fund to purchase an annuity to provide an income for the rest of his life. The fund may not be used in any other way (except for a small portion, which may be taken in cash).

**computer** An electronic tool that manipulates information in accordance with a predefined sequence of instructions. Computers have a simple 'brain', called the *central processing unit (CPU), that can do arithmetic and take decisions based on the results, and a *memory, which stores the instructions and information. Strictly speaking, all other parts of a computer system are *peripheral devices, but in practice all computers have at least: an *input device, such as a keyboard, where the information is fed in; an *output device, such as a screen or printer, where they show the results of their work; and extra memory in the form of *magnetic disks or *magnetic tape, called *backing store. Inside a computer, both information and instructions are represented by binary numbers (*see* binary notation) and all processing is done using these numbers. A computer needs to be given detailed instructions, called a program, before it can perform even the simplest task. The art of composing these instructions is called *computer programming. The general term for the programs that a computer needs in order to operate is *software, while the computer and devices attached to it are called *hardware.

Computers come in three main sizes. The largest is the *mainframe, used typically for large-scale corporate data processing; and the smallest is the *microcomputer, designed for a single user. Between these extremes is the *minicomputer. These distinctions are becoming blurred, as minicomputers and microcomputers become more powerful.

Formerly, this type of machine was called a **digital computer** to distinguish it from other computing devices, such as the *analog computer. The term 'computer' is now synonymous with 'digital computer', which is rarely used.

**computer-aided design (CAD)** The use of computers for a variety of design projects, including determining the contours of car bodies, investigating the behaviour of bridges in windy conditions, and the creation of shoes and home interiors. There are two elements to CAD. On-the-screen drawing, using light pens or similar devices, allows the designer a similar versatility to that given by word-processing to the typist. Designs can be amended easily, rotated and looked at from different angles, and printed out as working drawings. Other programs can systematically analyse the design and test it against appropriate technical data. The final output of CAD is often transferred directly to *computer-aided manufacturing systems.

**computer-aided manufacturing (CAM)** The use of computers to control industrial processes, such as brewing, chemical manufacture, oil refining, and steel making. They are also used to control automatic machines that can be programmed to carry out different tasks, especially in the car-manufacturing industry.

**computer-assisted telephone interviewing** A system in which a telephone interviewer conducts a sales or marketing interview, using a computer and a computerized questionnaire. This system reduces the number of errors as the interviewer keys in the respondent's answers as they are given and the computer follows a complex questionnaire routing efficiently, enabling the required statistics to be extracted automatically.

**computer-assisted trading (CAT)** The use of computers by brokers and traders on a market, such as a stock exchange or foreign-exchange market, to facilitate trading by displaying prices, recording deals, etc.

**computer-based training (CBT)** The use of computers and video films in training. Exercises are shown on the video screen and participants have to key in answers to questions, which are assessed by the computer.

**computer output microfilm (COM)** A system that produces computer output on microfilm or microfiche. The information is recorded on film either by a camera that photographs a miniature screen displaying the computer output, or by a laser beam recorder, which writes directly onto the film. COM is used where there is a large volume of output, particularly where immediate reference is not required: for example, to produce updated copies of a bank's customer account at the end of each day's business.

**computer programming** The process of writing the list of instructions that a computer must follow to solve a problem. The list of instructions, called the **program**, must be complete in every detail, since the computer cannot think for itself. The steps involved in writing a program are: understanding the problem, planning the solution, preparing the program, testing the program and removing errors, and documenting the program. A *systems analyst will often help the programmer understand the problem and produce a detailed specification of the required program. This specification is further resolved into a sequence of logical steps, often by drawing a *flowchart. The program itself is a translation of these steps into a *programming language, which can be fed into the computer. Once the program is complete the errors, or *bugs, must be located and corrected. Finally, a manual for the users of the program must be written.

**concealed unemployment** *See* hidden unemployment.

**concentrated segmentation (niche marketing)** A comparatively small segment of a market that has been identified, usually by a small company, in which to concentrate their efforts. For example, a company might decide to offer a service converting cine films to video tapes; as long as this activity does not threaten the profitability of the manufacturers of either video tapes or cine films, it can establish a profitable niche in the market.

**concept test** A technique used in *marketing research to assess the reactions of consumers to a new product or a proposed change to an existing product (*see* new product development). Before an organization invests in production facilities for a new product it writes a **concept statement** describing the proposed product and commissions a market researcher to interview a small number of potential consumers either in *group discussion or *depth interview. Respondents are shown the concept statement and their reactions are explored in considerable detail. The results of these interviews help the organization to understand what it should do with the proposed product if it is to be successful.

**concert-party agreements** Secret agreements between apparently unconnected shareholders to act together to manipulate the share price of a company or to influence its management. The Companies Act (1981) laid down that the shares of the parties to such an agreement should be treated as if they were owned by one person, from the point of view of disclosing interests in a company's shareholding.

**conciliation** *See* Advisory Conciliation and Arbitration Service (ACAS); mediation.

**condition 1.** A major term of a contract that, if unfulfilled, constitutes a fundamental *breach of contract and may invalidate it. *Compare* warranty. **2.** A provision that does not form part of a contract but either suspends the contract until a specified event has happened (**condition precedent**) or brings it to an end in specified circumstances (**condition subsequent**). An example of a condition precedent is an agreement to buy a particular car if it passes its MOT test; an example of a condition subsequent is an agreement that entitles the purchaser of goods to return them if he is dissatisfied with them.

**conditional bid** *See* takeover bid.

**conditionality** The terms under which the *International Monetary Fund (IMF) provides balance-of-payments support to member states. The principle is that support will only be given on the condition that it is accompanied by steps to solve the underlying problem. Programmes of economic reform are agreed with the member; these emphasize the attainment of a sustainable balance-of-payments position and boosting the supply side of the economy. The most recent review of general guidelines of conditionality principles was undertaken in 1988. Lending by commercial banks is frequently linked to IMF conditionality.

**conditional sale agreement** A contract of sale under which the price is payable by instalments and ownership does not pass to the buyer (although he is in possession of the goods) until specified conditions relating to the payment have been fulfilled. The seller retains ownership of the goods as security until he is paid in full. A conditional sale agreement is a form of *consumer credit regulated by the Consumer Credit Act (1974) if the buyer is an individual, the credit does not exceed £15,000, and the agreement is not otherwise exempt.

**Confederation of British Industry (CBI)** An independent non-party organization formed in 1965, by a merger of the National Association of British Manufacturers, the British Employers Confederation, and the Federation of British Industry, to promote prosperity in British industry and to represent industry in dealings with the government. Membership, which totals approximately 50,000 companies, is voluntary. The governing body is the Confederation of British Industry Council, which meets monthly; there are 13 Regional Councils that deal with local industrial problems.

**conference lines** *See* shipping conference.

**conferencing** *See* audioconferencing; videoconferencing.

**confidentiality clause** A clause in a *contract of employment that details certain types of information the employee will acquire on joining the firm that may not be passed on to anyone outside the firm.

**confirmed letter of credit** *See* letter of credit.

**confirming house** An organization that purchases goods from local exporters on behalf of overseas buyers. It may act as a principal or an agent, invariably pays for the goods in the exporters' own currency, and purchases on a contract that is enforceable in the exporters' own country. The overseas buyer, who usually pays the confirming house a commission or its equivalent, regards the confirming house as a local buying agent, who will negotiate the best prices on its behalf, arrange for the shipment and insurance of the goods, and provide information regarding the goods being sold and the status of the various exporters.

**conflict of interests** A situation that can arise if a person (or firm) acts in two or more separate capacities and the objectives in these capacities are not identical. The conflict may be between self-interest and the interest of a company for which a person works or it could arise when a person is a director of two companies, which find themselves competing. The proper course of action in the case of a conflict of interests is for the person concerned to declare his interests, to make known the way in which they conflict, and to abstain from voting or sharing in the decision-making procedure involving these interests.

**Confravision** *See* videoconferencing.

**conglomerate** A group of companies merged into one entity, although they are active in totally different fields. A conglomerate is usually formed by a company wishing to diversify so that it is not totally dependent on one industry. Many tobacco firms and brewers have diversified in this way.

**consequential-loss policy** *See* business-interruption policy.

**conservatism** A prudent and not overoptimistic view of the state of affairs of a company or other organization. Because it is regarded as imprudent to distribute to shareholders profits that may not materialize, it is a general principle of accounting not to anticipate profits before they are realized but to anticipate losses as soon as they become foreseeable. This view, which may be pessimistic, is generally accepted to be a *true and fair view. *See also* prudence concept.

**consideration 1.** A promise by one party to a *contract that constitutes the price for buying a promise from the other party to the contract. A consideration is essential if a contract, other than a *deed, is to be valid. It usually consists of a promise to do or not to do something or to pay a sum of money. **2.** The money value of a contract for the purchase or sale of securities on the London Stock Exchange, before commissions, charges, stamp duty, and any other expenses have been deducted.

**consignee 1.** Any person or organization to whom goods are sent. **2.** An agent who sells goods, usually in a foreign country, on *consignment on behalf of a principal (consignor).

**consignment 1.** A shipment or delivery of goods sent at one time. **2.** Goods sent **on consignment** by a principal (consignor) to an agent (consignee), usually in a foreign country, for sale either at an agreed price or at the best market price. The agent, who usually works for a commission, does not normally pay for the goods until they are sold and does not own them, although he will usually have possession of them. The final settlement, often called a **consignment account**, details the cost of the goods, the expenses incurred, the agent's commission, and the proceeds of the sale.

**consignment note** A document accompanying a consignment of goods in transit. It is signed by the *consignee on delivery and acts as evidence that the goods have received. It gives the names and addresses of both consignor and consignee, details the goods, usually gives their gross weight, and states who has responsibility for insuring them while in transit. It is not a negotiable document (*compare* bill of lading) and in some circumstances is called a **way bill.**

**consignor 1.** Any person or organization that sends goods to a *consignee. **2.** A principal who sells goods on *consignment through an agent (consignee), usually in a foreign country.

**consistency concept** Applying the same accounting principles to successive accounting periods. It is possible to quantify profit in a variety of ways. If different methods are used in different periods, one period cannot be compared with another and no clear trend can emerge. It is therefore a principle that similar methods should be maintained from year to year. This may still make it difficult to compare one company with another, a problem to which accounting standards are a partial solution (*see* statements of standard accounting practice).

**consolidated accounts** The combined accounts of a *group of companies. Although a parent company and its subsidiaries (the companies its owns and controls) are separate companies, it is customary to combine their results in a single set of accounts, which eliminates inter-company shareholdings and inter-company indebtedness; it also aggregates the assets and liabilities of all the companies. Parent companies of groups are required by law to prepare and file consolidated accounts in addition to individual accounts of the subsidiary companies. *See also* holding company.

**consolidated annuities** *See* Consols.

**Consolidated Fund** The Exchequer account, held at the Bank of England and controlled by the Treasury, into which taxes are paid and from which government expenditure is made. It was formed in 1787 by the consolidation of several government funds.

**consolidation** An increase in the *nominal price of a company's shares, by combining a specified number of lower-price shares into one higher-priced share. For example five 20p shares may be consolidated into one £1 share. In most cases this can be done by an ordinary resolution at a general meeting of the company.

**Consols** Government securities that pay interest but have no *redemption date. The present bonds, called **consolidated annuities** or **consolidated stock**, are the result of merging several loans at various different times going back to the 18th century. Their original interest rate was 3% on the nominal price of £100; most now pay 2½% and therefore stand at a price that makes their annual *yield comparable to long-dated *gilt-edged securities, e.g. at £273/4 they yield about 9%.

**consortium** A combination of two or more large companies formed on a temporary basis to quote for a large project, such as a new power station or dam. The companies would then work together, on agreed terms, if they were successful in obtaining the work. The purpose of forming a consortium may be to eliminate competition between the members or to pool skills, not all of which may be available to the individual companies.

**consortium relief** A means of enabling companies owned by a consortium to transfer to members of the consortium (or vice versa) the benefit of their tax losses or certain other payments for which they are unable to obtain tax relief in their own right. The relief is not available for all losses and the requirements for consortium status are rigidly defined.

**constructive dismissal** A situation that arises when an employer's behaviour towards an employee is so intolerable that the employee is left with no option but resignation. In these circumstances the employee can still claim compensation for *wrongful dismissal.

**constructive total loss** A loss in which the item insured is not totally destroyed but is so severely damaged that it is not financially worth repairing. The Marine Insurance Act (1906) defines a constructive total loss as one in which "the subject matter insured is reasonably abandoned on account of its actual total loss appearing to be unavoidable, or because it could not be preserved from actual total loss without an expenditure which would exceed its value when the expenditure had been incurred."

**consulage** *See* consular invoice.

**consular invoice** An export invoice that has been certified in the exporting country by the consul of the importing country. This form of invoice is required by the customs of certain countries (especially South American countries) to enable them to charge the correct import duties. A *certificate of origin may also be required. The fee charged by the consul for this or any other commercial service is called the **consulage**.

**consumer advertising** The *advertising of goods or services specifically aimed at the potential end-user, rather than at an intermediary in the selling chain. *Compare* trade advertising.

**consumer credit** Short-term loans to the public for the purchase of goods. The most common forms of consumer credit are credit accounts at retail outlets, personal loans from banks and finance houses, *hire purchase, and *credit cards. Since the Consumer Credit Act (1974), the borrower has been given greater protection, particularly with regard to regulations establishing the true rate of interest being charged when loans are made (*see* annual percentage rate). The Act also made it necessary for anyone giving credit in a business (with minor exceptions) to obtain a licence. *See* consumer-credit register.

**consumer-credit register** The register kept by the Director General of Fair Trading (*see* consumer protection), as required by the Consumer Credit Act (1974), relating to the licensing or carrying on of consumer-credit businesses or consumer-hire businesses. The register contains particulars of undetermined applications, licences that are in force or have at any time been suspended or revoked, and decisions given by the Director under the Act and any appeal from them. The public is entitled to inspect the register on payment of a fee.

**consumer durable** *See* consumer goods.

**consumer goods** Goods that are purchased by members of the public (*compare* capital goods). **Consumer durables** are consumer goods, such as cars, refrigerators, and television sets, whose useful life extends over a relatively long period. The purchase of a consumer durable falls between *consumption and *investment, which tends to complicate *macroeconomic analysis of these two variables. **Consumer non-durables** or **disposables** are goods that are used up within a short time after purchase. They include food, drink, newspapers, etc. Personal services, such as those provided by doctors, hairdressers, etc., are also classed by economists as consumer goods.

**consumer market** The market for *consumer goods, as opposed to the industrial market, in which buyers of goods are not the end users.

**consumer non-durable** *See* consumer goods.

**consumer preference** The way in which consumers in a free market choose to divide their total expenditure in purchasing goods and services; these preferences constitute the basis of *consumer theory. Using a limited number of assumptions, an individual's preferences can be built up into a *utility function. Applying *price theory to the utility functions of individuals enables a model to be constructed of the behaviour of markets in an economy.

**consumer price index** The name for the *Retail Price Index in the USA and some other countries.

**consumer profile** A profile of a typical consumer of a product, in terms of age, sex, social class, and other characteristics.

**consumer protection** The protection, especially by legal means, of consumers. It is the policy of current UK legislation to protect consumers against unfair contract terms. In particular they are protected by the Unfair Contract Terms Act (1977) and the Sale of Goods Act (1979) against terms that attempt to restrict the seller's implied undertakings that he has a right to sell the goods, that the goods conform with either description or sample (*see* trade description), and that they are of merchantable quality and fit for their particular purpose. There is also provision for the banning of unfair consumer trade practices in the Fair Trading Act (1973). This Act provides for a **Director General of Fair Trading**, who is responsible for reviewing commercial activities relating to the supply of goods to consumers and discovering any practices against the economic interest of the consumer. He may refer certain practices to the **Consumer Protection Advisory Committee** or take legal action himself. Consumers (including individual businessmen) are also protected when obtaining credit by the Consumer Credit Act (1974). *See also* consumer-credit register. There is provision for the imposition of standards relating to the safety of goods under the Consumer Protection Act (1987), which also makes the producer of a product liable for any damage it causes (*see* products liability).

**consumer research** Any form of *marketing research undertaken among the final consumers of a product or service. For example, a manufacturer supplying man-made fibres to a shirt factory might undertake consumer research, interviewing purchasers of shirts, in order to establish the merits, or otherwise, of his fibres. *Compare* industrial marketing research.

**Consumers' Association (CA)** A charitable organization formed in 1957 to provide independent and technically based guidance on the goods and services available to the public. The Consumers' Association tests and investigates products and services and publishes comparative reports on performance, quality, and value in its monthly magazine *Which?*. It also publishes *Holiday Which?* and *Gardening From Which?* as well as various books, including *The Legal Side of Buying a House*, *Starting Your Own Business*, and *The Which? Book of Saving and Investing*.

**consumers' expenditure** *See* consumption.

**consumer theory** The theory explaining the choices made by individuals and households in terms of the concept of *utility. Consumers are assumed to be able to order their preferences (*see* consumer preference) in such a way that they can choose a basket of goods that maximize their utility, subject to the constraint that their income is limited. The application of *price theory and *demand theory to this problem forms the basis of *microeconomics (together with the theory of the firm) and also of *macroeconomics.

**consumption 1.** The using up of *consumer goods and services for the satisfaction of the present needs of individuals, organizations, and governments. **2.** The amount of money spent by the whole of an economy on consumer goods and services. Economists contrast this amount of money (sometimes called **consumers' expenditure**) with the amount spent on *investment, which provides for future consumption. In most economies about 80% of national income is spent on consumption, the balance going to investment. However, the distinction between consumption and investment is not always clear as some goods, such as consumer durables, provide for both present and future consumption.

**containerization** The use of large rectangular containers for the shipment of goods. Goods are packed into the containers at the factory or loading depot and transported by road in these containers to the port of shipment, where they are loaded direct onto the ship without unpacking. At the port of destination they can again be transported by road to the final user. It is usual for such shipments to be covered by a **container bill of lading** (**combined-transport bill of lading**).

**contango** The former practice of carrying the purchase of stocks and shares over from one account day on the London Stock Exchange to the next. *See* cash and new.

**contempt of court** An act that hinders the course of justice or that constitutes disrespect to the lawful authority of the court. Contempt may be divided into acts committed in court (for instance, unseemly behaviour or refusing to answer a question as a witness) and acts committed out of court (such as intimidating a witness or refusing to obey a court order). Contempt of court is punishable by fine or imprisonment or both.

**contemptuous damages** *See* damages.

**contestable markets theory** The theory that prices in *oligopolies may be close to the perfectly competitive level due to the threat of entry by maverick firms. For example, in the airline industry although there are a relatively small number of competitors, prices tend to remain low, since as soon as the airlines try to raise their prices *entrepreneurs enter the market and undercut existing fares. This leads to the suggestion that governments need not try to encourage perfect competition in markets as long as they ensure that they are contestable.

**contingency insurance** An insurance policy covering financial losses occurring as a result of a specified event happening. The risks covered by policies of this kind are various and often unusual, such as a missing documents indemnity, the birth of twins, or *pluvial insurance.

**contingency plan** A plan that is formulated to cope with some event or circumstances that may occur in the future. For example, the contingency may be an increase in sales, in which case the plan would include means of increasing production very quickly.

**contingent annuity (reversionary annuity)** An annuity in which the payment is conditional on a specified event happening. The most common form is an annuity purchased jointly by a husband and wife that begins payment after the death of one of the parties (*see* joint-life and last-survivor annuities).

**contingent-interest** *See* vested interest.

**contingent liability** A liability that, at a balance sheet date, can be anticipated to arise if a particular event occurs. Typical examples include a court case pending against the company, the outcome of which is uncertain, or loss of earnings as a result of a customer invoking a penalty clause in a contract that may not be completed on time. Under the Companies Act (1985), such liabilities must be explained by a note on the company balance sheet.

**continuation** An arrangement between an investor and a stockbroker in which the broker reduces his *commission for a series of purchases by that investor of the same stock over a stated period.

**continuous stationery** Headed stationery (notepaper, invoices, etc.) printed in a continuous roll, suitable for use on a computer output device.

**contra** A book-keeping entry on the opposite side of an account to an earlier entry, with the object of cancelling the effect of the earlier entry. *See also* per contra.

**contraband** Illegally imported or exported goods, i.e. goods that have been smuggled into or out of a country.

**contract** A legally binding agreement. Agreement arises as a result of an *offer and *acceptance, but a number of other requirements must be satisfied for an agreement to be legally binding. There must be *consideration (unless the contract is by *deed); the parties must have an intention to create legal relations; the parties must have capacity to contract (i.e. they must be competent to enter a legal obligation, by not being a minor, mentally disordered, or drunk); the agreement must comply with any formal legal requirements; the agreement must be legal (*see* illegal contract); and the agreement must not be rendered void either by some common-law or statutory rule or by some inherent defect.

In general, no particular formality is required for the creation of a valid contract. It may be oral, written, partly oral and partly written, or even implied from conduct. However, certain contracts are valid only if made by deed (e.g. transfers of shares in statutory *companies, transfers of shares in British ships, legal *mortgages, certain types of *lease) or in writing (e.g. *hire-purchase agreements, *bills of exchange, promissory notes, contracts for the sale of land made after 21 September 1989), and certain others, though valid, can only be enforced if evidenced in writing (e.g. guarantees, contracts for the sale of land made before 21 September 1989). *See also* contract of employment. Certain contracts, though valid, may be liable to be set aside by one of the parties on such grounds as *misrepresentation or the exercise of undue influence. *See also* affirmation of contract; breach of contract.

**contract guarantee insurance** An insurance policy designed to guarantee the financial solvency of a contractor during the performance of a contract. If the contractor becomes financially insolvent and cannot complete the work the insurer makes a payment equivalent to the contract price, which enables another contractor to be paid to complete the work. *See also* credit insurance.

**contracting out** *See* State Earnings-Related Pension Scheme.

**contract note** A document sent by a stockbroker or commodity broker to his client as evidence that he has bought (in which case it may be called a **bought note**) or sold (a **sold note**) securities or commodities in accordance with the client's instructions. It will state the quantity of securities or goods, the price, the date (and sometimes the time of day at which the bargain was struck), the rate of commission, the cost of the transfer stamp and VAT (if any), and – finally – the amount due and the settlement date.

**contract of employment (contract of service)** A legally enforceable agreement entered into orally or in writing by an employer and an employee. There is no requirement in UK law that a contract of employment must be in writing; however, the basic terms and conditions of employment are normally written and signed by both parties to ensure that each knows their rights and obligations. The Employment Protection (Consolidation) Act (1978) states that every employee must be given, within 13 weeks of joining, a note giving the date of joining, job title, salary, pay day, holiday entitlement, sick pay, pensions, notice requirements, and disciplinary procedures. Most employers expand this statement to provide as detailed a contract as possible in order to avoid misunderstandings. Larger organizations provide a company handbook giving a detailed account of company policies, disciplinary rules, and any other relevant matters. A contract can be varied by mutual agreement but neither an employee nor an employer can contract out of obligations covered by normal common-law rights.

**contra proferentem** A rule of interpretation primarily applying to documents. If any doubt or ambiguity arises in the interpretation of a document, the rule requires that the doubt or ambiguity should be resolved against the party who drafted it or who uses it as a basis for a claim against another. For instance, a plaintiff who sues for breach of a written contract can expect that any ambiguity in the terms of the contract will be resolved against him. The expression derives from the Latin: verba chartarum fortuis accipiuntur contra proferentum, the words of a contract are construed more strictly against the person proclaiming them.

**contribution 1.** The amount that, under *marginal-costing principles, a given transaction produces to cover fixed overheads and to provide profit. The contribution is normally taken to be the selling price of a given unit of merchandise, less the variable costs of producing it. Once the total contributions exceed the fixed overheads, all further contribution represents pure profit. **2.** The sharing of claim payments between two or more insurers who find themselves insuring the same item, against the same risks, for the same person. As that person is not entitled to claim more than the full value of the item once, each insurer pays a share. For example, if a coat was stolen from a car, it might be insured under both a personal-effects insurance and a motor policy. As the policyholder is only entitled to the value of the coat (he cannot profit from the theft), each insurer contributes half of the loss.

**contributory** Any person who is liable to contribute towards the assets of a company on liquidation. The list of contributories will be settled by the liquidator or by the court. This list will include all shareholders, although those who hold fully paid-up shares will not be liable to pay any more.

**contributory pension** A *pension in which the employee as well as the employer contribute to the pension fund. *Compare* non-contributory pension.

**control accounts** Accounts in which the balances are designed to equal the aggregate of the balances on a substantial number of subsidiary accounts. Examples are the sales ledger control account (or total debtors account), in which the balance equals the aggregate of all the individual debtors' accounts, the purchase ledger control account (or total creditors account), which performs the same function for creditors, and the stock control account, whose balance should equal the aggregate of the balances on the stock accounts for each item of stock. This is achieved by entering in the control accounts the totals of all the individual entries made in the subsidiary accounts. The purpose is twofold: to obtain total figures of debtors, creditors, stock, etc., at any given time, without adding up all the balances on the individual records, and to have a cross-check on the accuracy of the subsidiary records.

**controlling interest** An interest in a company that gives a person control of it. For a shareholder to have a controlling interest in a company, he would normally need to own or control more than half the voting shares. However, in practice, a shareholder might control the company with considerably less than half the shares, if the shares that he does not own or control are held by a large number of people. For legal purposes, a director is said to have a controlling interest in a company if he alone, or together with his wife, minor children, and the trustees of any settlement in which he has an interest, owns more than 20% of the voting shares in a company or in a company that controls that company.

**convenience store** A store that trades primarily on the convenience it offers to customers. The products stocked may be influenced by local tastes or ethnic groups and the stores are often open long hours as well as being conveniently placed for customers in local shopping parades. The most successful are members of a multiple chain.

**conversational systems** *See* interactive.

**conversion** The tort (civil wrong) equivalent to the crime of theft. It is possible to bring an action in respect of conversion to recover damages, but this is uncommon.

**convertibility** The extent to which one currency can be freely exchanged for another. Since 1979 sterling has been freely convertible. The *International Monetary Fund encourages free convertibility, although many governments try to maintain some direct control over foreign-exchange transactions involving their own currency, especially if there is a shortage of hard-currency foreign-exchange reserves.

**convertible (conversion issue) 1.** A security, usually a *bond or *debenture, that can be converted into the ordinary shares or preference shares of the company at a fixed date or dates at a fixed price. In effect, **convertible loan stock** is equivalent to a bond plus a stock *option. **2.** A government security in which the holder has the right to convert a holding into new stock instead of obtaining repayment.

**convertible term assurance** A *term assurance that gives the policyholder the option to widen the policy to become a *whole of life policy or an *endowment assurance policy, without having to provide any further evidence of good health. All that is required is the payment of the extra premium. The risks of AIDS has meant that policies of this kind are no longer available, as insurers are not now prepared to offer any widening of life cover without evidence of good health.

**cooling-off period** The 14 days that begins when a life-assurance policy is received, during which a new policyholder can change his mind about taking out the assurance. During this period the policyholder can elect to cancel the policy, in which case he is entitled to receive a full refund of premiums.

**co-operative 1. (worker co-operative)** A type of business organization common in labour-intensive industries, such as agriculture, and often associated with communist countries. Agricultural co-operatives are encouraged in the developing countries, where individual farmers are too poor to take advantage of expensive machinery and large-scale production. In this case several farms pool resources to jointly purchase and use agricultural machinery. In recent times the principle has been extended to other industries in which factory employees have arranged a worker buy-out in order to secure threatened employment. The overall management of such co-operatives is usually vested in a committee of the employee-owners. **2. (consumer co-operative)** A movement launched in 1844 by 28 Rochdale weavers who combined to establish retail outlets where members enjoyed not only the benefits of good-quality products at fair prices but also a share of the profits (a dividend) based on the amount of each member's purchases.

**copy** Matter that is to be printed. It includes text, tables, and illustrations.

**copyright** The exclusive right to reproduce or authorize others to reproduce artistic, dramatic, literary, or musical works. It is conferred by the Copyright Act (1988), which also extends to sound broadcasting, cinematograph films, and television broadcasts. Copyright lasts for the author's lifetime plus 50 years from the end of the year in which he died (or from the end of the year in which a film or broadcast was made); it can be assigned or transmitted on death. The principal remedies for breach of copyright (*see* piracy) are an action for *damages and account of profits or an *injunction. It is a criminal offence to make or deal in articles that infringe a copyright.

**copywriter** A person who writes the text for advertisements or other promotional material. Copywriters are usually employed by an *advertising agency, although in the case of highly technical advertising matter they are often employed by the company manufacturing or distributing the product.

**core** An old-fashioned type of computer *memory, made from doughnut-shaped cores about 0.01 inch in diameter. These could be magnetized to store information, each holding one binary digit. Core memories were superseded in the 1970s by memories that use *integrated circuits, often called silicon chips.

The term is also used as a synonym for *main store memory of any type.

**corner** A *monopoly established by an organization that succeeds in controlling the total supply of a particular good or service. It will then force the price up until further supplies or substitutes can be found. This objective has often been attempted, but rarely achieved, in international *commodity markets. Because it is undesirable and has antisocial effects, a corner can now rarely be attempted as government restrictions on monopolies and antitrust laws prevent it.

**corporate image** The image that a company projects of itself. To gain a benevolent image for the way a company treats its employees or the environment, for example, can be as important to its sales as its individual brand images. In recent years there has been an increase in advertising to create an acceptable corporate image. For example, Shell has spent considerable sums advertising its concern for the unspoilt countryside.

**corporate plan** *See* business plan.

**corporate raider** A person or company that buys a substantial proportion of the equity of another company (the target company) with the object of either taking it over or of forcing the management of the target company to take certain steps to improve the image of the company sufficiently for the share price to rise enough for the raider to sell his holding at a profit.

**corporate venturing** The provision of venture capital by one company, either directly or by means of a venture-capital fund, for another company; the objectives are usually either to obtain information about the activities of the company requiring the venture capital or its markets or as a preliminary step towards acquiring that company. It may often be a means of moving into a fresh market cheaply and without needing to acquire the necessary expertise and personnel required to do so on its own.

**corporation** A succession of persons or body of persons authorized by law to act as one person and having rights and liabilities distinct from the individuals forming the corporation. The artificial personality may be created by royal charter, statute, or common law. The most important type is the registered *company formed under the Companies Act. **Corporations sole** are those having only one individual forming them; for example, a bishop, the sovereign, the Treasury Solicitor. **Corporations aggregate** are composed of more than one individual, e.g. a limited company. They may be formed for special purposes by statute; the BBC is an example. Corporations can hold property, carry on business, bring legal actions, etc., in their own name. Their actions may, however, be limited by the doctrine of *ultra vires. *See also* public corporation.

**corporation tax** A tax levied on the trading profits and other income of companies and other incorporated bodies. The rate of corporation tax is set in the UK by the Chancellor in his Budget; it is currently 35%, although companies with profits below a fixed amount (£200,000 in 1990–91) pay a reduced rate of 25%. For profits between £200,000 and £1M p.a. there is an increasing rate. Thus no company with profits below £1M pays the full rate of 35%. Corporation tax is paid in two parts: *advance corporation tax and mainstream corporation tax. Of particular interest is the relationship between corporation taxes on companies and income taxes on their individual shareholders. In 'classical' corporation-tax systems, corporation tax is levied on the company and then, from what remains, dividends are paid to shareholders, who are again taxed in full by means of income tax. In *imputation systems, such as that used in the UK, part of the company's corporation tax is effectively treated as a payment on account of the shareholders' income tax on their dividends. *See also* franked investment income.

**corruption** The introduction of errors into computer data through mechanical accident or malfunction. All forms of electronic or magnetic data storage are vulnerable to corruption, which can occur for no discernable reason. Corruption commonly occurs when data is sent over telephone lines between communicating computers, the bad quality of the line causing the data received to differ from that sent.

**cost 1.** An expenditure, usually of money, for the purchase of goods or services. **2.** An expenditure, usually of money, incurred in achieving a goal, e.g. producing certain goods, building a factory, or closing down a branch. *See also* economic cost; opportunity cost.

**cost accountant** An *accountant whose principal function is to gather and manipulate data on the costs and efficiency of industrial processes and thus to advise management on the profitability of ventures. It is the cost accountant who operates *budgetary control of departments, estimates *unit costs, and provides the information required for preparing *tenders.

**cost and freight** *See* c & f.

**cost-benefit analysis** A method of deciding whether or not a particular project should be undertaken, by comparing the relevant economic costs and the potential benefits. It can be used for private investment projects, calculating outlays and returns, and estimating the *net present value of the project: if this is positive the project would be profitable. Cost-benefit analysis is also frequently used by governments in an attempt to evaluate all the social costs and benefits of a project (e.g. road building), which is much more problematic, involving such considerations as *externalities, *public goods, macroeconomic consequences, etc. *See also* shadow price.

**cost centre** A unit of a business or other organization that generates identifiable expenditure for cost-accounting purposes (*see* cost accountant); for example, a department in a school or a single machine in a workshop could be regarded as cost centres. Each cost centre is frequently given its own budget, which is subsequently compared with the actual costs incurred. In some organizations central overheads, which the cost centre cannot control, are apportioned appropriately to cost centres. This may be necessary when one cost centre services other cost centres and it is required to assess the total cost of products or services for pricing purposes.

**cost effectiveness** **1.** Achieving a goal with the minimum of expenditure. **2.** Achieving a goal with an expenditure that makes the achievement viable in commercial terms. The M25, for example, could have been built as an eight-lane, rather than a six-lane, motorway but this was erroneously thought not to be cost effective.

**cost function** A function that defines the costs of different combinations of inputs for each level of output. In *producer theory, by making certain assumptions, a unique cost-minimizing combination of inputs for each level of output can be derived from the cost function.

**cost, insurance, and freight** *See* c.i.f.

**cost minimization** The behavioural assumption that an individual or firm will seek to purchase a given amount of goods or inputs at the least cost, *ceteris paribus. In *producer theory, making certain assumptions, there will exist a single cost-minimizing combination of inputs for any level of output. Thus, assuming that firms or *entrepreneurs choose to minimize costs, their behaviour can be predicted. It can be shown that the profit-maximizing level of output for a firm (*see* profit maximization) is also cost minimizing; however, it need not be assumed that firms actually maximize profits, which is a different behavioural assumption. *See also* managerial theory of the firm; satisficing behaviour.

**cost-plus** **1.** Contract terms in which a supplier provides goods and services at cost plus an agreed fee or percentage. These terms are sometimes used if it is difficult to estimate costs in advance. They are, however, unpopular because the supplier is motivated to increase the costs rather than to reduce them and the buyer has to enter into an open-ended commitment. There are, however, circumstances in which such a contract is inevitable, for example if the costs of materials cannot be ascertained in advance or if it is not possible to estimate the amount of work to be done. This form of contract is sometimes used in the building trade. **2.** A method of pricing work in which the cost of doing the work is marked up by a fixed percentage to cover either overheads or profit or both.

**cost-push inflation** An increase in the prices of goods caused by increases in the cost of inputs (especially wages and raw materials). As an explanation of *inflation, cost-push theories became popular in the 1970s when they appeared to explain the rapid inflation of that period, which followed on from very rapid rises in wages and the increases in oil prices. However, the theory is also widely criticized as: (a) it describes only changes in relative prices (e.g. oil) rather than rises in the general price level (which is how inflation is defined); and (b) most economists would now agree that price rises can only continue if there is an accompanying increase in the *money supply.

**cost unit** An item of production or a specific service provided to which costs can be attributed. In a company making refrigerators a cost unit might be one refrigerator although in a company making refrigerator parts, each part might be a cost unit.

**Council for Mutual Economic Assistance (CMEA** *or* **COMECON)** An eastern-bloc organization formed in 1949 to promote the development of the national economies and the science and technology of its member states. The members are Bulgaria, Cuba, Czechoslovakia, German Democratic Republic, Hungary, Mongolia, Poland, Romania, USSR, and Vietnam. Yugoslavia participates in the work of some CMEA bodies.

**counterbid 1.** A bid made in an auction that exceeds a bid already made (which then becomes the **underbid**). **2.** A second bid made in reply to a *counteroffer.

**counteroffer** A reply made to a *bid. If a seller makes an *offer of goods on specified terms at a specified price, the buyer may accept it or make a bid against the offer. If the seller finds the bid unacceptable he may make a counteroffer, usually on terms or at a price that are a compromise between those in the offer and bid. If the buyer still finds the counteroffer unacceptable he may make a *counterbid.

**countertrading** The practice in international trading of paying for goods in a form other than by hard currency. For example, a South American country wishing to buy aircraft may countertrade (usually through a third party) by paying in coffee beans.

**countervailing credit** *See* back-to-back credit.

**countervailing duty** An extra import duty imposed by a country on certain imports. It is usually used to prevent *dumping or to counteract export subsidies given by foreign countries.

**coupon 1.** One of several dated slips attached to a bond, which must be presented to the agents of the issuer or the company to obtain an interest payment or dividend. They are usually used with *bearer bonds; the **coupon yield** is the *yield provided by a bearer bond. **2.** The rate of interest paid by a fixed-interest bearer bond. A 5% coupon implies that the bond pays 5% interest.

**coupon yield** *See* coupon.

**covenant** A promise made in a deed under seal. Such a promise can be enforced by the parties to it as a contract, even if the promise is gratuitous: for example, if A covenants to pay B £100 per month, B can enforce this promise even though he has done nothing in return. Covenants may also be used to minimize income tax, by transferring income from higher rate taxpayers to non-taxpayers (such as children or charities). However, since the Finance Act (1988), only covenants made to charities offer much scope for tax planning.

Covenants may be entered into concerning the use of land, frequently to restrict the activities of a new owner or tenant (e.g. a covenant not to sell alcohol or run a fish-and-chip shop). Such covenants may be enforceable by persons deriving title from the original parties. This is an exception to the general rule that a contract cannot bind persons who are not parties to it. If the land is leasehold, a covenant "touching and concerning land" may be enforced by persons other than the original parties if there is "privity of estate" between them, i.e. if they are in the position of *landlord and tenant. If the land is freehold, the benefit of any covenant (i.e. the rights under it) may be assigned together with the land. The burden of the covenant (i.e. the duties under it) will pass with the land only if it is a restrictive covenant. This means that it must be negative in nature, such as a covenant not to build on land.

**cover 1.** The security provided by *insurance or *assurance against a specified risk. **2.** *See* dividend cover. **3.** Collateral given against a loan or credit, as in option dealing. **4.** A hedge purchased to safeguard an open position in commodity futures or currency dealing.

**covernote** Temporary proof of cover issued prior to the main insurance-policy documents. Motor-insurance covernotes are temporary versions of the *certificate of insurance; as motor insurers are rarely able to issue the full policy and certificate immediately, a covernote is issued as an abbreviated version for a short period (usually 30 or 60 days).

**CPA** Abbreviation for certified public accountant. *See* certified accountant.

**cpi** Abbreviation for characters per inch.

**CP/M** *Trademark* Acronym for Control Program for Microcomputers. CP/M is an *operating system developed by Digital Research Inc. in 1976 and is now widely used on *microcomputers. There are hundreds of programs available to computers using CP/M.

**CPP accounting** Abbreviation for *current purchasing power accounting.

**CPU** Abbreviation for *central processing unit.

**craft union** *See* trade union.

**crash 1.** A rapid and serious fall in the level of prices in a market. **2.** A breakdown of a computer system. A program is said to crash if it terminates abnormally. A computer is said to crash either if it suffers a mechanical failure, or if one of the programs running on it misbehaves in a way that causes the computer to stop. Generally, the computer must then be switched off, any necessary repairs made, and restarted. This usually results in the loss of the information held in the memory when the system crashed.

**crawling peg (sliding peg)** A method of exchange-rate control that accepts the need for stability given by fixed (or pegged) exchange rates, while recognizing that fixed rates can be prone to serious misalignments, which in turn can cause periods of financial upheaval. Under crawling peg arrangements, countries alter their pegs by small amounts at frequent intervals, rather than making large infrequent changes. This procedure provides flexibility and, in conjunction with the manipulation of interest rates, reduces the possibility of destabilizing speculative flows of capital. However, it is exposed to the criticism made against all fixed-rate regimes, that they are an inefficient alternative to the free play of market forces. At the same time, the crawling peg loses a major advantage of fixed rates, which is to inject certainty into the international trading system.

**CRC** Abbreviation for *camera-ready copy.

**credit 1.** The reputation and financial standing of a person or organization. **2.** The sum of money that a trader allows his customer before requiring payment. **3.** The ability of members of the public to purchase goods with money borrowed from finance companies, banks, and other money lenders. **4.** An entry on the right-hand side of an *account in double-entry bookkeeping, showing a positive asset.

**credit account** *See* charge account.

**credit brokerage** *See* ancillary credit business.

**credit call** A national or international telephone call that is charged to the caller's credit card.

**credit card** A plastic card issued by a bank or finance organization to enable holders to obtain credit in shops, hotels, restaurants, petrol stations, etc. The retailer or trader receives monthly payments from the credit-card company equal to its total sales in the month by means of that credit card, less a service charge. The customer also receives monthly statements from the credit-card company, which he may pay in full within a certain number of days with no interest charged, or he may make a specified minimum payment and pay a high rate of interest (usually between 24% and 30% p.a.) on the outstanding balance. In the UK the main cards are **Barclaycard**, **Access**, **American Express**, and **Diners Club**. *See also* debit card; gold card.

**credit control** Any system used by an organization to ensure that its outstanding debts are paid within a reasonable period. It involves establishing a **credit policy**, *credit rating of clients, and chasing accounts that become overdue. *See also* factoring.

**credit guarantee** *See* credit insurance.

**credit insurance 1.** An insurance policy that continues the repayments of a particular debt in the event of the policyholder being financially unable to do so because of illness, death, redundancy, or any other specified cause. **2.** A form of insurance or **credit guarantee** against losses arising from bad debts. This is not usually undertaken by normal insurance policies but by specialists known as factors (*see* factoring). *See also* Export Credits Guarantee Department.

**credit note** A document expressing the indebtedness of the organization issuing it, usually to a customer. When goods are supplied to a customer an invoice is issued; if the customer returns all or part of the goods the invoice is wholly or partially cancelled by a credit note.

**creditor** One to whom an organization or person owes money. The *balance sheet of a company shows the total owed to creditors and a distinction has to be made between creditors who will be paid during the coming accounting period and those who will not be paid until later than this.

**creditors' committee** A committee of creditors of an insolvent company or a bankrupt individual, which represents all the creditors. They supervise the conduct of the administration of a company or the bankruptcy of an individual or receive reports from an administrative *receiver.

**creditors' voluntary liquidation (creditors' voluntary winding-up)** The winding-up of a company by special resolution of the members when it is insolvent. A **meeting of creditors** must be held within 14 days of such a resolution and the creditors must be given seven days' notice of the meeting. Notices must also be posed in the *Gazette* and two local newspapers. The creditors also have certain rights to information before the meeting. A *liquidator may be appointed by the members before the meeting of creditors or at the meeting by the creditors. If two different liquidators are appointed, an application may be made to the court to resolve the matter.

**credit rating** An assessment of the creditworthiness of an individual or a firm, i.e. the extent to which they can safely be granted credit. Traditionally, banks have provided confidential trade references, but recently **credit-reference agencies** have grown up, which gather information from a wide range of sources, including the county courts, bankruptcy proceedings, hire-purchase companies, and professional debt collectors. This information is then provided, for a fee, to interested parties. The consumer was given some protection from such activities in the Consumer Credit Act (1974), which allows an individual to obtain a copy of all information held on him by such agencies, as well as the right to correct any discrepancies.

**credit-reference agency** *See* ancillary credit business; banker's reference; commercial agency; credit rating.

**credit sale agreement** *See* hire purchase.

**credit squeeze** A government measure, or set of measures, to reduce economic activity by restricting the money supply. Measures used include increasing the interest rate (to restrain borrowing), controlling moneylending by banks and others, and increasing down payments or making other changes to hire-purchase regulations.

**credit transfer (bank giro)** A method of settling a debt by transferring money through a bank or post office. The debtor completes written instructions naming the creditor, his address, and account number. Several creditors may be listed and settled by a single transaction. The popularity of this system led the banks to introduce a credit-clearing system in 1961 and the post office in 1968. It is especially useful for persons who do not have bank accounts.

**creeping takeover** The accumulation of a company's shares, by purchasing them openly over a period on a stock exchange, as a preliminary to a takeover (*see* takeover bid). Under the Securities and Exchange Commission (SEC) regulations, in the USA once a 5% holding in another company has been acquired this must be disclosed to the SEC within 10 days (according to section 13(d) of the US Securities and Exchange Act). The **section-13(d) window**, however, enables more than 5% of the stock to be accumulated before a declaration is made.

**critical-path analysis** A technique for planning a complicated operation so that it can be completed as quickly and cheaply as possible. The total operation (e.g. building a power station) is broken down into separate steps, each represented by arrows on a chart; the length of the arrow is related to the time anticipated that each step will take. Many arrows will be parallel, but there will also be points of intersection. The chain of arrows representing the longest time is the critical path, as this determines the total time that must elapse before the job is completed. The chart is arranged to show how each step dovetails with others, when supplies of materials are required, when plant and equipment is needed, as well as details of the managers and workforce involved. The technique can be applied to many commercial operations.

**crore** In India, Pakistan, and Bangladesh, 100 *lakhs, i.e. 10 million. It usually refers to a number of rupees.

**crossed cheque** *See* cheque.

**crowding-out** The concept in economics suggesting that a reduction in private expenditure results from an increase in government expenditure. For example, if the government stimulates *aggregate demand by maintaining a *budget deficit, the increase in government borrowing will raise interest rates, which will inhibit private *investment. Keynesians tend to believe that the effect of crowding-out is limited, while new classical macroeconomists believe it is complete, i.e. that every additional £1 of government expenditure causes a fall of £1 in private expenditure.

**crystallization** *See* charge; floating charge.

**cum-** *See* ex-.

**cum-dividend** *See* ex-.

**cum-new** Denoting a share that is offered for sale with the right to take up any *scrip issue or *rights issue. *Compare* ex-new.

**cumulative preference share** A type of *preference share that entitles the owner to receive any dividends not paid in previous years. Companies are not obliged to pay dividends on preference shares if there are insufficient earnings in any particular year. Cumulative preference shares guarantee the eventual payment of these dividends in arears before the payment of dividends on ordinary shares, provided that the company returns to profit in subsequent years.

**currency** **1.** Any kind of money that is in circulation in an economy. **2.** Anything that functions as a *medium of exchange, including coins, banknotes, cheques, *bills of exchange, promissory notes. etc. **3.** The money in use in a particular country. *See* foreign exchange. **4.** The time that has to elapse before a bill of exchange matures.

**current account** **1.** An active account at a bank or building society into which deposits can be paid and from which withdrawals can be made by cheque (*see also* cheque account). The bank or building society issues cheque books free of charge and supplies regular statements listing all transactions and the current balance. Banks sometimes make charges for current accounts, based on the number of transactions undertaken, but modern practice is to waive these charges if a certain credit balance has been maintained for a given period. Building societies usually make no charges and pay interest on balances maintained in a current account. Banks, in order to remain competitive, are following this practice, although they have traditionally not paid interest on current-account balances. **2.** The part of the *balance of payments account that records non-capital transactions. **3.** An account in which intercompany or interdepartmental balances are recorded. **4.** An account recording the transactions of a partner in a partnership that do not relate directly to his capital in the partnership (*see* capital account).

**current assets** Assets that form part of the *circulating capital of a business and are turned over frequently in the course of trade. The most common current assets are stock in trade, debtors, and cash. *Compare* capital asset.

**current-cost accounting (CCA)** A method of accounting, recommended by the *Sandilands Committee, to deal with the problem of showing the effects of inflation on business profits. Instead of showing assets at their historical cost (i.e. their original purchase price), less depreciation where appropriate, the assets are shown at their current cost (*see* replacement cost) at the time of producing the accounts. This method of accounting was used considerably in the UK in the late 1970s and early 1980s, when inflation was high; it was not popular, however, and as inflation reduced many companies abandoned it.

**current liabilities** Amounts due to the creditors of an organization that are due to be paid within twelve months.

**current purchasing power accounting (CPP accounting)** A method of accounting designed to deal with the problem of showing the effects of inflation on business profits. In this method the historical cost of an asset (i.e. its original purchase price) is increased by *indexation using, for example, the retail price index. The method was not adopted because the *Sandilands Committee recommended *current-cost accounting as preferable. However, current-cost accounting does not strictly show the effects of inflation because it can confuse the effects of inflation with price changes caused by other factors.

**current ratio** The proportion of the *current assets to the *current liabilities of an organization. This ratio is used in the analysis of *balance sheets to gauge the likelihood that an organization can pay its debts regularly. There is no absolute figure for the ratio that is desirable, although clearly an excess of current liabilities over current assets would be a cause for concern; different ratios might be appropriate to different sorts of business. In analysing successive balance sheets, trends might be more important than the absolute figures.

**current yield** *See* yield.

**curriculum vitae** (Latin: the course of one's life) An account of one's education, qualifications, and career prepared by a candidate for a job and given to a prospective employer. It should be typed and give full details. It is usual to add a section on outside interests. In the USA it is often called a **resumé**.

**cursor** The symbol on a computer screen that shows where the next character will appear. The cursor is typically a winking square of light, an underline, or a hollow square.

**customer service** The services an organization offers to its customers, especially of industrial goods and expensive consumer goods, such as computers or cars. Customer services cover a wide variety of forms, including after-sales servicing, such as a repair and replacement service, extended guarantees, regular mailings of information, and, more recently, freephone telephone calls in case of complaints. The appeal of a company's products are greatly influenced by the customer services it offers.

**custom of the trade** A practice that has been used in a particular trade for a long time and is understood to apply by all engaged in that trade. Such customs or practices may influence the way in which a term in a contract is interpreted and courts will generally take into account established customs of a trade in settling a dispute over the interpretation of a contract.

**Customs and Excise** *See* Board of Customs and Excise.

**customs entry** The record kept by the Customs of goods imported into or exported from a country. In the UK a *bill of entry is used for either imports (**entry in**) or exports (**entry out**). If no duty is involved with a consignment it is given **free entry**.

**customs tariff** A listing of the goods on which a country's government requires customs duty to be paid on being imported into that country, together with the rate of customs duty applicable. For the UK the Customs Tariff is published by the *Stationery Office.

**customs union** A union of two or more states to form a region in which there are no import or export duties between members but goods imported into the region bear the same import duties. The *European Community is an example.

**cut and paste** A technique used in wordprocessing (*see* wordprocessor). The term describes the way a section of text, such as a paragraph, can be moved within a document: first, the section is 'cut' from its original position; then, it is 'pasted' into the new position.

**cwo** Abbreviation for cash with order.

# D

**D/A** Abbreviation for *documents against acceptance.

**DAGMAR** Abbreviation for defining advertising goals for measured advertising results. This is the principle, developed in the USA in the 1960s, that the function of advertising is to communicate and that its success or failure should be measured against the specific objectives defined for it.

**daisywheel printer** A type of printer that carries its font on a plastic disk with characters on the end of stalks radiating from the centre; this 'daisywheel' looks a little like a flower. The wheel is rotated until the required letter is in position, when a hammer strikes it against the paper through an inked ribbon. The wheel moves along the line as it prints. These printers produce a high-quality print as a reasonable cost, but they are not as fast as other types of printers and cannot produce graphics. The daisywheels are removeable, and different ones can be used to provide alternative fonts, print styles (e.g. italic), and character sets. *Compare* dot matrix printer; laser printer; line printer; thermal printer.

**damages** Compensation, in monetary form, for a loss or injury, breach of contract, tort, or infringement of a right. Damages refers to the compensation awarded, as opposed to damage, which refers to the actual injury or loss suffered. The legal principle is that the award of damages is an attempt, as far as money can, to restore the injured party to the position he was in before the event in question took place; i.e. the object is to provide restitution rather than profit. Damages are not assessed in an arbitrary fashion but are subject to various judicial guidelines. In general, damages capable of being quantified in monetary terms are known as **liquidated damages**. In particular, liquidated damages include instances in which a genuine pre-estimate can be given of the loss that will be caused to one party if a contract is broken by the other party. If the anticipated breach of contract occurs this will be the amount, no more and no less, that is recoverable for the breach. However, liquidated damages must be distinguished from a *penalty. Another form of liquidated damages is that expressly made recoverable under a statute. These may also be known as **statutory damages** if they involve a breach of statutory duty or are regulated or limited by statute. **Unliquidated damages** are those fixed by a court rather than those that have been estimated in advance.

**General damages** represents compensation for general damage, which is the kind of damage the law presumes to exist in any given situation. It is recoverable even without being specifically claimed and is awarded for the usual or probable consequences of the wrongful act complained of. For example, in an action for medical negligence, pain and suffering is presumed to exist, therefore if the action is successful, general damages would be awarded as compensation even though not specifically claimed or proved. Loss of earnings by the injured party, however, must be specifically claimed and proved, in which case they are known as **specific damages**.

**Nominal** and **contemptuous damages** are those awarded for trifling amounts. These are awarded either when the court is of the opinion that although the plaintiff's rights have been infringed he has not suffered any real loss, or, although actual loss has resulted, the loss has been caused by the conduct of the plaintiff himself. The prospect of receiving only nominal or contemptuous damages prevents frivolous actions being brought. The award is usually accompanied by an order that each party bears his own legal costs. **Exemplary damages**, on the other hand, are punitive damages awarded not merely as a means of compensation but also to punish the party responsible for the loss or injury. This usually occurs when the party causing the damage has done so wilfully or has received financial gain from his wrongful conduct. Exemplary damages will be greater than the amount that would have been payable purely as compensation. **Prospective damages** are awarded to a plaintiff, not as compensation for any loss he has in fact suffered at the time of a legal action but in respect of a loss it is reasonably anticipated he will suffer at some future time. Such an injury or loss may sometimes be considered to be too remote and therefore not recoverable. *See* remoteness of damage.

**dandy note** A delivery order issued by an exporter and countersigned by HM Customs and Excise, authorizing a bonded warehouse to release goods for export.

**danger money** *See* occupational hazard.

**data** The information that is processed, stored, or produced by a computer.

**database** An organized collection of information held on a computer. A special computer program, called a **database management system (DBMS)**, is used to organize the information held in the database according to a specified schema, to update the information, and to help users find the information they seek. There are two kinds of DBMS: simple DBMS, which are the electronic equivalents of a card index; and programmable DBMS, which provide a *programming language that allows the user to analyse the data held in the database. On large computer systems, other programs can generally communicate with the DBMS and use its facilities. The term **data bank** is used for a collection of databases.

**data file** A *file on a computer system that contains *data, contrasted with one that contains a program (*see* computer programming). A data file is usually subdivided into *records and *fields.

**Datapost** A Royal Mail fast service for packages weighing up to 27.5 kg. The same-day door-to-door service by radio-controlled motorcycles and vans is more expensive than the overnight service, which guarantees next-day delivery to any point in the UK. There is also a Datapost International Service, which operates to many countries.

**data processing (DP)** The class of computing operations that manipulate large quantities of information. In business, these operations include book-keeping, printing invoices and mail shots, payroll calculations, and general record keeping. Data processing forms a major use of computers in business, and many firms have full-time data-processing departments.

**data protection** Safeguards relating to personal data in the UK, i.e. personal information about individuals that is stored on a computer. The principles of data protection, the responsibilities of data users, and the rights of data subjects are governed by the Data Protection Act (1984).

The principles of data protection include the following:

(1) The information to be contained in personal data shall be obtained, and personal data shall be processed, fairly and lawfully.

(2) Personal data shall be held only for specified and lawful purposes and shall not be used or disclosed in any manner incompatible with those purposes.

(3) Personal data held for any purpose shall be relevant to that purpose.

(4) Personal data shall be accurate and, where necessary, kept up to date.

(5) Personal data held for any purpose shall not be kept longer than necessary for that purpose.

(6) Appropriate security measures shall be taken against unauthorized access to, or alteration, disclosure, or destruction of personal data and against accidental loss or destruction of personal data.

Data users must register their activities with the **Data Protection Registrar** by means of a registration form obtained from a post office. This requires the data user to give: a description of the personal data it holds and the purposes for which the data is held; a description of the sources from which it intends or may wish to obtain the data or the information to be contained in the data; a description of any persons to whom it intends or may wish to disclose the data; the names or a description of any countries or territories outside the UK to which it intends or may wish directly or indirectly to transfer from data subjects for access to the data. A data user who fails to register is guilty of the offence of failing to register.

An individual is entitled to be informed by any data user whether he holds personal data of which that individual is the subject. He is also entitled to obtain a printout from a registered data user of any personal data held by him and to demand that any inaccurate or misleading information is corrected or erased. If a court is satisfied on the application of a data subject that personal data held by a data user concerning him is inaccurate it may order the rectification or erasure of the data. Additionally it may order the rectification or erasure of any data held by the data user that contains an expression of opinion that appears to the court to be based on the inaccurate data.

**dated security** A stock that has a fixed *redemption date.

**date stamp** A date stamped on the packaging of prepacked perishable food to indicate either the date by which it must be sold by a retailer (**sell-by date**) or the date by which it must be consumed by the consumer (**consume-by date**). This is a legal requirement for prepacked perishable food sold in the UK.

**dawn raid** An attempt by one company or investor to acquire a significant holding in the equity of another company by instructing brokers to buy all the shares available in that company as soon as the stock exchange opens, usually before the target company knows that it is, in fact, a target. The dawn raid may provide a significant stake from which to launch a *takeover bid. The conduct of dawn raids is now restricted by the *City Code on Takeovers and Mergers.

**day books** *Books of prime entry that provide a record of series of similar documents; for example the *sales day book records details of invoices rendered to customers, while the *purchase day book records invoices issued to the organization by suppliers. Other day books might record credit notes issued by suppliers or credit notes issued to customers. These books are then used as a source for making double-entry postings (*see* post) to the individual accounts of customers or suppliers as appropriate, while periodic totals are posted to the sales or the purchases account as appropriate.

**day order** An order to a stockbroker, commodity broker, etc., to buy or sell a specified security or commodity at a fixed price or within fixed limits; the order is valid for the day on which it is given and automatically becomes void at the close of trading on that day.

**days of grace** The extra time allowed for payment of a *bill of exchange or insurance premium after the actual due date. With bills of exchange the usual custom is to allow 3 days of grace (not including Sundays and *Bank Holidays) and 14 days for insurance policies.

**DBMS** Abbreviation for database management system. *See* database.

**DCF** Abbreviation for *discounted cash flow.

**dead-cat bounce** A temporary recovery on a stock exchange, caused by *short covering after a substantial fall. It does not imply a reversal of the downward trend.

**dead freight** *Freight charges incurred by a shipper for space reserved but not used.

**deadweight cargo** Any cargo, such as minerals and coal, for which the *freight is charged on the basis of weight rather than volume.

**deadweight debt** A debt that is incurred to meet current needs without the security of an enduring asset. It is usually a debt incurred by a government; the *national debt is a deadweight debt incurred by the UK government during the two World Wars.

**deadweight loss** The loss to society arising from an inefficient allocation of resources. Two typical examples are monopoly and taxation. However, in the case of taxation the loss may be justified by reference to some other (say, political) principle. Economists frequently attempt to measure deadweight losses when they are thought to exist although their estimates are usually controversial.

**deadweight tonnage** *See* tonnage.

**dealer 1.** A trader of any kind. **2.** A person who deals for himself as a principal, such as a *market maker on a stock exchange, a commodity merchant, etc., rather than as a broker or agent.

**dealer brand** Any product on which a middleman, usually a retailer, puts his own brand name. For example, St Michael is the dealer brand for Marks and Spencer's products.

**dear money (tight money)** A monetary policy in which loans are difficult to obtain and only available at high rates of interest. *Compare* cheap money.

**death duties** Taxes levied on a person's estate at the time of his death. It has long been thought appropriate to levy a tax on a person when he has no further use for his assets. The principal death duty in the UK was *estate duty, which was introduced in 1894. This became *capital-transfer tax in 1974, which itself became *inheritance tax in 1986. Both these taxes also tax life-time gifts, which estate duty itself had begun to do, otherwise any form of death duty can be avoided by giving away all or part of one's estate before death.

**death-valley curve** A curve on a graph showing how the venture capital invested in a new company falls as the company meets its start-up expenses before its income reaches predicted levels. This erosion of capital makes it difficult for the company to interest further investors in providing additional venture capital. *See also* maximum slippage.

**debenture 1.** The most common form of long-term loan taken by a company. It is usually a loan repayable at a fixed date, although some debentures are *irredeemable securities; these are sometimes called *perpetual debentures. Most debentures also pay a fixed rate of interest, and this interest must be paid before a *dividend is paid to shareholders. Most debentures are also secured on the borrower's assets, although some, known as **naked debentures** or *unsecured debentures, are not. In the USA debentures are usually unsecured, relying only on the reputation of the borrower. In a *secured debenture, the bond may have a *fixed charge (i.e. a charge over a particular asset) or a *floating charge. If debentures are issued to a large number of people (for example in the form of **debenture stock** or **loan stock**) trustees may be appointed to act on behalf of the debenture holders. There may be a premium on redemption and some debentures are *convertible, i.e. they can be converted into ordinary shares on a specified date, usually at a specified price. The advantage of debentures to companies is that they carry lower interest rates than, say, overdrafts and are usually repayable a long time into the future. For an investor, they are usually saleable on a stock exchange and involve less risk than *equities. **2.** A *deed under seal setting out the main terms of such a loan.

**debit** An entry on the left-hand side of an *account in double-entry book-keeping, showing an amount owed by the organization keeping the book. In the case of a bank account, a debit shows an outflow of funds from the account.

**debit card** A plastic card issued by a bank or building society to enable its customers with cheque accounts to pay for goods or services at certain retail outlets by using the telephone network to debit their cheque accounts directly. The retail outlets, such as petrol stations and some large stores, need to have the necessary computerized input device, into which the card is inserted; the customer may be required to tap in his *personal identification number before entering the amount to be debited. Some debit cards also function as *cheque cards and *cash cards.

**debit note** A document sent by an organization to a person showing that the recipient is indebted to the organization for the amount shown in the debit note. Debit notes are rare as invoices are more regularly used; however, a debit note might be used when an invoice would not be appropriate, e.g. for some form of inter-company transfer other than a sale of goods or services.

**debt** A sum owed by one person to another. In commerce, it is usual for debts to be settled within one month of receiving an invoice, after which *interest may be incurred. A long-term debt may be covered by a *bill of exchange, which can be a *negotiable instrument. *See also* debenture.

**debt adjusting** *See* ancillary credit business.

**debt collection agency** An organization that specializes in collecting the outstanding debts of its clients, charging a commission for doing so. *See also* ancillary credit business.

**debt counselling** *See* ancillary credit business.

**debt neutrality** *See* Ricardian equivalence theorem.

**debtor** One who owes money to another. In *balance sheets, debtors are those who owe money to the organization and a distinction has to be made between those who are expected to pay their debts during the next accounting period and those who will not pay until later.

**debt service ratio (DSR)** The proportion of annual export earnings needed to service a country's external debts, including both interest payments and repayment of principal. The DSR is an important statistic, indicating the severity of a country's indebtedness. The effect of rescheduling programmes can be examined by comparing pre- and post-rescheduling DSRs.

**debug** *See* bug.

**decimal currency** A currency system in which the standard unit is subdivided into 100 parts. Following the example of the USA in 1792, most countries have introduced a decimal system. However, it was not until 15 February, 1971 that decimalization was introduced in the UK, following the recommendations of the Halesbury Committee of 1961. The UK now has eight decimal coins, the 1p, 2p, 5p, 10p, 20p, 50p, £1 (introduced in 1983), and £2 (introduced in 1989). The ½p was introduced to ease the transition from a system based on 240 units to one based on 100 units but was abandoned in 1984.

**decimal notation** The everyday system of writing numbers, using the ten digits 0 to 9. The positions of the digits, from the right, in a decimal number refer to increasing multiples (or powers) of 10. *See also* binary notation.

**decision-making unit (DMU)** The informal group of individuals within an organization that decides which items the organization should buy. Commercial buying is undertaken by a group of people, rather than by individuals; it is important for the seller to discover the composition of this group within each of his potential customers, and to recognize that membership changes periodically. The group's composition varies according to the cost and complexity of the item being bought, but might comprise: the company's purchasing manager, the proposed user of the item (the **internal user**), the **influencer** (one or more people, such as the production scheduler, indirectly associated with the use of the item), and the **decider** (one or more people, such as a director, who authorize the purchase).

**decision tree** A diagram used to map the various possible courses of action that flow from a decision and the subsequent decisions that have to be made as a result of it. It consists of a series of levels at each of which the possible courses of action are represented by branches arising from decision points. It is often used in analysing financial situations and possible investments.

**deck cargo** Cargo that is carried on the deck of a ship rather than in a hold. This may increase the insurance premium – depending on the nature of the cargo.

**declaration day** The last day but one of an account on the London Stock Exchange, on which options must be declared, i.e. the owner of the option must state whether or not he wishes to exercise his option to purchase (call) or sell (put) the securities concerned.

**declaration of compliance** *See* company formation.

**declaration of solvency** A declaration made by the directors of a company seeking voluntary liquidation that it will be able to pay its debts within a specified period, not exceeding 12 months from the date of the declaration. It must contain a statement of the company's assets and liabilities, and a copy must be sent to the Registrar of Companies. A director who participates in a declaration of solvency without reasonable grounds will be liable to a fine or imprisonment on conviction. *See* members' voluntary liquidation.

**decreasing returns to scale** *See* returns to scale.

**decreasing term assurance** A form of *term assurance in which the amount to be paid in the event of the death of the *life assured reduces with the passage of time. These policies are usually arranged in conjunction with a cash loan or mortgage and are designed to repay the loan if the *life assured dies. As the amount of the loan decreases with successive repayments the sum assured reduces at the same rate.

**deductible** The amount deducted from a claim in an *excess policy; whereas "excess" is the usual word in motor-vehicle and householders' policies, "deductible" is used in large commercial insurances. It is an amount that is deducted from every claim that is paid. If a claim is made for a figure below the deductible no payment is made. Deductibles are usually applied to policies in return for a premium reduction.

**deductions at source** A method of tax collection in which a person paying income to another deducts the tax on the income and is responsible for paying it to the authorities. Tax authorities have found that, in general, it is easier to collect tax from the payer rather than the recipient of income, especially if paying the tax is made a condition of the payer's obtaining tax relief for the payment. The payee receives a credit against his tax liability for the tax already suffered. Examples of this in the UK tax system are *PAYE, sharehold dividends, interest on government securities, deeds of covenant, trust income, and sub-contractors in the building industry. Normally, tax is deducted at the basic rate of income tax only, although in certain cases, such as PAYE and payments from discretionary trusts, other rates might be used.

**deed** A document that has been signed, sealed, and delivered. The seal and the delivery make it different from an ordinary written agreement. The former use of sealing wax and a signet to effect the seal is now usually replaced by using a small paper disc; delivery may now be informal, i.e. by carrying out some act to show that the deed is intended to be operative. Some transactions, such as conveyances of land, must be carried out by deed to be effective.

**deed of arrangement** *See* arrangement.

**deed of assignment** *See* assignment.

**deed of covenant** A legal document, which must be in a specified form, used to transfer income from one person to another with a view to making a saving in tax. It authorizes regular annual payments to be made, which must normally be at least six (except in the case of payments to charities, when it can be three). The person making the payment deducts income tax at the basic rate from the payment, in most cases obtaining his tax relief on it. Any recipient who is exempt from tax (e.g. a charity) can reclaim the tax deducted. In certain cases, such as payments to charities, tax relief at higher rates may be available to the payer; this does not apply to student children.

**deed poll** A *deed having a straight edge at the top, as opposed to an *indenture. A deed poll was used when only one party was involved in an action, e.g. when a person declared that he wished to be known by a different name. Deeds commonly now have straight edges and are used for all purposes.

**deep-discount bond** A fixed-interest security paying little or no interest (in the latter case it may be called a *zero-coupon bond). Because it provides little or no income it is offered at a substantial discount to its *redemption value, providing a large capital gain in place of income. This may have tax advantages in certain circumstances.

**de facto** (Latin: in fact) Denoting that something exists as a matter of fact rather than by right. For example, a plaintiff may have de facto control of a property. This compares to **de jure** (Latin: in law), which denotes that something exists as a matter of legal right. For example, a planning authority may have acted de jure in refusing a planning application. In international law, one government may recognize another de facto, i.e. acknowledge that it is in control of the country even though it has no legal right to be. De jure recognition acknowledges that it has a legal right to govern. The basis of the distinction, however, is more political than legal.

**default 1.** Failure to do something that is required by law, especially failure to comply with the rules of legal procedure. **2.** Failure to comply with the terms of a contract. A seller is in default if he fails to supply the right quality goods at the time he has contracted to do so. A buyer is in default if he fails to take up documents or pay for goods when he has contracted to do so. Before taking legal action against a defaulter a **default notice** must be served on him.

**deferred annuity** An *annuity in which payments do not start at once but either at a specified later date or when the policyholder reaches a specified age.

**deferred asset** An asset the realization of which is likely to be considerably delayed. An example might be a payment of *advance corporation tax (ACT), which can be used to offset a future payment of *corporation tax. If there is no possibility of a liability to corporation tax in the near future, the ACT is a deferred asset rather than an actual asset.

**deferred liability** A prospective liability that will only become a definite liability if some future event occurs. *See also* contingent liability.

**deferred ordinary share 1.** A type of ordinary share, formerly often issued to founder members of a company, in which dividends are only paid after all other types of ordinary share have been paid. They often entitle their owners to a large share of the profit. **2.** A type of share on which little or no dividend is paid for a fixed number of years, after which it ranks with other ordinary shares for dividend.

**deferred-payment agreement** *See* hire purchase.

**deferred rebate** A rebate offered by a supplier of goods or services to customers on the understanding that further goods and services are purchased from the same supplier. The rebate is usually paid periodically, after the supplier has seen that he has the customer's continued support. Some shipping companies offer a deferred rebate to shippers.

**deferred taxation** A sum set aside for tax in the *accounts of an organization that will become payable in a period other than that under review. It arises because of timing differences between tax rules and accounting conventions. The principle of **deferred-tax accounting** is to re-allocate a tax payment to the same period as that in which the relevant amount of income or expenditure is shown. Historically, the timing difference has arisen in company accounts because the percentages used for the calculation of capital allowances have differed from those used for depreciation.

**deficit financing** The creation of a government *budget deficit for the purpose of influencing economic activity.

**deflation** A general fall in the *price level; the opposite of *inflation. As with inflation, a general change in the price level should, in theory, have no real effect. However, if traders are holding goods whose prices fall, they may suffer such large losses that they are forced into bankruptcy. However, agents holding money are simultaneously better off, although there may be a lag in increasing expenditure, during which a *recession may occur. The only major deflation in this century occurred during the Great Depression in the 1920s and 1930s. Since then, governments have avoided deflation wherever possible. *See also* disinflation.

**defunct company** A company that has been wound up and has therefore ceased to exist.

**degearing** The process in which some of the fixed-interest loan stock of a company is replaced by *ordinary share capital. *See* capital gearing.

**de-industrialization** A substantial fall in the importance of the manufacturing sector in the economy of an industrialized nation as it becomes uncompetitive with its neighbours. This may result from bad industrial relations, poor management, inadequate investment in capital goods, or short-sighted government economic policies. In many cases each of these factors contributes to de-industrialization.

**de jure** *See* de facto.

**del credere agent** A selling *agent who guarantees to pay for any goods he sells on his principal's behalf if his customer fails to do so. He charges an extra commission for covering this risk.

**delegatus non potest delegare** (Latin: a delegate cannot further delegate) The rule that a person to whom a power, trust, or authority is given to act on behalf, or for the benefit of, another, cannot delegate his obligation unless he is expressly authorized to do so. For instance, an auditor who has been appointed to audit the accounts of a company cannot delegate his task to another unless he has been expressly allowed to do so. If express authorization has not been granted he will have acted *ultra vires.

**delivered price** A quoted price that includes the cost of packing, insurance, and delivery to the destination given by the buyer.

**delivery note** A document, usually made out in duplicate, that is given to the consignee of goods when they are delivered to him. The consignee, or his representative, signs one copy of the delivery note as evidence that the goods have been received. *See also* advice note.

**delivery order** A written document from the owner of goods to the holder of the goods (e.g. a warehouse company) instructing them to release the goods to the firm named on the delivery order or to the bearer (if made out to "bearer"). A delivery order backed by a dock or warehouse *warrant may be accepted by a bank as security for a loan.

**delta stocks** *See* alpha stocks.

**demand curve** A curve on a graph relating the quantity of a good demanded to its price. Economists usually expect the demand curve to slope downwards, i.e. an increase in the price of a good brings a lower level of demand. Demand curves are useful in developing theories describing the way in which an economy behaves. *See also* demand theory; price theory; supply curve.

**demand for money** The existence of a stable demand for money has been the core of *monetarism. If this is accepted, it can be shown that *fiscal policy is neutral, i.e. when government expenditure pushes up interest rates private investment is reduced accordingly. Furthermore, changes in the supply of money are a necessary and sufficient condition for changes in the nominal value of the *gross domestic product or for inflation. However, econometric evidence has failed to establish whether or not the demand for money is, in fact, stable.

**demand function** A function relating a good or service demanded to the preferences of an individual (*see* consumer preference). *See also* Hicksian demand function; Marshallian demand function.

**demand management** The use of economic policy *instruments by government to influence the level of *aggregate demand. These instruments include government expenditure, tax cuts (*see* fiscal policy), interest rates, and the money supply (*see* monetary policy). Demand management may consist of expanding aggregate demand when, say, *unemployment is rising or inhibiting aggregate demand when there is *inflation. Demand management is favoured by Keynesians, but became discredited to some extent during the 1970s when governments seemed unable to prevent prices and unemployment rising simultaneously (*see* stagflation).

**demand-pull inflation** A rise in prices caused by an excess of demand over supply in the economy as a whole. When the labour force and all resources are fully employed extra demand will only disappear as a result of rising prices. Popular in the 1960s and 1970s as a 'Keynesian theory' of *inflation, the demand-pull theory appeared to be supported by the *Phillips curve. This turned out not to be the case, however, and alternative, particularly monetarist, theories of inflation have since dominated. *See also* cost-push inflation; quantity theory of money.

**demand theory** A theory that concerns the relationship between the demand for goods and their prices; it forms the core of *microeconomics (*see also* price theory). By plotting the quantities that an individual would purchase at different prices, a *demand curve can be drawn. Summing the demand curves of individuals will yield a market demand curve, while summing the demands for all goods will in turn give an aggregate demand curve for an economy. In this way, a *macroeconomic model can be built up from microeconomic data.

**demarcation dispute** An industrial dispute between trade unions or between members of the same union regarding the allocation of work between different types of tradesmen or workers. The **Demarcation Dispute Tribunal** set up by the TUC has effectively dealt with many of these disputes.

**demarketing** The process of discouraging consumers from either buying or consuming a particular product, such as cigarettes. It may also be used if a product is found to be faulty and the producers do not wish to risk their reputation by continuing to sell it.

**demise charter** *See* chartering.

**demography** The study of human populations, including their size, composition (by age, sex, occupation, etc.), and sociological features (birth rate, death rate, etc.).

**demurrage 1.** Liquidated *damages payable under a charterparty (*see* chartering), at a specified daily rate for any days (**demurrage days**) required for completing the loading or discharging of cargo after the *lay days have expired. **2.** Unliquidated damages to which a shipowner is entitled if, when no lay days are specified, the ship is detained for loading or unloading beyond a reasonable time. **3.** Liquidated damages included in any contract to compensate one party if the other is late in fulfilling his obligations. This occurs frequently in building contracts. Even if the loss caused by the delay is less than the demurrage, the demurrage must be paid in full.

**denationalization** *See* privatization.

**Department of Employment** The UK government department responsible for: the working of the labour market; helping the unemployed find work; encouraging small firms; and encouraging the training of workers in industry. *See also* Advisory Conciliation and Arbitration Service (ACAS); Health and Safety Commission; Training Commission.

**Department of Energy** The UK government department responsible for: policy in all energy matters; the Atomic Energy Authority; the nuclear power industry; the development of North Sea oil and gas resources; coordinating energy efficiency; the development of new energy sources; and government functions in the publicly owned coal and electricity industries.

**Department of Trade and Industry** The UK government department responsible for: international trade policy; the promotion of exports (under the direction of the *British Overseas Trade Board); industrial policy; competition policy and *consumer protection, including relations with the *Office of Fair Trading and the *Monopolies and Mergers Commission; policy on scientific research and development; company legislation and the Companies Registration Office (*see* Companies House); patents and the *Patent Office; the insolvency service; and the regulation of the insurance industry.

**deposit 1.** A sum of money paid by a buyer as part of the sale price of something in order to reserve it. Depending on the terms agreed, the deposit may or may not be returned if the sale is not completed. **2.** A sum of money left with an organization, such as a bank, for safekeeping or to earn interest or with a broker, dealer, etc., as a security to cover any trading losses incurred. **3.** A sum of money paid as the first instalment on a *hire-purchase agreement. It is usually paid when the buyer takes possession of the goods.

**deposit account** An *account with a bank from which money cannot be withdrawn by cheque (*compare* cheque account). The interest paid will depend on the current rate of interest and the notice required by the bank before money can be withdrawn, but it will always be higher than that on a *current account.

**deposit bonds** National Savings Deposit Bonds, introduced by the Department for *National Savings in 1983. They offer a premium rate of interest on lump sums between £100 and £100,000. Interest is taxable, but not deducted at source.

**depreciation 1.** An amount charged to the *profit and loss account of an organization to represent the wearing out or diminution in value of an asset. The amount charged is normally based on a percentage of the value of the asset as shown in the books; however, the way in which the percentage is used reflects different views of depreciation. **Straight-line depreciation** allocates a given percentage of the cost of the asset each year, thus suggesting an even spread of the cost of the asset over its useful life. **Reducing- (diminishing-)balance depreciation** applies a constant percentage reduction first to the cost of the asset and subsequently to the cost as reduced by previous depreciations. In this way reducing amounts are charged periodically to the profit and loss account; by this method the depreciated value of the asset in the balance sheet may approximate more nearly to its true value, in that many assets depreciate more quickly early and more slowly later in their life. Thus depreciation is principally a means of allocating the cost of an asset over its useful life. *See also* accumulated depreciation. **2.** A fall in the value of a currency with a *floating exchange rate relative to another. Depreciation can refer both to day-to-day movements and to long-term realignments in value. For currencies with a *fixed exchange rate a *devaluation or *revaluation of currency is required to change the relative value. *Compare* appreciation.

**depression (slump)** An extended or severe period of *recession. Depressions occur infrequently and some economists believe they occur in long (about 50-year) cycles (*see* Kondratieff waves). The most recent Great Depression occurred in the 1930s; prior to that they occurred in the periods 1873–96, 1844–51, and 1810–17. Depressions are usually associated with falling prices (*see* deflation) and large-scale *involuntary unemployment. They are often preceded by major financial crashes, e.g. the Wall Street crash of 1929. Keynes' *general theory (1936) attempted to explain how depressions can occur and he advocated fiscal *reflation to resolve them by raising employment. (Note that Keynes advocated these policies for depressions, not for the recessions of the ordinary *business cycle.) Monetarists have claimed that excessively restrictive *monetary policy causes depressions and that monetary expansion will alleviate depressions.

**depth interview** An unstructured interview that explores a marketing issue for purposes of *marketing research. The interviewer, a specialist acting on behalf of a client, will have previously compiled a topic guide that identifies the points to be explored; the respondent is part of a *sample chosen to match certain criteria (e.g. if the problem concerned tea all respondents would be tea drinkers). The interview is conducted informally and the interviewer adopts a passive role; he encourages the respondent to talk and ask questions, while ensuring that all the points on the topic guide are covered. After a minimum of ten such interviews, the interviewer reports back to his client with his marketing recommendations. Depth interviews are a qualitative procedure (*see* qualitative marketing research).

**deregulation** The removal of controls imposed by governments on the operation of markets. Many economists and politicians believe that during this century governments have imposed controls over markets that have little or no justification in economic theory; some have even been economically harmful. For example, in the post-war era, as a result of the Bretton Woods agreements, many governments imposed controls on the flow of capital between countries. In the belief that this was harmful to economic growth, many governments have recently eliminated these restrictions. However, most economists still argue that certain markets should be regulated (*see* regulation), particularly if a *monopoly is involved.

**derivative action** A legal action brought by a shareholder on behalf of a company, when the company cannot itself decide to sue. A company will usually sue in its own name but if those against whom it has a cause of action are in control of the company (i.e. directors or majority shareholders) a shareholder may bring a derivative action. The company will appear as defendant so that it will be bound by, and able to benefit from, the decision. The need to bring such an action must be proved to the court before it can proceed.

**desk research** A *marketing research study using mainly external published data and material but also including some internal reports, company records, etc. *See also* off-the-peg research.

**desktop evaluation** The process of deciding whether a computer system or program can perform a particular task by testing it in realistic, or desktop, circumstances. This contrasts with evaluation on a theoretical basis, using technical data supplied by the manufacturer.

**desktop publishing (DTP)** An application of computers that enables small companies and individuals to produce reports, advertising, magazines, etc., to near-typeset quality. A typical system comprises a *microcomputer, using DTP software, and a *laser printer. The capabilities of the software vary with price, although all offer basic page formatting and the ability to use several *founts. More elaborate systems enable graphics to be incorporated into the text and simulate many of the functions of professional typesetting systems. A common feature is the ability to preview each page on the computer's screen before it is printed; many DTP systems therefore require a computer with a superior graphics capability. The laser printer is usually capable of printing text and graphics at a resolution of 300 dots per inch, although some programs are capable also of driving typesetting machines, which use resolutions of over 1000 dots per inch.

**devaluation** A fall in the value of a currency relative to gold or to other currencies. Governments engage in devaluation when they feel that their currency has become overvalued, for example through high rates of inflation making exports uncompetitive or because of a substantially adverse *balance of trade. The intention is that devaluation will make exports cheaper and imports dearer, although the loss of confidence in an economy forced to devalue invariably has an adverse effect. Devaluation is a measure that need only concern governments with a *fixed exchange rate for their currency. With a *floating exchange rate, devaluation or revaluation takes place continuously and automatically (*see* depreciation; revaluation of currency).

**developing countries** Countries that often have abundant natural resources but lack the capital and entrepreneurial and technical skills required to develop them. The average income per head and the standard of living in these countries is therefore far below that of the industrial nations. Often known as the **third world**, these countries are being supported by various United Nations organizations as well as by western and eastern bloc nations, both of whom wish to influence their political development.
The developing countries, in which some 70% of the world's population lives, are characterized by poverty, poor diet, the prevalence of disease, high fertility, overpopulation, illiteracy, poor educational facilities, and an agricultural economy. Many depend on a single product for their exports and are therefore vulnerable in world markets. The third world consists of most of Africa (except the Republic of South Africa), most of Asia (except Japan and the USSR), and much of South America.

**diacritical marks** Accents, umlauts, cedillas, etc., set above or below a letter in foreign languages to modify the sound of the letter.

**diaeresis** Two dots, similar to an umlaut, set above the second of two adjacent vowels in a word to show that both are to be pronounced separately, e.g. Noël.

**diary panel** A group of shops and shoppers who keep a regular record of all purchases or purchases of selected products, for the purpose of *marketing research.

**dies non** (Latin: short for *dies non juridicus*, a non-juridical day) A day on which no legal business can be transacted; a non-business day.

**differentiated marketing** Marketing in which provision is made to meet the special needs of consumers. For example, weight watchers require diet drinks and left-handed people require left-handed scissors.

**diffusion of innovation** The process by which the sale of new products and services spreads among customers. Initially, only those with confidence in the new product or who like taking risks will try it out. Once the innovators have accepted the product, a larger group of early adopters will come into the market. These **opinion leaders** will in turn bring about a wider acceptance by consumers. The diffusion process can be speeded up by making new products more attractive, for example by giving away free samples or by special introductory prices.

**digital computer** *See* computer.

**dilapidations** Disrepair of leasehold premises. The landlord may be liable to repair certain parts of domestic premises (e.g. the structure and exterior, and the sanitary appliances) under the Landlord and Tenant Act (1985) if the lease is for less than seven years. Otherwise, the lease will usually contain a covenant by either the landlord or the tenant obliging them to keep the premises in repair.

Under the Landlord and Tenant Act (1985), a landlord cannot enforce a repairing covenant against a tenant by ending the lease prematurely unless he first serves a notice on him specifying the disrepair and giving time for the repairs to be carried out. If there is no covenant in the lease, the tenant is under a common-law duty not to damage the premises and must keep them from falling down.

**dilution of equity** An increase in the number of ordinary shares in a company without a corresponding increase in its assets or profitability. The result is a fall in the value of the shares as a result of this dilution.

**diminishing-balance depreciation** *See* depreciation.

**diminishing returns** *See* returns to scale.

**dinkie** Denoting an affluent married couple who may be expected to be extensive purchasers of consumer goods. The word is formed from *d*ouble-*i*ncome *n*o-*k*ids.

**direct access** A method of extracting information from a computer memory. Information stored in a type of memory that supports direct access, such as *random-access memory and *magnetic disk, can be retrieved, or accessed, immediately regardless of its location within the memory. *Compare* sequential access.

**direct costs** Costs that would not be incurred but for the production of a particular *cost unit. These are the costs, such as materials and labour, that can be attributed directly to a cost unit, usually without the necessity of apportionment. The labour cost would be measured as the length of time needed to produce the cost unit at a given rate per hour. *Compare* overhead costs.

**direct debit** A form of *standing order given to a bank by an account holder to pay regular amounts from his cheque account to a third party. Unlike a normal standing order, however, the amount to be paid is not specified; the account holder trusts the third party to claim from his bank an appropriate sum.

**direct labour** **1.** Employees of an organization who carry out work for that organization, rather than employees of a contractor used to carry out the work. For example, some councils use direct labour to collect refuse, while others use a contractor. **2.** The labour involved in producing goods or providing services, rather than the ancillary costs of supporting the producers and providers. *Compare* indirect labour.

**direct letter of credit** *See* letter of credit.

**direct-mail selling** A form of *direct marketing in which sales literature or other promotional material is mailed directly to selected potential purchasers. The seller may be the producer of the products or a business that specializes in this form of marketing. The seller may build up his own list of potential customers or he may buy or rent a list (*see* list renting). *See also* mail-order house.

**direct marketing** Selling by means of dealing directly with consumers rather than through retailers. Methods include mail order (*see* mail-order house), *direct-mail selling, *cold calling, *telephone selling, door-to-door calling, etc.

**director** A person appointed to carry out the day-to-day management of a company. A public company must have at least two directors, a private company at least one. The directors of a company, collectively known as the **board of directors**, usually act together, although power may be conferred (by the *articles of association) on one or more directors to exercise executive powers; in particular there is often a *managing director with considerable executive power.

The first directors of a company are usually named in its articles of association or are appointed by the subscribers; they are required to give a signed undertaking to act in that capacity, which must be sent to the *Registrar of Companies. Subsequent directors are appointed by the company at a general meeting, although in practice they may be appointed by the other directors for ratification by the general meeting. Directors may be discharged from office by an ordinary resolution with special notice at a general meeting, whether or not they have a *service contract in force. They may be disqualified for *fraudulent trading or *wrongful trading or for any conduct that makes them unfit to manage the company.

Directors owe duties of honesty and loyalty to the company (fiduciary duties) and a duty of care; their liability in *negligence depends upon their personal qualifications (e.g. a chartered accountant must exercise more skill than an unqualified man). Directors need no formal qualifications. Directors may not put their own interests before those of the company, may not make contracts (other than service contracts) with the company, and must declare any personal interest in work undertaken by the company. Their formal responsibilities include: presenting to members of the company, at least annually, the *accounts of the company and a *directors' report; keeping a register of directors, a register of directors' shareholdings, and a register of shares; calling an *annual general meeting; sending all relevant documents to the Registrar of Companies; and submitting a statement of affairs if the company is wound up (*see* liquidator).

Directors' remuneration consists of a salary and in some cases **directors' fees**, paid to them for being a director, and an expense allowance to cover their expenses incurred in the service of the company. Directors' remuneration must be disclosed in the company's accounts and shown separately from any pension payments or *compensation for loss of office. *See also* executive director.

**Director General of Fair Trading** *See* consumer-credit register; consumer protection.

**directors' report** An annual report by the directors of a company to its shareholders, which forms parts of the company's *accounts required to be filed with the Registrar of Companies under the Companies Act (1985). The information that must be given includes the principal activities of the company, a fair review of the developments and position of the business with likely future developments, details of research and development, significant issues on the sale, purchase, or valuation of assets, recommended dividends, transfers to reserves, names of the directors and their interests in the company during the period, employee statistics, and any political or charitable gifts made during the period. *See also* medium-sized company; small company.

**direct taxation** Taxation, the effect of which is intended to be borne by the person or organization that pays it. Economists distinguish between direct taxation and indirect taxation. The former is best illustrated by *income tax, in which the person who receives the income pays the tax and his income is thereby reduced. The latter is illustrated by *VAT, in which the tax is paid by traders but the effects are borne by the consumers who buy the trader's goods. In practice these distinctions are rarely clear-cut. Corporation tax is a direct tax but there is evidence that its incidence can be shifted to consumers by higher prices or to employees by lower wages. Inheritance tax could also be thought of as a direct tax on the deceased, although its incidence falls on the heirs of the estate.

**direct utility function** A type of *utility function that expresses the preferences of individuals (*see* consumer preference) in terms of the goods and services that they choose to purchase. *Compare* *indirect utility function.

**dirty bill of lading (foul *or* claused bill of lading)** A *bill of lading carrying a clause or endorsement by the master or mate of the ship on which goods are carried to the effect that the goods (or their packing) arrived for loading in a damaged condition.

**dirty float** A technique for managing the exchange rate in which a government publicly renounces direct intervention in the foreign exchange markets while continuing to engage in intervention surreptitiously. This technique was widely used after the collapse of the Bretton Woods fixed exchange rate system in the early 1970s as governments were unable to agree programmes of explicitly managed floating but were not prepared to accept fully floating rates.

**disbursement** A payment made by a professional person, such as a solicitor or banker, on behalf of a client. This is claimed back when the client receives an account for the professional services.

**discharge** To release a person from a binding legal obligation by agreement, by the performance of an obligation, or by law. For example, the payment of a debt discharges the debt; similarly, a judicial decision that a contract is frustrated discharges the parties from performing it.

**disclosure 1.** The obligation, in contract law, that each party has to the other to disclose all the facts relevant to the subject matter of the contract. *See* utmost good faith. **2.** The obligation, in company law, that a company has to disclose all relevant information and results of trading to its shareholders. *See* directors' report.

**discount 1.** A deduction from a *bill of exchange when it is purchased before its maturity date. The party that purchases (discounts) the bill pays less than its face value and therefore makes a profit when it matures. The amount of the discount consists of interest calculated at the *bill rate for the length of time that the bill has to run. *See* discount market. **2.** A reduction in the price of goods below list price, for buyers who pay cash (**cash discount**), for members of the trade (**trade discount**), for buying in bulk (**bulk** or **quantity discount**), etc. **3.** The amount by which the market price of a security is below its *par value. A £100 par value loan stock with a market price of £95 is said to be at a 5% discount.

**discount broker** *See* bill broker.

**discounted cash flow (DCF)** A method of appraising capital-investment projects by comparing their income in the future and their present and future costs with the current equivalents. The current equivalents take account of the fact that future receipts are less valuable than current receipts, in that interest can be earned on current receipts; on the other hand future payments are less onerous than current payments, as interest can be earned on money retained for future payments. Accordingly, future receipts and payments are discounted to their present values by applying discount factors, taking account of interest that could be earned for the relevant number of years to the date of payment or receipt. *See also* net present value.

**discount house 1.** A shop that is open to members of the public, or in some cases to members of a trade, that sells goods, usually consumer durables, at prices that are close to wholesale prices. **2.** A company or bank on the *discount market that specializes in discounting *bills of exchange, especially *Treasury bills.

**discounting back** Reducing a future payment or receipt to its present equivalent by taking account of the interest, which when added to the present equivalent for the relevant number of years would equate to the future payment or receipt. *See also* discounted cash flow.

**discount market** The part of the *money market consisting of banks, *discount houses, and *bill brokers. By borrowing money at short notice from commercial banks or discount houses, bill brokers are able to *discount bills of exchange, especially Treasury bills, and make a profit. The loans are secured on unmatured bills.

**discount rate 1.** *See* bill rate. **2.** The rate of interest charged by the US Federal Reserve Banks when lending to other banks.

**discretionary order 1.** An order given to a stockbroker, commodity broker, etc., to buy or sell a stated quantity of specified securities or commodities, leaving the broker discretion to deal at the best price. **2.** A similar order given to a stockbroker in which the sum of money is specified but the broker has discretion as to which security to buy for his client.

**discretionary trust** A trust in which the shares of each beneficiary are not fixed by the settlor in the trust deed but may be varied at the discretion of some person or persons (often the trustees). In an **exhaustive discretionary trust** all the income arising in any year must be paid out during that year, although no beneficiary has a right to any specific sum. In a **nonexhaustive discretionary trust**, income may be carried forward to subsequent years and no beneficiary need receive anything. Such trusts are useful when the needs of the beneficiaries are likely to change, for example when they are children.

**discriminating monopoly** A monopoly in which the supplier sells his products or services to consumers at different prices; by dividing the market into segments and charging each market segment the price it will bear, the monopolist increases his profit. An example is the different domestic and industrial tariffs operated by suppliers of electricity in many countries.

**discriminating tariff** A tariff that is not imposed equally by a country or group of countries on all its trading partners. The abolition of discriminating tariffs is one of the purposes of the *General Agreement on Tariffs and Trade and other tradings blocs.

**discrimination** The illegal practice of treating some people less favourably than others because they are of a different sex (**sexual discrimination**), race (**racial discrimination**), or religion (**religious discrimination**). *See* equal pay.

**diseconomies of scale** *See* economies of scale.

**dishonour 1.** To fail to accept (*see* acceptance) a *bill of exchange (**dishonour by non-acceptance**) or to fail to pay a bill of exchange (**dishonour by non-payment**). A dishonoured foreign bill must be protested (*see* protest). **2.** To fail to pay a cheque when the account of the drawer does not have sufficient funds to cover it. When a bank dishonours a cheque it marks it "refer to drawer" and returns it to the payee through his bank.

**disinflation** A gentle form of *deflation, to restrain *inflation without creating unemployment. Disinflationary measures include restricting consumer spending by raising the *interest rate, imposing restrictions on *hire-purchase agreements, and introducing price controls on commodities in short supply.

**disintermediation** The elimination of financial intermediaries, such as brokers and bankers, from transactions between borrowers and lenders or buyers and sellers in financial markets. An example of disintermediation is the *securitization of debt. Disintermediation has been a consequence of improved technology and *deregulation (*see also* globalization). Disintermediation allows both parties to a financial transaction to reduce costs by eliminating payments of commissions and fees. Disintermediation often occurs when governments attempt to impose direct controls on the banking system, such as reserve asset ratios and lending ceilings. In response, the market develops new instruments and institutions that are not covered by the direct controls. When these controls are relaxed, funds may return to the normal banking system, i.e. there may be **reintermediation**.

**disinvestment** A reduction in the capital stock of an economy, usually following an economic depression during which there has not been sufficient investment to match the loss in the value of *capital goods caused by normal wear and tear.

**disk** *See* magnetic disk.

**disk capacity** The amount of data that can be stored on a computer *magnetic disk. Usually disk capacity is described in terms of *bytes, where one byte will hold one character of the computer's *character set.

**disk drive** A computer *peripheral device that transfers information to or from a *magnetic disk. The disk is turned at high speed under a read/write head that is similar to the head on a tape recorder. Information is stored on the disk as a pattern of magnetic spots, which are read or written as the appropriate part of the disk passes under the head. *See also* floppy disk; hard disk.

**disk operating system (DOS)** An *operating system designed for use on a computer that has one or more *disk drives.

**displacement tonnage** *See* tonnage.

**display advertising** Advertising, such as a full-page or quarter-page advertisement, often containing a logo or illustration, rather than simple *classified advertising.

**disposable income 1.** The income a person has available to spend after payment of taxes, National Insurance contributions, and other deductions, such as pension contributions. **2.** In *national income accounts, the total value of income of individuals and households available for consumer expenditure and savings, after deducting income tax, National Insurance contributions, and remittances overseas.

**disposables** *See* consumer goods.

**distrain** To seize goods as a security for the performance of an obligation, especially the seizure of goods by a landlord because a tenant is in arrears with his rent.

**distributable profits** The profits of a company that are legally available for distribution as *dividends. They consist of a company's accumulated realized profits after deducting all realized losses, except for any part of these net realized profits that have been previously distributed or capitalized. *Public companies, however, may not distribute profits to such an extent that their net assets are reduced to less than the sum of their called-up capital (*see* share capital) and their undistributable reserves (*see* capital reserves).

**distributable reserves** The retained profits of a company that it may legally distribute by way of *dividends. *See* distributable profits.

**distributed logic** A computer system that supplements the main computer with remote terminals capable of doing some of the computing, or with electronic devices capable of making simple decisions, distributed throughout the system. *See also* distributed processing.

**distributed processing** A system of processing data in which several computers are used at various locations within an organization instead of using one central computer. The computers may be linked to each other in a *network, allowing them to cooperate, or they may be linked to a larger central computer, although a significant amount of the processing is done without reference to the central computer.

**distribution 1.** A payment by a company from its *distributable profits, usually by means of a *dividend. **2.** A dividend or quasi-dividend on which *advance corporation tax is payable. **3.** In economics, the allocation of resources among agents in an economy. Distribution has been a fundamental question in economics since earliest times. The *Classical school investigated the distribution of *wealth between classes, workers, capitalists, and landlords. Economists of the *neoclassical school have tended to accept initial endowments of wealth and to analyse the distribution of *income resulting from the *production process, operating through the laws of supply and demand (*see* market forces). Mainstream economists do not prejudge the issue of whether or not income and wealth should be redistributed, providing instead theories of optimal *taxation, to establish how any redistribution could be achieved efficiently. **4.** In statistics, a representation of the possible values that can be taken by a random variable. A distribution can be thought of as a curve in which each point represents the probability of the random variable taking that particular value. A **distribution function** must define all the possible values that a random variable may take, so that the sum of all probabilities of that function will be one. Although there is a potentially infinite number of distributions, certain families of distributions are the most interesting as they are easy to use and seem to occur most frequently in nature. Of these the best known is the **normal distribution**, in which the largest probabilities are clustered around a central point called the *mean, while the probabilities become smaller as the distance from the mean increases; this gives the curve a bell-like shape. **5.** The allocation of goods to consumers by means of wholesalers and retailers. **6.** The division of property and assets according to law, e.g. of a bankrupt person or a deceased person.

**distribution centre** A warehouse, usually owned by a manufacturer, that receives goods in bulk and despatches them to retailers.

**distribution channel** The network of firms necessary to distribute goods or services from the manufacturers to the consumers; it therefore primarily consists of wholesalers and retailers.

**distributor** An intermediary, or one of a chain of intermediaries (*see* distribution channel), that specializes in transferring a manufacturer's goods or services to the consumers.

**diversification 1.** Movement by a manufacturer or trader into a wider field of products. This may be achieved by buying firms already serving the target markets or by expanding existing facilities. It is often undertaken to reduce reliance on one market, which may be diminishing (e.g. tobacco), to balance a seasonal market (e.g. ice cream), or to provide scope for general growth. **2.** The spreading of an investment portfolio over a wide range of companies to avoid serious losses if a recession is localized to one sector of the market.

**dividend 1.** The distribution of part of the earnings of a company to its shareholders. The dividend is normally expressed as an amount per share on the *par value of the share. Thus a 15% dividend on a £1 share will pay 15p. However, investors are usually more interested in the **dividend yield**, i.e. the dividend expressed as a percentage of the share value; thus if the market value of these £1 shares is now £5, the dividend yield would be 1/5 × 15% = 3%. The size of the dividend payment is determined by the board of directors of a company, who must decide how much to pay out to shareholders and how much to retain in the business; these amounts may vary from year to year. In the UK it is usual for companies to pay a dividend every six months, the largest portion (the **final dividend**) being announced at the company's AGM together with the annual financial results. A smaller **interim dividend** usually accompanies the interim statement of the company's affairs, six months before the AGM. In the USA dividends are usually paid quarterly. *See also* dividend cover; yield. Interest payments on *gilt-edged securities are also sometimes called dividends although they are fixed. **2.** A payment made by a *co-operative society out of profits to its members. It is usually related to the amount the member spends and is expressed as a number of pence in the pound.

**dividend cover** The number of times a company's *dividends to ordinary shareholders could be paid out of its *net profits after tax in the same period. For example, a net dividend of £400,000 paid by a company showing a net profit of £1M is said to be covered 2½ times. Dividend cover is a measure of the probability that dividend payments will be sustained (low cover might make it difficult to pay the same level of dividends in a bad year's trading) and of a company's commitment to investment and growth (high cover implies that the company retains its earnings for investment in the business). Negative dividend cover is unusual, and is taken as a sign that a company is in difficulties. *See also* price-dividend ratio.

**dividend equalization reserve** A reserve formerly created to smooth out fluctuations in the incidence of taxation so that dividends could be maintained. Such reserves are now normally referred to as **deferred-tax accounts** (*see* deferred taxation).

**dividend limitation (dividend restraint)** An economic policy in which the dividends a company can pay to its shareholders are limited by government order. It is usually part of a *prices and income policy to defeat *inflation, providing a political counterpart to a *wage freeze.

**dividend mandate** A document in which a shareholder of a company notifies the company to whom dividends are to be paid.

**dividend stripping (bond washing)** The practice of buying *gild-edged securities after they have gone *ex-dividend and selling them cum-dividend just before the next dividend is due. This procedure enables the investor to avoid receiving dividends, which in the UK are taxable as income, and to make a tax-free *capital gain. This activity has mainly been indulged in by high-rate taxpayers but has now become of little interest since the rules regulating the taxation of accrued interest were changed.

**dividend waiver** A decision by a major shareholder in a company not to take a dividend, usually because the company cannot afford to pay it.

**dividend warrant** The cheque issued by a company to its shareholders when paying *dividends. It states the tax deducted and the net amount paid. This document must be sent by non-taxpayers to the Inland Revenue when claiming back the tax.

**dividend yield** *See* dividend.

**division of labour** The specialization of workers in the processes of production (or any other economic activity). Division of labour was identified by Adam Smith (1723–90) in *The Wealth of Nations* as one of the greatest contributions to the advancement of national wealth then (early in the Industrial Revolution) being experienced in the UK. The idea that specialization permits higher production and therefore improved economic welfare is the basis for one of the fundamental principles of economics, the theory of *comparative advantage, and for the almost universal support amongst economists for free trade.

**DMU** Abbreviation for *decision-making unit.

**dock receipt (wharfinger's receipt)** A receipt for goods given by a dock warehouse or wharf, acknowledging that the goods are awaiting shipment. A more formal document is a **dock warrant** or wharfinger's warrant (*see* warrant), which gives the holder title to the goods.

**documentary bill** A *bill of exchange attached to the shipping documents of a parcel of goods. These documents include the *bill of lading, insurance policy, dock warrant, invoice, etc.

**documentary credit** A credit arrangement in which a bank agrees to accept *bills of exchange drawn by an exporter of goods on the foreign buyer for a stated sum, provided that the bill of exchange has specified shipping documents attached to it.

**document merge** The process of combining two or more documents to produce a single document. This is a common operation in wordprocessing (*see* wordprocessor). An example is the combining of data produced by a *spreadsheet program into a document produced by a wordprocessing program.

**documents against acceptance (D/A)** A method of payment for goods that have been exported in which the exporter sends the shipping documents with a *bill of exchange to a bank or agent at the port of destination. The bank or agent releases the goods when the bill has been accepted by the consignee. *Compare* cash against documents.

**documents against presentation (D/P)** *See* cash against documents.

**dogs** Goods with a low share of a market, especially those in new or slow markets, which are therefore unlikely to yield attractive profits.

**dollar stocks** US or Canadian securities.

**domicile (domicil)** The country or place of a person's permanent home, which may differ from that of his nationality or place of residence. Domicile is determined by both the physical fact of residence and the continued intention of remaining there. For example, a citizen of a foreign country who is resident in the UK is not necessarily domiciled there unless he intends to make it his permanent home. Under the common law, it is domicile and not residence or nationality that determines a person's civil status, including the capacity to marry. A corporation may also have a domicile, which is determined by its place of registration.

**donor** A person making a gift or transferring property to another person (the **donee**).

**dormant company** A company that has had no significant accounting transactions for the accounting period in question. Such a company need not appoint auditors.

**DOS 1.** Abbreviation for *disk operating system. **2.** *See* MS-DOS.

**dot matrix printer** A type of printer used with a computer that prints characters as a pattern of dots produced by fine needles striking the paper through a normal typewriter ribbon. There are usually seven or nine needles on the print head; as they move across the paper they build up the required matrix to form each character. Using this system, it is possible to build up a wide range of foreign-language and mathematical symbols, as well as the normal letters of the alphabet. Some dot matrix printers can also produce graphs and other illustrations. They are fast and cheap, but the output is not of the highest quality. *Compare* daisywheel printer; laser printer; thermal printer.

**double-entry book-keeping** A method of recording the transactions of a business in a set of *accounts, such that every transaction has a dual aspect and therefore needs to be recorded in at least two accounts. For example, when a person (debtor) pays cash to a business for goods he has purchased, the cash held by the business is increased and the amount due from the debtor is decreased by the same amount; similarly, when a purchase is made on credit, the stock is decreased and the amount owing to creditors is increased by the same amount. This double aspect enables the business to be controlled because all the *books of accounts must balance.

**double option** *See* option.

**double-page spread** Two facing pages in a magazine or newspaper used in advertising as if they were a single page.

**double taxation** Taxation that falls on the same source of income in more than one country. Taxation is normally levied on a person's worldwide income in his country of residence. He may also be taxed in other countries in which he has a permanent trading establishment. Because this would inhibit trade, arrangements are normally made to mitigate or abolish this double taxation. This is often achieved by double-taxation treaties between countries; it may also be imposed by a country unilaterally. The principal methods of **double-taxation relief** are: (1) inclusion of the income in one country after deduction of the tax levied in the other; (2) agreement between countries that only one of them will tax the income; and (3) double-tax credits, enabling one country to allow a credit against its own tax for the tax paid in the other country.

**Dow Jones Industrial Average** An index of security prices issued by Dow Jones & Co. (a US firm providing financial information), used on the New York Stock Exchange. It is a narrowly based index, comparable to the London Financial Times Ordinary Share Index (*see* Financial Times Shares Indexes), having 30 constituent companies. The index was founded in 1884, based then on 11 stocks (mostly in railways), but was reorganized in 1928 when it was given the value of 100. Its lowest point was on 2 July, 1932, when it reached 41. In 1987 it exceeded 2400. There are three other Dow Jones indexes, representing price movements in US home bonds, transportation stocks, and utilities. *Compare* Standard and Poor's 500 Stock Index.

**down time** The period during which a computer is out of action, usually because of a fault or for routine maintenance work.

**DP** Abbreviation for *data processing.

**D/P** Abbreviation for *documents against presentation.

**draft 1.** *See* bank draft. **2.** Any order in writing to pay a specified sum, e.g. a *bill of exchange. **3.** A preliminary version of a document, before it has been finalized.

**drawback** The refund of import duty by the Customs and Excise when imported goods are re-exported. Payment of the import duty and claiming the drawback can be avoided if the goods are stored in a *bonded warehouse immediately after unloading from the incoming ship or aircraft until re-export.

**drawee 1.** The person on whom a *bill of exchange is drawn (i.e. to whom it is addressed). The drawee will accept it (*see* acceptance) and pay it on maturity. **2.** The bank on whom a cheque is drawn, i.e. the bank holding the account of the individual or company that wrote it. **3.** The bank named in a *bank draft. *Compare* drawer.

**drawer 1.** A person who signs a *bill of exchange ordering the *drawee to pay the specified sum at the specified time. **2.** A person who signs a cheque ordering the drawee bank to pay a specified sum of money on demand.

**drip-feed** To fund a new company in stages rather than by making a large capital sum available at the start.

**drop-dead fee** A fee paid by an individual or company that is bidding for another company to the organization lending the money required to finance the bid. The fee is only paid if the bid fails and the loan is not required. Thus, for the price of the drop-dead fee, the bidder ensures that he only incurs the interest charges if the money is required.

**drop lock** A new form of issue in the bond market that combines the benefits of a bank loan with the benefits of a bond. The borrower arranges a variable-rate bank loan on the understanding that if long-term interest rates fall to a specified level, the bank loan will be automatically refinanced by a placing of fixed-rate long-term bonds with a group of institutions. They are most commonly used on the international market.

**DSR** Abbreviation for *debt service ratio.

**DTP** Abbreviation for *desktop publishing.

**dual-capacity system** A system of trading on a stock exchange in which the functions of *stockjobber and *stockbroker are carried out by separate firms. In a **single-capacity system** the two functions can be combined by firms known as *market makers. Dual capacity existed on the *London Stock Exchange prior to October, 1986 (*see* Big Bang), since when a single-capacity system has been introduced, bringing London into line with most foreign international stock markets. The major advantage of single capacity is that it cuts down on the costs to the investor, although it can also create more opportunity for unfair dealing (*see* Chinese wall).

**due date** The date on which a debt is due to be settled, such as the maturity date of a *bill of exchange.

**dump 1.** To transfer the contents of a computer's main memory onto a backing memory. This provides a *back-up copy of the main memory for security reasons. **2.** A printout of the contents of computer memory, or of a *file, used to diagnose such problems as *crashes.

**dumping** The selling of goods abroad at prices below their marginal cost, which implies that the seller is making a loss. It has often been argued that developing countries wishing to establish an industry should use tariffs and quotas to ensure a monopoly for producers at home, while selling goods cheaply overseas. However, this practice inevitably leads to claims of dumping by other countries. The *General Agreement on Tariffs and Trade allows countries to prevent dumping by imposing tariffs on goods that are being dumped, although it is always difficult to establish conclusively that a particular price level constitutes dumping. Dumping is prohibited in the EEC.

**duopoly** A market in which there are only two producers or sellers of a particular product or service and many buyers. The profits in such an imperfect form of competition are in practice usually less than could be achieved if the two suppliers merged to form a *monopoly but more than if the two allowed competition to force them into *marginal costing. *See also* collusive duopoly.

**durables** *See* consumer goods.

**Dutch auction** An *auction sale in which the auctioneer starts by calling a very high price and reduces it until he receives a bid.

**Dutch disease** The deindustrialization of an economy as a result of the discovery of a natural resource. So named because it occurred in Holland after the discovery of North Sea gas; it has also been applied to the UK since the discovery of North Sea oil. The discovery of such a resource lifts the value of the country's currency, making manufactured goods less competitive; exports therefore decline and imports rise.

**duty** A government tax on certain goods or services. *See* death duties; excise duty; import duty; stamp duty; tariff.

**duty of care** *See* negligence.

# E

**E & OE** Abbreviation for errors and omissions excepted. These letters are often printed on invoice forms to safeguard the sender in case he has made an error in the recipient's favour.

**ear** The advertsing space at the top left or right corner of a newspaper's front page.

**early bargains** *See* after-hours deals.

**earned income** Income generally acquired by the personal exertion of the taxpayer as distinct from such passive income as dividends from investments. It is often thought by tax theorists that earned income should be taxed at a lower rate than unearned income, since the latter accrues without the expenditure of the taxpayer's time. This has been reflected in different ways in the UK, with such measures as earned-income relief, wife's earned-income relief, and investment-income surcharge. Earned income consists primarily of wages and salaries, business profits, royalties, and some pensions. Apart from wife's earned-income relief, there are currently no differences between the rates of taxation for earned and unearned income.

**earnings per share (eps)** The earnings of a company over a stated period (usually one year) divided by the number of ordinary shares issued by the company. The earnings (sometimes called **available earnings**) are calculated as annual profits, after allowing for tax and any *exceptional items. **Fully diluted earnings per share** include any shares that the company is committed to issuing but has not yet issued (e.g. through *convertibles). *See also* price-earnings ratio.

**earnings-related pension** *See* pension; State Earnings-Related Pension Scheme.

**earnings yield** *See* yield.

**easement** A right, such as a right of way, right of water, or right of support, that one owner of one piece of land (the dominant tenement) may have over the land of another (the servient tenement). The right must benefit the dominant tenement and the two pieces of land must be reasonably near each other. The right must not involve expenditure by the owner of the servient tenement and must be analogous to those rights accepted in the past as easements. An easement may be granted by *deed or it may be acquired by prescription (lapse of time, during which it is exercised without challenge); it may also be acquired of necessity (for example, if A sells B a piece of land that B cannot reach without crossing A's land) or when 'continuous and apparent' rights have been enjoyed with the part of the land sold before it was divided. Existing easements over the land of third parties pass with a conveyance of the dominant tenement.

**easy money** *See* cheap money.

**EC** Abbreviation for *European Community.

**ECGD** Abbreviation for *Export Credits Guarantee Department.

**econometrics** The study of economic phenomena based on observed data. Its aim is sometimes described as the statistical verification of economic theories, in much the same way as experiments are used to verify theories in the natural sciences. However, this is only a partial analogy as the object of the economist's research is not always amenable to experiment. In addition, the results of econometric analysis are usually open to more than one interpretation.

The main tool of the econometrician is regression analysis. This establishes the statistical relationship between two or more sets of economic data. These results can then be used for forecasting. To achieve this the economic theorist relates one economic variable (the dependent variable) to one or more other economic variables (the independent variables). The regression establishes the nature of the relationship by yielding coefficients of the relevant parameters. One of the problems of econometrics is to ensure that data adequately reflects the underlying variable in question. For example, a monetarist will claim that the demand for money is determined by, amongst other things, the interest rate. The econometrician can regress some measure of the money stock on some particular interest rate, but must be aware that the data may not exactly represent the variables suggested by the theory.

**economic cost** In economics, the total sacrifice involved in doing something. It will include the *opportunity cost and therefore is greater than the accounting cost, which is restricted to the total outlay of money.

**Economic Development Council** *See* National Economic Development Office.

**economic effects of taxation** The ways in which taxation can affect the taxpayer's behaviour. When a tax is imposed, some people will alter their behaviour to minimize the effects the tax has on them. They may work harder or less hard, they may buy different sorts of goods from those they bought previously, or they may even emigrate. Taxes that most distort consumer choices are said to impose the heaviest 'excess burden'.

**economic good** *See* good.

**economic growth** The expansion of the output of an economy, usually expressed in terms of the increase of national income. Nations experience different rates of economic growth mainly because of differences in population growth, investment, and technical progress.

**economic profit** *See* profit.

**economic rate of substitution** *See* marginal rate of substitution.

**economic rent** *See* rent.

**economics** A social science concerning behaviour in the fields of production, *consumption, *distribution, and *exchange. Economists analyse the processes involved and investigate the consequences for the individual, such organizations as the firm, and society as a whole.

There are many competing schools of thought in economics: the main division has been between the *Classical and *neoclassical schools. Adam Smith (1723–90), the founder of the Classical school, emphasized primarily the concept of economic value and the distribution of wealth between the classes – workers, capitalists, and landlords. The Marxist school of thought is one of its offshoots. The neoclassical school, now the mainstream of western economic thought, emphasizes the role of allocating scarce resources between competing ends. The founders of this school, W S Jevons (1835–82) and M E L Walras (1834–1910), were known as *Marginalists. Neoclassical economics is itself divided into two broad areas of research, *microeconomics – analysing the relationship between individual economic units (the consumer, firm, etc.) – *macroeconomics, which analyses the connection between economic aggregates, money, total employment, and government. Both fields place a heavy emphasis on the individual or household as the basic unit of analysis, rather than the classes.

**economic sanctions** Action taken by one country or group of countries to harm the economic interest of another country or group of countries, usually to bring about pressure for social or political change. Sanctions normally take the form of restrictions on imports or exports, or on financial transactions. They may be applied to specific items or they may be comprehensive trade bans. There is considerable disagreement over their effectiveness. Critics point out that they are easily evaded and often inflict more pain on those they are designed to help than on the governments they are meant to influence. They can also harm the country that imposes sanctions, through the loss of export markets or raw material supplies. In addition the target country may impose retaliatory sanctions. *See also* embargo.

***Economic Trends*** A monthly publication of the UK *Central Statistical Office devoted to economic statistics.

**economies of scale (scale effect)** Reductions in the average cost of production, and hence in the unit costs, when output is increased. If the average costs of production rise with output, this is known as **diseconomies of scale**. Economies of scale can enable a producer to offer his product at more competitive prices and thus to capture a larger share of the market. **Internal economies of scale** occur when better use is made of the factors of production and by using the increased output to pay for a higher proportion of the costs of marketing, financing, and development, etc. Internal diseconomies can occur when a plant exceeds its optimum size, requiring a disproportionate unwieldy administrative staff. **External economies** and diseconomies arise from the effects of a firm's expansion on market conditions and on technological advance.

**ECSC** Abbreviation for *European Coal and Steel Community.

**ECU** Abbreviation for *European Currency Unit.

**edition 1.** The version of a book published on a particular date. **2.** The number of copies of that version printed at that time.

**editorial advertisement** An advertisement in a newspaper or magazine written in the form of an editorial feature. Such advertisements must, however, be clearly labelled "advertisement".

**EEC** Abbreviation for *European Economic Community.

**effective demand** Demand for goods and services for which money is available to convert them into actual purchases. In Keynesian theory (*see* Keynesianism), it is often argued that in a *recession or a *depression there will be inadequate effective demand as unemployed workers demand goods but have no means to pay for them. Similarly, firms would demand labour if only there was someone to buy their goods. *Pump priming can add to effective demand and move the economy towards full employment by means of the *multiplier process.

**effective tax rate** The average tax rate that is applicable in a given circumstance. In many cases the actual rate of tax applying to an amount of income or to a gift may not, for various reasons, be the published rate of the tax; these reasons include the necessity to gross up, the complex effects of some reliefs, and peculiarities in scales of rates. The effective rate is therefore found by dividing the additional tax payable as a result of the transaction by the amount of the income, gift, or whatever else is involved in the transaction.

**efficiency 1. (technical efficiency)** A measure of the ability of a manufacturer to produce the maximum output of acceptable quality with the minimum of inputs. One company is said to be more efficient than another if it can produce the same output as the other with less inputs, irrespective of the price factor. **2. (economic efficiency)** A measure of the ability of an organization to produce and distribute its product at the lowest possible cost. A firm can have a high technical efficiency but a low economic efficiency because their prices are too high to meet competition.

**efficiency variance** A variance arising when the actual productivity of an organization is greater or less than that budgeted for.

**efficiency wage theory** The theory that the productivity of a worker increases with the wages he is paid. It was first applied in the late 1950s to developing economies; in this context it became clear that higher wages enable poor workers to improve their diets and thus to become more productive. Recently, however, the argument has been applied in developed economies to explain *involuntary unemployment. Higher wages raise morale and company loyalty, attract better quality workers, and result in less shirking. Thus, even in a recession, when some workers are being made redundant, firms will be unwilling to reduce wages levels to reflect the balance of supply and demand.

**EFTA** Abbreviation for *European Free Trade Association.

**EFTPOS** Abbreviation for *electronic funds transfer at point of sale.

**EftPos UK Ltd** *See* Association for Payment Clearing Services.

**EGM** Abbreviation for *extraordinary general meeting.

**EIB** Abbreviation for *European Investment Bank.

**elasticity** The percentage change in a variable resulting from unit change of another variable. The concept was first introduced into economics by Alfred Marshall (1842–1924) and has been widely used ever since. The price **elasticity of demand**, for example, represents the percentage change in demand for a good resulting from a change in its price; this is usually negative, reflecting the shape of the demand curve. However, by convention elasticities are usually represented as positive numbers. For an elasticity of zero the relationship is said to be perfectly inelastic; if the elasticity is infinite, it is said to be perfectly elastic; if it is one it is said to be unit elastic. Elasticities greater than one are called relatively elastic; less than one, relatively inelastic.

**electronic funds transfer at point of sale (EFTPOS)** The automatic debiting of a purchase price from the customer's bank or credit-card account by a computer link between the checkout till and the bank or credit-card company. This system is still experimental in the UK, and can only work when the customer has a *debit card or *credit card recognized by the retailer. The system is in use in some petrol filling stations and large stores. In addition to a debit card or credit card, the user may also be required to have a *personal identification number (PIN) with which to identify himself to the computer.

**electronic mail** The transfer of correspondence, such as letters and memos, from one computer to another. The computers are connected by cables or telephone lines (using a *modem). Often a central computer acts as a post office, providing each user with a space in its memory, called a **mailbox**, where messages can be left. The users periodically contact the mailbox to check for messages.

**electronic point of sale (EPOS)** A computerized method of recording sales in retail outlets, using a laser scanner at the checkout till to read *bar codes printed on the items' packages. Other advantages over conventional checkout points include a more efficient use of checkout staff time and the provision of a more detailed receipt to the customer. With some 80% of goods sold in grocery outlets now being bar-coded, retailers are increasingly converting to EPOS.

**electronic transfer of funds (ETF)** The transfer of money from one bank account to another by means of computers and communications links. Banks routinely transfer funds between accounts using computers; another variety of ETF is the telebanking service enabling the *Viewdata network to be used for banking in a customer's home. *See also* electronic funds transfer at point of sale.

**eligibility** Criteria that determine which bills the Bank of England will discount, as *lender of last resort. Such bills, known as **eligible paper**, include Treasury bills, short-dated gilts, and first-class trade bills.

**ellipsis** A printing symbol consisting of three full stops in a row, indicating that a word or passage has been omitted from the printed matter.

**em** A unit of length used in printing. The 12-point em is a standard unit in typography, equal to 1/6 of an inch. An **em rule** is a horizontal line one em long. *Compare* en.

**embargo** A ban on some or all of the trade with one or more countries. A trade embargo is a form of *economic sanction. Prominent examples include the international ban on trade in arms with South Africa and on certain high-technology products with the Eastern bloc. Full embargos are rare and difficult to apply in practice.

**embezzlement** A form of theft in which an employee dishonestly appropriates money or property given to him on behalf of his employer. The special offence of embezzlement ceased to exist in 1969.

**employee buy-out** The acquisition of a controlling interest in the equity of a company by its employees. This may occur if the company is threatened with closure and the employees wish to secure their jobs. By obtaining financial backing, the employees acting as a group of individuals, or by means of a trust, can acquire a majority of the shares by offering existing shareholders more than their *break-up value. An employee buy-out has also taken place on *privatization of a company.

**employee participation** **1.** The encouragement of motivation in a workforce by giving shares in the company to employees. Employee shareholding (*see* employee share-ownership plan) is now an important factor in improving industrial relations. **2.** The appointment to a board of directors of a representative of the employees of a company, to enable the employees to take part in the direction of the company.

**employee share-ownership plan (ESOP)** A method of giving employees shares in the business for which they work. Various such plans came into existence in the UK after their announcement in 1989; in 1990, in order to encourage their growth, company owners were given *roll-over relief from capital-gains tax for sales of shares through ESOPs.

**employers'-liability insurance** An insurance policy covering an employer's legal liability to pay compensation to any of his employees suffering or contracting injury or disease during the course of their work. This type of insurance is compulsory by law for anyone who employs another person (other than members of their own family) under a contract of service. A *certificate of insurance must be displayed at each place of work, confirming that employer's liability insurance is in force and giving details of the policy number and the insurer's name and address.

**employment agency** An organization that introduces suitable potential employees to employers, charging the employer a fee, usually related to the initial salary, for the service. Employment agencies also provide temporary staff, in which they are the employers, the temporary staff member being charged out at an hourly rate. Employment agencies that specialize in finding suitable managers and executives for a firm, or finding suitable jobs for executives who want a change, are often known as **head hunters**. Such agencies will often provide a short list of candidates in order to save their client's time in personnel selection. Employment agencies run by the government are called *job centres.

**employment protection** The safeguarding of an employee's position with regard to his employment. According to the Employment Protection (Consolidation) Act (1978), an employer must give an employee, within 13 weeks of the start of the employment, a contract stating the rate of pay, hours of work, holiday entitlement, details of sick pay and pension scheme (if any), and the length of notice to be given by either side to terminate the contract. *See also* discrimination; redundancy; unfair dismissal.

**EMS** Abbreviation for *European Monetary System.

**en** Half of an *em. An **en rule** is half as long as an em rule.

**encumbrance (incumbrance)** A charge or liability, such as a *mortgage or registered judgment, to which land is subject.

**endorsement (indorsement)** **1.** A signature on the back of a *bill of exchange or cheque, making it payable to the person who signed it. A bill can be endorsed any number of times, the presumption being that the endorsements were made in the order in which they appear, the last named being the holder to receive payment. If the bill is **blank endorsed**, i.e. no endorsee is named, it is payable to the bearer. In the case of a **restrictive endorsement** of the form "Pay X only", it ceases to be a *negotiable instrument. A **special endorsement**, when the endorsee is specified, becomes payable **to order**, which is short for 'in obedience to the order of'. **2.** A signature required on a document to make it valid in law. **3.** An amendment to an *insurance policy or cover note, recording a change in the conditions of the insurance.

**endowment assurance** An assurance policy that pays a specified amount of money on an agreed date or on the death of the *life assured, whichever is the earlier. As these policies guarantee to make a payment (either to the policyholder or his or her dependants) they offer both life cover and a reasonable investment. A *with-profits policy will also provide bonuses in addition to the sum assured. These policies are often used in the repayment of a personal *mortgage or as a form of saving, although they lost their tax relief on premiums in the Finance Act (1984).

**Engel curve** A curve relating the expenditure on a good as income rises. The relationship was first analysed by the 19th-century statistician C L E Engel (1821–96). **Engel's law** states that the proportion of expenditure on food will fall as income rises, i.e. food is a *necessary good. Engel curves are useful for separating the effect of income on demand from the effects of changes in relative prices.

**Engel's law** *See* Engel curve.

**Enterprise Allowance Scheme** A scheme, administered by the Department of Employment, to provide a £40-per-week grant, for up to 52 weeks, to enable the unemployed to set up their own businesses. To be eligible for the scheme applicants must: be receiving unemployment benefit; have been unemployed for at least 8 weeks; be able to invest at least £1000 in the business during the first 12 months; be between 18 and 65 years of age; agree to work full-time in the business; and have the business approved by the Department of Employment.

**enterprise zone** An area, designated as such by the government, in which its aim is to restore private-sector activity by removing certain tax burdens and by relaxing certain statutory controls. Benefits, which are available for a 10-year period, include: exemption from rates on industrial and commercial property; 100% allowances for corporation- and income-tax purposes for capital expenditure on industrial and commercial buildings; exemption from industrial training levies; and a simplified planning regime.

**entrepôt trade** Trade that passes through a port, district, airport, etc., before being shipped on to some other country. The entrepôt trade may make use of such a port because it is conveniently situated on shipping lanes and has the warehouses and customs facilities required for re-export or because that port is the centre of the particular trade concerned and facilities are available there for sampling, testing, auctioning, breaking bulk, etc. Much entrepôt trade used to pass through London, but since the decline of the London docks other European ports, such as Rotterdam, have taken its place. Singapore and Hong Kong are also centres of the entrepôt trade.

**entrepreneur** An individual who undertakes (from the French *entreprendre* to undertake) to supply a good or service to the market for his own profit. The entrepreneur will usually invest his own capital in the business and take on the risks associated with the investment. To an economist, an entrepreneur is simply a representative capitalist whose behaviour is to be studied. More recently the view has been expressed that it is the initiative of entrepreneurs that creates a society's wealth and that governments should therefore establish conditions in which they will thrive.

**environmental scanning** An examination of the environment of a business in order to identify its marketing opportunities, competition, etc.

**EPOS** Abbreviation for *electronic point of sale.

**EPP** Abbreviation for *executive pension plan.

**eps** Abbreviation for *earnings per share.

**equal pay** The requirement of the Equal Pay Act (1970), expressing the principle of **equal opportunities**, that men and women in the same employment must be paid at the same rate for like work or work rated as equivalent. Work is rated as equivalent when the employer has undertaken a study to evaluate his employees' jobs in terms of the skill, effort, and responsibility demanded of them and the woman's job is given the same grade as the man's or when an independent expert appointed by an *industrial tribunal evaluates the two jobs as of equal value.

**equilibrium** A situation in which the forces acting on an economic variable are exactly balanced, so that there is no tendency for that variable to change. Equilibrium is a fundamental tool of economic analysis, since without it there is no posssibility of predicting the value that a variable will take. Most of economics is concerned with defining the forces that act on a variable and establishing a process, such as supply and demand, by which equilibrium is reached.

**equitable interest** An interest in, or ownership of, property that is recognized by *equity but not by the *common law. A beneficiary under a trust has an equitable interest. Any disposal of an equitable interest (e.g. a sale) must be in writing. Some equitable interests in land must be registered or they will be lost if the legal title to the land is sold. Similarly, equitable interests in other property will be lost if the legal title is sold to a bona fide purchaser for value who has no notice of the equitable interest. In such circumstances the owner of the equitable interest may claim damages from the person who sold the legal title.

**equities** The ordinary shares of a company, especially those of a publicly owned quoted company. In the event of a liquidation, the ordinary shareholders are entitled to share out the asssets remaining after all other creditors (including holders of *preference shares) have been paid out. Investment in equities on a stock exchange represents the best opportunity for capital growth, although there is a high element of risk as only a small proportion (if any) of the investment is secured. Although equities pay relatively low profit-related dividends, unlike *fixed-interest securities, they are popular in times of inflation as they tend to rise in value as the value of money falls.

**equity** **1.** A beneficial interest in an asset. For example, a person having a house worth £100,000 with a mortgage of £20,000 may be said to have an equity of £80,000 in the house. **2.** The net assets of a company after all creditors (including the holders of *preference shares) have been paid off. **3.** The amount of money returned to a borrower in a mortgage or hire-purchase agreement, after the sale of the specified asset and the full repayment of the lender of the money. **4.** The ordinary share capital of a company (*see* equities; equity capital). **5.** The system of law developed by the medieval chancellors and later by the Court of Chancery. It is distinguished from the *common law, which was developed by the king's courts, having originated from the residual jurisdiction delegated by the king to the chancellor. A citizen dissatisfied by the common law could petition the chancellor, who might grant him relief on an ad-hoc basis. In time this developed into a complementary but separate system of law providing remedies unavailable at common law, such as specific performance of a contract rather than damages. Until 1873 equity was applied and administered by the Court of Chancery, and equitable remedies were not available in the common law courts (and vice versa). However, the Judicature Acts 1873 and 1875 merged the two systems so that any court may now apply both common law and equity. If the two sets of rules contradict each other, the equitable rule prevails. Equity has been particularly important in the development of the law of trusts, land law, administration of estates, and alternative remedies for breach of contract.

**equity accounting** The practice of showing in a company's accounts a share of the undistributed profits of another company in which it holds a share of the *equity (usually a share of between 20% and 50%). The share of profit shown by the equity-holding company is usually equal to its share of the equity in the other company. Although none of the profit may actually be paid over, the company has a right to this share of the undistributed profit.

**equity capital** The part of the share capital of a company owned by ordinary shareholders, although for certain purposes, such as *pre-emption rights, other classes of shareholders may be deemed to share in the equity capital and therefore be entitled to share in the profits of the company or any surplus assets on winding up. *See also* A shares.

**equity dilution** A reduction in the percentage of the *equity owned by a shareholder as a result of a new issue of shares in the company, which rank equally with the existing voting shares.

**equity-linked policy** An insurance or assurance policy in which a proportion of the premiums paid are invested in equities. The surrender value of the policy is therefore the selling price of the equities purchased; as more premiums are paid the portfolio gets larger. Although investment returns may be considerably better on this type of policy than on a traditional *endowment policy, the risk is greater, as the price of equities can fall dramatically reducing the value of the policy. With *unit-linked policies, a much wider range of investments can be achieved and the risk is correspondingly reduced.

**ERM** Abbreviation for Exchange Rate Mechanism. *See* European Monetary System.

**ERNIE** Abbreviation for electronic random number indicating equipment. *See* premium bonds.

**erratum slip** A slip of paper containing last-minute corrections to a book or other printed document. It is stuck into the front of the book or to the appropriate page.

**errors and omissions excepted** *See* E & OE.

**escalation clause** A clause in a contract authorizing the contractor to increase the price in specified conditions of all or part of the services or goods he has contracted to supply. Escalation clauses are common in contracts involving work over a long period in times of high inflation. The escalation may refer to either or both labour and materials and may or may not state the way in which the price is permitted to escalate. An escalation clause does not convert a contract into a *cost-plus contract, but it represents a move in that direction.

**escape clause** A clause in a contract releasing one party from all or part of his contractual obligations in certain specified circumstances.

**escrow** A *deed that has been signed and sealed but is delivered on the condition that it will not become operative until some stated event happens. It will become effective as soon as that event occurs and it cannot be revoked in the meantime.

**ESOP** Abbreviation for *employee share-ownership plan.

**estate 1.** The sum total of a person's assets less his liabilities (usually as measured on his death for the purposes of *inheritance tax). **2.** A substantial piece of land, usually attached to a large house.

**estate duty** A former tax on the estate of a deceased person at the time of his death. This tax applied in the UK from 1894 to 1974, when it was converted to *capital-transfer tax. The latter tax became *inheritance tax in 1986, when it reverted to a form of taxation similar to the former estate duty. The tax is levied on the assets less the liabilities of the deceased, taking into account (to forestall avoidance) certain gifts made in a defined period before the death.

**estate in land** The length of time for which a piece of land will be held; for example, a life estate would last only for the life of the owner. *See* fee simple.

**estoppel** **1.** A rule of evidence by which a person is prevented from denying that a certain state of affairs exists if he has previously asserted that it does. **2.** (**promissory estoppel**) The rule that if a person has declared that he will not insist upon his strict legal rights under a contract, he will not later be allowed to insist upon them if the other party has relied on that declaration. He may, however, be allowed to enforce his strict legal rights, on giving reasonable notice, if this would not be inequitable. **3.** (**proprietary estoppel**) The rule that if one person allows or encourages another person to act to his detriment in respect of land, he will not later be able to refuse to grant something that he allowed the other person to expect. For example, if A encourages B to build a garage, which can only be reached by driving over A's land, saying that he will grant a right of way, he will not later be able to refuse to do so.

**ETF** Abbreviation for *electronic transfer of funds.

**Euratom** Abbreviation for *European Atomic Energy Community.

**Euro-ad** An advertisement designed to be used in all countries of the European Community. Certain products, particularly those that have been in use for long enough for national traditions to build up, are not suitable for multinational advertisements. Food is an example: there would be no point in advertising tinned baked beans in France. On the other hand, advertisements for cars, a more recent product, can have an equal appeal in all European countries.

**eurobond** A *bond issued in a *eurocurrency, which is now one of the largest markets for raising money (it is much larger than the UK stock exchange). The reason for the popularity of the eurobond market is that *secondary market investors can remain anonymous, usually for the purpose of avoiding tax. For this reason it is difficult to ascertain the exact size and scope of operation of the market. Issues of new eurobonds normally take place in London, largely through syndicates of US and Japanese investment banks; they are *bearer securities, unlike the shares registered in most stock exchanges and interest payments are free of any *withholding taxes. There are various kinds of eurobonds. An ordinary bond, called a **straight**, is a fixed-interest loan of 3 to 8 years duration; others include **floating-rate notes**, which carry a variable interest rate based on the *LIBOR; and perpetuals, which are never redeemed. Some carry *warrants and some are *convertible. *See also* note issuance facility; swap; zero-coupon bonds.

**eurocheque** A cheque drawn on a European bank, which can be cashed at any bank or bureau de change in the world that displays the EC sign (of which there are some 200 000). It can also be used to pay for goods and services in shops, hotels, restaurants, garages, etc., that display the EC sign (over 4 million). The cheques are blank and are made out for any amount as required, usually in the local currency. They have to be used with a **Eurocheque Card**, which guarantees cheques for up to about £100. In most cases, a commission of 1.25% is added to the foreign currency value of the cheque before it is converted to sterling and there is a 30p cheque charge (for cheques drawn on UK banks).

**eurocurrency** A currency held in a European country other than its country of origin. For example, dollars deposited in a bank in Switzerland are *eurodollars, yen deposited in Germany are **euroyen**, etc. Eurocurrency is used for lending and borrowing; the eurocurrency market often provides a cheap and convenient form of liquidity for the financing of international trade and investment. The main borrowers and lenders are the commercial banks, large companies, and the central banks. By raising funds in eurocurrencies it is possible to secure more favourable terms and rates of interest, and sometimes to avoid domestic regulations and taxation. Most of the deposits and loans are on a short-term basis but increasing use is being made of medium-term loans, particularly through the raising of *eurobonds. This has to some extent replaced the syndicated loan market, in which banks lent money as a group in order to share the risk. *Euromarkets emerged in the 1950s.

**eurodollars** Dollars deposited in financial institutions outside the USA. The eurodollar market evolved in London in the late 1950s when the growing demand for dollars to finance international trade and investment coincided with a greater supply of dollars. The prefix 'euro' indicates the origin of the practice but it now refers to all dollar deposits made anywhere outside the USA. *See also* eurocurrency.

**euromarket** **1.** A market that emerged in the 1950s for financing international trade. Its principal participants are *commercial banks, large companies, and the central banks of members of the EC. Its main business is in *eurobonds issued in *eurocurrencies. The largest euromarket is in London, but there are smaller ones in Paris and Brussels. **2.** The European Community, regarded as one large market for goods.

**European Atomic Energy Community (Euratom)** The organization set up by the six members of the *European Coal and Steel Community in 1957; the *European Economic Community was established at the same time. Euratom was formed to create the technical and industrial conditions necessary to produce nuclear energy on an industrial scale. The UK, Denmark, and Ireland joined in 1973; Greece joined in 1981; and Spain and Portugal joined in 1986.

**European Coal and Steel Community (ECSC)** The first of the *European Communities (EC), founded in 1953. The ECSC created a common market in coal, steel, iron ore, and scrap between the original six members of the EC (Belgium, France, West Germany, Italy, Luxembourg, and the Netherlands). These six countries, in 1957, signed the Treaty of Rome setting up the *European Economic Community.

**European Commission (Commission of the European Communities)** The single executive body formed in 1967 from the three separate executive bodies of the *European Coal and Steel Community, the *European Atomic Energy Community, and the *European Economic Community. It now consists of 17 Commissioners: two each from the UK, France, West Germany, Spain, and Italy; and one each from Belgium, the Netherlands, Luxembourg, Ireland, Denmark, Greece, and Portugal. The Commissioners accept joint responsibility for their decisions, which are taken on the basis of a majority vote. The Commission initiates action in the *European Community and mediates between member governments.

**European Community (EC)** The twelve nations (Belgium, Denmark, France, West Germany, Greece, Ireland, Italy, Luxembourg, the Netherlands, Portugal, Spain, and the UK) that joined together to form an economic community, with some common monetary, political, and social aspirations. The community grew from the *European Coal and Steel Community, the *European Atomic Energy Community, and the *European Economic Community. The Commission of the European Communities (*see* European Commission) was formed in 1967 with the Council of the European Communities. The community policy emerges from a dialogue between the Commission, which initiates and implements the policy, and the Council, which takes the major policy decisions. The European Parliament, formed in 1957, exercises democratic control over policy, and the European Court of Justice imposes the rule of law on the community, as set out in its various treaties.

**European Currency Unit (ECU)** A currency medium and unit of account created in 1979 to act as the reserve asset and accounting unit of the *European Monetary System. The value of the ECU is calculated as a weighted average of a basket of specified amounts of *European Community (EC) currencies; its value is reviewed periodically as currencies change in importance and membership of the EC expands. It also acts as the unit of account for all EC transactions. It has some similarities with the *Special Drawing Rights of the *International Monetary Fund; however, ECU reserves are not allocated to individual countries but are held in the *European Monetary Cooperation Fund. Private transactions using the ECU as the denomination for borrowing and lending have proved popular. It has been suggested that the ECU will be the basis for a future European currency to replace all national currencies.

**European Economic Community (EEC; Common Market)** The European common market set up by the six member states of the *European Coal and Steel Community in 1957. At the same time the *European Atomic Energy Community was set up; the controlling bodies of these three communities were merged in 1967 to form the Commission of European Communities (*see* European Commission) and the Council of European Communities. The European Parliament and the European Court of Justice were formed in accordance with the Treaty of Rome in 1957. The treaty aimed to forge a closer union between the countries of Europe by removing the economic effects of their frontiers. This included the elimination of customs duties and quotas between members, a common trade policy to outside countries, the abolition of restrictions on the movement of people and capital between member states, and a *Common Agricultural Policy. In addition to these trading policies, the treaty envisaged a harmonization of social and economic legislation to enable the Common Market to work (*see also* European Investment Bank). The UK, Ireland, and Denmark joined in 1973, Greece joined in 1979, and Portugal and Spain became members in 1986, making a total of 12 nations. *See also* European Community; European Monetary System.

**European Free Trade Association (EFTA)** A trade association formed in 1960 between Austria, Denmark, Norway, Portugal, Sweden, Switzerland, and the UK. Finland and Iceland joined later while the UK, Denmark, and Portugal left on joining the *European Community (EC). EFTA is a looser association than the EC, dealing only with trade barriers rather than generally coordinating economic policy. All tariffs between EFTA and EC countries were abolished finally in 1984. EFTA is governed by a council in which each member has one vote; decisions must normally be unanimous and are binding on all member countries.

**European Investment Bank (EIB)** A bank set up under the Treaty of Rome in 1958 to finance capital-investment projects in the *European Economic Community (EEC). It grants long-term loans to private companies and public institutions for projects that further the aims of the Community. The twelve members of the *European Community subscribed to the Bank's capital of 28,000M ECU but most of the funds lent by the bank are borrowed on the international capital markets. The bank is non-profit making and charges interest at a rate that reflects the rate at which it borrows. Its headquarters are in Luxembourg.

**European Monetary Cooperation Fund (FECOM)** A fund organized by the *European Monetary System in which members of the *European Community deposit reserves to provide a pool of resources to stabilize exchange rates and to finance *balance of payments support. In return for depositing 20% of their gold and gross dollar reserves, member states have access to a wide variety of credit facilities, denominated in ECU, from the fund.

**European Monetary System (EMS)** A European system of exchange-rate stabilization involving the countries of the *European Community. There are two elements: the **Exchange Rate Mechanism (ERM)**, under which participating countries commit themselves to maintaining the value of their currencies within agreed narrow limits, and a *balance of payments support mechanism, organized through the *European Monetary Cooperation Fund. The ERM is generally regarded as having helped to maintain exchange-rate stability and to have encouraged the coordination of macroeconomic policy. It operates by giving each currency a value in ECUs and drawing up a **parity grid** giving exchange values in ECUs for each pair of currencies. If market rates differ from this parity by more than a permitted percentage (currently 2.25%), the relevant governments have to take action to correct the disparity. Some, however, including the UK, which does not yet participate in the ERM (although it is committed to do so when certain conditions have been met), argue that it imposes too great a constraint on domestic monetary policy. The ultimate goal of the EMS is also controversial. To some its function is to facilitate monetary cooperation; to others, it is the first step towards a single European currency and a European central bank. *See also* European Currency Unit.

**European option** *See* option.

**Europort** Any of the main European ports, especially Rotterdam.

**Eurotunnel** *See* Channel Tunnel.

**euroyen** *See* eurocurrency.

**evergreen fund** A fund that provides capital for new companies and supports their development for a period with regular injections of capital.

**ex-** (Latin: without) A prefix used to exclude specified benefits when a security is quoted. A share is described as **ex-dividend** (xd or ex-div) when a potential purchaser will no longer be entitled to receive the company's current dividend, the right to which remains with the vendor. Government stocks go ex-dividend 36 days before the interest payment. Similarly, **ex-rights**, **ex-scrip**, **ex-coupon**, **ex-capitalization** (**ex-cap**), and **ex-bonus** mean that each of these benefits belongs to the vendor rather than the buyer. **Ex-all** means that all benefits belong to the vendor. **Cum-** (Latin: with) has exactly the opposite sense, meaning that the dividend or other benefits belong to the buyer rather than the seller. The price of a share that has gone ex-dividend will usually fall by the amount of the dividend, while one that is **cum-dividend** will usually rise by this amount. However, in practice market forces usually mean that these falls and rises are often slightly less than expected.

**ex-all** *See* ex-.

**ex-bonus** *See* ex-.

**ex-capitalization (ex-cap)** *See* ex-.

**excepted perils** *See* peril.

**exceptional items** Costs or income affecting a company's *profit and loss account that arise from the normal activities of the company but are of exceptional magnitude, either large or small. These should be disclosed separately in arriving at the trading profit or loss but not after the normal trading profit or loss has been shown (*compare* extraordinary items).

**excess capacity** The part of the output of a plant or process that is not currently being utilized but which, if it could be, would reduce the average cost of production. The excess capacity is thus the amount by which the present output must be increased to reduce the average cost per unit to a minimum.

**excess demand** *See* excess supply (or demand).

**excess policy** An insurance policy in which the insured is responsible for paying a specified sum (the **excess**) of each claim and cannot make claims of a lower value than this excess. For example, a £100 excess on a motor-insurance policy means that the insured has to pay the first £100 of any claim and cannot make a claim on the policy for less than £100. This arrangement enables the insurer to offer the insurance at a lower premium than would otherwise be the case as he avoids the administrative cost of small claims and also makes a saving on claims paid out. *See also* deductible; franchise.

**excess supply (***or* **demand)** The amount by the quantity of a good supplied (or demanded) in the market exceeds the quantity demanded (or supplied). It is a fundamental tenet of economics, since at least the time of Adam Smith (1723–90), that excess supply (or demand) will be eliminated by falling (or rising) prices in a free market. *Keynesianism challenged the belief that this mechanism would be effective in the labour market but a convincing justification for this view has yet to be established.

**exchange** The economic process enabling values to be traded between individuals or groups. Values include physical goods, services, information, and even promises (e.g. *options). The arena in which exchange takes place is a *market; it is the focus of economic analysis. The achievement of the *neoclassical school of economics has been to prove that under certain restrictive conditions the free exchange of values between individuals will yield a Pareto-optimal outcome (*see* Pareto optimality), one that could not be improved upon by any other method of allocation. Much of modern economics is concerned with investigating the consequences of relaxing the restrictive conditions to take into account such real-world problems as *monopoly, limited *information, and the absence of certain markets.

**exchange control** Restrictions on the purchase and sale of foreign exchange. It is operated in various forms by many countries, in particular those who experience shortages of *hard currencies; sometimes different regulations apply to transactions that would come under the capital account of the *balance of payments. There has been a gradual movement toward dismantling exchange controls by many countries in recent years. The UK abolished all form of exchange control in 1979.

**exchange equalization account** An account set up in 1932 and managed by the Bank of England on behalf of the government. It contains the official gold and foreign-exchange reserves (including *Special Drawing Rights) of the UK and is used as a buyer of foreign exchange to support the value of sterling. Although all *exchange controls were abolished in 1979, the Bank of England still makes use of this account to help to stabilize rates of exchange.

**exchange of contracts** A procedure adopted in the sale and purchase of land in which both parties sign their copies of the contract, having satisfied themselves as to the state of the property, etc., and agreed that they wish to be bound. There need be no physical exchange of documents; the parties or their advisers can exchange contracts by agreeing to do so orally (for example, by telephone). From that moment the contract is binding and can normally be enforced by specific performance. Contracts for the sale of land must be in writing.

**exchange rate** *See* rate of exchange.

**Exchange Rate Mechanism (ERM)** *See* European Monetary System.

**Exchequer stocks** *See* gilt-edged security.

**excise duty** A duty or tax levied on certain goods, such as alcoholic drinks and tobacco products, produced and sold within the UK, unlike customs duty (*see* customs tariff), which is levied on imports. Both excise and customs duties are collected by the *Board of Customs and Excise.

**exclusive economic zone** *See* territorial waters.

**ex-coupon** *See* ex-.

**ex-dividend** *See* ex-.

**executive director (working director)** A *director of a company who is also an employee (usually full-time) of that company. An executive director will often have a specified role in the management of the company, such as finance director, marketing director, production director, etc. A **non-executive director** is a member of the board of directors and is therefore involved in planning and policy-making, but not in the day-to-day management of the company. A non-executive director is often employed for prestige (if he is well known), for his experience or contacts, or for his specialist knowledge, which may only be required occasionally.

**executive pension plan (EPP)** A pension for a senior executive or director of a company in which the company provides a tax-deductible contribution to the premium. An executive pension plan may be additional to any group pension scheme provided by the company, as long as the pension limit of two-thirds of the working salary is not exceeded.

**executor** A person named in a will of another person to gather in the assets of his estate, pay his liabilities, and distribute any residue to the beneficiaries in accordance with the instructions contained in the will.

**exemplary damages** *See* damages.

**exempt gilts** Government gilt-edged securities that pay interest gross, unlike ordinary gilts, on which tax is deducted from interest payments.

These gilts are of particular interest to foreign buyers and others, such as institutions, who do not pay income tax.

**exercise notice** Formal notification from the owner of an *option to the person of the firm that has written it that he wishes to exercise the option to buy (for a call option) or sell (for a put option) at the *exercise price.

**exercise price (striking price)** The price per share at which a traded *option entitles the owner to buy the underlying security in a call option or to sell it in a put option. *See also* exercise notice.

**ex factory** *See* ex works.

**ex gratia** (Latin: as of grace) Denoting a payment made out of gratitude, moral obligation, kindness, etc., rather than to fulfil a legal obligation. When an ex gratia payment is made, no legal liability is admitted by the payer.

**ex growth** (of a share or a company) Having had substantial growth in the past but now not holding out prospects for immediate growth of earnings or value.

**Eximbank** *See* Export-Import Bank.

**ex-new** Describing a share that is offered for sale without the right to take up any *scrip issue or *rights issue. *Compare* cum-new.

**expenditure** A sum spent for goods or services. Expenditure may or may not become an *expense in the *profit and loss account, depending on whether any residual value remains for the purchasing organization.

**expenditure tax (outlay tax)** A tax on the expenditure of individuals or households. This form of taxation, of which VAT is an example, is often preferred by tax theorists to income taxes, as it does not distort the incentive to work. It is also argued that it taxes what persons take out of the communal pot rather than what they put into it.

**expense 1.** A sum spent for goods or services, which therefore no longer represents an asset of the purchasing organization. Expenses are normally shown as charge against profit in the *profit and loss account. **2.** A sum of money spent by an employee during the course of his work. The employee records this expenditure in an **expense account**, for refund by the company and submission to the Inland Revenue on a P11D form for assessment of any taxable element.

**expert system** A computer program that simulates the knowledge and experience of an expert in a particular field enabling it to solve problems in that field. It can be questioned by a non-expert and will give the answer that the expert would give. Also called **intelligent knowledge-based systems**, expert systems are an application of *artificial intelligence techniques. They are finding increasing use in commerce for such tasks as evaluating loan applications and buying stocks and shares.

**expiry date 1.** The date on which a contract expires. **2.** The last day on which an *option expires. In a European option the option must be taken up or allowed to lapse on this date. In an American option the decision can be taken at any time up to the expiry date.

**Export Credits Guarantee Department (ECGD)** A UK government department, responsible to the Secretary of State for Trade and Industry, that operates under the Export Guarantees and Overseas Investment Act (1978). It encourages exports from the UK by making export credit insurance available to exporters and guaranteeing repayment to UK banks that provide finance for exports on credit terms of two years or more. It also insures British private investment overseas against war risk, expropriation, and restrictions on the making of remittances.

**Export-Import Bank (Eximbank)** A US bank established by the US government to foster trade with the USA. It provides export credit guarantees and guarantees loans made by commercial banks to US exporters.

**export incentive** An incentive offered by a government to exporters. They can include subsidies, grants, tax concessions, credit facilities, etc., but are now discouraged by the EC.

**export licence** A licence required before goods can be exported from a country. Export licences are only required in the UK for certain works of art, antiques, etc., and certain types of arms and armaments.

**exports** Goods or services sold to foreign countries. In terms of the *balance of payments, goods are classified as *visibles, while such services as banking, insurance, and tourism are treated as invisibles. The UK has traditionally relied on its invisibles to achieve its trade balance as it tends to spend more on imports than it receives in exports. *See also* Export Credits Guarantee Department.

**ex quay** Delivery terms for goods in which the seller pays all freight charges up to the port of destination, unloading onto the quay, and loading onto road or rail vehicles. Thereafter all transport charges must be paid by the buyer.

**ex-rights** *See* ex-; ex-new.

**ex-scrip** *See* ex-; ex-new.

**ex ship (free overboard; free overside)** Delivery terms for goods shipped from one place to another. The seller pays all charges up to the port of destination, including unloading from the ship. All subsequent charges, e.g. lighterage, loading charges, etc., are paid by the buyer.

**extended guarantee** A servicing and maintenance cover that a manufacturer offers to a customer for a specified additional amount of time beyond the original guarantee. Extended guarantees usually have to be paid for.

**external account** A sterling bank account held by someone who resides outside the sterling area.

**externalities** Those economic effects of a business that are not recorded in its accounts as they do not arise from individual transactions of the business. For example, local overcrowding may arise because a large number of employees have been attracted to the neighbourhood, thus incurring extra costs for roads, schools, health care, etc. More generally, an externality results from an economic choice that is not reflected in market prices. For example, siting a railway station close to a housing estate represents an externality to householders on that estate if they are not asked to pay for it. It is an external economy if the householders benefit from greater freedom to travel and an external diseconomy if the noise of trains keeps them awake at night. It is usually argued that governments should internalize external diseconomies, such as pollution, by means of taxation.

**external storage** *See* backing store.

**extraordinary general meeting (EGM)** Any general meeting of a company other than the *annual general meeting. Most company's articles give the directors the right to call an EGM whenever they wish. Members have the right to requisition an EGM if they hold not less than 10% of the paid-up share capital; a resigning auditor may also requisition a meeting. Directors must call an EGM when there has been a serious loss of capital. The court may call an EGM if it is impracticable to call it in any other way. Those entitled to attend must be given 14 days' notice of the meeting (21 days if a special resolution is to be proposed). *See also* agenda; order of business.

**extraordinary items** Costs or income affecting a company's *profit and loss account that do not derive from the normal activities of the company and, if undisclosed, would distort the normal trend of profits. Such items are therefore disclosed after the normal trading profit or loss has been shown. Extraordinary items are to be distinguished from *exceptional items, which arise from the normal activities of the company but are of exceptional magnitude.

**extraordinary resolution** A resolution submitted to a general meeting of a company; 14 days' notice of such a resolution is required, and the notice should state that it is an extraordinary resolution. 75% of those voting must approve the resolution for it to be passed.

**extrapolation** *See* interpolation.

**ex warehouse** Delivery terms for goods that are available for immediate delivery, in which the buyer pays for the delivery of the goods but the seller pays for loading them onto road or rail transport. *Compare* at warehouse.

**ex works (ex factory)** Delivery terms for goods in which the buyer has to pay for transporting them away from the factory that made them. In some cases, however, the seller will pay for loading them onto road or rail transport.

# F

**faa** Abbreviation for *free of all averages.

**face value 1.** The nominal value (*see* nominal price) printed on the face of a security. This is known also as the *par value. It may be more or less than the market value. **2.** The value printed on a banknote or coin.

**facing matter** An advertisement in a publication positioned to appear opposite an editorial page, especially when the advertisement is related to the text of the article.

**facsimile transmission** *See* Fax.

**fact book** A file containing information on the history of a product, including data on sales, distribution, competition, customers, and relevant market research undertaken, as well as a detailed record of the product's performance in relation to the marketing effort made on its behalf.

**factor 1.** A firm that engages in *factoring. **2.** An individual or firm that acts as an agent (often called a **mercantile agent**) in certain trades, usually receiving a **factorage** (commission or fee) based on the amount of sales achieved. Unlike some other forms of agent, a factor takes possession of the goods and sells them in his own name.

**factoring** The buying of the trade debts of a manufacturer, assuming the task of debt collection and accepting the credit risk, thus providing the manufacturer with working capital. **With service factoring** involves collecting the debts, assuming the credit risk, and passing on the funds as they are paid by the buyer. **With service plus finance factoring** involves paying the manufacturer up to 90% of the invoice value immediately after delivery of the goods, with the balance paid after the money has been collected. This form of factoring is clearly more expensive than with service factoring. In either case the factor, which may be a bank or finance house, has the right to select its debtors. *See also* undisclosed factoring.

**factors of production** The resources required to produce economic *goods. They are land (including all natural resources), labour (including all human work and skill), capital (including all money, assets, machinery, raw materials, etc.), and entrepreneurial ability (including organizational *management skills, inventiveness, and the willingness to take risks). For each of these factors there is a price, i.e. rent for land, wages for labour, interest for capital, and profit for the entrepreneur.

**factory costs** The costs of producing goods, which have been incurred in the area of the factory; these include materials, factory labour, costs of machinery, and costs of factory buildings, but not mark-up or profit.

**facultative reinsurance** A form of *reinsurance in which the terms, conditions, and reinsurance premium is individually negotiated between the insurer and the reinsurer. There is no obligation on the reinsurer to accept the risk or on the insurer to reinsure it if it is not considered necessary. The main differences between facultative reinsurance and *coinsurance is that the policyholder has no indication that reinsurance has been arranged. In coinsurance, the coinsurers and the proportion of the risk they are covering are shown on the policy schedule. Also, coinsurance involves the splitting of the premium charged to the policyholder between the coinsurers, whereas the reinsurers charge entirely separate reinsurance premiums.

**Faculty of Actuaries** *See* Institute of Actuaries.

**fair average quality** *See* faq.

**fair trading** *See* consumer protection.

**fall-back price** *See* Common Agricultural Policy.

**falsification of accounts** A dishonest entry in a firm's books of account, made by an employee with the object of covering up the theft of goods or money from the firm.

**family brand** A group of brand names for the products of a company, all of which contain the same word to establish their relationship in the minds of the consumers. *See also* product line.

**family life-cycle** The six stages of family life based on demographic data: (1) young single people; (2) young couples with no children; (3) young couples with youngest child under six years; (4) couples with dependent children; (5) older couples with no children at home; (6) older single people. These groups have been useful in marketing and advertising for defining the markets for certain goods and services, as each group has its own specific and distinguishable needs and interests.

**FAO** Abbreviation for *Food and Agricultural Organization.

**faq 1.** Abbreviation for fair average quality. This is a trade description of certain commodities that are offered for sale on the basis that the goods supplied will be equal to the average quality of the current crop or recent shipments rather than on the basis of a specification or quality sample. **2.** Abbreviation for free alongside quay. *See* free alongside ship.

**fas** Abbreviation for *free alongside ship.

**fast food 1.** Food that requires little or no preparation before serving. **2.** Food served in restaurants with limited and standardized menus. They are often self-service establishments offering the same quality of product throughout the country or, in some cases, internationally.

**fast-moving consumer goods (FMCG)** Products that move off the shelves of retail shops quickly, which therefore require constant replenishing. Fast-moving consumer goods include standard groceries, etc., sold in supermarkets as well as records and tapes sold in music shops.

**fate** Whether or not a cheque or bill has been paid or dishonoured. A bank requested by another bank to **advise fate** of a cheque or bill is being asked if it has been paid or not.

**Fax** A widely used method of communication between businesses, linked to the international telephone system. Firms purchase or hire a Fax machine and install a separate telephone line; this enables them to send copies of documents, diagrams, plans, etc., for the cost of a telephone call of equal duration (approximately 30–60 seconds per A4 page). The Fax machine itself has the appearance of a desk-top copier. Provided sender and recipient have equipment conforming to international standards, communication can be worldwide.

**FCA** Abbreviation for Fellow of the *Institute of Chartered Accountants.

**FCCA** Abbreviation for Fellow of the Chartered Association of Certified Accountants. *See* certified accountant; chartered accountant.

**FCII** Abbreviation for Fellow of the *Chartered Insurance Institute.

**feasibility study** A study of the financial factors involved in producing a new product, setting up a new process, etc. The study will analyse the technical feasibility with detailed costings of set-up expenses, running expenses, and raw-material costs, together with expected income. The capital required and the interest charges will also be analysed to enable an opinion to be given as to the commercial viability of product, process, etc.

**FECOM** *See* European Monetary Cooperation Fund.

**federal funding rate** The rate of interest charged by the US Federal Reserve System when lending money to the rest of the banking system.

**Federal Reserve System** The banking system in the USA that performs the functions of a central bank. The system consists of twelve Federal Reserve Districts, in each of which a Federal Reserve Bank acts as *lender of last resort. The activities of the twelve Reserve Banks are controlled from Washington by the Federal Reserve Board. The Federal Reserve System is used to implement the USA's monetary policy.

**fee simple** The most usual freehold *estate in land. Although all land in England is theoretically held by the Crown, the owner of a fee simple (or his heirs) will own the land forever and may dispose of it as he wishes both during his lifetime and by will. The land will revert to the Crown only if he dies without leaving a will and with no surviving relatives. This is the only type of freehold estate that can now exist in common law, as opposed to in equity.

**FIA** Abbreviation for Fellow of the *Institute of Actuaries.

**fiat money** Money that a government has declared to be legal tender, although it has no intrinsic value and is not backed by reserves. Most of the world's paper money is now fiat money. *Compare* fiduciary issue.

**fictitious asset** An asset, usually as shown in a balance sheet, that does not exist. The reasons for showing it may be fraudulent, the asset may have ceased to exist but has not been taken out of the accounts, or it may be an asset, such as goodwill, that no longer has any value. *Compare* intangible asset.

**fidelity guarantee** An insurance policy covering employers for any financial losses they may sustain as a result of the dishonesty of employees. Policies can be arranged to cover all employees or specific named persons. Because of the nature of the cover, insurers require full details of the procedure adopted by the organization in recruiting and vetting new employees and they usually reserve the right to refuse to cover a particular person without giving a reason.

**fiduciary** **1.** Denoting a person who holds property in trust or as an executor. Persons acting in a fiduciary capacity do so not for their own profit but to safeguard the interests of some other person or persons. **2.** Denoting a loan that is made on trust, rather than against some security.

**fiduciary issue** **1.** In the UK, the part of the issue of banknotes by the Bank of England that is backed by government securities, rather than by gold. Nearly the whole of the note issue is now fiduciary. *Compare* fiat money. **2.** Formerly, in the UK, banknotes issued by a bank without backing in gold, the value of the issues relying entirely on the reputation of the issuing bank.

**field** A subdivision of a *record in a computer *file. For example, club membership records might be divided into fields for name, address, membership number, and so on.

**FIFO** Abbreviation for *first in, first out.

**fifth-generation computer** A very advanced type of computer, still under development. These computers will use VLSI (very large-scale integration) silicon chips, containing millions of transistors on a single chip. Their input devices will be less cumbersome and more direct than keyboards; for example, they might recognize speech; they will also incorporate *artificial intelligence.

**file** In computer technology, a collection of related information held on *backing store that is treated as a unit. A file may contain a program, which can be copied into *main store memory and executed; it may contain modules from which programs are built; or it may contain data of any kind. A data file is often subdivided into *records. For example, the subscription information of a magazine might be kept in a subscriptions file, with a record for each customer.

**filmsetting** *See* phototypesetting.

**FIMBRA** Abbreviation for Financial Intermediaries, Managers and Brokers Regulatory Association Ltd. *See* Self-Regulatory Organization.

**final accounts** The accounts for a company produced at the end of its financial year (*see* annual accounts), as opposed to any **interim accounts** produced during the year, often after six months. Interim accounts are for the guidance of management and are often not audited, whereas final accounts must be audited and are open for inspection by the public.

**final dividend** *See* dividend.

**final invoice** An invoice that replaces a *proforma invoice for goods. The proforma invoice is sent before all the details of the goods are known. The final invoice contains any missing information and states the full amount still owing for the goods.

**finance** **1.** The practice of manipulating and managing money. **2.** The capital involved in a project, especially the capital that has to be raised to start a new business. **3.** A loan of money for a particular purpose, especially by a *finance house.

**finance house** An organization, many of which are owned by *commercial banks, that provides finance for *hire-purchase agreements. A consumer, who buys an expensive item (such as a car) from a trader and does not wish to pay cash, enters into a hire-purchase contract with the finance house, who collects the deposit and instalments. The finance house pays the trader the cash price in full, borrowing from the *commercial banks in order to do so. The finance house's profit is the difference between the low rate of interest it pays to the commercial banks to borrow and the high rate it charges the consumer. Most finance houses are members of the *Finance Houses Association.

**Finance Houses Association** A UK organization of *finance houses, set up in 1945 to regulate the trade of hire purchase and to negotiate with the government on acceptable terms and conditions. Its members control most UK hire-purchase agreements.

**financial accountant** An accountant whose primary responsibility is the management of the finances of an organization and the preparation of its annual accounts. *Compare* cost accountant; management accountant.

**financial adviser 1.** Anyone who offers financial advice to someone else, especially one who advises on *investments. **2.** An organization, usually a merchant bank, who advises the board of a company during a takeover (*see* takeover bid).

**financial futures** *See* futures contract; London International Financial Futures Exchange.

**Financial Intermediaries, Managers and Brokers Regulatory Association Ltd (FIMBRA)** *See* Self-Regulatory Organization.

**financial investment** *See* investment.

**Financial Ombudsman 1.** (**Banking Ombudsman**) An official in charge of a service set up in 1986 by 19 banks, which fund the service, to investigate complaints from bank customers against the service provided by member banks. **2.** (**Building Societies Ombudsman**) An official in charge of a service set up in 1987 by all the UK building societies, who fund the service, to investigate complaints from building-society customers against the services provided by the building societies. **3.** (**Insurance Ombudsman**) An official in charge of a service set up in 1981 by some 200 insurance companies, who fund the service, to settle disputes between insurance policyholders and member insurance companies. The Insurance Ombudsman's Bureau also runs the **Unit Trusts Ombudsman Scheme**, to settle disputes between unit-trust holders and unit-trust management companies. It was set up in 1988.

***Financial Statistics*** A monthly publication of the UK *Central Statistical Office giving a full account of financial statistics.

**Financial Times Share Indexes** A number of share indexes published by the *Financial Times*, daily except Sundays and Mondays, as a barometer of share prices on the London Stock Exchange. The **Financial Times Actuaries Share Indexes**, of which there are 54, are calculated by the *Institute of Actuaries as weighted arithmetic averages for various sectors of the market (capital goods, consumer goods, etc.) and divided into various industries. These are used widely by investors, portfolio managers, etc. The **Financial Times Ordinary Share Index** represents the movement of industrial shares on the basis of 30 market leaders. An unweighted geometric average, it is published twice daily (noon and the close of business) and gives an indication of market mood. The more recent (1984 = 1000) **Financial Times Stock Exchange 100 Index** (FTSE 100 Index, known as **FOOTSIE**) is based on the price of 100 securities and gives the best overall indication of market movements on a daily basis. *Compare* Dow Jones Industrial Average.

**financial year 1.** Any year connected with finance, such as a company's accounting period or a year for which budgets are made up. **2.** A specific period relating to corporation tax, i.e. the year beginning 1st April (the year beginning 1st April 1988 is the financial year 1988). Corporation-tax rates are fixed for specific financial years by the Chancellor in his budget; if a company's accounting period falls into two financial years the profits have to be apportioned to the relevant financial years to find the rates of tax applicable. *Compare* fiscal year.

**financier** A person who uses his own money to finance a business deal or venture or who makes arrangements for such a deal or venture to be financed by a merchant bank or other financial institution.

**financing gap** The difference between a country's foreign exchange requirements, for imports and the servicing of its debts, and what it has available from export receipts and overseas earnings. This gap must be filled either by raising further foreign exchange (donor aid, loans, etc.) or by cutting back the requirements, either by reducing imports or rescheduling the repayment of debts. Forecasting the financing gap and negotiating means of bridging it are major elements in helping countries with balance of payments problems.

**fine trade bill** A *bill of exchange that is acceptable to the Bank of England as security, when acting as *lender of last resort. It will be backed by a first-class bank or finance house.

**fine-tuning** The use by governments of small changes in *monetary and *fiscal policy to maintain a constant level of *aggregate demand, usually that associated with full employment. In the 1950s and 1960s it was widely believed that the government could raise employment during *recessions and curb *inflation during booms by the use of fine-tuning. With the emergence of *stagflation in the 1970s, however, this belief was largely undermined and the policy is now discredited. *See also* demand management.

**fire insurance** An insurance policy covering financial losses caused by damage to property by fires. Most fire insurance policies also include cover for damage caused by lightning and explosions of boilers or gas used for domestic purposes. On payment of an additional premium, fire insurers are usually prepared to widen the cover to include such special *perils as those arising from various weather-related or man-made causes.

**firm 1.** Any business organization. **2.** A business partnership.

**firm offer** An offer to sell goods that remains in force for a stated period. For example, an 'offer firm for 24 hours' binds the seller to sell if the buyer accepts the offer within 24 hours. If the buyer makes a lower *bid during the period that the offer is firm, the offer ceases to be valid. An offer that is not firm is usually called a *quotation in commercial terms.

**firm order** An order to a broker (for securities, commodities, currencies, etc.) that remains firm for a stated period or until cancelled. A broker who has a firm order from his principal does not have to refer back if he can execute the terms of the order in the stated period.

**firmware** Computer programs or data that are stored in a memory chip. These are often built into a computer to make it unnecessary to load the programs or data from disk or from the keyboard. Wordprocessing programs, for example, on *read-only memory (ROM) chips are built into some business computer systems. Firmware is also used where absolute reliability is required, for example in air-traffic control.

**first-class paper (fine paper)** A *bill of exchange, cheque, etc., drawn on or accepted or endorsed by a first-class bank, *finance house, etc.

**first in, first out (FIFO)** A method of charging homogeneous items of stock to production when the cost of the items has changed. It is assumed, both for costing and stock valuation purposes, that the earliest items taken into stock are those used in production although this may not necessarily correspond with the physical movement of the stock items. *Compare* last in, first out.

**first-loss policy** A property insurance policy in which the policyholder arranges cover for an amount below the full value of the items insured and the insurer agrees not to penalize him for under-insurance. The main use of these policies is in circumstances in which a total loss is virtually impossible. For example, a large warehouse may contain £2.5m worth of wines and spirits but the owner may feel that no more than £500,000 worth could be stolen at any one time. The solution is a first-loss policy that deals with all claims up to £500,000 but pays no more than this figure if more is stolen. First-loss policies differ from *coinsurance agreements with the policyholders because the insured is not involved in claims below the first-loss level and the premiums are not calculated proportionately. In the above example, the premium might be as much as 80–90% of the premium on the full value.

**first of exchange** *See* bills in a set.

**fiscal drag** A restraint on the expansion of an economy as a consequence of government taxation policy. In a progressive tax system, a rise in inflation will cause wage earners to pay a higher proportion of their income in tax, even though their real wages are unchanged; this is **nominal fiscal drag. Real fiscal drag** occurs when all taxpayers pay a higher proportion of their income in taxation, as a result of rising real wages. In these circumstances there is a rise in government tax revenues as a proportion of *gross domestic product. The most widely advocated remedy for fiscal drag is the *indexation of tax thresholds.

**fiscal policy** The use of government spending to influence macroeconomic conditions. Keynes advocated the encouragement of public works in order to create employment during *recessions, arguing that fiscal policy would be more effective than *monetary policy. Fiscal policy was actively pursued to sustain full employment in the post-war years; however, monetarists and others have claimed that this set off the inflation of the 1970s. In fact, they claim that every £1 spent by a government crowds out £1 spent by the private sector, leaving no real effect. Fiscal policy has remained 'tight' in most western countries in the 1980s, with governments actively attempting to reduce the level of public expenditure.

**fiscal year** The year beginning on 6th April in one year and ending on 5th April in the next (the fiscal year 1988/89 runs from 6th April 1988 to 5th April 1989). Income tax, capital-gains tax, and annual allowances for inheritance tax are calculated for fiscal years, and the UK Budget estimates refer to the fiscal year. In the USA it runs from 1st July to the following 30th June. The fiscal year is sometimes called the **tax year** or the **year of assessment.** *Compare* financial year.

**fixed asset** *See* capital asset.

**fixed capital** The amount of capital tied up in the *capital assets of an organization. *Compare* circulating capital.

**fixed capital formation** An investment over a given period, as used in the *national income accounts. It consists primarily of investment in manufacturing and housing. **Gross fixed capital formation** is the total amount of expenditure on investment, while **net fixed capital formation** includes a deduction for the *depreciation of existing capital.

**fixed charge (specific charge)** A *charge in which a creditor has the right to have a specific asset sold and applied to the repayment of his debt if the debtor defaults on any payments. The debtor is not at liberty to deal with the asset without the charge-holder's consent. *Compare* floating charge.

**fixed costs** *See* overhead costs.

**fixed debenture** A *debenture that has a *fixed charge as security. *Compare* floating debenture.

**fixed disk** A computer memory device in which *magnetic disks are permanently attached to the computer. In practice the term is almost synonymous with *hard disk. *Floppy disks are never fixed disks.

**fixed exchange rate** A *rate of exchange between one currency and another that is fixed by government and maintained by that government buying or selling its currency to support or depress its currency. *Compare* floating exchange rate.

**fixed-interest security** A type of *security that gives a fixed stated interest payment once or twice per annum. They include *gilt-edged securities, *bonds, *preference shares, and *debentures; as they entail less risk than *equities they offer less scope for capital appreciation. They do, however, often give a better *yield than equities.

The prices of fixed-interest securities tends to move inversely with the general level of interest rates, reflecting changes in the value of their fixed yield relative to the market. Fixed-interest securities tend to be particularly poor investments at times of high inflation as their value does not adjust to changes in the price level. To overcome this problem some gilts now give index-linked interest payments.

**fixed spot** A television advertising spot for which a premium is paid (normally 15%) to ensure that it is transmitted in a preselected commercial break during a particular programme.

**fixtures and fittings** Items normally forming part of the setting in which an organization conducts its business, as distinct from the *plant and machinery it uses in conducting the business.

**flag of convenience** The national flag of a small country, such as Panama, Liberia, Honduras, or Costa Rica, flown by a ship that is registered in one of these countries, although the ship is owned by a national of another country. The practice of registering ships in these countries, and sailing them under flags of convenience, grew up in the post-war years and still continues. The object for the shipowners is to avoid taxation and the more stringent safety and humanitarian conditions imposed on ships and their crews by the larger sea-going nations.

**flat yield** *See* yield.

**fleet rating** A single special premium rate quoted by an insurer for covering the insurance on a number of ships or vehicles owned by one person or company, rather than considering each one individually. The fleet need not consist of identical vehicles or vessels but common ownership is essential. A common method of fleet rating is to examine the claims history of the fleet against the total premium. In this way one fleet member may have several claims (which would individually merit a premium adjustment) but if the rest of the fleet is claim-free no adjustment need be made. Insurers vary on the minimum number constituting a fleet.

**flexitime (flexihours)** A system of working, especially in offices, in which employees are given a degree of flexibility in the hours they work. Provided they work an agreed number of hours per day, they may start or finish work at different times. The object is usually to reduce time spent in rush-hour travelling.

**floating charge** A *charge over the assets of a company; it is not a legal charge over its fixed assets but floats over the charged assets until crystallized by some predetermined event. For example, a floating charge may be created over all the assets of a company, including its trading stock. The assets may be freely dealt with until a crystallizing event occurs, such as the company going into liquidation. Thereafter no further dealing may take place but the debt may be satisfied from the charged assets. Such a charge ranks in priority after legal charges and after preferred creditors in the event of a winding-up. It must be registered (*see* register of charges).

**floating debenture** A *debenture that has a *floating charge as security. *Compare* fixed debenture.

**floating debt** The part of the *national debt that consists primarily of short-term *Treasury bills. *See also* funding operations.

**floating exchange rate** A *rate of exchange between one currency and others that is permitted to float according to market forces. *Compare* fixed exchange rate.

**floating policy** An insurance policy that has only one sum insured although it may cover many items. No division of the total is shown on the policy and the policyholder is often able to add or remove items from the cover without reference to the insurers, provided that the total sum insured is not exceeded. Cover for contractors' plant and machinery is often arranged on this basis, because it enables them to purchase specialist equipment for a particular contract, without having to contact the insurers on every occasion.

**floating-rate interest** An *interest rate on certain bonds, certificates of deposit, etc., that changes with the market rate in a predetermined manner, usually in relation to the *base rate.

**floating-rate note** *See* eurobond.

**floating warranty** A guarantee given by one person to another that induces this other person to enter into a contract with a third party. For example, a car dealer may induce a customer to enter into a hire-purchase contract with a finance company. If the car does not comply with the dealer's guarantee, the customer may recover damages from the dealer, on the basis of the hire-purchase contract, even though the dealer is not a party to that contract.

**floor 1.** The room in a stock exchange, commodity exchange, Lloyd's, etc., in which dealing takes place. Dealings are invariably restricted to *floor traders. **2.** *See* collar.

**floor trader** A member of a stock exchange, commodity market, Lloyd's, etc., who is permitted to enter the dealing room of these institutions and deal with other traders, brokers, underwriters, etc. Each institution has its own rules of exclusivity, but in many, computer dealing is replacing face-to-face floor trading.

**floppy disk** A removable flexible disk used with computers to store information. The disk is made of plastic coated with magnetic material. When in use the disk is inserted into a *disk drive unit. Floppy disks range in diameter from three to eight inches. They are slower and less reliable than *hard disks, but are much cheaper. They are used mainly in small microcomputer systems, for *back-up copies, and for transferring programs or information between computers. *See also* magnetic disk.

**flotation** The initial sale of a private (or government-owned) company's stock to the general public, usually by means of an *introduction, *issue by tender, *offer for sale, *placing, or *public issue. Once flotation has taken place, the shares can be traded on a *stock exchange. The purpose of a flotation is to raise new capital for the business or for the owners to realize their investment. In the UK, a flotation may be made on the *main market of the *London Stock Exchange, on the *unlisted securities market, the *over-the-counter market, or the *third market.

**flotsam** **1.** Items from a cargo or ship that is itself floating on the sea after a shipwreck. In British waters, it can be claimed by the ship's owners for 366 days after the shipwreck; thereafter it belongs to the Crown. *Compare* jetsam. **2.** In *marine insurance, items or rights that are lost as a result of breach of *warranty. In such cases the policyholder has no right to make an insurance claim.

**flowchart** A diagram representing the sequence of logical steps required to solve a problem. It is a useful tool for the computer programmer, being used to plan a program. There are a number of conventional symbols used in flowcharts. The important ones are the process box, which indicates a process taking place, and the decision lozenge, which indicates where a decision is needed.

**flurry** A burst of activity on a speculative market, especially on a financial market.

**flying picket** *See* picketing.

**flyposting** *See* poster.

**FMCG** Abbreviation for *fast-moving consumer goods.

**FOB** Abbreviation for free on board. This is the basis of an export contract in which the seller pays for sending the goods to the port of shipment and loading them on to the ship or aircraft. The seller also pays for the insurance up to this point. Thereafter the transport and insurance charges have to be paid by the buyer. If the goods are to travel by rail rather than by ship, the equivalent terms are **FOR (free on rail)**.

**FoC** Abbreviation for father of chapel. This is the person in charge of a trade-union chapel, usually a unit of the union within a particular printing or publishing firm.

**Food and Agricultural Organization (FAO)** A specialized agency of the United Nations with its headquarters in Rome. It is responsible for organizing world agriculture, especially in developing countries, to improve nutrition and avoid famine.

**FOOTSIE** *See* Financial Times Share Indexes.

**FOQ** Abbreviation for free on quay. *See* free alongside ship.

**FOR** Abbreviation for free on rail. *See* FOB.

**forced sale** A sale that has to take place because it has been ordered by a court or because it is necessary to raise funds to avoid bankruptcy or liquidation.

**forced saving** A government measure imposed on an economy with a view to increasing savings and reducing expenditure on *consumer goods. It is usually implemented by raising taxes, increasing interest rates, or raising prices.

**force majeure** (French: superior force) An event outside the control of either party to a contract (such as a strike, riot, war, act of God) that may excuse either party from fulfilling his contractual obligations in certain circumstances, provided that the contract contains a force majeure clause. If one party invokes the force majeure clause the other may either accept that it is applicable or challenge the interpretation. In the latter case an *arbitration would be involved.

**foreclosure** The legal right of a lender of money if the borrower fails to repay the money or part of it on the due date. The lender must apply to a court to be permitted to sell the property that has been held as security for the debt. The court will order a new date for payment in an order called a foreclosure nisi. If the borrower again fails to pay, the lender may sell the property. This procedure can occur when a mortgagor fails to pay the mortgagee (bank, building society, etc.) his mortgage instalments, in which the security is the house in which the mortgagor lives. The bank, etc., then forecloses the mortgage, dispossessing the mortgagor.

**foreign bill** *See* inland bill.

**foreign exchange** The currencies of foreign countries. Foreign exchange is bought and sold in *foreign exchange markets. Firms or organizations require foreign exchange to purchase goods from abroad or for purposes of investment or speculation.

**foreign-exchange broker** A *broker who specializes in arranging deals in foreign currencies on the *foreign-exchange markets. Most transactions are between commercial banks and governments. Foreign-exchange brokers do not normally deal direct with the public or with firms who require foreign currencies for buying goods abroad (who buy from commercial banks). Foreign-exchange brokers earn their living from the *brokerage paid on each deal.

**foreign-exchange dealer** A person who buys and sells *foreign exchange on a *foreign-exchange market, usually as an employee of a *commercial bank. Banks charge fees or commissions for buying and selling foreign exchange on behalf of their customers; dealers may also be authorized to speculate in forward exchange rates.

**foreign-exchange market** An international market in which foreign currencies are traded. It consists primarily of foreign-exchange dealers employed by *commercial banks (acting as principals) and *foreign-exchange brokers (acting as intermediaries). Although tight *exchange controls have been abandoned by many governments, including the UK government, the market is not entirely free in that the market is to some extent manipulated by the Bank of England on behalf of the government, usually by means of the *exchange equalization account. Currency dealing has a *spot currency market for delivery of foreign exchange immediately and a forward-exchange market (*see* forward-exchange contract) in which transactions are made for foreign currencies to be delivered at agreed dates in the future. This enables dealers, and their customers who require foreign exchange in the future, to hedge their purchases and sales. *Options on future exchange rates can also be traded in London through the *London International Financial Futures Exchange.

**foreign investment** Investment in the domestic economy by foreign individuals or companies. Foreign investment takes the form of either direct investment in productive enterprises or investment in financial instruments, such as a portfolio of shares. Countries receiving foreign investment tend to have a mixed attitude towards it. While the creation of jobs and wealth is welcome, there is frequently antagonism on the grounds that the country is being 'bought by foreigners'. Similarly, there is frequently resentment in the country whose nationals invest overseas, especially if there is domestic unemployment. Nevertheless, foreign investment is increasingly important in the economy of the modern world.

**foreign sector** The part of a country's economy that is concerned with external trade (imports and exports) and capital flows (inward and outward).

**foreign trade multiplier** The effect that an increase in home demand has on a country's foreign trade. The primary effect is to increase that country's imports of raw materials. A secondary effect may be an increase in exports because the increased home demand enables manufacturers to be more competitive and also because the countries supplying the increased quantities of imports have more foreign exchange available to increase *their* imports.

**foreign trade zone** *See* free port.

**forfeited share** A partly paid share in a company that the shareholder has to forfeit because he has failed to pay a subsequent part- or final payment.

**forgery** The legal offence of making a false instrument in order that it may be accepted as genuine, thereby causing harm to others. Under the Forgery and Counterfeiting Act (1981), an instrument may be a document or any device (e.g. magnetic tape) on which information is recorded. An instrument is considered false, for example, if it purports to have been made or altered by someone who did not do so, on a date or at a place when it was not, or by someone who does not exist. It is also an offence to copy or use a false instrument knowing it to be false or to make or possess any material meant to be used to produce such specified false instruments as money, cheques, share certificates, cheque cards, credit cards, passports, or registration certificates.

**formation of a company** *See* company formation.

**for the account** Denoting a transaction on the *London Stock Exchange, either a purchase or a sale, that the investor intends to close out with an equivalent sale or purchase within the same *account. In dealing for the account the investor will only be called upon for any losses he makes and not for the total cost of a purchase.

**FORTRAN** Acronym for FORmula TRANslation, a computer programming language. The first *high-level language to gain widespread acceptance for mathematical and scientific work, FORTRAN has been dominant in this field since the 1950s. A vast amount of software has been written in it, forming an investment that ensures its continued popularity despite advances in programming language design.

**forward-dated** *See* post-date.

**forward dealing** Dealing in commodities, securities, currencies, freight, etc., for delivery at some future date at a price agreed at the time the contract (called a **forward contract**) is made. This form of trading enables dealers and manufacturers to cover their future requirements by *hedging their more immediate purchases. *See* futures contract.

**forward delivery** Terms of a contract in which goods are purchased for delivery at some time in the future (*compare* spot goods). Commodities may be sold for forward delivery up to one year or more ahead, often involving shipment from their port of origin.

**forward-exchange contract** An agreement to purchase *foreign exchange at a specified date in the future at an agreed exchange rate. In international trade, with floating rates of exchange, the forward-exchange market provides an important way of eliminating risk on future transactions that will require foreign exchange. The buyer on the forward market gains by the certainty such a contract can bring; the seller, by buying and selling exchanges for future delivery, makes a market and earns his living partly from the profit he makes by selling at a higher price than that at which he buys and partly by speculation. There is also an active *option market in forward foreign-exchange rates. *See* foreign-exchange market.

**forwarding agent** *See* shipping and forwarding agent.

**forward integration** *See* integration.

**foul bill of lading** *See* dirty bill of lading.

**founders' shares** Shares issued to the founders of a company. They often have special rights to dividends. *See also* deferred ordinary share.

**fount (font)** A set of types for all the characters of one size and design of typeface, as used in printing.

**four-plus cover** An advertising campaign in which a particular advertisement is exposed to the public at least four times. It is based on the theory that four exposures is the most likely to persuade a person to buy a particular product.

**fourth-generation computer** The latest type of computer in use, incorporating VLSI (very large-scale integration) silicon chips that contain thousands of transistors on each chip. A microprocessor (a computer on a single chip) is used in small fourth-generation machines, called microcomputers. Fourth-generation computers are characterized by reliable technology, sophisticated programming, the use of laser printers, and peripherals for special applications, such as point-of-sale terminals. *Compare* fifth-generation computer.

**FPA** Abbreviation for free of particular average. *See* average.

**fractional banking** A banking practice that some governments impose on their banks, calling for a fixed fraction between cash reserves and total liabilities.

**franchise 1.** A licence given to a manufacturer, distributor, trader, etc., to enable them to manufacture or sell a named product or service in a particular area for a stated period. The holder of the licence (**franchisee**) usually pays the grantor of the licence (**franchisor**) a royalty on sales, often with a lump sum as an advance against royalties. The franchisor may also supply the franchisee with finance and technical expertise. **2.** A clause in an insurance policy, often a *marine-insurance policy, that excludes the payment of claims up to a specified level but agrees to pay in full all claims above it. For example, a £25 franchise would mean that insurers would not deal with claims below £25 but a claim for £26 would be met in full. Franchise clauses are used by insurers to avoid administratively expensive small claims, while still providing the policyholder with full cover for larger claims. *See also* excess policy.

**franco** (French: free) Export delivery terms in which the seller of goods pays for all transport and insurance to the buyer's warehouse, even when this is in a foreign country. In some countries the term **rendu** is used.

**franked investment income** Dividends and other distributions from UK companies that are received by other companies. The principle of the *imputation system of taxation is that once one company has paid corporation tax, any dividends it pays can pass through any number of other companies without carrying a further corporation-tax charge, hence the term 'franked'. Thus franked investment is exempt from corporation tax in the recipient company; moreover, the amount of tax credit included in the franked investment income can reduce the amount of *advance corporation tax that the recipient company has to pay on its own dividends.

**franked payment** A dividend or other distribution from a UK company together with the amount of *advance corporation tax (ACT) attributable to the dividend. Thus, if the basic rate of income tax is 30%, a franked payment is the dividend actually paid plus 3/7 of it, i.e. it is a grossed-up dividend. In any accounting period ACT is actually paid on franked payments less *franked investment income at the basic rate of income tax.

**franking machine** A machine supplied by the Post Office to businesses who wish to frank their own mail. Instead of taking mail to a post office and using gummed stamps, the business uses the machine to produce slips of paper that state the postage and the date. The machine records the amount of money spent, which has to be remitted to the Post Office.

**fraud** A false representation by means of a statement or conduct, in order to gain a material advantage. A contract obtained by fraud is voidable on the grounds of fraudulent *misrepresentation. If a person uses fraud to induce someone to part with money he would not otherwise have parted with, this may amount to theft. *See also* fraudulent conveyance; fraudulent preference; fraudulent trading.

**fraudulent conveyance** The transfer of property to another person for the purposes of putting it beyond the reach of creditors. For example, if A transfers his house into the name of his wife because he realizes that his business is about to become insolvent, the transaction may be set aside by the court under the provisions of the Insolvency Act (1986).

**fraudulent preference** Putting a creditor of a company into a better position than he would have been in, at a time when the company was unable to pay its debts. If this occurs because of an act of the company within six months of winding-up (or two years if the preference is given to a person connected with the company), an application to the court may be made to cancel the transaction. The court may make any order that it thinks fit, but no order may prejudice the rights of a third party who has acquired property for value without notice of the preference.

**fraudulent trading** The carrying on of the business of a company with intent to defraud creditors or for any other fraudulent purpose. This includes accepting money from customers when the company is unable to pay its debts and cannot meet its obligations under the contract. The liquidator of a company may apply to the court for an order against any person who has been a party to fraudulent trading to make such contributions to the assets of the company as the court thinks fit. Thus an officer of the company may be made personally liable for some of its debts. 'Fraudulent' in this context implies actual dishonesty or real moral blame; this definition has limited the usefulness of the remedy as fraud is notoriously difficult to prove. *See* wrongful trading.

**free alongside ship (fas)** The basis of an export contract in which the seller pays for sending the goods to the port of shipment but not for loading them onto the ship. He must cover all insurance up to this point and any *lighterage required. This may be the same as **free on quay (FOQ)** or **free alongside quay (faq)** as long as the ship can reach the dock. If it is unable to, the buyer has to pay for lighterage.

**free capital** Capital in the form of cash. *See also* liquid assets.

**free competition (free economy)** An economy in which the market forces of supply and demand control prices, incomes, etc., without government interference. In such an economy private enterprise is dominant, the public sector being very small.

**free depreciation** A method of granting tax relief to organizations by allowing them to charge the cost of fixed assets against taxable profits in whatever proportions and over whatever period they choose. This gives businesses considerable flexibility, enabling them to choose the best method of depreciation depending on their anticipated cash flow, profit estimates, and taxation expectations.

**free docks** The basis of an export contract in which the exporter pays to deliver the goods to the shipping dock, but the buyer arranges and pays for loading and shipping.

**free economy** *See* free competition.

**free enterprise** *See* private enterprise.

**Freefone** A telephone service provided to businesses by British Telecom in which the business pays for incoming calls. A customer dials the operator and asks to be connected to a specified Freefone number knowing that he will not have to pay for the call. This service is much used by *mail-order houses.

**free good** *See* good.

**freehold** An *estate in land that is now usually held in *fee simple. Land that is not freehold will be *leasehold.

**free in and out** Denoting a selling price that includes all costs of loading goods (into a container, road vehicle, ship, etc.) and unloading them (out of the transport).

**free issue** *See* scrip issue.

**free list** A list, issued by the Customs and Excise, of the goods that can be imported into the UK without paying duty.

**free lunch** Economists' jargon for a nonexistent benefit. It derives from a 19th century tavern that advertised free food, it being clearly understood that anyone attempting to exploit this offer without buying a drink would be thrown out. The phrase is now used to reflect the economist's belief that wherever there appears to be a free benefit, someone, somewhere, always pays for it.

**free market 1.** A market that is free from government interference, prices rising and falling in accordance with supply and demand. **2.** A security that is widely traded on a stock exchange, there being sufficient stock on offer for the price to be uninfluenced by availability. **3.** A foreign-exchange market that is free from pegging of rates by governments, rates being free to rise and fall in accordance with supply and demand.

**free of all averages (faa)** Denoting a marine insurance that covers only a total loss, general *average and particular average losses being excluded.

**free of capture** Denoting a *marine insurance in which capture, seizure, and mutiny are excluded. This usually applies to policies taken out in wartime.

**free of particular average (FPA)** *See* average.

**free on board** *See* FOB.

**free on board and trimmed** Denoting an *FOB contract in the coal trade in which the seller is responsible for delivering the coal to the ship, paying for it to be loaded, and ensuring that it is correctly stowed.

**free on quay (FOQ)** *See* free alongside ship.

**free on rail (FOR)** *See* FOB.

**free overside (free overboard)** *See* ex ship.

**free port** A port, such as Bremerhaven, Gdansk, Rotterdam, or Singapore, that is free of customs duties. The area around the port, known as a **foreign trade zone** or **free zone**, specializes in *entrepôt trade, as goods can be landed and warehoused before re-export without the payment of customs duties.

**Freepost** A UK business reply service provided by the Post Office. Members of the public reply to advertisements and the advertiser pays the cost of postage after the reply is delivered.

**freesheets** Local newspapers and magazines that are published and distributed without charge. As they are entirely dependent upon advertising revenue they must ensure high circulation and low costs to survive. For this reason, the advertising to editorial ratio is exceptionally high.

**free trade** The flow of goods and services across national frontiers without the interference of laws, tariffs, quotas, or other restrictions. Although the mercantilists advocated trade restrictions, since the time of the *physiocrats and Adam Smith (1723–90) most economists have believed that free trade is economically beneficial (*see also* comparative advantage). However, in the political world, free trade has tended to be a fashion that has come and gone according to which party is in power. Before the 19th century most countries restricted the flow of trade across its borders; in the 19th century itself, free trade became quite widely espoused, partly as a result of the work of Adam Smith and David Ricardo (1772–1823). After World War I political pressure in most countries led to major restrictions on trade, although since World War II, under the auspices of GATT, trade barriers have continuously, if slowly, fallen.

**freeze-out** Pressure applied to minority shareholders of a company that has been taken over, to sell their stock to the new owners.

**free zone** *See* free port.

**freight 1.** The transport of goods by sea (**sea freight**) or air (**air freight**). **2.** The goods so transported. **3.** The cost of shipping goods for a particular voyage by sea or by air. Freight is charged on the basis of weight (*see* deadweight cargo) or the volume occupied, described as **weight or measurement**. Usually the rate is quoted per tonne or per cubic metre, the freight paid being whichever is the greater amount. Air freights are usually quoted on the basis of per kilo or per 7000 cubic centimetres, whichever is the greater. Certain cargoes are charged on an *ad valorem basis, expressed as a percentage of the FOB value. Freight is normally paid when the goods are delivered for shipment but in some cases it is paid at the destination (**freight forward**).

**freight forward** Denoting a shipment of goods in which the *freight is payable by the buyer at the port of destination.

**freight insurance** A form of marine or aviation insurance in which a consignor or a consignee covers loss of sums paid in *freight for the transport of goods.

**Freightliner** A door-to-door container service for the transport of goods within the UK, provided by British Rail. The standard containers are owned by BR and delivered by road to the consignor's factory, where they are packed by the consignor. BR collects the container by road, delivers it to a Freightliner terminal, whence it is sent by rail to another terminal near the consignee. BR then delivers the container by road to the consignee.

**freight note** An invoice from a shipowner to a shipper showing the amount of *freight due on a shipment of goods.

**freight release** An endorsement on a *bill of lading made by a shipowner or his agent, stating that *freight has been paid and the goods may be released on arrival. Sometimes the freight release is a separate document providing the same authority.

**frequency** In advertising, the number of times an average person in a target audience is to be exposed to an advertising message within a specified period.

**frictional unemployment** The amount of *unemployment consistent with the efficient operation of the economy. It is therefore equivalent to the *natural rate of unemployment (NAIRU). Frictional unemployment exists because of lags between workers leaving one job and taking up another and because there are times of the year when many new workers (such as school leavers) enter the labour market; in these circumstances there is some delay in finding them all jobs.

**Friendly Society** A UK non-profitmaking association registered as such under the Friendly Society Acts (1896–1955). Mutual insurance societies, dating back to the 17th century, they were widespread in the 19th and 20th centuries, until many closed in 1946 after the introduction of National Insurance. Some developed into trade unions and some large insurance companies are still registered as Friendly Societies. Since 1984 their tax-free investment plans have been restricted by government regulations.

**fringe benefits** **1.** Non-monetary benefits offered to the employees of a company in addition to their wages or salaries. They include company cars, expense accounts, the opportunity to buy company products at reduced prices, private health plans, canteens with subsidized meals, luncheon vouchers, cheap loans, social clubs, etc. Some of these benefits, such as company cars, do not escape the tax net. **2.** Benefits, other than dividends, provided by a company for its shareholders. They include reduced prices for the company's products or services, Christmas gifts, and special travel facilities.

**front door** *See* back door.

**front-end load** The initial charge made by a unit trust, life-assurance company, or other investment fund to pay for administration and commission for any introducing agent. The investment made on behalf of the investor is, therefore, the total of his initial payment less the front-end load.

**front running** The practice by *market makers on the London Stock Exchange of dealing on advance information provided by their brokers and investment analysis department, before their clients have been given the information. *See also* Chinese wall.

**frozen assets** Assets that for one reason or another cannot be used or realized. This may happen when a government refuses to allow certain assets to be exported.

**frustration of contract** The termination of a contract as a result of an unforseen event that makes its performance impossible or illegal. A contract to sell an aircraft could be frustrated if it crashed before the contract was due to be implemented. Similarly an export contract could be frustrated if the importer was in a country that declared war on the country of the exporter.

**FT Cityline** A telephone service giving information on the Financial Times Ordinary Share Index (*see* Financial Times Share Indexes). The information is updated seven times each day.

**FT Share Indexes** *See* Financial Times Share Indexes.

**full-cost pricing** A method of pricing goods that takes account of their full cost, i.e. cost including fixed costs in addition to *direct costs (*compare* marginal cost). Clearly a business must ultimately cover its full costs to be profitable, but there are some circumstances in which goods sold at less than their full cost could make some contribution, whereas if they remain unsold at a higher price they make no contribution.

**full employment** The state in which all the economic resources of a country (and more specifically, its labour force) are fully utilized. Since Keynes, governments have tended to see full employment as the ultimate goal of economic policy, although in recent years there has been some debate on exactly what is meant by full employment and whether or not governments can achieve it. M Friedman, for example, has argued that there is a *natural rate of unemployment that the economy will reach, consistent with a level of *frictional unemployment, which cannot be eliminated. New classical macroeconomists have argued that the government cannot influence the level of unemployment. Post-Keynesians, however, still claim that *involuntary unemployment can be eliminated by government policy. *See also* over-full employment.

**fully paid share capital** *See* share capital.

**functions of money** In economics, money fulfils the functions of acting as a *medium of exchange, a *unit of account, and a *store of value. None of the functions occur in an economy based on barter.

**fund** A reserve of money or investments held for a specific purpose, e.g. to provide a source of pensions (*see* pension funds) or to sell as units (*see* fund manager; unit trust).

**fundamental term** A term in a contract that is of such importance to the contract that to omit it would make the contract useless. Fundamental terms cannot be exempted from a contract, whereas lesser terms, known as conditions and *warranties, can be exempted by mutual consent.

**funded debt** The part of the *national debt that the government is under no obligation to repay by a specified date. This consists mostly of *Consols. *See also* funding operations.

**funding operations 1.** The replacement of short-term fixed-interest debt (*floating debt) by long-term fixed-interest debt (*funded debt). This is normally associated with the government's handling of the National Debt through the operations of the Bank of England. The bank buys Treasury bills and replaces them with an equal amount of longer-term government bonds, thus lengthening the average maturity of government debt. This has the effect of tightening the monetary system, as Treasury bills are regarded by the commercial banks as liquid assets while bonds are not. *See also* overfunding. **2.** A change in the *capital gearing of a company, in which short-term debts, such as overdrafts, are replaced by longer-term debts, such as *debentures.

**fund manager (investment manager)** An employee of one of the larger institutions, such as an insurance company, *investment trust, or *pension fund, who manages its investment fund. The fund manager decides which investments the fund shall hold, in accordance with the specified aims of the fund, e.g. high income, maximum growth, etc.

**fund of funds** A *unit trust belonging to an institution in which most of its funds are invested in a selection of other unit trusts owned by that institution. It is designed to give maximum security to the small investor.

**futures** *See* futures contract.

**Futures and Options Exchange** *See* London FOX.

**futures contract** An agreement to buy or sell a fixed quantity of a particular commodity, currency, or security for delivery at a fixed date in the future at a fixed price. Unlike an *option, a futures contract involves a definite purchase or sale and not an option to buy or sell; it therefore may entail a potentially unlimited loss. However, **futures** provide an opportunity for those who must purchase goods regularly to hedge against changes in price. For *hedging to be possible there must be speculators willing to offer these contracts; in fact trade between speculators usually exceeds the amount of hedging taking place by a considerable amount. In London, futures are traded in a variety of markets. Financial futures are traded on the *London International Financial Futures Exchange; the *Baltic Exchange deals with shipping and agricultural products; *London FOX deals with cocoa, coffee, and other foodstuffs; the *London Metal Exchange with metals; and the International Petroleum Exchange with oil. In these **futures markets**, in many cases actual goods (*see* actuals) do not pass between dealers, a bought contract being cancelled out by an equivalent sale contract, and vice versa; money differences arising as a result are usually settled through a *clearing house (*see also* International Commodities Clearing House). In some futures markets only brokers are allowed to trade; in others, both dealers and brokers are permitted to do so. *See also* forward-exchange contract.

**futures market** *See* futures contract.

**future value** The value that a sum of money (the **present value**) invested at compound interest will have in the future. If the future value is $F$, and the present value is $P$, at an annual rate of interest $r$, compounded annually for $n$ years, $F = P(1 + r)^n$. Thus a sum with a present value of £1000 will have a future value of £1973.82 at 12% p.a., after 6 years.

# G

**G3; G5; G7; G10; G77** Abbreviations for *Group of Three, *Group of Five, *Group of Seven, etc.

**GA** Abbreviation for general average. *See* average.

**GAFTA** Abbreviation for *Grain and Free Trade Association.

**galley (galley proof)** A set of proofs of printed matter set on long sheets of paper before it has been broken down into separate pages. After the galleys have been read and corrected the text is set on individual pages and the *page proofs are then read. In automatic typesetting, text is very often set directly into pages and the galley stage is omitted.

**galloping inflation** *See* hyperinflation.

**game theory** A mathematical theory, developed by J von Neumann (1903–57) and O Morgenstern (1902– ) in 1944, concerned with predicting the outcome of games of strategy (rather than games of chance) in which the participants have incomplete information about the others' intentions. Under *perfect competition there is no scope for game theory, as individual actions are assumed not to influence others significantly; under *oligopoly, however, this is not the case. Game theory has been increasingly applied to economics in recent years, particularly in the theory of industrial organizations.

**gaming contract** A contract involving the playing of a game of chance by any number of people for money. A **wagering contract** involves only two people. In general, both gaming contracts and wagering contracts are solid and no action can be brought to recover money paid or won under them.

**gamma stocks** *See* alpha stocks.

**gap analysis** A methodical tabulation of all the known requirements of consumers in a particular category of products, together with a cross-listing of all the features provided by existing products to satisfy these requirements. Such a chart shows up any gaps that exist and therefore provides a pointer to any new products that could supply an unfulfilled demand.

**garnishee order** An order made by a judge on behalf of a *judgment creditor restraining a third party (often a bank), called a **garnishee**, from paying money to the judgment debtor until sanctioned to do so by the court. The order may also specify that the garnishee must pay a stated sum to the judgment creditor, or to the court, from the funds belonging to the judgment debtor.

**gatekeeper** A manager in a large company who controls the flow of information. It is the gatekeeper who decides what information shall be passed upwards to a parent company and downwards to a subsidiary.

**GATT** Abbreviation for *General Agreement on Tariffs and Trade.

**gazump** To raise the price of, or accept a higher offer for, land, buildings, etc., on which a sale price has been verbally agreed but before the *exchange of contracts has taken place. The intending purchaser with whom the sale price had been verbally agreed has no remedy, even though he may have incurred expenditure on legal fees, surveys, etc. Proposals to reform the law in this respect have been put forward but not enacted.

**gazunder** To reduce an offer on a house, flat, etc., immediately before exchanging contracts, having previously agreed a higher price. In a market in which house prices are falling, the unscrupulous buyer is aware that the seller will be extremely anxious to sell, having incurred legal expenses and perhaps having bought another property. Gazundering is the dishonest buyer's opportunity in a falling market, as *gazumping was the dishonest seller's opportunity in a rising market.

**GDP** Abbreviation for *gross domestic product.

**GDP deflator** The factor by which the value of GDP (*see* gross domestic product) at current prices must be reduced (deflated) to express GDP in terms of the prices of some base year (e.g. 1980). The GDP deflator is thus a measure of *inflation.

**gearing** *See* capital gearing; degearing; leverage.

**gearing adjustment** An adjustment in *current-cost accounting to allow for the fact that in inflationary times profits may accrue to a company from its fixed-interest capital, so that the whole cost of capital maintenance need not fall on the profits available to the ordinary shareholders.

**General Agreement on Tariffs and Trade (GATT)** A trade treaty that has been in operation since 1948, to which 95 nations are party and a further 28 nations apply its rules de facto; thus some 90% of world trade is governed by GATT regulations. Its objectives are to expand world trade and to provide a permanent forum for international trade problems. Special attention is given to the trade problems of developing countries. The *Tokyo round, concluded in 1979, agreed many tariff reductions, non-tariff measures, and a revised anti-dumping code. The **Uruguay round**, begun in 1986, is conducting negotiations on non-tariff measures, subsidies, safeguards, etc. The GATT office is in Geneva.

**general average** *See* average.

**General Commissioners** A body of local businessmen appointed by the Lord Chancellor to hear appeals against assessments to income tax, corporation tax, and capital-gains tax. They sit with a clerk to give them guidance on legal and technical matters. *Compare* Special Commissioners.

**general crossing** *See* cheque.

**general damages** *See* damages.

**general equilibrium analysis** The simultaneous analysis of the behaviour of all agents in an economy, including all possible feedback effects. The best-known result from this form of analysis was published in 1959 by

G Debreu, based on the work of M E L Walras (1834–1910), which proved that under certain conditions a perfectly competitive economy (*see* perfect competition) will produce a Pareto-optimal outcome (*see* Pareto optimality). This result has been used by neoclassical economists to advocate the extension of competition within and between economies. Extending this work has been very difficult because of the complexity of the analysis. Thus, while a *partial equilibrium analysis is less rigorous, it has often been more productive than general equilibrium analysis.

**general offer** An offer of sale made to the general public rather than to a restricted number of people. For example, an object displayed in a shop window with a price tag is a general offer; the shopowner must sell to anyone willing to pay the price.

**general partner** *See* partnership.

**general strike** *See* strike.

**general theory** The general theory of employment, interest, and money by J M Keynes (1883–1946), which represents the first attempt to distinguish *macroeconomics from *microeconomics; it is arguably the most influential economic text of the 20th century. In it, Keynes attempted to synthesize the behaviour of *consumption, *investment, and the money markets and to provide a rationale for *involuntary unemployment. Published in 1936, during the Depression, the theory offered prescriptions for eliminating unemployment, which were widely accepted and pursued by economists and governments after World War II. However, since the 1970s the new classical macroeconomists have successfully challenged many of Keynes' assumptions, although debate still continues as to what Keynes really meant.

**general union** *See* trade union.

**generic** A product that is not branded (*see* brand; own brand). While identical to its branded equivalent, a generic is packaged more simply and is not promoted by the manufacturer; it is therefore cheaper. Generics are found in such fields as groceries and drugs; for example, Panadol (Trademark) is a branded version of the generic drug paracetamol.

**generic advertising** Advertising a type of product or service rather than a branded product or service. Wool and fabric conditioners are both products that have benefitted from generic advertising.

**generic name** The name of a class or category of products, such as videos or pens. Sometimes the name of an extensively promoted and successful brand comes to be used as a generic name; examples include Biro, Hoover, Levi, and Walkman, which are then often written without an initial capital letter.

**geodemographic segmentation** *See* market segmentation.

**geometric mean** An average obtained by calculating the *n*th root of a set of *n* numbers. For example the geometric mean of 7, 100, and 107 is $\sqrt[3]{74\,900} = 42.15$, which is considerably less than the *arithmetic mean of 71.3.

**Giffen good** A good for which demand falls at the same time as its price falls (and vice versa). This upsets the usual relationship between price and demand (*see* demand theory); it is still hotly debated whether or not such goods exist. The usual example is that of potatoes during the Irish potato famine in the 19th century. At that time, potatoes were the main expenditure of a typical Irish family. As the price of potatoes rose, the demand also rose because no close substitutes were available. This meant that even less was spent on other forms of food, such as meat.

Other examples are hard to find. The phenomenon was named by Sir Robert Giffen (1837–1910).

**gift** The transfer of an asset from one person to another for no consideration. Gifts have importance for tax purposes; if they are sufficiently large they may give rise to charges under *capital-transfer tax or *inheritance tax. *See* inter vivos gifts.

**gift with reservation** A gift in which the donor retains some benefit for himself (e.g. the gift of a house in which the donor continues to reside). In general, the donor is treated as not having parted with the asset until he releases the reservation.

**gilt-edged market makers** *See* gilts primary dealers.

**gilt-edged security (gilt)** A *fixed-interest security or stock issued by the British government in the form of **Exchequer stocks** or **Treasury stocks.** Gilts are among the safest of all investments, as the government is unlikely to default on interest or on principal repayments. They may be irredeemable (*see* Consols) or redeemable. **Redeemable gilts** are classified as: **long-dated gilts** or **longs** (not redeemable for fifteen years or more), **medium-dated gilts** or **mediums** (redeemable in five to fifteen years), or **short-dated gilts** or **shorts** (redeemable in less than five years).

Like most fixed-interest securities, gilts are sensitive not only to interest rates but also inflation rates. This led the government to introduce **index-linked gilts** in the 1970s, with interest payments moving in a specified way relative to inflation.

Most gilts are issued in units of £100. If they pay a high rate of interest (i.e. higher than the current rate) a £100 unit may be worth more than £100 for a period of its life, even though it will only pay £100 on *redemption.

**gilts primary dealers** The 19 *market makers approved and supervised by the Bank of England for dealing directly with the Bank of England in gilt-edged securities. They have to some extent taken over the role of the *government broker.

**Gini coefficient** *See* Lorenz curve.

**giro** A banking arrangement for settling debts that has been used in Europe for many years. In 1968 the Post Office set up the UK **National Girobank** (now Girobank plc) based on a central office in Bootle, Merseyside. Originally a system for settling debts between people who did not have bank accounts, it now offers many of the services provided by *commercial banks, with the advantage that there are many more post offices, at which Girobank services are provided, than there are bank branches. Also the post offices are open for longer hours than banks. Girobank also offers banking services to businesses, including an **automatic debit transfer** system, enabling businesses to collect money from a large number of customers at regular intervals for a small charge.

The **Bank Giro** is a giro system operated in the UK, independently of Girobank, by the clearing banks. It has no central organization, being run by bank branches. The service enables customers to make payments from their accounts by credit transfer to others who may or may not have bank accounts.

**Bancogiro** is a giro system in operation in Europe, enabling customers of the same bank to make payments to each other by immediate book entry.

**globalization** The process that has enabled investment in financial markets to be carried out on an international basis. It has come about as a result of improvements in technology and *deregulation; as a result of globalization, for example, investors in London can buy shares or bonds directly from Japanese brokers in Tokyo rather than passing through intermediaries. *See also* disintermediation.

**global product** A product that is marketed throughout the world with the same *brand name, such as Coca Cola, Guinness, Levi, and McDonalds. The advantage of a global product is that it usually enables an advertisement or image to be used worldwide. However, one drawback is that some advertising slogans do not travel well. For example, 'things come alive with Pepsi', when translated into Chinese, led the populace to believe that their ancestors would be brought back from the dead; the Vauxhall Nova encountered similar problems in Spain, where Nova means 'no go'!

**global search** A wordprocessing operation in which a complete document is searched by the computer, in order to locate every occurrence of a particular word or phrase. In **global search and replace**, the word or phrase is replaced automatically by some other designated word or phrase.

**GmbH** Abbreviation for *Gesellschaft mit beschränkter Haftung*. It appears after the name of a West German company, being equivalent to the British abbreviation Ltd (i.e. denoting a private limited company). *Compare* AG.

**GNP** Abbreviation for *gross national product.

**godfather offer** A *tender offer pitched so high that the management of the target company is unable to discourage shareholders from accepting it.

**going-concern concept** A principle of accounting practice that assumes businesses to be going concerns, unless circumstances indicate otherwise. It assumes that an enterprise will continue in operation for the foreseeable future, i.e. that the accounts assume no intention or necessity to liquidate or significantly curtail the scale of the enterprise's operation. The implications of this principle are that assets are shown at cost, or at cost less depreciation, and not at their break-up values; it also assumes that liabilities applicable only on liquidation are not shown.

**gold card** A *credit card that entitles its holder to various benefits (e.g. an unsecured overdraft, some insurance cover, a higher limit) in addition to those offered to standard card holders.

**gold clause** A clause in a loan agreement between governments stipulating that repayments must be made in the gold equivalent of the currency involved at the time either the agreement or the loan was made. The purpose is to protect the lender against a fall in the borrower's currency, especially in countries suffering high rates of inflation.

**gold coins** Coins made of gold ceased to circulate after World War I. At one time it was illegal to hold more than four post-1837 gold coins and there have been various restrictions on dealing in and exporting gold coins at various times. Since 1979 (Exchange Control, Gold Coins Exemption, Order) gold coins may be imported and exported without restriction, except that gold coins more than 50 years old with a value in excess of £8000 cannot be exported without authorization from the Department of Trade and Industry. *See also* Britannia coins; Krugerrand.

**golden handcuffs** *See* golden hello.

**golden handshake** *See* compensation for loss of office.

**golden hello** A financial incentive paid by a securities firm (e.g. a *market maker) to a newly employed specialist, such as a dealer or investment analyst, who leaves another firm. **Golden handcuffs** are financial inducements used to persuade specialists to stay in a particular firm. Golden hellos and golden handcuffs became popular at the time of *Big Bang, when the major stockbroking firms were attempting to gain a large share of the market by employing the leading specialists.

**golden parachute** A clause in the employment contract of a senior executive in a company that provides for financial and other benefits if the executive is sacked or decides to leave as the result of a takeover or change of ownership.

**golden share** A share in a company that controls at least 51% of the voting rights. A golden share has been retained by the UK government in some *privatization issues to ensure that the company does not fall into foreign or other unacceptable hands.

**gold market** *See* bullion.

**gold pool** An organization of eight countries (Belgium, France, Italy, Netherlands, Switzerland, UK, USA, and West Germany) that between 1961 and 1968 joined together in an attempt to stabilize the price of gold.

**gold standard** A former monetary system in which a country's currency unit was fixed in terms of gold. In this system currency was freely convertible into gold and free import and export of gold was permitted. The UK was on the gold standard from the early 19th century until it finally withdrew in 1931. Most other countries withdrew soon after. *See also* International Monetary Fund.

**gold tranche** *See* reserve tranche.

**good** A *commodity or *service that is regarded by economists as satisfying a human need. An **economic good** is one that is both needed and sufficiently scarce to command a *price. A **free good** is also needed but it is in abundant supply and therefore does not need to be purchased; air is an example. However, a commodity or service that is free but has required an effort to produce or obtain is not a free good in this sense. *See also* capital good; consumer goods.

**Goodhart's Law** The law in economics that any attempt by a government to control an economic variable will distort that variable so as to render the government's control ineffective. Formulated by the monetary economist C A E Goodhart (1936– ), it was first applied to measures of the *money supply, such as M0, M1, and £M3, that monetarist governments (*see* monetarism) attempted to control in order to reduce inflation. Goodhart's reasoning is that once the public become aware of the government's attempt to control an economic variable, it will attempt to evade the control, which will distort the government's plans.

**goods-in-transit insurance** An insurance policy covering property against loss or damage while it is in transit from one place to another or being stored during a journey. Policies often specify the means of transport to be used, which may include the postal service. Goods shipped solely by sea are not covered by this type of policy, in which case a *marine insurance policy would be used.

**goodwill** An *intangible asset that normally represents the excess of the value of a business over the value of its *tangible assets. This excess value is largely attributable to the fact that the business generates profits in excess of the return to be expected from investing a sum equivalent to the value of the tangible assets alone. The goodwill is a saleable asset when a business is sold and is sometimes shown as such in the balance sheet. However, for limited companies, the Companies Act (1981) stipulates that goodwill purchased in this way must be written off by charges to the *profit and loss account over a period not exceeding its economic life. *See* amortization.

**go public** To apply to a stock exchange to become a *public limited company. *See* flotation.

**go slow** A form of *industrial action in which employees work at a deliberately slow pace. *See also* work-to-rule.

**government actuary** A government officer with a small team of *actuaries, with the responsibility of estimating future expenditure on the basis of observed population trends. He provides a consulting service to government departments and to Commonwealth governments, advising on social security schemes and superannuation arrangements and on government supervision of insurance companies and *Friendly Societies.

**government broker** The stockbroker appointed by the government to sell government securities on the London Stock Exchange, under the instructions of the Bank of England. He is also the broker to the National Debt Commissioners (*see* national debt). Until October 1986 (*see* Big Bang) the government broker was traditionally the senior partner of Mullins & Co. Since then, when the Bank of England started its own gilt-edged dealing room, the government broker has been appointed from the gilt-edged division of the Bank of England, although some of his functions are now undertaken by the *gilts primary dealers.

**government security (government stock)** *See* gilt-edged security.

**Government Statistical Service (GSS)** A service of statistical information and advice provided to the UK government by specialist staffs employed in the statistics divisions of individual government departments. The statistics collected are made generally available through such publications as *Social Trends*, published for the Central Statistical Office by HM Stationery Office.

**Grain and Free Trade Association (GAFTA)** A commodity association that controls dealings in grain and other commodities on London's *Baltic Exchange.

**granny bond** An index-linked savings certificate (*see* National Savings). They were formerly only available to persons over retirement age, hence the name.

**grant-in-aid** Any grant from central government to a local authority for particular services, other than the rate-support grant.

**grant of probate** An order from the High Court in the UK authorizing the executors of a will to deal with and distribute the property of the deceased person. If the person died intestate or did not appoint executors, the administrator of the estate has to obtain *letters of administration.

**graphics** Any computer output except *alphanumeric characters, especially the drawings, graphs, and symbols that can be produced on most modern computers. Most business software packages include a graphics program that can produce graphs, histograms, and piecharts. More powerful graphics programs are used for *computer-aided design and animation. There are also hardware graphics aids, such as light pens that allow freehand drawing on the computer screen, and plotters and printers that produce paper versions of designs.

**Green Book** An informal name for the book entitled *Unlisted Securities Market*, issued by the London Stock Exchange. It sets out the terms and conditions for admission to the USM and the subsequent obligations of the companies involved.

**green card system** An international system that is designed to provide proof that a motor insurance policy issued in one country satisfies the insurance regulations of another. The card on which the system is based is carried by motorists travelling abroad. It is green and contains details of the policy cover in the language of every country subscribing to the system. Although the first European Community directive on motor insurance means that green cards are not essential, they are the easiest and most convenient method of proving adequate insurance cover is held. Contrary to popular belief, however, holding a green card does not automatically imply that the full cover of the policy operates in other countries. It serves only as a proof of cover. The representative of the green card system in the UK is the *Motor Insurers Bureau.

**green currencies** The currencies of members of the European (Economic) Community using artificial rates of exchange for the purposes of the *Common Agricultural Policy (CAP). Their object is to protect farm prices in the member countries from the wide variations due to fluctuations in the real rates of exchange. Green currencies used as their bases the European Unit of Account (EUA; this was later replaced by the *European Currency Unit or ECU).

The **green pound** is the popular name for the British pound sterling used as a green currency. It is used to calculate payments due by or to the UK to or from the fund of the CAP, i.e. when the value of the pound sterling differs from the value of the green pound.

**green-field project** A project that starts from scratch, e.g. building a factory on a virgin site in the country.

**greenmail** The purchase of a large block of shares in a company, which are then sold back to the company at a premium over the market price in return for a promise not to launch a bid for the company. This practice is not uncommon in the USA, where companies are much freer than in the UK to buy their own shares. Although the morality of greenmail is dubious, it can be extremely profitable.

**green pound** *See* green currencies.

**green product** A product that its manufacturers claim will not cause damage to the environment, especially a new product or formulation that is being launched to replace one that is said to cause environmental damage.

**Gresham's Law** The law, usually attributed to the Elizabethan financier Thomas Gresham (1519–79), that bad money will drive out good. For example, if there are gold coins in circulation and the government tries to save money by lowering the gold content of a new coinage, the old coins will fall out of circulation, as no one will exchange the good (old) for the bad (new). Instead the good coins will be melted down and sold for their gold value.

**grey knight** In a takeover battle, a counter bidder whose ultimate intentions are undeclared. The original unwelcome bidder is the *black knight, the welcome counter bidder for the target company is the *white knight. The grey knight is an ambiguous intervener whose appearance is unwelcome to all.

**grey market 1.** Any market for goods that are in short supply. It differs from a *black market in being legal; a black market is usually not. **2.** A market in shares that have not been issued, although they are due to be issued in a short time. Market makers will often deal with investors or speculators who are willing to trade in anticipation of receiving an allotment of these shares or are willing to cover their deals after flotation. This type of grey market provides an indication of the market price (and premium, if any) after flotation.

**grey wave** A company that is thought to be potentially profitable and ultimately a good investment, but that is unlikely to fulfil expectations in the near future. The fruits of an investment in the present should be available when the investor has grey hair.

**gross domestic product (GDP)** The monetary value of all the goods and services produced by an economy over a specified period. It is measured in three ways:

(1) on the basis of expenditure, i.e. the value of all goods and services bought, including consumption, capital expenditure, increase in the value of stocks, government expenditure, and exports less imports;

(2) on the basis of income, i.e. income arising from employment, self-employment, rent, company profits (public and private), and stock appreciation;

(3) on the basis of the value added by industry, i.e. the value of sales less the costs of raw materials.

In the UK, statistics for GDP are published monthly by the government on all three bases, although there are large discrepancies between each measure. Economists are usually interested in the real rate of change of GDP to measure the performance of an economy, rather than the absolute level of GDP. *See also* GDP deflator; gross national product (GNP); net national product (NNP).

**gross income 1.** The income of a person or an organization before the deduction of the expenses incurred in earning it. **2.** Income that is liable to tax but from which the tax has not been deducted. For many types of income, tax may be deducted at source (*see* deductions at source) leaving the taxpayer with a net amount.

**grossing up** Converting a net return into the equivalent gross amount. For example, if a net dividend ($d$) has been paid after deduction of $r$% tax, the gross equivalent ($g$) is given by: $g = 100d/(100 - r)$.

**gross interest** The amount of interest applicable to a particular loan or deposit before tax is deducted. Interest rates may be quoted gross (as they are on government securities) or net (as in most building society deposits or bank deposits). The gross interest less the tax deducted at the basic rate of income tax gives the net interest. Any tax suffered is usually, but not always, available as a credit against tax liabilities.

**gross investment** *See* net investment.

**gross margin** For a retailer, the difference between the retail price of goods offered for sale and the wholesale price at which they are purchased. This is often expressed as a percentage of the retail price, whereas the **mark-up** is often expressed as a fraction of the wholesale (purchase) price. For example a mark-up of ⅓ on a price of £75, will give a retail price of £100, which is a 25% gross margin.

**gross national product (GNP)** The *gross domestic product (GDP) with the addition of interest, profits, and dividends received from abroad by UK residents. The GNP better reflects the welfare of the population in monetary terms, although it is not as accurate a guide as to the productive performance of the economy as the GDP. *See also* net national product (NNP).

**gross profit** The total sales revenue of an organization, less the cost of the goods sold. The cost of the goods sold includes their purchase price and costs of bringing them to a state to be sold but not the costs of distribution, general administration, or finance costs.

**gross receipts** The total amount of money received by a business in a specified period before any deductions for costs, raw materials, taxation, etc. *Compare* net receipts.

**gross tonnage (register tonnage)** *See* tonnage.

**gross weight 1.** The total weight of a package including its contents and the packing. The **net weight** is the weight of the contents only, i.e. after deducting the weight of the packing (often called the **tare**). **2.** The total weight of a vehicle (road or rail) and the goods it is carrying, as shown by a weighbridge. The **net weight** is obtained by deducting the tare of the vehicle.

**gross yield** The *yield on a security calculated before tax is deducted. This yield is often quoted for the purpose of comparison even on ordinary shares, where dividends have tax deducted before they are paid. The yield after tax is called the **net yield**.

**ground rent** Rent payable under a lease that has been granted or assigned for a capital sum (premium). Normally, long leases on offices, flats, etc., are granted for such a premium, payable when the lease is first granted; in addition the leaseholder pays the landlord a relatively small annual ground rent. *See also* rent.

**group accounts** *See* group of companies.

**group discussion** A *marketing research technique that brings between six and eight respondents together for at least an hour to discuss a marketing issue under the guidance of an interviewer. It has been found that, once such a group is relaxed, it will fully explore issues in a manner that shows what is important to it, using its own rather than marketers' terminology. The interviewer, who is a specialist acting on behalf of a client, first draws up a **topic guide** that identifies the points to be explored. A *sample of respondents – who must not know one another – is recruited to match certain criteria (e.g. if the problem concerned tea all respondents would be tea drinkers). The group meets in the interviewer's home or in a hotel; with the discussion being tape-recorded, the interviewer describes the problem and then adopts a passive role, allowing the group to discuss their views and interjecting further questions only if some aspect of the problem is not being explored adequately. After a minimum of four group discussions, the interviewer will have sufficient material to write a report and recommend a particular action. Group discussions are a qualitative procedure (*see* qualitative marketing research) frequently used to determine attitudes to particular products or advertisements.

**group income** A means of enabling dividends and other payments to be made between members of groups of companies in specified circumstances without accounting for *advance corporation tax. The dividends when received are not *franked investment income but are referred to as group income. Like franked investment income they are not liable to corporation tax in the recipient company.

**group life assurance** A life-assurance policy that covers a number of people, usually a group of employees or the members of a particular club or association. Often a single policy is issued and premiums are deducted from salaries or club-membership fees. In return for an agreement that all employees or members join the scheme, insurers are prepared to ask only a few basic questions about the health of a person joining. However, with the advent of AIDS insurers are no longer prepared to waive all health enquiries, as they were a few years ago.

**group of companies** A holding (parent) company together with its subsidiaries. A company is a **subsidiary** of another company if the parent company holds more than half of the nominal value of its *equity capital or holds some shares in it and controls the composition of its board of directors. If one company has subsidiaries, which themselves have subsidiaries, all the companies involved are members of the same group. Groups of companies are required by the UK Companies Act (1985) to file **group accounts**, which normally consist of a consolidated balance sheet and a consolidated profit and loss account for the whole group. In certain circumstances the Department of Trade and Industry will permit groups to publish separate accounts for each subsidiary without consolidation or to omit certain subsidiaries from group accounts. The reasons for doing so, however, have to be bona fide.

**Group of Five (G5)** The five countries France, Japan, UK, USA, and West Germany who have agreed to stabilize their exchange rates by acting together to overcome adverse market forces.

**Group of Seven (G7)** The seven leading industrial nations outside the communist bloc: USA, Japan, West Germany, France, UK, Italy, and Canada. This group evolved from the first economic summit held in 1976 and is now an annual meeting attended by heads of state. The original aim was to discuss economic coordination but the agenda has since broadened to include political issues. However, increasing enthusiasm for international economic cooperation in the 1980s has led to collective action, for example on exchange rates as a result of meetings of G7 finance ministers. The *Group of Ten (G10) and, more recently, the *Group of Five (G5) have also been important forums for economic cooperation.

**Group of Seventy Seven (G77)** The developing countries of the world.

**Group of Ten (G10; The Paris Club)** The ten relatively prosperous industrial nations that agreed in 1962 to lend money to the *International Monetary Fund (IMF). They are Belgium, Canada, France, Italy, Japan, Netherlands, Sweden, West Germany, UK, and USA. They inaugurated *Special Drawing Rights.

**Group of Three (G3)** The three largest western industrialized economies, i.e. the USA, West Germany, and Japan.

**group relief** A means of enabling a company within a group of companies to transfer to another company in the same group the benefit of its own tax losses or certain other payments for which it is unable to obtain tax relief in its own right. This relief is not available for all losses and the requirements for group status are rigidly defined. There are other possible tax reliefs available within groups, e.g. surrender of *advance corporation tax, group roll-overs (*see* roll-over relief), and *group income, although none of these qualify as group relief.

**growth** An increase in the value of an asset. If growth is sought in an investment, it is an increase in its capital value that is required. *See also* growth stocks.

**growth curve** A curve on a graph in which a variable is plotted as a function of time. The curve thus illustrates the growth of the variable. This may be used to show the growth of a population, sales of a product, price of a security, etc.

**growth industry** Any industry that is expected to grow faster than others.

**growth recession** *See* recession.

**growth stocks** Securities that are expected to offer the investor sustained *capital growth. Investors and investment managers often distinguish between growth stocks and income stocks. The former are expected to provide *capital gains; the latter, high income. The investor will usually expect a growth stock to be an ordinary share in a company whose products are selling well and whose sales are expected to expand, whose capital expenditure on new plant and equipment is high, whose earnings are growing, and whose management is strong, resourceful, and investing in product development and long-term research.

**GSS** Abbreviation for *Government Statistical Service.

**guarantee 1.** *See* warranty. **2.** A promise made by a third party (**guarantor**), who is not a party to a contract between two others, that he will be liable if one of the parties fails to fulfil his contractual obligations. For example, a bank may make a loan to a person, provided that a guarantor is prepared to repay the loan if the borrower fails to do so. The banker may require the guarantor to provide some *security to support his guarantee. *See also* bank guarantee.

**guaranteed-income bond** A bond issued by a life-assurance company that guarantees the purchaser a fixed income as well as a guaranteed return of capital at the end of the term. *See also* single-premium assurance.

**guaranteed minimum pension** The earnings-related component of a state pension that a person would have been entitled to as an employee of a company, had he not contracted out of the *State Earnings-Related Pension Scheme (SERPS). Any private pension contract must pay at least the guaranteed minimum pension if it is to be an acceptable replacement of a SERPS pension.

**guaranteed stocks** Stocks issued by UK nationalized industries on which the income is guaranteed by the government.

**guarantor** A person who guarantees to pay a debt incurred by someone else if that person fails to repay it. A person who acts as a guarantor for a bank loan, for example, must repay the loan if the borrower fails to repay it when it becomes due.

**guide price** *See* Common Agricultural Policy.

**guinea** A former British coin, originally made from gold from Guinea. First issued in 1663, it had various values; during most of the 18th century it stood at 21 shillings. Although it was replaced in 1817 by the sovereign, it persisted, not as a coin but as a unit of 21 shillings for charging for some professional services, until decimalization (1971).

# H

**half-commission man** A person who is not a member of a stock exchange but works for a *stockbroker, introducing clients in return for half, or some other agreed share, of the commission.

**half-tone** A photograph printed in a book by means of a process in which light and shade are represented by dots on the printed page. The image is obtained by photographing it through a screen.

**hallmark** A series of marks stamped on articles made of gold, silver, or platinum in the UK to indicate the maker, the hall or assay office making the mark, the quality of the metal, and the date of assay. Each of the seven halls (London, Birmingham, Sheffield, Chester, Glasgow, Edinburgh, and Dublin) have distinguishing marks (e.g. a leopard for London, an anchor for Birmingham). The quality of gold is indicated by a *carat mark (22, 18, 14, and 9 carats; Dublin also has a 12-carat mark). The quality mark for silver in England is a lion (passant) and in Scotland a thistle or lion (rampant). The date is indicated by a letter in a specifically shaped shield. Hallmarks on precious metals are accepted internationally and there are heavy penalties for forging them or selling articles made of these metals (except those exempt) without them.

**hall test** A marketing research test conducted in a room or hall close to a shopping centre. Consumers, selected at random, are invited to come to the hall to participate in the test, which involves answering questions about their buying habits, purchases, etc.

**hammering** An announcement on the *London Stock Exchange that a broker is unable to meet his obligations. It was formerly (until 1970) introduced by three blows of a hammer by a waiter and followed by the broker's name.

**handbill** A form of printed advertisement delivered into the hands of likely customers; handbills may also be attached to their cars or put through their letter boxes.

**Hang Seng Index** An arithmetically weighted index based on the capital value of 33 stocks on the Hong Kong Stock Exchange.

**hard copy** A printed copy of computer output, such as a report or a program listing.

**hard currency** A currency that is commonly accepted throughout the world; they are usually those of the western industrialized countries although other currencies have achieved this status, especially within regional trading blocs. Holdings of hard currency are valued because of their universal purchasing power. Countries with *soft currencies go to great lengths to obtain and maintain stocks of hard currencies, often imposing strict restrictions on their use by the private citizen.

**hard disk** A type of computer memory that uses a rigid magnetic disk enclosed in a sealed container to store information. Almost always non-removable (*see* fixed disk), they are used in various sizes, ranging from packs containing a stack of disks in a plastic hood, to *Winchester disks used on small business computers. Hard disks can hold more information than *floppy disks, are more reliable, and operate more quickly. *See also* magnetic disk.

**hard sell** The use of selling methods that emphasize the supposed virtues of a product, repeat its name, and attempt to force it upon the consumer by forceful and unsubtle means. *Compare* soft sell.

**hardware** The electronic and mechanical parts of a computer system; for example, the central processing unit, disk drive, screen, and printer. *Compare* software.

**harvesting strategy** Making a short-term profit from a particular product shortly before withdrawing it from the market. This is often achieved by reducing the marketing support it enjoys, such as advertising, on the assumption that the effects of earlier advertising will still be felt and the product will continue to sell.

**haulage** The charge made by a **haulier (haulage contractor)** for transporting goods, especially by road. If the goods consist of a large number of packages (e.g. 100 tonnes of cattlefood packed in 2000 bags each weighing 50 kilograms) there will be a separate charge for loading and unloading the vehicle.

**head hunter** *See* employment agency.

**head lease** The main or first *lease, out of which **sub-leases** may be created. For example, if A grants a 99-year lease to B and B then grants a 12-year lease of the same property to

C, the 99-year lease is the head lease and the 12-year lease is a sub-lease.

**Health and Safety Commission** A commission appointed by the Secretary of State for Employment to look after the health, safety, and welfare of people at work; to protect the public from risks arising from work activities; and to control the use and storage of explosives and other dangerous substances. It is composed of representatives from trade unions, employers, and local authorities with a full-time chairman. The **Health and Safety Executive** is a statutory body that advises the Commission and carries out its policies through 20 area offices. It includes HM Factory Inspectorate and a Medical Division, which itself includes the Employment Medical Advisory Service.

**hedging** An operation undertaken by a trader or dealer who wishes to protect an *open position, especially a sale or a purchase of a commodity, currency, security, etc., that is likely to fluctuate in price over the period that the position remains open. For example, a manufacturer may contract to sell all that he can produce of his product for the next six months. If his product depends on a raw material that fluctuates in price, and if he does not have sufficient raw material in stock, he will have an open position. He may decide to hedge his sales by buying the raw material he requires on a *futures contract; if he has to pay for it in a foreign currency he may also hedge his currency needs by also buying that foreign currency forward. Operations of this type do not offer total protection because the prices of *spot goods and futures do not always move together, but it is possible to reduce the vulnerability of an open position substantially by hedging. *See also* option.

**Hedging against inflation** is protecting one's capital against the ravages of inflation by buying *equities or making other investments that are likely to rise with the general level of prices.

**hedonic demand theory** An economic theory that the price an individual will pay for a good reflects the sum of the characteristics of that good. This implies that demand for the characteristics is the true object of *demand theory, not the goods themselves. While this is appealing in principle, it is very difficult to enumerate all the characteristics of goods in practice.

**hereditament** Real property. Originally, it was property that would be inherited by the heir on intestacy. **Corporeal hereditaments** are physical real property, such as land, buildings, trees, and minerals. **Incorporeal hereditaments** are intangible rights, such as easements or profits à prendre, attached to land.

**hexadecimal notation** A number system that uses 16 symbols, the ten digits 0–9 and the letters A–F, to represent numbers. For example, the decimal number 26 is written as 1A in hexadecimal notation. Hexadecimal numbers are widely used in computer programming as they are easier to follow than *binary notation, yet are easy to convert to binary if required. *Compare* octal notation.

**Hicksian (*or* compensated) demand function** A function that expresses an individual's demand for goods by means of prices at a particular level of *utility. Unlike the *Marshallian demand function, a change in the price of a good will have only one effect on the Hicksian demand function; as the level of utility is held constant, this is the *substitution effect. The concept was introduced by Sir John Hicks (1904–   ).

**hidden reserve** Funds held in reserve but not disclosed on the balance sheet (they are also known as **off-balance sheet reserves** or **secret**

**reserves**). They arise when an asset is deliberately either undisclosed or undervalued. Such hidden reserves are permitted for some banking institutions but are not permitted for limited companies as they reduce profits and therefore the corporation tax liability of the company.

**hidden tax** A tax, the *incidence of which may be hidden from the person who is suffering it. An example could be a tax levied on goods at the wholesale level, which increases the retail price in such a way that the final customer cannot detect either that it has happened or the amount of the extra cost. Again, a government might introduce a hidden tax by artificially causing the price of a certain utility to be raised above its usual commercial value.

**hidden unemployment (concealed unemployment)** Unemployment that does not appear in government statistics, which are usually based on figures for those drawing unemployment benefit. In addition to these people there will be some who are doing very little productive work, although they are not claiming benefit. These include workers on short time for whom their employers are expecting to be able to provide work shortly; workers who are not usefully employed, although their employers think they are; and others not claiming benefit for a variety of reasons.

**hierarchy of effects** The steps in the persuasion process that lead a consumer to purchase a particular product. They are awareness, knowledge, liking, preference, conviction, and finally a decision to purchase.

**higher rates** Rates above the *standard (basic) rate in the UK income tax schedule of rates. The higher rate may apply to all income above a specified figure or there may be a series of higher-rate bands, the tax rate in each band being higher than its predecessor. *See also* progressive tax.

**high-involvement product** A product that involves the consumer in taking time and trouble before deciding on a purchase. This will include looking in several catalogues, shops, etc., to compare prices and the products themselves. High-involvement products include such major purchases as cars and houses, as well as hi-fi equipment and many other domestic products. *Compare* low-involvement product.

**high-level language** A type of computer-programming language. High-level languages are designed to reflect the needs of the programmer rather than the capabilities of the computer (*compare* low-level language). They use abstract data and control structures, and symbolic names for variables. There are a large number of high-level languages; *BASIC, *COBOL, *Pascal, *FORTRAN, *Algol, and *C are examples. Programs written in high-level languages must be translated into *machine code before they are run (*see* compiler; interpreter).

**high yielder** A stock or share that gives a high yield but is more speculative than most, i.e. its price may fluctuate.

**hire purchase (HP)** A method of buying goods in which the purchaser takes possession of them as soon as he has paid an initial instalment of the price (a **deposit**) and obtains ownership of the goods when he has paid all the agreed number of subsequent instalments. A **hire-purchase agreement** differs from a **credit-sale agreement** and **sale by instalments** (or a **deferred payment agreement**) because in these transactions ownership passes when the contract is signed. It also differs from a contract of hire, because in this case ownership never passes. Hire-purchase agreements were formerly controlled by government regulations stipulating the minimum deposit and the length of the repayment period. These controls were removed in 1982. Hire-purchase agreements were also formerly controlled by the Hire Purchase Act (1965), but most are now regulated by the Consumer Credit Act (1974). In this Act a hire-purchase agreement is regarded as one in which goods are bailed in return for periodical payments by the bailee; ownership passes to the bailee if he complies with the terms of the agreement and exercises his option to purchase.

A hire-purchase agreement often involves a *finance company as a third party. The seller of the goods sells them outright to the finance company, which enters into a hire-purchase agreement with the hirer.

**histogram** A graph of a frequency distribution constructed with rectangles, in which the areas of the rectangles are proportional to the frequencies.

**historical-cost accounting** The traditional form of accounting, in which assets are shown in balance sheets at their cost to the organization (historical cost), less any appropriate depreciation. In times of high inflation this method tends to overstate profits; as a result other forms of accounting (*see* current-cost accounting; current purchasing power accounting) were used. Most organizations have now reverted to historical-cost accounting.

**HMSO** Abbreviation for Her Majesty's *Stationery Office.

**holder** The person in possession of a *bill of exchange or promissory note. He may be the payee, the endorsee, or the bearer. When value (which includes a past debt or liability) has at any time been given for a bill, the holder is a **holder for value**, as regards the acceptor and all who were parties to the bill before value was given. A **holder in due course** is one who has taken a bill of exchange in good faith and for value, before it was overdue, and without notice of previous dishonour or of any defect in the title of the person who negotiated or transferred the bill. He holds the bill free from any defect of title of prior parties and may enforce payment against all parties liable on the bill.

**holding company (parent company)** A company in a *group of companies that holds shares in other companies (usually, but not necessarily, its subsidiaries).

**holiday and travel insurance** An insurance policy covering a variety of risks for the duration of a person's holiday. Although policies vary, an average policy covers the policyholder's baggage and personal effects for *all risks; compensation for delays while travelling to or from the holiday; refund of deposits lost if the holiday has to be cancelled for any of a number of specific causes; theft or loss of money, traveller's cheques, or credit cards; payment of medical expenses or costs of flying home in the event of illness or injury; and payment of legal compensation for injuring other people or damaging their property because of negligence.

**holistic evaluation** An evaluation of an advertising or marketing campaign as a whole, as distinct from an analysis of the results of its constituent parts.

**home audit** Marketing research conducted in the home by means of panels of householders who keep a diary on a regular basis recording the products they buy and when they buy them.

**home banking** Carrying out normal banking transactions by means of a home computer linked to a bank's computer.

**home service assurance** *See* industrial life assurance.

**horizontal integration** *See* integration.

**horizontal mobility** *See* mobility of labour.

**hostile takeover** A *takeover bid that is opposed by the target company.

**hot-metal typesetting** A method of setting type in which the type itself is cast from molten metal. Systems include the Monotype and Linotype machines. Hot-metal typesetting has largely been replaced by *phototypesetting, which can more easily be computer-controlled.

**hot money 1.** Money that moves at short notice from one financial centre to another in search of the highest short-term interest rates, for the purposes of *arbitrage, or because its owners are apprehensive of some political intervention in the money market, such as a *devaluation. Hot money can influence a country's *balance of payments. **2.** Money that has been acquired dishonestly and must therefore be untraceable.

**house brand** *See* own brand.

**household insurance 1.** An insurance policy that covers the structure of a home (often called **buildings insurance**). **2.** An insurance policy covering the personal goods and effects kept inside a home (**contents policy**). A **comprehensive householder's policy** will cover both the buildings and the contents.

**house journal** A journal produced by a large organization for the benefit of its employees. It usually seeks to keep all levels of employees informed of the general policy of the company, to function as a forum for employees, and often to provide social news. It often helps to create a corporate identity, especilly if employees are widely distributed geographically.

**HP** Abbreviation for *hire purchase.

**hull insurance** *Marine or aircraft insurance covering the structure of a ship, boat, hovercraft, or aeroplane and the equipment maintained permanently on board.

**human capital** The skills, general or specific, acquired by an individual in the course of training and work experience. The concept was introduced by Gary Becker in the 1960s in order to point out that wages reflect in part a return on human capital. This theory has been used to explain large variations in wages for apparently similar jobs and why even in a *recession a firm may retain its workers on relatively high wages, in spite of *involuntary unemployment. *See also* efficiency wage theory; implicit contract theory. *Compare* physical capital.

**hundredweight 1.** (Abbreviation: cwt) A former unit of weight used in the UK, equal to 112 lbs; 20 cwt equal 1 long ton (2240 lbs). It is being replaced by metric units; many bags that used to contain 1 cwt of material now contain 50 kilos. 1 cwt = 50.8 kilograms. According to the Weights and Measures Act (1985) it is no longer legal to use the hundredweight in trade. 112 lbs of a product may be sold, but it must be invoiced as 112 lbs, not 1 cwt. **2.** A unit of weight used in the USA, equal to 100 lbs; 20 hundredweights equal 1 short ton (2000 lbs).

**hyperinflation (galloping inflation)** Extremely high *inflation. It is usually associated with inflation rates of some 50% per month, within a period of only a few months, and with social disorder. Examples of hyperinflation occurred in Germany after World War I and in Greece at the end of World War II.

**hypermarket** A very large shop, usually having a selling area of at least 50 000 square feet (4645 square metres), that sells a wide range of products. Able to buy in bulk, hypermarkets are often located adjacent to towns and base their attraction on low prices and to car-owning consumers, who can make many of their purchases in one place where parking facilities are provided. Hypermarkets are becoming increasingly common, especially in grocery and do-it-yourself retailing.

**hypothecation** 1. An authority given to a banker, usually as a **letter of hypothecation**, to enable him to sell goods that have been pledged (*see* pledge) to him as security for a loan. The goods have often been pledged as security in relation to a documentary bill, the banker being entitled to sell the goods if the bill is dishonoured by non-acceptance or non-payment. **2.** A mortgage granted by a ship's master to secure the repayment with interest, on the safe arrival of the ship at her destination, of money borrowed during a voyage as a matter of necessity (e.g. to pay for urgent repairs). The hypothecation of a ship itself, with or without cargo, is called **bottomry** and is effected by a **bottomry bond**; that of its cargo alone is **respondentia** and requires a **respondentia bond**. The bondholder is entitled to a maritime *lien.

**hysteresis** A lag in a variable of a system, with respect to the effect causing the variation. In economics, it refers to the assumption that the present level of an economic variable depends on past levels. For example, when unemployment rises, a new classical economist (*see* new classical macroeconomics) would expect wages to fall and the demand for labour to rise, so that unemployment would quickly disappear. If there is hysteresis, however, this may not occur since workers, once excluded from the labour market, may lose skills and therefore be unable to compete with employed workers to push down wages. Hysteresis may be thought of as a *post-Keynesian theory, which may justify the use of government expenditure to alleviate unemployment (*see also* insider-outsider theory).

# I

**IBA** Abbreviation for *Independent Broadcasting Authority.

**IBRD** Abbreviation for *International Bank for Reconstruction and Development.

**ICC** Abbreviation for *International Chamber of Commerce.

**ICCH** Abbreviation for *International Commodities Clearing House.

**icon** A symbol on a computer screen to represent the action that a program can carry out; for example, a jar of money representing a budgeting program, or a calculator for an arithmetic program.

**ICSA** Abbreviation for *Institute of Chartered Secretaries and Administrators.

**IDA** Abbreviation for *International Development Association.

**IDC** Abbreviation for *industrial development certificate.

**IFC** Abbreviation for *International Finance Corporation.

**illegal contract** A contract prohibited by statute (e.g. one between traders providing for minimum resale prices) or illegal at common law on the grounds of *public policy. An illegal contract is totally void, but neither party (unless innocent of the illegality) can recover any money paid or property transferred under it, according to the maxim *ex turpi causa non oritur actio* (a right of action does not arise out of an evil cause). Related transactions may also be affected. A related transaction between the same parties (e.g. if X gives Y a promissory note for money due from him under an illegal contract) is equally tainted with the illegality and is therefore void. The same is true of a related transaction with a third party (e.g. if Z lends X the money to pay Y) if the original illegality is known to him.

**illegal partnership** A partnership formed for an illegal purpose and therefore disallowed by law. A partnership of more than 20 partners is illegal, except in the case of certain professionals, e.g. accountants, solicitors, and stockbrokers.

**ILO** Abbreviation for *International Labour Organization.

**ILU** Abbreviation for *Institute of London Underwriters.

**image** A composite mental picture formed by people about an organization or its products. One of the functions of advertising and public relations is to create a favourable image or to improve one that is unfavourable.

**IMF** Abbreviation for *International Monetary Fund.

**immediate-access store** A type of computer memory that can supply information with no significant delay. Usually composed of *integrated circuits, it is used for *main store memory.

**immediate annuity** An *annuity contract that begins to make payments as soon as the contract has come into force.

**immediate holding company** A company that has a *controlling interest in another company, even though it is itself controlled by a third company, which is the *holding company of both companies.

**immigrant remittances** Money sent by immigrants from the country in which they work to their families in their native countries. These sums can be a valuable source of foreign exchange for the native countries.

**impact day** The day on which the terms of a *new issue of shares are made public.

**impact printer** *See* printer.

**imperfect competition (imperfect market)** A market in which some of the assumptions of *perfect competition do not apply. For example, there may be incomplete *information about the market or firms may be able to influence the prices of goods sold by their output decisions. These departures from perfect competition may be considered individually or together (*see* monopoly; oligopoly). Using perfect competiton as a benchmark, economists are interested in analysing specific forms of imperfect competition in order to understand inefficiencies in the real world, such as large-scale unemployment and excessive volatility in financial markets.

**Imperial preference** *See* preferential duty.

**Imperial units** A system of units formerly widely used in the UK and the rest of the English-speaking world. It includes the pound (lb), quarter (qt), *hundredweight (cwt), and *ton (ton); the foot (ft), yard (yd), and mile (mi); and the gallon (gal), British thermal unit (btu), etc. These units are being replaced by metric units, especially *SI units, although Imperial units persist in some contexts, e.g. town gas is charged for in cubic feet and quoted for in btu, petrol is sold in litres but quoted for in gallons, beer is sold in pints, etc. *Compare* metric system; US Customary units.

**impersonal account** A ledger *account that does not bear the name of a person. These accounts normally comprise the *nominal accounts, having such names as motor vehicles, heat and light, and stock in trade.

**implicit contract theory** The theory, introduced by several economists in the 1970s, that wage contracts contain an element of insurance for workers. Thus firms provide an implicit contract guaranteeing stable wages and employment in return for lower average pay, much as an insurance company charges a premium. This theory explains why, even in a recession, employers are reluctant to reduce wages creating the possibility of *involuntary unemployment. Unfortunately it has subsequently been shown that overemployment is just as likely an outcome as underemployment as a result of implicit contracts. However, the theory has provided many new insights into the operation of labour markets.

**implied term** A provision of a contract not agreed to by the parties in words but either regarded by the courts as necessary to give effect to their presumed intentions or introduced into the contract by statute (as in the case of contracts for the sale of goods). An implied term may constitute either a *condition of the contract or a *warranty; if it is introduced by statute it often cannot be expressly excluded.

**import deposit** A method of restricting importation to a country, in which the importer is required to deposit a sum of money with the Customs and Excise authorities when goods are imported. The UK used this method in 1968–70, as an alternative to raising import duties, when it had an agreement with GATT not to do so.

**import duty** A tax or *tariff on import goods. Import duties can either be a fixed amount or a percentage of the value of the goods. They have been a major type of barrier used to protect domestic production against foreign competition and they have also been an important source of government revenue, especially in developing countries.

**import entry form** A form completed by a UK importer of goods and submitted to the Customs and Excise for assessment of the *import duty, if any. When passed by the Customs the form functions as a *warrant to permit the goods to be removed from the port of entry.

**import licence** A permit allowing an importer to bring a stated quantity of certain goods into a country. Import licences are needed when *import restrictions include *import quotas, currency restrictions, and prohibition. They also function as a means of exchange control, the licence both permitting importation and allowing the importer to purchase the required foreign currency.

**import quota** An *import restriction imposed on imported goods, to reduce the quantity of certain goods allowed into a country from a particular exporting country, in a stated period. The purpose may be to conserve foreign currency, if there is an unfavourable *balance of payments, or to protect the home market against foreign competition (*see* protectionism). Quotas are usually enforced by means of *import licences.

**import restrictions** Restrictions imposed on goods and services imported into a country, which usually need to be paid for in the currency of the exporting country. This can cause a serious problem to the importing country's *balance of payments, hence the need for restrictions, which include *tariffs, *import quotas, currency restrictions, and prohibition. Prohibition will also apply to preventing the importation of illegal goods (e.g. drugs, arms). The restrictions may also be imposed to protect the home industry against foreign competition (*see* protectionism) or during the course of political bargaining.

**imports** Goods or services purchased from another country. *See* import duty; import quota; import restrictions.

**import surcharge** An extra tax applied by a government on certain imports in addition to the normal *tariff. It is usually applied as a temporary measure by a government that is in difficulty with its *balance of payments, but does not wish to violate its international tariff agreements.

**imprest account** A means of controlling petty-cash expenditure in which a person is given a certain sum of money (float or imprest). When he has spent some of it he provides appropriate vouchers for the amounts spent and is then reimbursed so that the float is restored. Thus at any given time the person should have either vouchers or cash to a total of the amount of his float.

**impulse buying** The buying of a product by a consumer without previous intention and almost always without evaluation of competing brands. Impulse buying is encouraged by such factors as prominent shelf position; manufacturers are therefore keen to persuade retailers to give these advantages to their products.

**imputation system of taxation** A system of *corporation tax in which some, or all, of the corporation tax is treated as a tax credit on account of the income tax payable by the shareholders on their dividends. Such a system was introduced into the UK in 1972 and the imputation works through the system of *advance corporation tax.

**imputed cost** An estimate of the *opportunity cost of making use of a resource that is already owned and so has no formal price. In arriving at the true cost of a process, the imputed cost must be added to the actual outlay. For example, if one wishes to instal a commercial process in one's garage, the true cost is obtained by adding the interest one could earn on the money obtained by selling the garage to the cost of the plant, machinery, raw materials, etc.

**IMRO** Abbreviation for Investment Management Regulatory Organization. *See* Self-Regulatory Organization.

**in bond** Delivery terms for goods that are available for immediate delivery but are held in a *bonded warehouse. The buyer has to pay the cost of any Customs duty due, the cost of loading from the warehouse, and any further on-carriage costs.

**incentive compatibility** The compatibility of an economic rule or mechanism with the incentives of the individuals concerned. That non-incentive compatible rules are usually of limited use has important implications for the operation of government policy. For example, it is of little use for the government to ask individuals how much they value a good provided by the state, such as roads, and use this response as a basis for deciding how much to spend on roads, as all road users have an incentive to overstate their needs. Such a mechanism for determining expenditure on roads is not incentive compatible.

**incestuous share dealing** The buying and selling of shares in companies that belong to the same group, in order to obtain an advantage of some kind, usually a tax advantage. The legality of the transaction will depend on its nature.

**incidence of taxation** The impact of a tax on those who bear its burden, rather than those who pay it. For example, *VAT is paid by traders, but the ultimate burden of it falls on the consumer of the trader's goods or services. Again, a company may pay corporation tax but if it then raises its prices or reduces its employees' wages to recoup the tax it may be said to have shifted the incidence.

**income 1.** Any sum that a person or organization receives either as a reward for effort (e.g. salary or trading profit) or as a return on investments (e.g. rents or interest). From the point of view of taxation, income has to be distinguished from *capital. *See also* income profit. **2.** In economics, the flow of economic *value attributed to an individual or group of individuals over a particular period. The total of income throughout an individual's life, minus expenditure, is equal to that person's *wealth. Note that both income and wealth are generally considered to include more than simply money; for example, benefits in kind, services rendered by governments, and *human capital should all be included.

**income and expenditure account** An account, similar to a *profit and loss account, prepared by an organization whose main purpose is not the generation of profit. It records the income and expenditure of the organization and results in either a surplus of income over expenditure or of expenditure over income.

**income bond 1. (National Savings Income Bond)** A type of bond introduced by the Department for *National Savings in 1982. They offer monthly interest payments on investments between £2000 and £100,000. Interest is taxable but not deducted at source. The bonds have a guaranteed life of 10 years. Indexed income bonds were withdrawn in 1987. **2.** *See* guaranteed-income bond.

**income distribution** The payment by a *unit trust of its half-yearly income to unit holders, in proportion to their holdings. The income distributed is the total income less the manager's service charge and income tax at the standard rate.

**income effect** The change in the overall purchasing power of an individual as the price of a good changes (*compare* substitution effect). For example, if the price of butter falls, a consumer will be able to purchase the same quantity as before and still have some income remaining to spend on other goods. The income effect is usually thought to be negative with respect to the good whose price has changed; implying, in the above example, that expenditure on butter would rise.

**income in kind** *See* benefits in kind.

**income profit** Any sum accruing to a person or organization that represents pure income and is not partly a payment of capital. This term is especially used in a legal sense, when deciding whether annual payments are assessable to Schedule D Case III for income tax purposes. If they are pure income profit they are so assessable; if not there may be a capital element to be disentangled.

**income redistribution** The use of taxation to ensure that the incomes of the richer members of the community are reduced and those of the poorer members are increased. Tax theorists tend to favour ability-to-pay taxation and *progressive taxes for this purpose, combined with various forms of *income support for the poorer members of the community.

**incomes policy** A government policy aimed at controlling inflation and maintaining full employment by holding down increases in wages and prices by statutory or other means. In the 1960s and 1970s, when *Keynesianism, *demand-pull, and *cost-push theories of *inflation were popular, incomes policies were widely pursued in the developed world. This reflected the strong belief that inflation and unemployment were closely connected (*see* Phillips curve), although incomes policies were unpopular with workers, whose wages were held down, often to an extent that caused a fall in the purchasing power of their incomes, as a result of inflation. However, the emergence of *stagflation cast doubts on the benefits of these policies; with the rise of *monetarism and the increasing popularity of laissez-faire government (*see* laissez-faire economy) in the 1980s, incomes policies became much less attractive.

**income stock** A stock or share bought primarily for the steady and relatively high income it can be expected to produce. This may be a fixed-interest *gilt or an ordinary share with a good *yield record.

**income support** A government benefit for those with low incomes in the UK. It is primarily given to those who arc unemployed, over 60, bringing up children alone, or unable to work through disability or through caring for relatives. Those with capital or savings in excess of £6000 are not eligible.

**income tax** A direct tax on income. The principal direct tax in most countries, it is levied on the incomes of either individual taxpayers or households. It lends itself particularly well to levying ability-to-pay taxes and to progressive taxation. Its defects are that it may discourage work and risk-bearing investment. The tax is calculated on the taxpayer's *taxable income, i.e. his gross income less his *income-tax allowances. Below a certain taxable income no tax is payable; above this threshold, tax is payable at the *standard rate of income tax. However, high-income earners pay tax at above the standard rate on part of their incomes. *See also* PAYE.

**income-tax allowances** Statutory sums that may be deducted from gross income in calculating income tax before the scale of rates is applied. These allowances are primarily concerned with the family circumstances of the taxpayer, such as the married-man's allowance, the single-person's allowance, age relief, and dependent-relative relief.

**income tax year** *See* fiscal year.

**income velocity of circulation** *See* velocity of circulation.

**incompatibility** The inability of one computer to handle data and programs produced for a different type of computer. *Compare* compatibility.

**inconvertible paper money** Paper money that is not convertible into gold. Most paper money now falls into this category, although until 1931, in the UK, the Bank of England had an obligation to supply any holder of a bank note with the appropriate quantity of gold.

**incorporated company** *See* company.

**Incorporated Society of British Advertisers Ltd (ISBA)** A society founded in 1900 as the Advertisers Protection Society. It is open to any advertising company and is concerned with all matters relating to advertising.

**Incoterms** A glossary of terms used in international trade published by the *International Chamber of Commerce in Paris. It gives precise definitions to eliminate misunderstandings between traders in different countries.

**increasing capital** An increase in the number or value of the shares in a company to augment its authorized *share capital. If the *articles of association of the company do not permit this to be done (with the agreement of the members), the articles will need to be changed. A company cannot increase its share capital unless authorized to do so by its articles of association.

**increasing returns to scale** *See* returns to scale.

**incremental costs** Costs incurred purely because of the occurrence of a specific event. An incremental cost is similar to a *marginal cost; it tends to be used when the principal significance is the exclusion of all costs except those that would not have been incurred but for the specific project in view.

**indemnity** **1.** An agreement by one party to make good the losses suffered by another, usually by payment of money, repair, replacement, or reinstatement. This is the function of *indemnity insurance. **2.** An undertaking by a bank's client, who has lost a document (such as a share certificate or bill of lading), that the bank will be held harmless against any consequences of the document's absence if it proceeds to service the documents that have not been mislaid. The bank usually requires a *letter of indemnity to make sure that it suffers no loss.

**indemnity insurance** Any insurance designed to compensate a policyholder for a loss suffered, by the payment of money, repair, replacement, or reinstatement. In every case the policyholder is entitled to be put back in the same financial position as he was immediately before the event against which he was insured occurred. There must be no element of profit to the policyholder nor any element of loss. Most – but not all – insurance policies are *indemnity contracts. For example, personal accident and life-assurance policies are not contracts of indemnity as it is impossible to calculate the value of a lost life or limb (as the value of a car or other property can be calculated).

**indenture** **1.** A *deed, especially one creating or transferring an estate in land. It derives its name from the former practice of writing the two parts of a two-part deed on one piece of parchment and separating the two parts by an irregular wavy line. The two parts of the indenture were known to belong together if the indented edges fitted together. A deed with a straight edge was known as a *deed poll. **2.** *See* apprentice.

**Independent Broadcasting Authority (IBA)** The organization formed in 1972 from the Independent Television Authority, itself set up in 1954 to provide television services in addition to those of the BBC. The IBA appoints local radio and ITV programme companies, owns transmitters, supervises the programmes broadcast, and controls advertising on radio and television. Both ITV and local radio are financed mainly by the sale of advertising time.

**independent intermediary** A person who acts as a representative of a prospective policyholder in the arrangement of an insurance or assurance policy. In *life assurance and *pensions it is a person who represents more than one insurer and is legally bound to offer advice to his clients on the type of assurance or investment contracts best suited to their needs. In general insurance the independent intermediary can represent more than six insurers and is responsible for advising clients on policies that best suit their needs. They must, themselves, have *professional-indemnity insurance to cover any errors that they may make. Although, in both cases, intermediaries are the servants of the policyholder (and the insurer is therefore not responsible for their errors), they are paid by the insurer in the form of a commission, being an agreed percentage of the first or renewal premium paid by the policyholder.

**independent retailer** A small *retailer, generally owning between one and nine shops. They are now less important than *multiple shops in many areas of retailing.

**indexation** The policy of connecting such economic variables as wages, taxes, social security payments, or pensions to rises in the general price level (*see* inflation) This policy is often advocated by economists in the belief that it mitigates the effects of inflation. In practice, complete indexation is rarely possible, so that inflation usually leaves somebody worse off (e.g. lenders, savers) and somebody better off (borrowers). *See* Retail Price Index.

**index-linked gilts** *See* gilt-edged security; indexation.

**index-linked savings certificates** *See* indexation; National Savings.

**index number** A number used to represent the changes in a set of values between a *base year and the present. *See* Financial Times Share Indexes; Retail Price Index.

**Index of Industrial Production** A UK index produced by the Central Statistical Office to show changes in the volume of production of the main British industries and of the economy as a whole.

**indicator** A statistical measure representing an economic variable. For example, M3 is one measure of the *money supply. Indicators are useful both in the formulation of economic theory and in assessing the effectiveness of economic policy. However, an indicator is not the same as the underlying variable itself. For example, the *Retail Price Index is a sample of the prices of goods and therefore only approximates to the actual level of prices.

**indifference curve** A curve on a graph representing the amounts of different goods that will yield the same level of utility to an individual. Each indifference curve represents a different level of utility. In *consumer theory, making various assumptions (including the assumption that individuals are rational), it follows that there will be a single combination of goods that will create *maximization of utility for a given level of income (*see* budget constraint). This combination will be on the highest indifference curve of the individual. *Isoquants in *producer theory represent an equivalent concept.

**indirect costs** *See* overhead costs. *Compare* direct costs.

**indirect labour 1.** Employees of a contractor who carry out work for another organization. This organization employs the contractor to supply the labour for a particular task. For example, some councils use a contractor, and indirect labour, to collect refuse. **2.** The part of a work force in an organization that does not directly produce goods or provide the service that the organization sells. Indirect labour includes office staff, cleaners, canteen workers, etc. *Compare* direct labour.

**indirect taxation** Taxation that is intended to be borne by persons or organizations other than those who pay the tax (*compare* direct taxation). The principal indirect tax in the UK is VAT, which is paid by traders as goods or services enter into the chain of production, but which is ultimately borne by the consumer of the goods or services. One of the advantages of indirect taxes is that they can be collected from comparatively few sources while their economic effects can be widespread.

**indirect utility function** A type of *utility function that expresses the preferences of individuals (*see* consumer preference) in terms of the prices of goods and services and the individual's income. *Compare* direct utility function.

**indorsement** See endorsement.

**industrial action** Any form of coordinated action in an **industrial dispute** by employees, with or without the support of a trade union, that seeks to force employers to agree to their demands relating to wages, terms of employment, working conditions, etc. It may take the form of a *go slow, *overtime ban, *work-to-rule, *sit-down strike, or *strike.

**industrial bank** A relatively small *finance house that specializes in *hire purchase, obtaining its own funds by accepting long-term deposits, largely from the general public.

**industrial democracy** Any system in which workers participate in the management of an organization or in sharing its profits. Workers' participation in decision-making can improve *industrial relations and add to workers' job satisfaction and motivation. Various schemes have been tried, including worker directors on the boards of some nationalized industries, workers' councils, and *profit-sharing schemes.

**industrial development certificate (IDC)** A certificate required by an industrial organization wishing to build a new factory in the UK or to extend an existing one. An IDC has to accompany any application for planning permission for industrial property. The certificate is issued by the Department of the Environment.

**industrial disability benefit** Benefits paid by the Department of Social Security in the UK to compensate for disablement resulting from an industrial accident or from certain diseases due to the nature of a person's employment. A weekly payment is made on a scale of disablement ranging from 14% to 100%. No benefit is paid for less than 14% disablement. Claimants have to submit to a medical examination. The payment is made if the claimant is still suffering disability 15 weeks or more after the accident or onset of the disease.

**industrial dispute** *See* industrial action; trade dispute.

**industrial espionage** Spying on a competitor to obtain *trade secrets by dishonest means. The information sought often refers to the development of new products, innovative manufacturing techniques, commissioned market surveys, forthcoming advertising campaigns, research plans, etc. The means used include a wide range of telephone- and computer-tapping devices, infiltration of the competitor's workforce, etc.

**industrial estate** *See* trading estate.

**industrial life assurance** A life-assurance policy, usually for a small amount, the premiums for which are paid on a regular basis (weekly or monthly) and collected by an agent of the assurance company, who calls at the policyholder's home. Records of the premium payments are kept in a book, which – together with the policy document – has to be produced to make a claim. This type of assurance began in industrial areas (hence its name), where small weekly policies were purchased to help pay the funeral expenses of the policyholder. The company official, who calls to collect the premium is called an *agent or, in certain areas, a tally man. This type of insurance is now more widely known as **home service assurance**.

**industrial marketing research (business marketing research)** Any form of *marketing research undertaken among organizations that buy and add value to products and services but are not the final consumers. *Compare* consumer research.

**industrial medicine** The health care provided by employers for their employees. Apart from the provision of first aid and nursing care in larger organizations, its main aims are prophylactic: the prevention of accidents by providing adequate guards for machinery, the prevention of disease by controlling noxious fumes, dusts, etc., and the prevention of stress by providing the best possible working environment. In addition, regular check-ups are often provided to monitor the health of members of staff.

**industrial psychology** The study of behaviour at work and the best ways of minimizing stress in an industrial setting. Consultant industrial psychologists may be called upon to advise management on various aspects of *industrial relations.

**industrial relations** The relationship between the management of an organization and its workforce. If industrial relations are good, the whole workforce will be well motivated to work hard for the benefit of the organization and its customers. The job satisfaction in such an environment will itself provide some of the rewards for achieving good industrial relations. If the industrial relations are bad, both management and workers will find the workplace an uncongenial environment, causing discontent, poor motivation, and a marked tendency to take self-destructive *industrial action.

**industrial tribunal** Any of the bodies established under the UK employment protection legislation to hear disputes between employers and employees or trade unions relating to statutory terms and conditions of employment. For example, the tribunals hear complaints concerning *unfair dismissal, *redundancy, *equal pay, and *maternity rights. Tribunals may also hear complaints from members of trade unions concerning unjustifiable disciplining by their union. The tribunal usually consists of a legally qualified chairman and two independent laymen. It differs from a civil court in that it cannot enforce its awards (this must be done by separate application to a court) and it can conduct its proceedings informally. Strict rules of evidence need not apply and the parties can present their own case or be represented by anyone they wish at their own expense; legal aid is not available. The tribunal has wide powers to declare a party in breach of a contract of employment, to award compensation, and to order the reinstatement or re-engagement of a dismissed employee.

**industrial union** *See* trade union.

**industry 1.** An organized activity in which *capital and *labour are utilized to produce *goods. **2.** The sector of an economy that is concerned with manufacture.

**inertia selling** A form of selling in which goods are sent to a potential customer on a sale-or-return basis, without his prior consent or knowledge. This method is especially prevalent with book, record, and video clubs.

**infant industry** A new industry that may merit some protection against foreign competition in the short term. The argument in favour of providing protection, say in the form of a tariff on imported competitors' goods, is that it would provide a period during which the new industry could streamline its process and make the necessary economies in order for it to become truly competitive with all market producers.

**inferior good** A good for which demand falls in absolute terms as income rises. For example, as a household's income rises it may switch from buying margarine to butter, in which case margarine would be called an inferior good. More precisely, an inferior good is one for which the income *elasticity of demand is less than zero. *See also* Engel curve; luxury good; necessary good; normal good.

**inflation** A persistent rise in the level of prices and wages throughout an economy. If the rises in wages are sufficient to raise production costs, further rises in prices inevitably occur, creating an **inflationary spiral** in which the rate of inflation increases continuously and in some cases alarmingly (*see* hyperinflation). The causes of inflation are not simple and probably cannot be ascribed to any single factor, although according to *monetarism (*see also* quantity theory of money) it is an inevitable consequence of too rapid an increase in the *money supply, a view proposed by the US economist Milton Friedman (1912– ). Monetarists therefore believe that inflation can be restricted by judicious control of the money supply. They also tend to support the view that high unemployment is a deterrent to persistent wage claims (those in work being grateful for being paid at all) and therefore provides a curb on the inflationary spiral. It also reduces the total demand in the economy. Followers of the British economist John Maynard Keynes (1883–1946), on the other hand, believe that factors other than the money supply create inflation. According to this theory, low inflation and low unemployment can both be maintained by a rigidly enforced *incomes policy.

The rate of inflation in an economy is usually measured by means of price indexes, notably the *Retail Price Index. *See also* cost-push inflation; demand-pull inflation; hyperinflation.

**inflation accounting** A method of accounting that, unlike *historical-cost accounting, attempts to take account of the fact that a monetary unit (e.g. the pound sterling) does not have a constant value; because of the effects of inflation, successive accounts expressed in that unit do not necessarily give a fair view of the trend of profits. The principal methods of dealing with inflation have been *current-cost accounting and *current purchasing power accounting.

**inflationary gap 1.** The difference between the total spending in an economy (both private and public) and the total spending that would be needed to maintain full employment. **2.** Government expenditure in excess of taxation income and borrowing from the public. This excess will be financed by increasing the money supply, by printing paper money, or by borrowing from banks.

**inflationary spiral** *See* inflation.

**inflation tax** A tax that does not exist in the UK, although it has been proposed, principally by the Liberal Party, to deprive traders of some of the benefit they obtain by raising prices excessively or agreeing to excessive wage demands in a time of inflation.

**infopreneurial industry** The manufacture and sale of electronic office and factory equipment for the distribution of information. *See* information technology.

**information** In economics, the data that determines the expectation and choices of individuals. In recent years, techniques for analysing economic problems involving incomplete information have been developed. In general, there are two main problems concerning information. First, there is the problem of asymmetric information, in which some individuals involved in market transactions have more information than others (e.g. the second-hand car salesman knows more about the vehicle he is offering than the potential buyer). Secondly, there is the problem of incomplete information. In this case the range of possible outcomes is so great that it is difficult to form any expectations at all. Some examples of information problems are the measurement of *risk in financial markets, the study of incentives in labour and other markets (*see* moral hazard), and the study of *rational expectations.

**information technology** The use of computers and other electronic means to process and distribute information. Information can be transferred between computers using cables, satellite links, or telephone lines. Networks of connected computers can be used to send *electronic mail or to interrogate *databases, using such systems as *Viewdata and *Teletext. These systems also enable *electronic transfer of funds to be made between banks, as well as telebanking and teleshopping from the home. The same technology is used in the entertainment industry to provide cable and satellite television and videotape and laser disk films.

**infrastructure (social overhead capital)** The goods and services, usually requiring substantial *investment, considered essential to the proper functioning of an economy. For example, roads, railways, sewerage, and electricity supply constitute essential elements of a community's infrastructure. Since the infrastructure often possesses many of the characteristics of *public goods, it is often argued that they should be funded, partly if not wholly, by the government by means of taxation.

**inherent vice** A defect or weakness of an item, especially of a cargo, that causes it to suffer some form of damage or destruction without the intervention of an outside cause. For example, certain substances, such as jute, when shipped in bales, can warm up spontaneously, causing damage to the fibre. Damage by this cause is excluded from most cargo insurance policies as an excepted *peril.

**inheritance tax** A tax on amounts inherited, usually in excess of a specified amount that is free of tax (£128,000 in the UK, from 1990). A true inheritance tax would be based on the amounts inherited by an individual or possibly by a household. The inheritance tax introduced in 1986 in the UK is not strictly such a tax since it taxes the total estate left by deceased persons together with *gifts that those persons may have made in a given period before their deaths. The former *capital-transfer tax on lifetime gifts was replaced in 1986 by a tapered charge on gifts made within seven years of death (*see* inter vivos gifts). This charge is added to the cumulative total used to calculate the amount of inheritance tax due.

**initial public offering (IPO)** The US term for a *flotation.

**injunction** An order by a court that a person shall do, or refrain from doing, a particular act. This is an equitable remedy that may be granted by the High Court wherever 'just and convenient'. The county court also has a limited jurisdiction to grant injunctions. An **interlocutory injunction** lasts only until the main action is heard. **Interim injunctions**, lasting a short time only, may be granted on the application of one party without the other being present (an **ex parte interim injunction**) if there is great urgency. **Prohibitory injunctions** forbid the doing of a particular act; **mandatory injunctions** order a person to do some act. Failure to obey an injunction is contempt of court and punishable by a fine or imprisonment.

**inland bill** A *bill of exchange that is both drawn and payable in the UK. Any other bill is classed as a **foreign bill**.

**Inland Revenue** *See* Board of Inland Revenue.

**inland waterways** *See* British Waterways Board.

**innovation** Any new approach to designing, producing, or marketing goods that gives the innovator or his company an advantage over competitors. By means of *patents, a successful innovator can enjoy a temporary monopoly, although eventually competitors will find ways of supplying a valuable market. Some companies rely on bringing out new products based on established demand, while others develop technological innovations that open up new markets.

**input** The process of introducing data or programs into a computer; for example by typing on the computer keyboard. An **input device** feeds data or programs into a computer. The most common is a keyboard; others include a card reader, joystick, mouse, light pen, voice input, optical-character reader, and bar-code reader.

**inputs** The *factors of production (land, labour, capital, and entrepreneurial ability) required by an organization to enable it to provide its outputs (goods or services).

**input tax** The VAT included in the price a trader pays for the goods and services that he acquires for use in his business. This tax can be deducted from his *output tax in arriving at the amount he must account for to the Customs and Excise.

**inquiry test** A method of testing the response to an advertisement or a particular medium by comparing the number of inquiries received as a result of it.

**inscribed stock (registered stock)** Shares in loan stock, for which the names of the holders are kept in a register rather than by the issue of a certificate of ownership. On a transfer, a new name has to be registered, which makes them cumbersome and unpopular in practice.

**inside money** Bank deposits held by individuals that are offset by loans to other individuals. Using tenuous assumptions it can be shown that in a competitive *equilibrium, with inside money a change in the price level has no *wealth effect, since the increase in the value of bank deposits is offset by an increase in the value of loans. Inside money is the basis of the *neutrality of money proposition. *Compare* outside money.

**insider dealing (insider trading)** Dealing in company securities with a view to making a profit or avoiding a loss while in possession of information that, if generally known, would affect their price. Under the Companies Securities (Insider Dealing) Act (1985) those who are or have been connected with a company (e.g. the directors, the company secretary, employees, and professional advisers) are prohibited from such dealing on or, in certain circumstances, off the stock exchange if they acquired the information by virtue of their connection and in confidence. The prohibition extends to certain unconnected persons to whom the information has been conveyed.

**insider-outsider theory** The theory that employed workers (insiders) may be able to maintain high wages even in periods of high unemployment, as they can prevent outsiders from competing for jobs. This may result from the power of unions or from the level of skills and training required before workers become effective. This theory has been used to explain the persistence of high unemployment in the western economies in the 1970s and 1980s (*see also* hysteresis).

**insolvency** The inability to pay one's debts when they fall due. In the case of individuals this may lead to *bankruptcy and in the case of companies to *liquidation. In both of these cases the normal procedure is for a specialist, a trustee in bankruptcy or a liquidator, to be appointed to gather and dispose of the assets of the insolvent and to pay the creditors. Insolvency does not always lead to bankruptcy and liquidation, although it often does. An insolvent person may have valuable assets that are not immediately realizable.

**inspection and investigation of a company** An inquiry into the running of a company made by inspectors appointed by the Department of Trade and Industry. Such an inquiry may be held to supply company members with information or to investigate fraud, unfair prejudice, nominee shareholdings, or *insider dealing. The inspectors' report is usually published.

**Inspector of Taxes** A civil servant responsible to the *Board of Inland Revenue for issuing tax returns and assessments, the conduct of appeals, and agreeing tax liabilities with taxpayers.

**instalment** One of a series of payments, especially when buying goods on *hire purchase, settling a debt, or buying a new issue of shares.

**instant** Of the current month. In commercial English it is common to refer to a date as, for example, the 5th instant, or more usually 5th inst., meaning the 5th of the current month. **Ultimo** is used to refer to the previous month, for example, the 5th ultimo, or more usually the 5th ult., means the 5th of last month. **Proximo** refers to the next month; the 5th proximo, or 5th prox., means the 5th of next month.

**Institute cargo clauses** Clauses issued by the *Institute of London Underwriters that are added to standard *marine insurance policies for cargo to widen or restrict the cover given. Each clause has a wording agreed by a committee of insurance companies and *Lloyd's underwriters. By attaching particular clauses to the policy the insurers are able to create an individual policy to suit the clients' requirements.

**Institute of Actuaries** One of the two professional bodies in the UK to which actuaries belong. To become an actuary it is necessary to qualify as a fellow of one or the other. The roots of the profession go back to 1756, when a Fellow of the Royal Society, James Dodson, produced the first table of premiums for life assurance, after having been turned down for an assurance policy on the grounds of his age. The Institute is in London; the other organization, the **Faculty of Actuaries**, is based in Edinburgh.

**Institute of Chartered Accountants** Any of the three professional accountancy bodies in the UK, the **Institute of Chartered Accountants in England and Wales**, the **Institute of Chartered Accountants of Scotland**, and the **Institute of Chartered Accountants in Ireland**. The institutes are separate but recognize similar codes of practice. The largest is the England and Wales institute, with some 90 000 members, who are identified by the letters ACA or FCA (as are members of the Ireland institute; in Scotland members use the letters CA). The institutes ensure high standards of education and training in accountancy, provide qualification by examination, and supervise professional conduct in the service of clients and of the public. They are members of the Consultative Committee of Accountancy Bodies, whose Accounting Standards Committee is responsible for drafting accounting standards. *See also* chartered accountant.

**Institute of Chartered Secretaries and Administrators (ICSA)** A professional body for secretaries and administrators in the UK. Founded in 1891 and granted a Royal Charter in 1902, the institute represents members' interests to government bodies on such matters as company law; publishes journals, reports, pamphlets, and papers; promotes the professional standing of members; and conducts the education and examination of members.

**Institute of Directors** An institution, founded in 1903 and granted a Royal Charter in 1906, to which any director of a UK company can apply to join. A non-political organization, it aims to further the cause of free enterprise and to assist its members in the functions of leadership throughout industry and commerce.

**Institute of London Underwriters (ILU)** An association of UK insurance companies that cooperate with each other in providing a market for *marine insurance and *aviation insurance. Although Lloyd's underwriters are not members of the institute, the two organizations work closely with each other. The ILU appoints agents to settle claims, provides certificates of insurance for cargo shippers insured by members, and is responsible for drawing up its own insurance contracts for the use of members; it also draws up the *Institute cargo clauses widely used in many marine and aviation policies.

**Institute of Management Services** An institution, founded in 1941, that is the professional and qualifying body for people in management services. Its main aims are to create and maintain professional standards for the practice of management services; to provide a system of qualifying examinations; and to encourage research and development in management services.

**Institute of Marketing** An association founded in 1911 to develop knowledge about marketing, to provide services for members and registered students in the field of marketing, and to promote the principles and practices of marketing throughout industry.

**Institute of Personnel Management (IPM)** A professional organization for personnel managers in the UK, founded in 1913. The Institute, an independent and non-political body, aims to encourage and assist the development of personnel management by promoting investigation and research and establishing standards of qualification and performance.

**Institute of Practitioners in Advertising (IPA)** The professional organization, established in 1917, for UK advertising agencies. Its members handle some 95% of all UK advertising.

**Institute of Public Relations (IPR)** The professional organization, established in 1948, for UK public relations consultancies. It is the largest in Europe.

**Institute of Sales Promotion (ISP)** The professional organization for UK sales-promotion agencies. Its purpose is to raise the status of the role of sales promotion in the UK. The Institute was formed in 1978, following the reorganization of the Sales Promotion Executives Association, itself founded in 1969.

**institutional investor (institution)** A large organization, such as an insurance company, unit trust, bank, trade union, or a pension fund of a large company, that has substantial sums of money to invest on a stock exchange, usually in the UK, in both gilts and equities. Institutions usually employ their own investment analysts and advisors; they are usually able to influence stock exchange sentiment more profoundly than private investors and their policies can often affect share prices. Because institutions can build up significant holdings in companies, they can also influence company policy, usually by making their opinions known at shareholders' meetings, especially during *takeover-bid negotiations.

**instrument** **1.** A tool that is used by a government in achieving its macroeconomic *targets. For example, interest rates and the *money supply may be considered instruments in the pursuit of stable prices, while government expenditure and taxation may be considered instruments in the pursuit of full employment. Debate in macroeconomics centres on which instruments are suitable for pursuing particular targets and whether or not the instruments are effective. *See also* indicator. **2.** A formal legal document. *See also* negotiable instrument.

**insurable interest** The legal right to enter into an insurance contract. A person is said to have an insurable interest if the event insured against could cause that person a financial loss. For example, anyone may insure their own property as they would incur a loss if an item was lost, destroyed, or damaged. If no financial loss would occur, no insurance can be arranged. For example, a person cannot insure his next-door neighbour's property. The limit of an insurable interest is the value of the item concerned, although there is no limit on the amount of life assurance a person can take out on his own life, or that of his spouse, because the financial effects of death cannot be accurately measured.

Insurable interest was made a condition of all insurance by the Life Assurance Act (1774). Without an insurable interest, an insured person is unable to enforce an insurance contract (or life-assurance contract) as it is the insurable interest that distinguishes insurance from a bet or wager.

**insurable risk** *See* risk.

**insurance** A legal contract in which an insurer promises to pay a specified amount to another party, the *insured, if a particular event (known as the *peril), happens and the insured suffers a financial loss as a result. The insured's part of the contract is to promise to pay an amount of money, known as the *premium, either once or at regular intervals. In order for an insurance contract to be valid, the insured must have an *insurable interest. It is usual to use the word 'insurance' to cover events (such as a fire) that may or may not happen, whereas *assurance refers to an event (such as death) that must occur at some time (*see also* life assurance). The main branches of insurance are: *accident insurance, *fire insurance, *holiday and travel insurance, *household insurance, *liability insurance, *lifestock and bloodstock insurance, loss-of-profit insurance (*see* business-interruption policy), *marine insurance, *motor insurance, National Insurance, *pluvial insurance, *private health insurance, and *property insurance. *See also* reinsurance.

**insurance broker** A person who is registered with the *Insurance Brokers Registration Council to offer advice on all insurance matters and arrange cover, on behalf of the client, with an insurer. An insurance broker's income comes from commission paid to him by insurers, usually in the form of an agreed percentage of the first premium or on subsequent premiums. *See also* Lloyd's.

**Insurance Brokers Registration Council** A statutory body established under the Insurance Brokers Registration Act (1977). It is responsible for the registration and training of insurance brokers and for laying down rules relating to such matters as accounting practice, staff qualifications, advertising, and the orderly conduct and discipline of broking businesses.

**Insurance Ombudsman** *See* Financial Ombudsman.

**insurance policy** A document that sets out the terms and conditions of an *insurance contract, stating the benefits payable and the *premium required. *See also* life assurance.

**insurance premium** *See* insurance; premium.

**insurance tied agent** An agent who represents a particular insurance company or companies. In life and pensions insurance, a tied agent represents only one insurer and is only able to advise the public on the policies offered by that one company. In general insurance (motor, household, holiday, etc.), a tied agent represents no more than six insurers, who are jointly responsible for the financial consequences of any failure or mistake the agent makes. In both cases the agent receives a commission for each policy that is sold and a further commission on each subsequent renewal of the policy. The commission is calculated as an agreed percentage of the total premium paid by the policyholder. The distinction between the two forms of insurance tied agent was a consequence of the Financial Services Act (1987) and the General Insurance Selling Code (1989) of the *Association of British Insurers.

**insured** A person or company covered by an *insurance policy. In some policies that cover death, the alternative word *assured may be used for the person who receives the payment in the event of the assured's death.

**intangible asset (invisible asset)** An asset that can neither be seen nor touched. The most common of these are *goodwill, *patents, *trademarks, and *copyrights. Goodwill is probably the most intangible and invisible of all assets as no document provides evidence of its existence and its commercial value is difficult to determine. However, it frequently does have very substantial value as the capitalized value of future profits, not attributable purely to the return on *tangible assets. While goodwill is called either an intangible asset or an invisible asset, such items as insurance policies and less tangible overseas investments are usually called invisible assets. *Compare* fictitious asset.

**integrated circuit** An electronic circuit built on a chip of semiconductor material, usually silicon. Typical chips are 1.5 mm square by 0.2 mm thick. All the circuit components – transistors, diodes, resistors, and capacitors, together with their interconnecting conductors – are formed on the chip at the time of manufacture. The number of components on a chip has increased dramatically since integrated circuits were first introduced; it is now commonplace for chips to contain millions of components. The term small-scale integration (SSI) describes chips holding less than 20 components; medium-scale integration (MSI), 20–200 components; large-scale integration (LSI), 200–1000 components; and very large-scale integration (VLSI), more than 1000 components.

**integrated software** A group of computer programs that are coordinated into a compatible interrelated whole. Wordprocessing (*see* wordprocessor), *database, *spreadsheet, *graphics, and communications programs are integrated into one package, and it is possible to transfer data between the programs. For example, data produced by the spreadsheet can be transferred to the graphics program and output as a chart, and data from the database can be incorporated into a letter produced using the wordprocessor.

**integration** The combination of two or more companies under the same control for their mutual benefit, by reducing competition, saving costs by reducing overheads, capturing a larger market share, pooling technical or financial resources, cooperating on research and development, etc. In **horizontal** (*or* **lateral**) **integration** the businesses carry out the same stage in the production process or produce similar products or services; they are therefore competitors. In a *monopoly, horizontal integration is complete, while in an *oligopoly there is considerable horizontal integration. In **vertical integration** a company obtains control of its suppliers (sometimes called **backward integration**) or of the concerns that buy its products or services (**forward integration**).

**intellectual property** An *intangible asset, such as a *copyright, *patent, or *trademark. *See also* royalty.

**intelligent knowledge-based systems** *See* expert system.

**intelligent terminal** A computer terminal that is able to process data without the help of the computer to which it is connected. An example is a point-of-sale terminal that adds up the cost of the items purchased and also informs the main stock computer that they are no longer in stock.

**inter-account dealing** Speculative transactions on the *London Stock Exchange in which all individual purchases and sales within an *account cancel each other out, so that only differences have to be settled. *See also* cash and new.

**interactive** Denoting a computer system in which the operator can communicate with the computer while it is running a program. The computer program prompts the operator when it needs information and stops running until the information is supplied. Interactive computer systems are also called **conversational systems**. *Compare* batch processing.

**inter-bank market 1.** The part of the London *money market in which banks lend to each other and to other large financial institutions. The *London Inter Bank Offered Rate (LIBOR) is the rate of interest charged on inter-bank loans. **2.** The market between banks in foreign currencies, including spot currencies and forward *options.

**inter-dealer broker** A member of the *London Stock Exchange who is only permitted to deal with *market makers, rather than the public.

**interdict** An *injunction in Scottish law.

**interest** The charge made for borrowing a sum of money. The **rate of interest** is the charge made, expressed as a percentage of the total sum loaned, for a stated period of time (usually one year). Thus, a rate of interest of 15% per annum means that for every £100 borrowed for one year, the borrower has to pay a charge of £15, or a charge in proportion for longer or shorter periods. In **simple interest**, the charge is calculated on the sum loaned only, thus $I = Prt$, where $I$ is the interest, $P$ is the principal sum, $r$ is the rate of interest, and $t$ is the period. In **compound interest**, the charge is calculated on the sum loaned plus any interest that has acrued in previous periods. In this case $I = P\,[(1 + r)^n - 1]$, where $n$ is the number of periods for which interest is separately calculated. Thus, if £500 is loaned for 2 years at a rate of 12% per annum, compounded quarterly, the value of $n$ will be $4 \times 2 = 8$ and the value of $r$ will be $12/4 = 3\%$. Thus, $I = 500\,[(1.03)^8 - 1]$ = £133.38, whereas on a simple-interest basis it would be only £120.

In general, rates of interest depend on the money supply, the demand for loans, government policy, the risk of nonrepayment as assessed by the lender, and the period of the loan. In economics, interest has two functions to perform:

(i) to make the amount saved by households equal the amount that firms wish to borrow for investment;

(ii) to make the amount of credit demanded equal the supply of credit.

The rate of interest that achieves this equilibrium is known as the **natural rate of interest**. First defined by K Wicksell (1851–1926), it implies that an actual interest rate below the natural rate will cause a rise in the prices of consumer goods (*see* time preference), which will fuel inflation and lead to an inadequate rate of savings. The general theory of Maynard Keynes, built around Wicksell's concepts, saw a role for governments in controlling credit by means of restricting the *money supply (*see* interest-rate policy). *See also* loanable-funds theory; term structure of interest rates.

**interest arbitrage** Transactions between financial centres in foreign currencies that take advantage of differentials in interest rates between the two centres and the difference between the forward and spot exchange rates. In some circumstances it is possible to make a profit by buying a foreign currency, changing it into the currency of the home market, lending it for a fixed period, and buying back the foreign currency on a forward basis.

**interest-only yield** *See* yield.

**interest-rate policy** The policy by which governments influence *interest rates through the supply of bonds. Higher rates of interest will discourage *investment and reduce the *demand for money, giving a downward impetus to both employment and prices. Lower interest rates will have the opposite effect. The type of interest-rate policy advocated by an economist or government depends on their economic beliefs: monetarists generally aim to control inflation, while Keynesians are anxious to raise employment. However, new classical macroeconomists have argued that the *neutrality of money combined with *rational expectations restrict interest-rate policy to influencing prices.

**interim accounts** *See* final accounts.

**interim dividend** *See* dividend.

**interim report** Any report other than a final report. In commercial terms, an interim report normally refers to the statements that a company makes halfway through its financial year, setting out the state of its profits for the first half year. It may or may not be used to justify its interim *dividend. An interim report is a requirement for a company seeking a quotation on the London Stock Exchange.

**intermediary** Any person or organization that acts as an *agent or *broker between the parties to a transaction. In the *money markets **intermediation** is the process of borrowing money at one rate of interest and lending it at a higher rate. *See also* disintermediation.

**intermediate good** A *good in the course of production, i.e. neither a raw material nor a finished product.

**internal audit** An *audit that an organization carries out on its own behalf, normally to ensure that its own internal controls are operating satisfactorily. Whereas an external audit is almost always concerned with financial matters, this may not necessarily be the case with an internal audit; internal auditors may also concern themselves with such matters as the observation of the safety and health at work regulations or of the equal opportunities legislation. It may also be used to detect any theft or fraud (*see also* internal control).

**internal control** The measures an organization employs to ensure that opportunities for fraud or misfeasance are minimized. Examples range from requiring more than one signature on certain documents, security arrangements for stock-handling, division of tasks, keeping of *control accounts, use of special passwords, handling of computer files, etc. It is one of the principal concerns of an *internal audit to ensure that internal controls are working properly so that the external auditors can have faith in the accounts produced by the organization. Internal control should also reassure management of the integrity of its operations.

**internal rate of return (IRR)** The rate of return that discounts the *net present value of a project to zero. This is a method used in conjunction with *discounted cash flow in order to discover what rate of return on the outlay is represented by the cash flows from the project. If the IRR exceeds the market rate of interest then the project is profitable. The IRR is usually considered less reliable than the actual net present value as a means of appraising a project. First, if returns in different periods fluctuate from positive to negative, a unique

IRR will not exist. Secondly, the IRR tends to favour projects with large returns early on, even if they have a low net present value. *See also* payback period.

**Internal Revenue Service (IRS)** The US federal organization that assesses and collects all personal and business federal taxes.

**internal storage** *See* main store.

**International Bank for Reconstruction and Development (IBRD)** A specialized agency working in coordination with the United Nations, established in 1945 to help finance post-war reconstruction and to help raise standards of living in developing countries, by making loans to governments or guaranteeing outside loans. It lends on broadly commercial terms, either for specific projects or for more general social purposes; funds are raised on the international capital markets. The Bank and its affiliates, the *International Development Association and the *International Finance Corporation, are often known as the **World Bank**; it is owned by the governments of 151 countries. Members must also be members of the *International Monetary Fund. The headquarters of the Bank are in Washington, with a European office in Paris and a Tokyo office.

**International Chamber of Commerce (ICC)** An international business organization that represents business interests in international affairs. Its office is in Paris. *See also* Incoterms.

**International Commodities Clearing House (ICCH)** A *clearing house established in 1888, as the **London Produce Clearing House**, to provide futures markets with clearing and guaranteeing services. Membership is at the discretion of the directors and rules for admission and codes of conduct are established by the ICCH. The ICCH is an independent body, owned by six major UK commercial banks. Exchanges making use of ICCH facilities include the *London International Financial Futures Exchange, *London FOX, the *London Metal Exchange, the *Baltic Exchange, the *London Traded Options Market, and the *International Petroleum Exchange.

**International Development Association (IDA)** An affiliate of the *International Bank for Reconstruction and Development (IBRD) established in 1960 to provide assistance for poorer developing countries (those with a GNP of less than $835 per head). With the IBRD it is known as the **World Bank**. It is funded by subscription and transfers from the net earnings of IBRD. The headquarters of the IDA are in Washington, with offices in Paris and Tokyo, administered by IBRD staff.

**International Finance Corporation (IFC)** An affiliate of the *International Bank for Reconstruction and Development (IBRD) established in 1956 to provide assistance for private investment projects. Although the IFC and IBRD are separate entities, both legally and financially, the IFC is able to borrow from the IBRD and reloan to private investors. The headquarters of the IFC are in Washington. Its importance increased in the 1980s after the emergence of the debt crisis and the subsequent reliance on the private sector.

**International Fund for Agricultural Development (IFAD)** A fund, proposed by the 1974 World Food Conference, that began operations in 1977 with the purpose of providing additional funds for agricultural and rural development in developing countries. The headquarters of the IFAD are in Rome.

**International Labour Organization (ILO)** A special agency of the United Nations, established under the Treaty of Versailles in 1919, with the aim of promoting lasting peace through social justice. The ILO establishes international labour standards, runs a programme of technical assistance to developing countries, and attacks unemployment through its World Employment Programme. The ILO is financed by contributions from its member states. Its headquarters, The International Labour Office, is in Geneva.

**International Monetary Fund (IMF)** A specialized agency of the United Nations established in 1945 to promote international monetary cooperation and expand international trade, stabilize exchange rates, and help countries experiencing short-term balance of payments difficulties to maintain their exchange rates. The Fund assists members by supplying the amount of foreign currency it wishes to purchase in exchange for the equivalent amount of its own currency. The member repays this amount by buying back its own currency in a currency acceptable to the Fund, usually within three to five years (*see also* Special Drawing Rights). The Fund is financed by subscriptions from its members, the amount determined by an estimate of their means. Voting power is related to the amount of the subscription – the higher the contribution the higher the voting rights. The head office of the IMF is in Washington. *See also* conditionality.

**International Petroleum Exchange (IPE)** An exchange, founded in London in 1980, that deals in *futures contracts and *options (including traded options) in oil (gas oil, Brent crude oil, and heavy fuel oil) with facilities for making an exchange of futures for physicals (*see* actuals). There are 35 floor members of the exchange, which is located in St Katharine Dock and shared with *London FOX.

**International Standards Organization (ISO)** An organization founded in 1946 to standardize measurements and other standards for industrial, commercial, and scientific purposes. The *British Standards Institution is a member.

**International Stock Exchange of the UK and Republic of Ireland Ltd** *See* London Stock Exchange.

**International Telecommunication Union (ITU)** A special agency of the United Nations founded in Paris in 1865 as the International Telegraph Union. It sets up international regulations for telegraph, telephone, and radio services; promotes international cooperation for the improvement of telecommunications; and is concerned with the development of technical facilities. It is responsible for the allocation and registration of radio frequencies for communications. Its headquarters are in Geneva.

**International Union of Credit and Investment Insurers** *See* Berne Union.

**interpolation** Estimation of the value of an unknown quantity that lies between two of a series of known values. For example, one can estimate by interpolation the population of a country at any time between known 10-year census figures. This would normally be done on a graph, the shape of the curve between known points enabling a good estimate to be made of the interpolated value. **Extrapolation** involves estimating an unknown quantity that lies outside a series of known values. Thus one could obtain a figure for the population of a country five years after its last census, by extending the population curve after the last known value.

**interpreter** A computer program that executes a program written in a *programming language without first translating it into *machine code. The program runs as if it had been so translated (*see* compiler), but much more slowly. An interpreter usually allows the operator to interrupt a program and inspect or alter the values of variables.

**intervention price** *See* Common Agricultural Policy.

**inter vivos gifts** Gifts made between living people. In the UK, inter vivos gifts were subject to *capital-transfer tax until March 1986, when this tax was replaced by *inheritance tax. Inter vivos gifts, known for inheritance-tax purposes as **lifetime transfers**, are taken into account in calculating the inheritance tax payable. Exempt transfers include those between spouses, a specified annual sum, small gifts (less than £250), certain marriage gifts, gifts to charities and political parties, and those made for national purposes. A tapering relief is given for lifetime transfers made over a seven-year period. Gifts made more than seven years before death escape tax.

**intestacy** The state of a person who has died without having left a will. Such a person is said to have died **intestate**. Intestacy can be either total (if no will is left at all) or partial (if not all the deceased's property is left by his will). The Administration of Estates Act contains a table showing the destination of the property in an intestacy. If the intestate has no close relatives but his spouse survives, then the spouse takes everything. If children or grandchildren survive him, the spouse takes £40,000 plus a life interest in half of the remaining estate. The children or grandchildren take the rest. If he is survived by a spouse and other close relatives but not by children or grandchildren, the spouse receives £85,000 plus half the rest of the estate, and the remainder goes to the nearest of the surviving relatives. If there are no relatives, the estate goes to the Crown as *bona vacantia*.

**in-the-money option** *See* intrinsic value.

**intrinsic value 1.** The value something has because of its nature, before it has been processed in any way. **2.** The difference between the market value of the underlying security in a traded *option and the *exercise price. An option with an intrinsic value is said to be **in the money**; an option with zero intrinsic is **at the money**; one with less intrinsic value than zero is **out of the money**. *See also* time value.

**introduction** A method of issuing shares on the *London Stock Exchange in which a broker or issuing house takes small quantities of the company's shares and issues them to clients at opportune moments. It is also used by existing public companies that wish to issue additional shares. *Compare* issue by tender; offer for sale; placing; public issue.

**inventory 1.** A list of articles, usually of items in a house or belonging to a person, with their values. **2.** The US term for *stock-in-trade.

**inventory and valuation policy** An insurance policy, usually on household goods and personal effects, that is based on a written schedule of every item in the premises. This schedule, which is prepared by professional valuers, shows the value of each item. In such policies, insurers agree that in the event of a claim they will not require any further proof of value, which they would require without a valued schedule. Inventory and valuation policies are rare because of the time and expense involved in obtaining a valuation. There are also the problems of having to add to the cover any new items purchased and of inflationary rises in the value of objects.

**inventory investment** Investment in stock-in-trade. This is the preferred term in the USA.

**investment 1.** The purchase of capital goods, such as plant and machinery in a factory in order to produce goods for future consumption. This is known as **capital investment**; the higher the level of capital investment in an economy, the faster it will grow. **2.** The purchase of assets, such as securities, works of art, bank and building-society deposits, etc., with a primary view to their financial return, either as income or capital gain. This form of **financial investment** represents a means of saving. The level of financial investment in an economy, will be related to such factors as the rate of interest, the extent to which investments are likely to prove profitable, and the general climate of business confidence.

**investment analyst** A person employed by stockbrokers, banks, insurance companies, unit trusts, pension funds, etc., to give advice on the making of investments. Many pay special attention to the study of *equities in the hope of being able to advise their employers to make profitable purchases of ordinary shares. To do this they use a variety of techniques, including a comparison of a company's present profits with its future trading prospects; this enables the analyst to single out the companies likely to outperform the general level of the market.

**investment bank** A US bank that fulfils many of the functions of a UK *merchant bank. It is usually one that advises on mergers and acquisitions and provides finance for industrial corporations by buying shares in a company and selling them in relatively small lots to investors.

**investment bond** A single-premium life-assurance policy in which an investment of a fixed amount is made (usually over £1000) in an *asset-backed fund. Interest is paid at an agreed rate and at the end of the period the investment is returned with any growth. Investment bonds confer attractive tax benefits in some circumstances. *See also* single-premium assurance; top slicing.

**investment club** A group of investors who, by pooling their resources, are able to make more frequent and larger investments on a stock exchange, often being able to reduce brokerage and to spread the risk of serious loss. The popularity of investment clubs has waned with the rise of *unit trusts and *investment trusts, both of which have the advantages of professional management.

**investment company** *See* investment trust.

**investment grant (investment incentive)** A grant made to a company by a government to encourage investment in plant, machinery, buildings, etc. The incentives may take various forms, some being treated as a deduction for tax purposes, others being paid whether or not the company makes a profit.

**investment income 1.** A person's income derived from investments. **2.** The income of a business derived from its outside investments rather than from its trading activities.

**Investment Management Regulatory Organization** *See* Self-Regulatory Organization.

**investment manager** *See* fund manager.

**investment portfolio** *See* portfolio.

**investment trust (investment company)** A company that invests the funds provided by shareholders in a wide variety of securities. It makes its profits from the income and capital gains provided by these securities. The investments made are usually restricted to securities quoted on a stock exchange, but some will invest in unquoted companies. The advantages for shareholders are much the same as those with *unit trusts, i.e. spreading the risk of investment and making use of professional managers. Investment trusts, which are not usually *trusts in the usual sense, but private or public limited companies, differ from unit trusts in that in the latter the investors buy units in the fund but are not shareholders. *See also* accumulation unit; unitization.

**invisible assets** *See* intangible asset.

**invisible balance** The *balance of payments between countries that arises as a result of transactions involving services, such as insurance, banking, shipping, and tourism (often known as **invisibles**), rather than the sale and purchase of goods. Invisibles can play an important part in a nation's current account, although they are often difficult to quantify. The UK relies on a substantial invisible balance in its balance of payments.

**invisible hand** The unseen forces by which the pursuit of rational self-interest (*see* rationality) achieves a socially desirable outcome. Conceived by Adam Smith (1723–90) to describe the operation of a market economy, the idea of the invisible hand forms the basis of neoclassical theories about general *equilibrium and *Pareto optimality.

**invisibles** *See* invisible balance.

**invitation to treat** *See* offer.

**invoice** A document stating the amount of money due to the organization issuing it for goods or services supplied. A **commercial invoice** will normally give a description of the goods and state how and when the goods were dispatched by the seller, who is responsible for insuring them in transit, and the payment terms.

**involuntary unemployment** Unemployment in which workers who would be willing to work for lower wages than those in employment are still unable to find work. J M Keynes (1883–1946) argued that recessions are characterized by involuntary unemployment because firms may be unwilling or unable to cut the wages of workers they employ (*see* efficiency wage theory). Although neoclassical economists have found difficulty accepting this concept in recent years, a number of theories (including the *implicit contract theory and the efficiency wage theory) have been suggested to explain it. The emergence of these theories reflects the need to explain the high and persistent levels of unemployment that began in the 1980s.

**IOU** A written document providing evidence of a debt, usually in the form "I owe you . . .". It is not a *negotiable instrument or a *promissory note and requires no stamp (unless it does include a promise to pay). It can, however, be used as legal evidence of a debt.

**IPA** Abbreviation for *Institute of Practitioners in Advertising.

**IPE** Abbreviation for *International Petroleum Exchange.

**IPM** Abbreviation for *Institute of Personnel Management.

**IPO** Abbreviation for *initial public offering.

**IPR** Abbreviation for *Institute of Public Relations.

**IRR** Abbreviation for *internal rate of return.

**irrational behaviour** *See* rationality.

**irredeemable securities (irredeemables)** Securities, such as some government loan stock (*see* Consols) and some *debentures, on which there is no date given for the redemption of the capital sum. The price of fixed-interest irredeemables on the open market varies inversely with the level of interest rates.

**irrevocable documentary acceptance credit** A form of irrevocable confirmed *letter of credit in which a foreign importer of UK goods opens a credit with a UK bank or the UK office of a local bank. The bank then issues an irrevocable letter of credit to the exporter, guaranteeing to accept *bills of exchange drawn on it on presentation of the shipping documents.

**irrevocable letter of credit** *See* letter of credit.

**IRS** Abbreviation for *Internal Revenue Service.

**ISBA** Abbreviation for *Incorporated Society of British Advertisers Ltd.

**ISLM model** A model that shows how the *equilibrium levels of *interest and national output are determined. It involves manipulation of the IS (investment-savings) curve and the LM (liquidity-money) curve. First formulated by John Hicks (1904– ), the model works by simultaneously finding the equilibrium values of these variables in the money market and the savings-investment market. The aim of the model was to provide a framework for contrasting Keynes' general theory with the standard neoclassical (later monetarist) theory and a basis for policy analysis. Although the ISLM model has been the standard textbook analysis for some fifty years, it lacks proper *microfoundations and has failed to resolve the major issues in macroeconomics.

**ISO** Abbreviation for *International Standards Organization.

**isocost** The set of combinations of goods that have the same total cost; this can be represented by a curve on a graph. In *producer theory isocosts refer to inputs and, making certain assumptions, there will be a unique minimum cost of producing a given level of output, which can be represented as the point at which a particular *isoquant curve touches the lowest possible isocost curve.

**isoquant** The combinations of inputs that can be used to produce a given level of output; this can be represented as a curve on a graph. The concept is analogous to that of an *indifference curve in *consumer theory. Given the prices of inputs faced by a firm, by making certain additional assumptions, the firm or *entrepreneur will be able to choose a single combination of inputs that minimizes costs for a given level of output (i.e. on a particular isoquant); this combination will also maximize profits. *See also* marginal rate of substitution.

**ISP** Abbreviation for *Institute of Sales Promotion.

**issue 1.** The number of shares or the amount of stock on offer to the public at a particular time. *See also* new issue; rights issue; scrip issue. **2.** The number of banknotes distributed by the Bank of England at a particular time.

**issue by tender** A method of issuing shares on the *London Stock Exchange in which an *issuing house asks investors to *tender for them. The stocks or shares are then allocated to the highest bidders. It is usual for the tender documents to state the lowest price acceptable. This method may be used for a *new issue or for loan stock (*see* debenture), but is not frequently employed. *Compare* introduction; offer for sale; placing; public issue.

**issued (*or* subscribed) share capital** *See* share capital.

**issue price** The price at which a *new issue of shares is sold to the public. Once the issue has been made the securities will have a market price, which may be above (at a premium on) or below (at a discount on) the issue price (*see also* stag). In an *introduction, *offer for sale, or *public issue, the issue price is fixed by the company on the advice of its stockbrokers and bankers; in an *issue by tender the issue price is fixed by the highest bidder; in a *placing the issue price is negotiated by the *issuing house or broker involved.

**issuing house** A financial institution, usually a *merchant bank, that specializes in the *flotation of private companies on a *stock exchange. In some cases the issuing house will itself purchase the whole issue (*see* underwriter), thus ensuring that there is no uncertainty in the amount of money the company will raise by flotation. It will then sell the shares to the public, usually by an *offer for sale, *introduction, *issue by tender, or *placing.

# J

**J-curve effect** An increase in the value of imports or a fall in the value of exports caused by a depreciation of a domestic currency. In the long term, it is expected that the falling value of the currency will make exports more competitive, causing them to increase, and imports more expensive, causing demand for them to fall. However, in the short term import and export volumes will usually be slow to adjust, so that imports will cost more in the depreciated currency and exports will earn less; the resulting decline in the *balance of payments is represented by the hook of the J. The subsequent fall in imports and rise in exports causing the balance of payments to rise sharply is represented by the upstroke of the J.

**jerque note** A certificate issued by a UK customs officer to the master of a ship, stating that he is satisfied that the cargo on his ship has been correctly entered on the *manifest and that no unentered cargo has been found after searching the ship.

**jetsam** Items of a ship's cargo that have been thrown overboard and sink below the surface of the water. *Compare* flotsam.

**jettisons** Items of a ship's cargo that have been thrown overboard to lighten the ship in dangerous circumstances. If the items are insured they constitute a general average loss (*see* average).

**job analysis** A detailed study of a particular job, the tools and equipment needed to do it, and its relation to other jobs in an organization. The analysis should also provide the information needed to say how the job should best be done and the qualifications, experience, or aptitudes of the person best suited to doing it.

**jobber 1.** A former dealer in stocks and shares, who had no contact with the general public, except through a *stockbroker. Jobbers were replaced by *market makers on the *London Stock Exchange in the Big Bang of October, 1986. **2.** A dealer who buys and sells commodities, etc., for his own account.

**jobbing backwards** Looking back on a transaction or event and thinking about how one might have acted differently, had one known then what one knows now.

**job centre** A UK government employment agency that seeks work for those out of a job, assists employers to find suitable employees, and provides training facilities for trades in which there are shortages of skilled workers. Most centres also offer occupational advice on retraining.

**job costing** The attribution of separate costs to each job undertaken. This method of costing is particularly used in engineering, when a variety of different jobs are undertaken and there is no homogeneous product.

**job description** *See* personnel selection.

**job evaluation** An assessment of the work involved in a particular job, the responsibilities borne by the person who does it, and the skills, experience, or qualifications required, all with a view to evaluating the appropriate remuneration for the job and the differentials between it and other work in the same organization.

**job sharing** Dividing the work of one full-time employee between two or more part-time employees.

**job specification** *See* personnel selection.

**joint account** A *bank account or a building-society account held in the names of two or more people, often husband and wife. On the death of one party the balance in the account goes to the survivor(s), except in the case of partnerships, executors' accounts, or trustees' accounts. It is usual for any of the holders of a joint account to operate it alone.

**joint and several liability** A liability that is entered into as a group, on the understanding that if any of the group fail in their undertaking the liability must be shared by the remainder. Thus if two people enter into a joint and several guarantee for a bank loan, if one becomes bankrupt the other is liable for repayment of the whole loan.

**Joint Industry Committee for National Readership Surveys** A committee formed in 1968 to run readership surveys, although continuous research and surveys have been undertaken since 1956. The committee comprises members of the *Institute of Practitioners in Advertising, the *Incorporated Society of British Advertisers Ltd, and the Newspaper and Periodical Contributors Committee. The committee reports twice yearly on the reading habits of 28 000 individuals, covering 200 titles.

**Joint Industry Committee for Radio Audience Research** A committee composed of representatives of the *Institute of Practitioners in Advertising, the *Incorporated Society of British Advertisers Ltd, and independent radio companies. The committee provides information on the listening habits of radio audiences.

**joint investment** A security purchased by more than one person. The certificate will bear the names of all the parties, but only the first named will receive notices. To dispose of the holding all the parties must sign the *transfer deed.

**joint-life and last-survivor annuities** Annuities that involve two people (usually husband and wife). A joint-life annuity begins payment on a specified date and continues until both persons have died. A last-survivor annuity only begins payment on the death of one of the two people and pays until the death of the other. *Compare* single-life pension.

**joint-stock bank** A UK bank that is a *public limited company rather than a private bank (which is a partnership). During the 19th century many private banks failed; the joint-stock banks became stronger, however, largely as a result of amalgamations and careful investment. In the present century, they became known as the *commercial banks or high-street banks.

**joint-stock company** A *company in which the members pool their stock and trade on the basis of their joint stock. This differs from the earliest type of company, the merchant corporations or regulated companies of the 14th century, in which each member traded with his own stock, subject to the rules of the company. Joint-stock companies originated in the 17th century and some still exist, although they are now rare.

**joint supply (complementary supply)** The supply of two or more separate commodities that are produced by the same process; examples include milk and butter, wool and mutton, and petrol and heavy oil. If the demand for one increases, the supply of the other will also increase, but its price will fall unless its demand also increases.

**joint tenants** Two or more persons who jointly own an interest in land. When one dies, the other(s) takes his share by right of survivorship. No one tenant has rights to any particular part of the property and no tenant may exclude another.

**journal** A book of prime entry in which transfers to be made from one *account to another are recorded. It is used for transfers not recorded in any other of the books of prime entry, such as the sales *day book or the *cash book.

**joystick** A computer input device resembling a small aircraft control stick that is used with computer games, and some computer-aided design programs, to move graphics symbols on the screen.

**judgment creditor** The person in whose favour a court decides, ordering the **judgment debtor** to pay the sum owed. If the judgment debtor fails to pay, the judgment creditor must return to the court asking for the judgment to be enforced.

**judgment debtor** *See* judgment creditor.

**junk bond** A *bond that offers a high rate of interest because it carries a higher than usual probability of default. The issuing of junk bonds to finance the takeover of large companies in the USA is a practice that has developed rapidly over recent years and has spread elsewhere. *See* leveraged buyout.

# K

**K** *See* kilobyte.

**kaffirs** An informal name for shares in South African gold-mining companies on the London Stock Exchange.

**kangaroos** An informal name for Australian shares, especially in mining, land, and tobacco companies, on the London Stock Exchange.

**keelage** A fee paid by the owners of a ship for being permitted to dock in certain ports and harbours.

**kerb market** **1.** The former practice of trading on the street after the formal close of business of the London Stock Exchange. **2.** Dealing on the London Metal Exchange in any metal, after the formal *ring trading in specified metals. **3.** Any informal market, such as one for dealing in securities not listed on a stock exchange.

**keyed advertisement** An advertisement designed to enable the advertiser to know where a respondent saw it, for instance by including a code number of a particular department in the return address.

**key-man assurance** An insurance policy on the life of a key employee (male or female) of a company, especially the life of a senior executive in a small company, whose death would be a serious loss to the company. In the event of the key man or woman dying, the benefit is paid to the company. In order that there should be an *insurable interest, a loss of profit must be the direct result of the death of the key man or woman.

**Keynesianism** An approach to *macroeconomics based on the work of J M Keynes (1883–1946). Failures of coordination between markets, even if these are internally efficient, may generate recession and mass unemployment. For example, unemployed workers may be unable to find jobs because there is no demand for the goods they produce although these same workers would demand the goods if they were employed and earned wages. In this view, governments have a role to play by putting money into workers' pockets through public works, i.e. creating the demand and raising the level of production through the *multiplier process. This view of the economy was widely accepted between the late 1940s and the 1960s, when it came under attack first by the monetarists and more recently by the new classical macroeconomists.

**key punch** A keyboard-operated device used to punch holes in cards to represent data for processing by computer.

**kilobyte** A measure of computer memory capacity, often abbreviated to K. One kilobyte is 1024 *bytes, or characters; a 64K memory therefore holds 64 × 1024 = 65 536 bytes or characters.

**kite** An informal name for an *accommodation bill. **Kite-flying** or **kiting** is the discounting of a kite (accommodation bill) at a bank, knowing that the person on whom it is drawn will dishonour it.

**Kite mark** *See* British Standards Institution.

**knock-for-knock agreement** An agreement between motor insurers that they will pay for their own policyholders' accident damage without seeking any contribution from the other insurers, irrespective of blame but provided that the relevant policies cover the risk involved. The agreement works on the principle that, having paid for their own policyholder's damage, insurer A refrains from claiming the payment back from insurer B, even though it may be clear that A's client was to blame for the accident. On another occasion the roles might be reversed; insurer B then refrains from claiming from A even though A's client was responsible.

Before the knock-for-knock agreement, insurers found they were exchanging roughly similar amounts in payment of these claims. The agreement has cut down the expensive administration of claim and counterclaim, which has helped to reduce the cost of premiums.

**knocking copy** Advertising copy that attacks a rival product.

**knot** A measure of the speed of a ship or aircraft; one nautical mile per hour, i.e. 1.15 miles per hour.

**Kondratieff waves** Long cycles in economic activity, whose existence were predicted by the Russian economist N D Kondratieff (1892–1931), who disappeared during Stalin's purges. The cycles were thought to last about 40 years, the last downswing having been the Depression of the 1930s. Careful inspection of the data, however, provides little evidence to support the existence of these cycles.

**Krugerrand** A South African coin containing 1 troy ounce of gold. Minted since 1967 for investment purposes, it enables investors to evade restrictions on holding gold. Since 1975 an import licence has been required to bring them into the UK. In the 1980s their popularity waned for political reasons and because other countries have produced similar coins. *See* Britannia coins.

# L

**labour** In economics, a factor of production usually combined with *land and *capital to produce outputs. Economists of the *Classical school and *Marxists attach a special importance to labour as the prime mover of the production process, a view encapsulated in the *labour theory of value. Economists of the *neoclassical school view labour as simply one input among several, whose value is determined by supply and demand, in the same way as any other commodity.

**labour-intensive** Denoting an industry or firm in which the remuneration paid to employees represents a higher proportion of the costs of production than the cost of raw materials or capital equipment (in the reverse case the industry or firm is said to be **capital-intensive**). While publishing is a labour-intensive industry, printing is capital-intensive.

**labour market** The market that determines who has jobs and the rate of pay for a particular job. It is a theoretical concept used in macroeconomic analysis.

**labour rate variance** A *variance arising when the rate at which labour is paid differs from the standard rate.

**labour theory of value** The theory that the value of an item is dependent on the quantity of labour that is required to make it. First developed by Adam Smith (1723–90), this theory was at the core of the *Classical school of economics and has subsequently been the basis of *Marxist economic theory. In practice, the theory has few useful applications and was superseded by the *Marginalist theory of value, which associates the value of a good with its utility and its scarcity.

**labour turnover rate** The ratio, usually expressed as a percentage, of the number of employees leaving an organization or industry in a stated period to the average number of employees working in that organization or industry during the period.

**labor union** US name for a *trade union.

**laches** (Norman French *lasches*: negligence) Neglect and unreasonable delay in enforcing an equitable right. If a plaintiff with full knowledge of the facts takes an unnecessarily long time to bring an action (e.g. to set aside a contract obtained by fraud) the court will not assist him; hence the maxim "the law will not help those who sleep on their rights". No set period is given but if the action is covered by *limitation-of-actions legislation, the period given will not be shortened. Otherwise the time allowed depends on the circumstances.

**Lady Macbeth strategy** A strategy used in takeover battles in which a third party makes a bid that the target company would favour, i.e. it appears to act as a *white knight, but subsequently changes allegiance and joins the original bidder.

**laesio enormis** (Latin: extraordinary injury) A doctrine of Roman law that is included in some European legal systems but not in English law. It states that the price given in a contract by way of *consideration must be fair and reasonable or the contract may be rescinded.

**Laffer curve** A curve on a graph, popularized by the US economist Arthur Laffer (1940– ), in which the amount of government tax revenue is plotted against the percentage rate of taxation. Supply-side economists (*see* supply-side economics) believe that this curve is hill-shaped, so that at very high rates of taxation a cut in percentage rates will actually increase government tax revenues. This belief is based on the view that lower tax rates will induce individuals to work harder and earn more income, enabling more tax to be raised at the lower rate. Although the Laffer curve has been very popular politically, there is little empirical evidence to support it.

**LAFTA** Abbreviation for *Latin American Free Trade Area.

**LAIA** Abbreviation for *Latin American Integration Association.

**laissez-faire economy** An economy in which government intervention is kept to a minimum and market forces are allowed to rule. The term, attributed to the French merchant J Gourlay (1712–59), is best translated as "let people do as they think best". Laissez-faire policies were popular with western governments until the time of Keynes, when mass *unemployment caused them to become discredited. During the *stagflation of the late 1960s and 1970s they became, once again, fashionable.

**lakh** In India, Pakistan, and Bangladesh 100,000. It often refers to this number of rupees. *See also* crore.

**lame duck** A company that has national prestige and is a large employer of labour but, being unable to meet foreign competition, is unable to survive without government support.

**LAN** Abbreviation for *local area network.

**land 1.** In economics, a factor of production usually combined with *labour and *capital to produce outputs. Land, which includes air, sea, minerals, etc., as well as land for building on, is unusual amongst factors of production in that it costs nothing to produce and is in fixed supply. However, the application of capital to land (e.g. irrigation) can considerably alter its characteristics and value. **2.** In law, the part of the earth's surface that can be owned, including all buildings, trees, minerals, etc., attached to or forming part of it. It also includes the airspace above the land necessary for its reasonable enjoyment. In law, a first-floor flat can constitute a separate piece of land, although it has no contact with the soil, just as a cellar may be a separate *estate in land.

**land bank 1.** The amount of land a developer owns that is awaiting development. **2.** *See* Agricultural Bank.

**land certificate** A document that takes the place of a *title deed for registered land. It is granted to the registered proprietor of the land and indicates ownership. It will usually be kept in the *Land Registry if the land is subject to a mortgage.

**land charges** Incumbrances, such as contracts, payments, or other matters, to which *land is subject; they must either be dealt with before the land is sold or they will bind the purchaser. Formerly, they bound a purchaser who had notice of them. Now most of them must be registered if they are to remain effective, either under the Land Charges Act (1972; applicable to unregistered land) or under the Land Registration Act (1925; applicable to registered land).

**Land Charges Register 1.** The register of *land charges on unregistered land, kept at the *Land Registry. Charges are registered against the name of the estate owner. The kinds of charges that must be registered are set out in the Land Charges Act (1972) and include estate contracts, equitable easements, restrictive covenants, and Inland Revenue charges. Registration constitutes actual notice to any person of the existence of the charge. If the purchaser of the land is unable to discover the charge (for example, because it is registered against the name of an estate owner that does not appear on the title deeds available to the purchaser), compensation may be payable to the purchaser at public expense. If a charge is not registered it will not bind a purchaser for money or money's worth of a legal estate in the land, even if he had notice of the charge. Some charges, if not registered, will not bind the purchaser for value of any interest (whether legal or equitable) in the land. **2.** (**Local Land Charges Register**) A register kept by each London borough, the City of London, and each district council. Public authorities may register statutory restrictions on land, e.g. under town and country planning legislation. These are registered against the land itself, not the estate owner. If restrictions are not registered, the purchaser is still bound by them, but will be entitled to compensation.

**landed** Denoting an export price in which the exporter pays all the charges to the port of destination and also pays the *landing charges but not any dock dues or onward carriage charges to the importer's factory or warehouse. *Compare* c.i.f.; franco.

**landing account** An account prepared by a public warehousing company to the owner of goods that have recently been landed from a ship and taken into the warehouse. The account, which is usually accompanied by a *weight note, states the date on which the warehouse rent begins and also the quantity of goods and their condition (from the external appearance of the packages).

**landing charges** The charges incurred in disembarking a cargo, or part of it, from a ship at the port of destination.

**landing order** A document given to an importer by the UK Customs and Excise, enabling him to remove newly imported goods from a ship or quay to a *bonded warehouse, pending payment of duty or their re-export.

**landlord and tenant** The parties to a *lease. The landlord grants the lease to the tenant. The landlord may be the freehold owner of the land or may himself be a tenant of a superior landlord (*see* head lease). The law of landlord and tenant is that governing the creation, termination, and regulation of leases. Many leases, both business and residential, are subject to statutory control. This means that the right of the landlord to end a lease and evict tenants may be restricted and the rent may be controlled. It is therefore essential to consider statutory rights when considering the relation between landlord and tenant, as well as those given by the lease.

**land registration** The system of registration of title introduced by the Land Registration Act (1925). Land is registered by reference to a map rather than against the name of any estate owner. Three registers are kept: (a) the **property register**, which records the land and estate; it refers to a map and mentions any interests benefiting the land, e.g. easements and restrictive covenants; (b) the **proprietorship register**, which sets out the name of the owner, the nature of the title (e.g. freehold or leasehold), and any restrictions on the way the land may be dealt with (e.g. if the land is subject to a trust); (c) the **charges register**, which contains notice of rights adverse to the land, such as mortgages and land charges as defined in the Land Charges Act (1972). The owner's title is guaranteed by registration in most circumstances. Rights in registered land may be a registered interest, which amounts to ownership of the freehold or leasehold estate; an overriding interest, which binds a purchaser without needing to be registered; or a minor interest, which needs to be protected by an entry on the register (e.g. interest under a trust or restrictive covenant). The purpose of the system was to simplify conveyancing, so that instead of checking title deeds, a search of the register would reveal all relevant matters. Unfortunately the simplicity of the system is breached by the difficulties arising from overriding interests. Eventually, all land in England and Wales will be registered, enabling unregistered conveyancing to disappear. *See also* rectification of register; registered land certificate.

**Land Registry** The official body that keeps the various land registers (*see* land registration). The main registry is in Lincoln's Inn Fields in London, but there are also regional registries. The Land Registry is responsible for implementing the policy of registering title to all land in England and Wales. *See also* rectification of register.

**land waiter** A UK customs officer who examines goods at the ports for export or import and ensures that the correct taxes or duties are levied.

**language laboratory** A room used, especially by business people, to study a foreign language. The object is to acquire a proficiency in the spoken language as quickly as possible and to this end extensive use is made of tape recorders and prerecorded tapes. A supervisor can plug into tapes made by students, to monitor their progress and correct errors, especially of pronunciation.

**laser disk** A disk with a silvery surface, on which information is stored and read by a laser. The disk surface is covered by circular tracks made up of tiny pits that hold the information. A higher-power laser beam is used to burn the pits into the disk when recording the information. The information is read by shining a laser light onto the tracks as the disk rotates. Some laser disks are read-only or write-once disks; however, erasable disks are available. Compact disks, used for high-quality recording of music, and video disks are common examples of laser disks. They are also used to store computer data, when they are usually called **optical disks** (*compare* magnetic disk), and for the publication of large information databases.

**laser printer** A type of high-speed printer used with computers. The characters to be printed are formed by a low-power laser, which alters the electrical charge on appropriate areas of the surface of a photoconductive rotating drum, as in *xerography. The drum is selectively coated with a toner powder, which is attracted by the charge produced by the laser but repelled by that on the rest of the drum. The powder is transferred to paper and fused into place by heated rollers. Laser printers are very fast and silent, producing high-quality lettering and graphics. They are often used with *desktop publishing equipment to produce reports, catalogues, and newsletters.

**LASH** Acronym for *l*ighter *a*board *sh*ip. This is a method of cargo handling in which cargo-carrying barges (lighters) are lifted on and off ocean-going ships by the ship's own crane, thus reducing to a minimum the need for port facilities.

**last in, first out (LIFO)** A method of charging homogeneous items of stock to production when the cost of the items has changed. It is assumed, both for costing and stock valuation purposes, that the latest items taken into stock are those used in production although this may not necessarily correspond with the physical movement of the goods. The LIFO method is not acceptable to the Inland Revenue in the UK. *Compare* first in, first out.

**last-survivor policy 1.** An assurance policy on the lives of two people, the sum assured being paid on the death of the last to die. *See also* joint-life and last-survivor annuities. **2.** A contract (formerly called a **tontine**) in which assurance is arranged by a group of people, who all pay premiums into a fund while they are alive. No payment is made until only one person from the group is left alive. At that point the survivor receives all the policy proceeds. Contracts of this kind are not available in the UK because of the temptation they provide to members to murder their fellows, in order to be the last survivor.

**last trading day** The last day on which commodity trading for a particular delivery period can be transacted.

**lateral integration** *See* integration.

**Latin American Free Trade Area (LAFTA)** An economic grouping that became the *Latin American Integration Association in 1981.

**Latin American Integration Association (LAIA)** An economic grouping of South American countries with headquarters in Montevideo. It took over the Latin American Free Trade Area (LAFTA) in 1981. Its members are Argentina, Bolivia, Brazil, Chile, Colombia, Ecuador, Mexico, Paraguay, Peru, Uruguay, and Venezuela.

**laundering money** Processing money acquired illegally (as by theft, drug dealing, etc.) so that it appears to have come from a legitimate source. This is often achieved by paying the illegal cash into a foreign bank and transferring its equivalent to a UK bank.

**LAUTRO** Abbreviation for Life Assurance and Unit Trust Regulatory Organization. *See* Self-Regulatory Organization.

**law of one price** The economic rule that, in the absence of trade barriers or transport costs, competition will ensure that a particular good will sell at the same price in all countries. The theory of *purchasing-power parity is based on this law.

**law of statistical regularity** The basic assumption in statistics that a random sample taken from a larger group will reflect the characteristics of the larger group. The larger the size of the sample in relation to the whole group, the more accurately it will reflect the group.

**Law Society** The professional body for solicitors in England and Wales, incorporated by royal charter in 1831. It controls the education and examination of articled clerks and the admission of solicitors. It also regulates their professional standards and conduct. It has over 40 000 members. Its College of Law provides courses for its examinations.

**lay days** The number of days allowed to a ship to load or unload without incurring *demurrage. **Reversible lay days** permit the shipper to add to the days allowed for unloading any days he has saved while loading. Lay days may be calculated as *running days (all consecutive days), *working days (excluding Sundays and public holidays), or *weather working days (working days on which the weather allows work to be carried out).

**laying off** Suspending or terminating the employment of workers because there is no work for them to do. If the laying off involves a permanent termination of employment, *redundancy payments will be involved.

**lay-off pay** US for *redundancy payment.

**LBO** Abbreviation for *leveraged buyout.

**LCD** Abbreviation for *liquid crystal display.

**lead** The first named underwriting syndicate on a Lloyd's insurance policy. When a broker seeks to cover a risk he will first try to get a large syndicate to act as lead, which encourages smaller syndicates to cover a share of the risk. The premium rate is calculated by the lead; if others wish to join in the risk they have to insure at that rate. On a collective policy the lead insurer is the first insurer on the schedule of insurers; he issues the policy, collects the premiums, and distributes the proportions to the coinsurers.

**leakage 1.** The loss of a liquid from a faulty container. **2.** An allowance for loss of a liquid in transit, due to evaporation or other causes.

**lean back** A period of cautious inaction by a government agency before intervening in a market. For example, a central bank might allow a lean-back period to elapse to allow exchange rates to stabilize, before intervening in the foreign exchange market.

**learning curve** A graph that illustrates a trainee's progress in learning a task. If the task is relatively simple, the rate of progress might be measured in units produced in a fixed period plotted against time. Typically the curve will climb sharply to a plateau, while learning is consolidated, before rising to the final level, representing the output of the skilled worker. In complicated tasks more sophisticated means have to be used to measure progress.

**lease** The grant of an interest in *leasehold land. It must give exclusive possession of the land and be for a fixed term. To be a legal lease it must be created by deed, unless it is for less than three years, take effect on possession (i.e. start immediately), and be for the best rent obtainable without taking a *premium. A lease that is not a legal lease may, however, be valid in equity as an agreement for a lease. This must be registered or any rights under it may be lost on the sale of the freehold or of a superior lease. A lease generally constitutes a bargain between the *landlord and tenant, containing rights and obligations on both sides. It may come to an end on the expiry of the term; alternatively it may be ended earlier, by the tenant surrendering it to the landlord or by the landlord ending it for the breach of some condition (e.g. failure to pay the rent or leaving the premises in disrepair). A lease may be assigned to someone else; for example, a leasehold flat may be sold for a capital sum. The new tenant will then take over the responsibilities under the lease (*see also* head lease). A **repairing lease** is one in which the tenant is obliged to pay for all repairs and is usually bound to leave the property at the end of his lease in the same condition as he found it at the start of the lease.

**lease-back (renting back)** A method of raising finance in which an organization sells its land or buildings to an investor (usually an insurance company) on condition that the investor will lease the property back to the organization for a fixed term at an agreed rental. This releases capital for the organization, enabling it to be used for other purposes.

**leasehold land** Land held under a *lease. The land will eventually revert to the freehold owner, although there has been some statutory modification of this right to repossession (e.g. in the Rent Acts). This is the most common way for blocks of offices to be owned. The landlord maintains possession of the common parts and creates separate leases for each office. The ownership of each office may subsequently change as leases are assigned.

**leasing** Hiring equipment, such as a car or a piece of machinery, to avoid the capital cost involved in owning it. In some companies it is advantageous to use capital for other purposes and to lease some equipment, paying for the hire out of income. The equipment is then an asset of the leasing company rather than the lessor. Sometimes a case can be made for leasing rather than purchasing, on the grounds that some equipment quickly becomes obsolete.

**ledger** A collection of *accounts of a similar type. Traditionally, a ledger was a large book with separate pages for each account. In modern systems, ledgers may consist of separate cards or computer records. The most common ledgers are the **nominal ledger** containing the *impersonal accounts, the **sales** (*or* **debtors**) **ledger** containing the accounts of an organization's customers, and the **purchase** (*or* **creditors**) **ledger** containing the accounts of an organization's suppliers.

**legal reserve** The minimum amount of money that building societies, insurance companies, etc., are bound by law to hold as security for the benefit of their customers.

**legal tender** Money that must be accepted in discharge of a debt. It may be **limited legal tender**, i.e. it must be accepted but only up to specified limits of payment; or **unlimited legal tender**, i.e. acceptable in settlement of debts of any amount. Bank of England notes and the £2 and £1 coins are unlimited legal tender in the UK. Other Royal Mint coins are limited legal tender; i.e. debts up to £10 can be paid in 50p and 20p coins; up to £5 by 10p and 5p coins; and up to 20p by bronze coins.

**lender of last resort** A country's central bank with responsibility for controlling its banking system. In the UK, the Bank of England fulfils this role, lending to *discount houses, either by repurchasing *Treasury bills, lending on other paper assets, or granting direct loans, charging the *base rate of interest. *Commercial banks do not go directly to the Bank of England; they borrow from the discount houses.

**lessee** A person who is granted a *lease; tenant. *See* landlord and tenant.

**lessor** A person granting a *lease; landlord. *See* landlord and tenant.

**letter of allotment** *See* allotment.

**letter of credit** A letter from one banker to another authorizing the payment of a specified sum to the person named in the letter on certain specified conditions (*see* letter of indication). Commercially, letters of credit are widely used in the international import and export trade as a means of payment. In an export contract, the exporter may require the foreign importer to open a letter of credit at his local bank (the issuing bank) for the amount of the goods. This will state that it is to be negotiable at a bank (the negotiating bank) in the exporter's country in favour of the exporter; often, the exporter (who is called the beneficiary of the credit) will give the name of the negotiating bank. On presentation of the *shipping documents (which are listed in the letter of credit) the beneficiary will receive payment from the negotiating bank.

An **irrevocable letter of credit** cannot be cancelled by the person who opens it or by the issuing bank without the beneficiary's consent, whereas a **revocable letter of credit** can. In a **confirmed letter of credit** the negotiating bank guarantees to pay the beneficiary, even if the issuing bank fails to honour its commitments (in an **unconfirmed letter of credit** this guarantee is not given). A confirmed irrevocable letter of credit therefore provides the most reliable means of being paid for exported goods. However, all letters of credit have an expiry date, after which they can only be negotiated by the consent of all the parties.

A **circular letter of credit** is one that will be negotiated by all the agents of the issuing bank, while a **direct letter of credit** is addressed to only one agent.

**letter of hypothecation** *See* hypothecation.

**letter of indemnity 1.** A letter stating that the organization issuing it will compensate the person to whom it is addressed for a specified loss. *See also* indemnity. **2.** A letter written to a company registrar asking for a replacement for a lost share certificate and indemnifying the company against any loss that it might incur in so doing. It may be required to be countersigned by a bank. **3.** A letter written by an exporter stating that he will be responsible for any losses arising from faulty packing, short weight, etc., at the time of shipment. If this letter accompanies the *shipping documents, the shipping company will issue a clean *bill of lading, even if the packages are damaged, enabling the exporter to negotiate his documents and receive payment without trouble.

**letter of indication (letter of identification)** A letter issued by a bank to a customer to whom a *letter of credit has been supplied. The letter has to be produced with the letter of credit at the negotiating bank; it provides evidence of the bearer's identity and a specimen of his signature. It is used particularly with a circular letter of credit carried by travellers, although *traveller's cheques are now more widely used.

**letter of intent** A letter in which a person formally sets out his intentions to do something, such as signing a contract in certain circumstances, which are often specified in detail in the letter. The letter does not constitute either a contract or a promise to do anything, but it does indicate the writer's serious wish to pursue the course he has set out.

**letter of licence** A letter from a creditor to a debtor, who is having trouble raising the money to settle the debt. The letter states that the creditor will allow the debtor a stated time to pay and will not initiate proceedings against him before that time. *See also* arrangement.

**letter of regret** A letter from a company, or its bankers, stating that an application for an allotment from a *new issue of shares has been unsuccessful.

**letter of renunciation** **1.** A form, often attached to an *allotment letter, on which a person who has been allotted shares in a *new issue renounces his rights to them, either absolutely or in favour of someone else (during the **renunciation period**). **2.** A form on the reverse of some *unit-trust certificates, which the holder completes when he wishes to dispose of his holding. He sends the completed certificate to the trust managers.

**letter of set-off** *See* set-off.

**letter-quality printer** A computer printer that produces letters of a quality that is adequate for business correspondence. A *daisywheel printer is an example. Some *dot matrix printers also produce good-quality output but they are usually called **near letter quality** (**NLQ**), as the letters are slightly inferior to those produced by a daisywheel printer.

**letters of administration** An order authorizing the person named (the *administrator) to distribute the property of a deceased person, who has not appointed anyone else to do so. The distribution must be in accordance with the deceased's will, or the rules of *intestacy if he did not leave one.

**leverage** **1.** The US word for *capital gearing. **2.** The use by a company of its limited assets to guarantee substantial loans to finance its business.

**leveraged buyout (LBO)** A takeover ploy (*see* takeover bid) in which a small company, whose assets are limited, borrows heavily on these assets and the assets of the target company in order to finance a takeover of a larger company, often making use of *junk bonds.

**liability** *See* contingent liability; current liabilities; deferred liability; long-term liability; secured liability.

**liability insurance** A form of insurance policy that promises to pay any compensation and court costs the policyholder becomes legally liable to pay because of claims for injury to other people or damage to their property as a result of the policyholder's negligence. Policies often define the areas in which they will deal with liability, e.g. personal liability or *employers' liability.

**LIBOR** Abbreviation for *London Inter Bank Offered Rate.

**licence 1.** Official permission to do something that is forbidden without a licence (e.g. sell alcohol or own a TV or a firearm). Licences may be required for social reasons or simply to enable revenue to be collected. Since the Consumer Credit Act (1974) all businesses involved with giving credit to purchasers of goods must be licensed by the Office of Fair Trading. **2.** *See* franchise. **3.** Formal permission to enter or occupy land. Such licenses are of three kinds. (a) The simplest gives the licensee the permission of the landowner to be on land (e.g. the right of a visitor to enter a house). It may be revoked at any time as long as the licensee is given time to leave. (b) **Contractual licences** are permissions to be on land in the furtherance of some contractual right (e.g. the right of the holder of a cinema ticket to be in the cinema). This type of licence has been used to get round the Rent Acts (which apply to leases only): a licence to occupy a flat or house may be granted, which is said to be revocable at any time. It has been held that if the licence gives the licensee exclusive possession of the property, it is in fact a lease, despite the fact that it is called a licence. The exact state of the law in this area is uncertain. It is also unclear whether a contractual licence can be made irrevocable and binding on those who were not a party to the contract (e.g. purchasers of the land). It was originally held that a licence could always be revoked, although damages might be payable. Recent cases have cast doubt on this proposition. (c) Licences coupled with an interest are those that go with a recognizable interest in the land of another, e.g. a *profit à prendre. Such licences are irrevocable and assignable. They bind successors in title in the same way as the interest in land to which they relate. **4.** A carrier's licence (*see* carrier).

**licensed dealer** A dealer licensed by the Department of Trade and Industry to provide investment advice and deal in securities, either as an agent or principal. Licensed dealers are not members of the *London Stock Exchange and are not covered by its compensation fund.

**licensed deposit taker** A category of financial institutions as defined by the Banking Act (1979), which divided banking into recognized banks, licensed deposit takers, and exempt institutions. To qualify for authorization, the licensed deposit taker had to satisfy the Bank of England that it conducted its business in a prudent manner. The aim of the Act was to bring more institutions under the supervision of the Bank of England. The Banking Act (1987), however, established a single category of authorized institutions eligible to carry out banking business.

**lien** The right of one person to retain possession of goods owned by another until the possessor's claims against the owner have been satisfied. In a **general lien**, the goods are held as security for all the outstanding debts of the owner, whereas in a **particular lien** only the claims of the possessor in respect of the goods held must be satisfied. Thus an unpaid seller may in some contracts be entitled to retain the goods until he receives the price, a carrier may have a lien over goods he is transporting, and a repairer over goods he is repairing. Whether a lien arises or not depends on the terms of the contract and usual trade practice. This type of lien is a **possessory lien**, but sometimes actual possession of the goods is not necessary. In an **equitable lien**, for example, the claim exists independently of possession. If a purchaser of the property involved is given notice of the lien it binds him; otherwise he will not be bound. Similarly a **maritime lien**, which binds a ship or cargo in connection with some maritime liability, does not depend on possession and can be enforced by arrest and sale (unless security is given). Examples of maritime liens are the lien of a salvor, those of seamen for their wages and of masters for their wages and outgoings, that of a bottomry or respondentia bondholder (*see* hypothecation), and that over a ship at fault in a collision in which property has been damaged.

**life assurance** An *insurance policy that pays a specified amount of money on the death of the *life assured or, in the case of an *endowment assurance policy, on the death of the life assured or at the end of an agreed period, whichever is the earlier. Life assurance grew from a humble means of providing funeral expenses to a means of saving for oneself or one's dependants, with certain tax advantages. *With-profits policies provide sums of money in excess of the sum assured by the addition of *bonuses. However, while this has done something to counter the effects of inflation, a better cover is now given by *unit-linked policies, which invest the premiums in funds of assets, by means of buying units in the funds. *See also* whole (of) life policy; qualifying policy.

**Life Assurance and Unit Trust Regulatory Organization (LAUTRO)** *See* Self-Regulatory Organization.

**life assured** The person upon whose death a life-assurance policy makes an agreed payment. The life assured need not be the owner of the policy.

**lifeboat** A fund set up to rescue dealers on an exchange in the event of a market collapse and the ensuing insolvencies.

**life-cycle hypothesis** The hypothesis that an individual's *consumption over any period is proportional to his expected *wealth. For example, an individual with a strong expectation of success in later life will borrow to finance consumption in early life. This hypothesis is now considered more acceptable than either Keynesian theories of consumption or the *permanent income hypothesis, which are both based on current income rather than wealth. Econometric investigation, however, has not tended to support any particular theory.

**life office** A company that provides *life assurance.

**lifetime transfers** *See* inter vivos gifts.

**LIFFE** Abbreviation for *London International Financial Futures Exchange.

**LIFO** Abbreviation for *last in, first out.

**light dues** A levy on shipowners for maintaining lighthouses, beacons, buoys, etc., collected in the UK by HM Customs and Excise on behalf of *Trinity House.

**lighter** A flat-bottomed cargo barge, usually without its own means of propulsion, that is towed by a tug for short distances, as between a ship and a quay or along a river from a port to a warehouse. The charge for transporting goods by lighter is the **lighterage**. *See also* LASH.

**limit** **1.** An order given by an invester to a stockbroker or commodity broker restricting a particular purchase to a stated maximum price or a particular sale to a stated minimum price. Such a **limit order** will also be restricted as to time; it may be given firm for a stated period or firm until cancelled. **2.** The *maximum fluctuations (up or down) allowed in certain markets over a stated period (usually one day's trading). In some volatile circumstances the market moves the limit up (or down). The movement of prices on the Tokyo Stock Exchange is limited in this way as it is on certain US commodity markets.

**limitation of actions** Statutory rules limiting the time within which civil actions can be brought. Actions in simple contract and tort must be brought within six years of the cause of action. There are several special rules, including that for strict liability actions for defective products, in which case the period is three years from the accrual of the cause of action or (if later) the date on which the plaintiff knew, or should have known, the material facts, but not later than ten years from the date on which the product was first put into circulation. The present UK law is contained in the Limitations Act (1980), the Latent Damage Act (1986), and the Consumer Protection Act (1987).

**limited by guarantee** *See* limited company.

**limited carrier** *See* carrier.

**limited company** A *company in which the liability of the members in respect of the company's debts is limited. It may be **limited by shares**, in which case the liability of the members on a winding-up is limited to the amount (if any) unpaid on their shares. This is by far the most common type of registered company. The liability of the members may alternatively be **limited by guarantee**; in this case the liability of members if limited by the memorandum to a certain amount, which the members undertake to contribute on winding-up. These are usually societies, clubs, or trade associations. Since 1980 it has not been possible for such a company to be formed with a share capital, or converted to a company limited by guarantee with a share capital. *See also* public limited company.

**limited liability** *See* limited company.

**limited market** A market for a particular security in which buying and selling is difficult, usually because a large part of the issue is held by very few people or institutions.

**limited partner** *See* partnership.

**limit order** *See* limit.

**limit price** The highest price that established sellers in a market can charge for a product, without inducing a new seller to enter the market. The limit price is usually lower than the monopoly price but higher than the competitive price. Limit pricing enables established sellers to make higher profits than they would if they sold at the competitive price and it also ensures that new firms will not enter the market and drive the prices down to competitive levels.

**line and staff management** A system of management used in large organizations in which there are two separate hierarchies; the **line management** side consists of **line managers** with responsibility for deciding the policy of and running the organization's main activities (such as manufacturing, sales, etc.), while the **staff management**, and its separate **staff managers**, are responsible for providing such supporting services as warehousing, accounting, transport, personnel management, and plant maintenance.

**line drawing** An illustration in a book or other printed matter in which the image is formed from black lines and tinted areas, with no variation in tone.

**line extending** Increasing a line of products by adding variations of an existing brand. For example, Coca Cola extended its line with Diet Coke. Line extending runs the risk of weakening the brand name. *Compare* line filling.

**line filling** Adding products to an existing line of products in order to leave no opportunities for competitors. Line filling can be horizontal or vertical. In horizontal line filling, a video manufacturer may produce machines with a variety of features, such as half-speed copying, multi-recording memory, etc., at the top end of the price range, and relatively cheap cost-effective machines at the opposite end; a competitor can therefore only compete on price. In vertical line-filling a manufacturer may produce a wide variety of brand names within a single product line; some detergent manufacturers do this. *Compare* line extending.

**line management** *See* line and staff management.

**line printer** A type of computer printer that prints a line of characters at a time, as opposed to a serial printer, which prints one character at a time. The most common types are chain printers and drum printers. The chain printer has embossed letters on a loop, containing several sets of characters, that travels across the paper. In a drum printer the letters are embossed on a drum that rotates. As the drum or chain rotates, all the As in the line are printed first, then the Bs, and so on until the line is complete. Line printers are very fast and are used with large computers.

**liner** A ship that operates on a scheduled service, as opposed to a tramp steamer. *Shipping conferences fix freight rates for liners on established routes.

**Linkline** A telephone service provided by British Telecom for advertisers of a wide variety of goods and services. Customers call Linkline, either free or at a local charge, and the rest of the charge is paid by the advertiser.

**liquid assets (liquid capital; quick assets; realizable assets)** Assets held in cash or in something that can be readily turned into cash (e.g. deposits in a bank current account, trade debts, marketable investments). The ratio of these assets to current liabilities provides an assessment of an organization's *liquidity or solvency. *See also* liquid ratio.

**liquidated damages** *See* damages.

**liquidation (winding-up)** The distribution of a company's assets among its creditors and members prior to its dissolution. This brings the life of the company to an end. The liquidation may be voluntary (*see* creditors' voluntary liquidation; members' voluntary liquidation) or by the court (*see* compulsory liquidation).

**liquidation committee** A committee set up by creditors of a company being wound up in order to consent to the *liquidator exercising certain of his powers. When the company is unable to pay its debts, the committee is usually composed of creditors only; otherwise it consists of creditors and *contributories.

**liquidator** A person appointed by a court, or by the members of a company or its creditors, to regularize the company's affairs on a *liquidation (winding-up). In the case of a *members' voluntary liquidation, it is the members of the company who appoint the liquidator. In a *creditors' voluntary liquidation, the liquidator may be appointed by company members before the **meeting of creditors** or by the creditors themselves at the meeting; in the former case the liquidator can only exercise his powers with the consent of the court. If two liquidators are appointed, the court resolves which one is to act. In a *compulsory liquidation, the court appoints a *provisional liquidator after the winding-up petition has been presented; after the order has been granted, the court appoints the *official receiver as liquidator, until or unless another officer is appointed.

The liquidator is in a relationship of trust with the company and the creditors as a body; if he is appointed in a compulsory liquidation, he is an officer of the court, is under statutory obligations, and may not profit from his position. A liquidator must be a qualified insolvency practitioner, according to the Insolvency Act (1986). Under this Act, insolvency practitioners must meet certain statutory requirements, including membership of an approved professional body (such as the Insolvency Practitioners' Association or the Institute of Chartered Accountants). On appointment, the liquidator assumes control of the company, collects the assets, pays the debts, and distributes any surplus to company members according to their rights. In the case of a compulsory liquidation, the liquidator is supervised by the court, the *liquidation committee, and the Department of Trade and Industry. He receives a *statement of affairs from the company officers and must report on these to the court.

**liquid capital** *See* liquid assets.

**liquid crystal display (LCD)** A type of display used in digital watches and calculators as well as some small televisions and computers. The display consists of a thin layer of a liquid crystal – a liquid with some of the properties of a crystal – sandwiched between thin pieces of polarizing glass with attached electrodes. The characters appear as black against a light background when a small electrical voltage is applied across the electrodes. A seven-segment pattern is used to produce the set of numbers.

**liquidity** The extent to which an organization's assets are liquid (*see* liquid assets), enabling it to pay its debts when they fall due and also to move into new investment opportunities.

**liquidity-preference theory** A theory associating the desire that people have to hold money with various motives; in particular, *transactions demand, *precautionary demand, and *speculative demand. First introduced by J M Keynes, this theory supplied a basis for many conclusions concerning the tendency of economies to suffer *recession and *depression, characterized by *involuntary unemployment. Liquidity-preference theory rejects the *loanable-funds theory and the natural rate of *interest. It also underplays the role of the rate of interest in ensuring that plans for present and future consumption are balanced (*see* time preference); this omission is now seen as a fundamental weakness in the theory.

**liquidity ratio** *See* cash ratio.

**liquidity trap** A situation that could arise in an economy in which interest rates have fallen so low that investors allow their preference for liquidity to prevent them from investing in bonds. The liquidity trap was described by J M Keynes in his *general theory to point out that traditional views (now associated with *monetarism) were inconsistent. As a result of the liquidity trap, falling investment could lead to falling aggregate demand as well as falling prices and wages, but with no tendency to restore equilibrium between aggregate supply and demand. Neoclassical economists were quick to point out that falling prices would increase the real value of consumers' wealth, thus increasing the demand for goods, which would lead to the restoration of (full-employment) equilibrium. Although the liquidity trap is still taught as part of the canon of economics, it is now considered to have little practical relevance.

**liquid ratio (acid-test ratio)** The ratio of the liquid assets of an organization to its *current liabilities. The liquid assets are normally taken to be trade debt and cash and any other assets that are readily marketable. The ratio gives an indication of the organization's ability to pay its debts without needing to make further sales. It therefore provides the ultimate (acid-test) proof of its solvency.

**listed company** A company that has a **listing agreement** (*see* listing requirements) with the London Stock Exchange and whose shares therefore have a *quotation on the *main market. These companies were formerly called **quoted companies**.

**listed security 1.** In general, a security that has a *quotation on a recognized *stock exchange. **2.** On the *London Stock Exchange, a security that has a quotation in the Official List of Securities of the *main market, as opposed to the unlisted securities market or the *third market. *See also* flotation; listing requirements; Yellow Book.

**listing requirements** The conditions that must be satisifed before a security can be traded on a stock exchange. To achieve a quotation in the Official List of Securities of the *main market of the *London Stock Exchange, the requirements contained in a **listing agreement** must be signed by the company seeking quotation. The two main requirements of such a listing are usually:
(i) that the value of the company's assets should exceed a certain value;
(ii) that the company publish specific financial information, both at the time of *flotation and regularly thereafter (*see* accounts; directors' report). Listing requirements are generally more stringent the larger the market. For example, the main market in London demands considerably more information from companies than the *unlisted securities market. The listing requirements are set out in the *Yellow Book.

**list price 1.** The retail price of a consumer good as recommended by the manufacturer and shown on his price list. If no price-maintenance agreements apply, a discount on the list price may be offered by the retailer to attract trade. **2.** A supplier's price as shown on an invoice to a retailer or wholesaler, before deduction of any discounts.

**list renting** The practice of renting a list of potential customers to an organization involved in *direct-mail selling or to a charity raising funds. When the terms of hiring restrict the use of the list to one mail shot, it is usual for the owner of the list to carry out the mailing, so that the hirer cannot copy it for subsequent use.

**lists closed** The closing of the application lists for a *new issue on the London Stock Exchange, after a specified time or after the issue has been fully subscribed.

**literal** A wrong letter in typeset matter, resulting from an error in keyboarding.

**litigation 1.** The taking of legal action. The person who takes it is called a **litigant**. **2.** The activity of a solicitor when dealing with proceedings in a court of law.

**livery company** One of some eighty chartered companies in the City of London that are descended from medieval craft guilds. Now largely social and charitable institutions, livery companies owe their name to the elaborate ceremonial dress (livery) worn by their officers. Several support public schools (e.g. Merchant Taylors, Haberdashers, Mercers) and although none are now trading companies, some still have some involvement in their trades (e.g. Fishmongers). In 1878 they joined together to form the City and Guilds of London Institute, which founded the City and Guilds College of the Imperial College of Science and Technology and has been involved in other forms of technical education.

**livestock and bloodstock insurance** An insurance policy covering the owners against financial losses caused by the death of an animal. Policies may be widened to include cover for treatment fees for certain specified diseases or lost profits for stud animals. Insurances can be arranged to cover single animals or for whole herds.

**Lloyd's** A corporation of underwriters (**Lloyd's underwriters**) and insurance brokers (**Lloyd's brokers**) that developed from a coffee shop in Tavern Street in the City of London in 1689. It takes its name from the proprietor of the coffee shop, Edward Lloyd. By 1774 it was established in the Royal Exchange and in 1871 was incorporated by act of parliament. It now occupies a new (1986) building in Lime Street (built by Richard Rogers). As a corporation, Lloyd's itself does not underwrite insurance business; all its business comes to it from some 260 Lloyd's brokers, who are in touch with the public, and is underwritten by some 350 *syndicates of Lloyd's underwriters, who are approached by the brokers and who do not, themselves, contact the public.
The 30 000 or so Lloyd's underwriters must each deposit a substantial sum of money with the corporation and accept unlimited liability before they can become members. They are grouped into syndicates, run by a syndicate manager or agent, but most of the members of syndicates are **names**, underwriting members of Lloyd's who take no part in organizing the underwriting business, but who share in the profits or losses of the syndicate and provide the risk capital. Lloyd's has long specialized in marine insurance but now covers almost all insurance risks. *See also* Lloyd's agent; Lloyd's List and Shipping Gazette; Lloyd's Register of Shipping.

**Lloyd's agent** A person situated in a port to manage the business of Lloyd's members, keeping the corporation informed of shipping movements and accidents, arranging surveys, helping in the settlement of marine insurance claims, and assisting masters of ships. A Lloyd's agent will be found in every significant port in the world.

**Lloyd's broker** *See* Lloyd's.

**Lloyd's List and Shipping Gazette** A daily newspaper published by Lloyd's, founded in 1734 and formerly known as **Lloyd's List**. It gives details of the movements of ships and aircraft, accidents, etc. **Lloyd's Loading List**, published weekly by Lloyd's, lists ships loading in British and continental ports, with their closing dates for accepting cargo. It also gives general news on the insurance market.

**Lloyd's Register of Shipping** A society formed by *Lloyd's in 1760 to inspect and classify all ocean-going vessels in excess of 100 tonnes. Ships are periodically surveyed by **Lloyd's surveyors** and classified according to the condition of their hulls, engines, and trappings. The society also provides a technical advice service. Its annual publication is called **Lloyd's Register of British and Foreign Shipping**. The Register enables underwriters to have instant access to the information they need to underwrite marine risks, even when the vessels may be thousands of miles away.

**Lloyd's underwriter** *See* Lloyd's.

**LME** Abbreviation for *London Metal Exchange.

**load line** One of a series of lines marked on the hull of a ship to show the extent to which the hull may be immersed in the water. Originally introduced by Samuel Plimsoll MP in 1874 and running right round the hull, the **Plimsoll Line** made a great contribution to safety at sea. The modern markings consist of a series of lines, usually painted on the hull amidships, applying to different conditions. The line marked TF applies to tropical fresh water, F fresh water,

T tropical sea water, S summer sea water, W winter sea water and WNA winter in the North Atlantic. Shipowners and masters who allow vessels to be overloaded face heavy penalties.

**loan** Money lent on condition that it is repaid, either in instalments or all at once, on agreed dates and usually that the borrower pays the lender an agreed rate of interest (unless it is an **interest-free loan**). *See also* balloon; bank loan; bridging loan; bullet; local loan; personal loan.

**loanable-funds theory** A theory that rates of interest must rise or fall until the amount that borrowers wish to raise in loans is exactly matched by the amount of funds lenders wish to lend in that period. First formulated by D H Robertson (1890–1963), it is based on the theory of the natural rate of *interest. J M Keynes's *liquidity-preference theory has recently lost ground to the loanable-funds theory.

**loan account** An account opened by a bank in the name of a customer to whom it has granted a loan, rather than an *overdraft facility. The amount of the loan is debited to this account and any repayments are credited; interest is charged on the full amount of the loan less any repayments. The customer's current account is credited with the amount of the loan. With an overdraft facility, interest is only charged on the amount of the overdraft, which may be less than the full amount of the loan.

**loanback** An arrangement in which an individual can borrow from the accumulated funds of his pension scheme. Usually a commercial rate of interest has to be credited to fund for the use of the capital. Some life assurance companies offer loan facilities on this basis of up to fifteen times the annual pension premium. Companies can also borrow from company pension schemes in the same way.

**loan capital** Money required to finance the activities of an organization that is raised by loans (*see* debenture), as distinct from its *share capital. The principal advantages of loan capital over share capital are that it can be readily repaid when the company has funds, interest charges are deductible for tax purposes, and there is no capital duty on the issue of loan stock. Unlike the share capital, which has a right to share in the organization's profits, loan capital is a fixed-interest return.

**Loan Guarantee Scheme** A UK government scheme that guarantees 70% of a company's overdraft for a 3% premium. The bank must accept the risk for the balance of 30%. Its purpose is to support small businesses.

**loan note** A form of loan stock (*see* debenture) in which an investor takes a note rather than cash as the result of a share offer to defer tax liability. The yield is often variable and may be linked to the *London Inter Bank Offered Rate. Loan notes are not usually marketable but are usually repayable on demand.

**loan stock** *See* debenture.

**local** A person who has a seat on a commodity futures market and who deals for himself.

**local area network (LAN)** A collection of linked computer stations (often microcomputers) restricted to a small local area, such as an office building or a university campus. Stations in the network are able to communicate with both a central computer and with each other. The network is used as a communications system as well as providing wide access to central computing facilities.

**local loan** A loan issued by a UK local government authority for financing capital expenditure.

**lock-out** A form of industrial action in which the employer refuses to allow employees access to their place of work unless they accept the employer's terms of employment.

**lock-up** An investment in assets that are not readily realizable or one that is specifically intended to be held for a long period – say, over ten years.

**loco** Denoting a price for goods that does not include any loading or transport charges; i.e. the goods are located in a specified place, usually the seller's warehouse or factory, and the buyer has to pay all the charges involved in loading, transporting, or shipping them to their destination. A **loco price** might be quoted in the form: *£100 per tonne, loco Islington factory*.

**locus poenitentiae** (Latin: an opportunity to repent) An opportunity for the parties to an illegal contract to reconsider their positions, decide not to carry out the illegal act, and so save the contract from being void. Once the illegal purpose has been carried out, no action in law is possible.

**logistics** The control of the movement of physical materials in a factory. It is usually subdivided into **materials management**, which is the control of the movement of materials in the factory, from the arrival of raw materials to the packaging of the product; and **distribution management** (*or* **marketing logistics**), which includes the storage of goods and their distribution to distributors and consumers.

**Lombard Street** The street in the City of London that is the centre of the money market. Many commercial banks have offices in or near Lombard Street, as do many bill brokers and discount houses. The Bank of England is round the corner.

**London Bankers' Clearing House** *See* Association for Payment Clearing Services.

**London Chamber of Commerce** The largest chamber of commerce in the UK. It provides the normal services of a chamber of commerce and in addition runs courses and examinations in business subjects.

**London Commodity Exchange** *See* London FOX.

**London FOX (Futures and Options Exchange)** A commodity exchange formed in 1987 from the **London Commodity Exchange**, which itself emerged after World War II as a successor to the London Commercial Sale Rooms. London FOX is located in a purpose-built exchange in St Katharine Dock, which it shares with the *International Petroleum Exchange. The commodities dealt in on the exchange are cocoa, coffee, raw sugar, and white sugar. The market is in futures (*see* futures contract) and *options, including traded options.

**London Inter Bank Offered Rate (LIBOR)** The rate of interest in the short-term wholesale market (*see* inter-bank market) in which banks lend to each other. The loans are for a minimum of £250,000 for periods from overnight up to five years. The importance of the market is that it allows individual banks to adjust their liquidity positions quickly, covering shortages by borrowing from banks with surpluses. This reduces the need for each bank to hold large quantities of liquid assets, thus releasing funds for more profitable lending transactions.

**London International Financial Futures Exchange (LIFFE)** A **financial futures** market opened in 1982, in London's Royal Exchange, to provide facilities within the European time zone for dealing in options and futures contracts, including those in government bonds, stock-and-share indexes, foreign currencies, and interest rates. A client wishing to buy or sell options or futures telephones a LIFFE broker, who instructs his booth on the floor of the exchange. The booth clerk hands a slip to the broker's trader in the *pit of the market who executes the transaction with another trader. The bargain details are passed to the *International Commodities Clearing House, who act as guarantors.

**London Metal Exchange (LME)** A central market for non-ferrous metals, established in London in 1877 to supply a central market for the import of large quantities of metal from abroad. The Exchange deals in copper, lead, zinc, aluminium (all in minimum lots of 25 tonnes), tin (minimum 5 tonnes), and nickel (minimum 6 tonnes). The official prices of the LME are used by producers and consumers worldwide for their long-term contracts. Dealings on the LME include *futures and *options contracts. Bargains are guaranteed by the *International Commodities Clearing House.

**London Stock Exchange** The market in London that deals in securities. Dealings in securities began in London in the 17th century. The name Stock Exchange was first used for New Jonathan's Coffee House in 1773, although it was not formally constituted until 1802. The development of the industrial revolution encouraged many other share markets to flourish throughout the UK, all the remnants of which amalgamated in 1973 to form The Stock Exchange of Great Britain and Ireland. After the *Big Bang in 1986 this organization became the **International Stock Exchange of the UK and Republic of Ireland Ltd**. Its major reforms included:

(i) allowing banks, insurance companies, and overseas securities houses to become members and to buy existing member firms;

(ii) abolishing scales of commissions, allowing commissions to be negotiated;

(iii) abolishing the division of members into jobbers and brokers, enabling a member firm to deal with the public, to buy and sell shares for their own account, and to act as *market makers;

(iv) the introduction of *SEAQ, a computerized dealing system that has virtually abolished face-to-face dealing on the floor of exchange.

In merging with members of the international broking community in London, the International Stock Exchange became a registered investment exchange and The Securities Association Ltd became a *Self-Regulatory Organization (SRO) complying with the Financial Services Act (1986).

The International Stock Exchange provides three markets for companies: the *main market for *listed companies, the *unlisted securities market (USM), and the *third market. It also offers a market in traded *options in equities, currencies, and indexes. *See* London Traded Options Market.

**London Traded Options Market (LTOM)** A subsidiary company of the International Stock Exchange (*see* London Stock Exchange). Self-financed by the sale of seats to floor members, it uses the floor of the old Stock Exchange Building for dealings in traded *options.

**long-dated gilt** *See* gilt-edged security.

**long position** A position held by a dealer in securities (*see* market maker), commodities, currencies, etc., in which his holding exceeds his sales, because he expects prices to rise enabling him to sell his longs at a profit. *Compare* short position.

**long run** A period of time that is sufficiently long for all the economic *factors of production to be varied by a firm to enable it to obtain the most efficient combination. *Compare* short run.

**longs 1.** *See* gilt-edged security. **2.** Securities, commodities, currencies, etc., held in a *long position.

**long-term debt** Loans and debentures that are not due for repayment for at least ten years.

**long-term liability** A sum owed that does not have to be repaid within the next accounting period of a business. In some contexts a long-term liability may be regarded as one not due for repayment within the next three, or possibly ten, years.

**long ton** *See* ton.

**loose insert** An advertising leaflet distributed with another publication and usually inserted loosely within its pages.

**Lorenz curve** A curve on a graph showing the degree of equality in the distribution among a population of some variable, usually income or wealth. It is drawn by plotting the cumulative percentage of the population (say households) against the cumulative percentage of, say, income, starting from the lowest income and ending with the highest. Complete equality would be represented by a 45° line between the axes. The degree of curvature away from the 45° line represents the degree of inequality. The **Gini coefficient** is a measure of this degree of inequality, being the ratio of the area between the actual curve and the 45° line to the area under the 45° line.

**loss adjuster** A person appointed by an insurer to negotiate an insurance claim. The loss adjuster, who is independent of the insurer, discusses the claim with both the insurer and the policyholder, producing a report recommending the basis on which the claim should be settled. The insurer pays a fee for this service based on the amount of work involved for the loss adjuster, not on the size of the settlement. *Compare* loss assessor.

**loss assessor** A person who acts on behalf of the policyholder in handling a claim. A fee is charged for this service, which is usually a percentage of the amount received by the policyholder. *Compare* loss adjuster.

**loss leader** A product or service offered for sale by an organization at a loss in order to attract customers. This practise was curbed in the UK by the Resale Prices Act (1976), although it still continues, especially in supermarkets.

**loss-of-profit policy** *See* business-interruption policy.

**loss ratio** The total of the claims paid out by an insurance company, underwriting *syndicate, etc., expressed as a percentage of the amount of premiums coming in in the same period. For example, if claims total £2m and premiums total £4m, the result is a 50% loss ratio. Insurers use this figure as a guide to the profitability of their business when they are reconsidering premium rates for a particular risk.

**lower case** The small letters of the alphabet, as compared to the capital letters, which are referred to as *upper case.

**low-involvement product** A product that involves the consumer in little or no trouble or deliberation when making a purchase. Such products are invariably cheap; manufacturers try to make them more interesting and appealing by means of advertising with the intention of developing a brand loyalty. The chimpanzees used in TV advertisements for PG Tips are an example of this form of advertising. *Compare* high-involvement product.

**low-level language** A type of computer language that is close to a form that the computer can understand directly. There are two kinds of low-level language: *assembly language and *machine code. Low-level languages are difficult to use but produce programs that run faster than programs written in the more convenient *high-level languages.

**LTOM** Abbreviation for *London Traded Options Market.

**Ltd** The usual abbreviation for *limited*. This (or the Welsh equivalent) must appear in the name of a private *limited company. *Compare* plc.

**Lucas critique** The view that government policies based on a particular model of the economy must take account of the way in which individuals change their behaviour if they become aware that the government is using that model. Formulated by the influential US macroeconomist Robert Lucas, the critique suggests that relationships between economic variables based on individual choices may not remain stable, making policy formulation difficult if not impossible. *See also* rational expectations.

**lump sum 1.** A sum of money paid all at once, rather than in instalments. **2.** A sum of money paid for freight, irrespective of the size of the cargo. **3.** An insurance benefit, such as a sum of money paid on retirement or redundancy or to the beneficiaries on the death of an insured person. Retirement *pensions can consist of a lump sum plus a reduced pension. **4.** A form of *damages; a **lump-sum award** is given in tort cases.

**lump-sum tax** A tax that calls on the taxpayer to pay a fixed amount, which is unrelated to any factor and which cannot be avoided by any action. Lump-sum taxes are said to be Pareto-efficient (*see* Pareto optimality), as they do not cause individuals to alter their pattern of consumption or employment, as most other taxes do. Examples of genuine lump-sum taxes are not easy to find in practice, although a *poll tax is often described as one.

**lump system** A system of employment used in the building trade in which workers are self-employed and paid lump sums for an agreed amount of work. This formerly enabled workers to avoid paying tax and insurance on their earnings; this kind of tax evasion has now been greatly reduced as employers now have to give to the Inland Revenue the details (including names and addresses) of all payments to self-employed workers.

**luncheon voucher (LV)** A printed ticket that can be used in payment or part-payment of a meal in certain restaurants. They are issued to employees by some firms that do not have canteens and are tax-free up to a fixed daily value. As this now stands at 15p per day, LVs have lost popularity, but can still be used as a *fringe benefit. They are sold to employers by luncheon-voucher firms, who pay the restaurants the face value of the tickets, making a commission on the transaction.

**Lutine Bell** A bell that hangs in the underwriting room at *Lloyd's and is rung for ceremonial occasions and to draw the attention of underwriters to an important announcement. It was formerly rung once if a ship sank and twice for good news. It was recovered from the *Lutine*, a ship that was insured by Lloyd's and sank in the North Sea in 1799, with a cargo of bullion (£1.4m value), most of which was lost.

**luxury good** A good for which demand increases more than proportionately as income rises (*compare* necessary good). Another way of saying this is that the income *elasticity of demand is greater than unity. *See also* Engel curve; inferior good; normal good.

**LV** Abbreviation for *luncheon voucher.

# M

**M0; M1; M2; M3; M4; M5** *See* money supply.

**machine code (machine language)** A code that a computer can understand directly. It is the most basic of the programming languages. Each action the processor can be instructed to perform is identified by a unique number, usually represented in *octal or *hexadecimal notation. A machine-code program is a series of such numbers, interspersed with other numbers, which are either parameters to control the actions or the data on which the actions are performed. All programs written in a *high-level language, such as BASIC or COBOL, must be converted into machine code by a *compiler before they can be run (or else executed by an *interpreter, which simulates this).

**machine-down time** The period during which a machine cannot be used, usually because of breakdown. If a machine is 'down', clearly production is not taking place, but it is customary in costing to attribute costs to the down time.

**machine-idle time** The period during which a machine is not being used. This is similar in effect to *machine-down time though it may be caused by lack of work rather than by a fault in the machine.

**machine-readable** Denoting any form of data that can be input directly into a computer. It includes data that has been stored on a magnetic medium (e.g. disks, tapes, etc.) and data that has been prepared by *OCR.

**macro** A sequence of computer instructions generated by a special piece of software called a **macrogenerator**. Instead of writing the same sequence of instructions every time it is needed, the programmer defines a macro by giving the sequence a symbolic name, which is used in the program. The macrogenerator (which may be part of a larger piece of software, such as an *assembler) automatically substitutes the sequence for all occurences of the name.

**macroeconomic model** A model of a country's economy that is based on *macroeconomic theory and *econometric analysis. These models make use of past data on such variables as output, employment, and consumption to forecast future values. In the UK, many such models are constructed by various groups of economists; the most influential is the model constructed by the Treasury. However, the quality and usefulness of the forecasts so obtained are still widely debated.

**macroeconomics** The study of economic aggregates and their relationships to, for example, money, employment, interest rates, government spending, investment, and consumption. J M Keynes (1883–1946) is widely credited with the foundation of macroeconomics since he sought to explain that even if the economy is operating efficiently at the *microeconomic level, unemployment and recession may still occur at the macroeconomic level, because of the lack of coordination between markets. Modern macroeconomics may be summarized as an attempt to assess the validity of this proposition (*see also* Keynesianism; monetarism) and to establish what role, if any, the government should play in the economy.

**magnetic disk** A flat circular disk coated with a magnetic material, used to store information in a computer system. The information is stored as a pattern of magnetized spots on the surface of the disk. The spots are arranged around the centre of the disk in concentric circles (tracks), which are divided into sectors. The disk is rotated in a *disk drive, under a read/write head, which can both

read the pattern of magnetized spots on the disk and write a pattern onto the disk.
Several kinds of disk are in use. Some (*fixed disks) are permanently fixed inside the disk drive, such as the *Winchester disk used in small computers. Others, such as the *floppy disk, can be removed from the disk drive. Large computers also use cartridge disks (in which the disks are held in a plastic cover) and disk packs (a stack of disks in a plastic hood).

**magnetic ink character recognition (MICR)** A type of magnetic ink used on cheques and other documents to enable them to be automatically sorted and the characters to be read and fed into a computer.

**magnetic tape** Plastic tape having a magnetic surface on which data can be stored. The data is stored as a series of magnetized spots, which run lengthwise along the tape. The data is read by moving the tape past a read/write head. Magnetic tape is used as *backing storage on large computers. *See also* magnetic disk.

**mailbox** *See* electronic mail.

**mailing list** A list of names and addresses used in *direct-mail selling, advertising, fund raising, etc. *See also* list renting.

**mailmerge** *See* wordprocessor.

**mail-order house** A firm that specializes in selling goods direct to customers by post. Orders are obtained from an illustrated catalogue supplied by the firm or by agents, who introduce the catalogue. The low costs of selling, especially the absence of retail premises, enable the mail-order houses to offer goods at competitive prices. *See also* direct-mail selling.

**mail shot** Selling, advertising, or fund-raising material sent to all the names on a mailing list.

**mail survey** Market research conducted by mail. As respondents have time to consider their answers, this form of research has advantages over both telephone surveys and face-to-face surveys. However, it does suffer from low returns.

**mainframe** The largest type of *computer, requiring an air-conditioned room and special staff, including operators, programmers, and system analysts, to run it. Used by large organizations, such as banks, they can handle vast amounts of information with ease and calculate at high speed. They can also handle many users simultaneously (*see* time sharing). *Compare* microcomputer; minicomputer.

**main market** The premier market for the trading of *equities on the *London Stock Exchange. For this market the *listing requirements are the most stringent and the liquidity of the market is greater than in the *unlisted securities market and the other junior markets. A company wishing to enter this market must have audited trading figures covering at least five years and must place 25% of its shares in public hands. The main market currently deals in over 2500 stocks. *Compare* over-the-counter (OTC) market; third market.

**main store (internal storage)** The main part of a computer *memory. Forming part of the *central processing unit, it is a high-speed or *immediate-access store, which holds programs while they are being executed and the intermediate results of calculations in progress. The main store is supplemented by *backing store.

**mainstream corporation tax (MCT)** The liability for corporation tax of a company for an accounting period after the relevant *advance corporation tax (ACT) has been deducted. In the imputation system of corporation tax in the UK, payments on account of ACT are paid when dividends are paid to shareholders.

Mainstream corporation tax is the balance remaining to be paid.

**making a price** On the *London Stock Exchange, the quoting by a *market maker of a price at which he will sell securities and a price at which he will buy them, usually without knowing whether the broker or other person asking him to make a price wishes to buy or sell. Having made a price, the market maker is bound to buy or sell at the prices he has quoted, though he may limit the quantity by saying so at the time he makes the price.

**making-up price** The price at which securities that have not been paid for on *account day (on the London Stock Exchange) are carried forward to the next account.

**managed currency (managed floating)** A currency in which the government controls, or at least influences, the exchange rate. This control is usually exerted by the central bank buying and selling in the foreign-exchange market. *See also* clean floating.

**managed fund** A fund, made up of investments in a wide range of securities, that is managed by a life-assurance company to provide low risk investments for the smaller investor, usually in the form of *investment bonds, *unit trusts, or unit-linked saving plans (*see* unit-linked policy). The *fund managers will have a stated investment policy favouring a specific category of investments.

**management 1.** The running of an organization, which in economic theory is sometimes regarded as a *factor of production, with land, labour, and capital. Management has two main components: an organizational skill, including the ability to delegate, and an entrepreneurial sense. The organizational skill, involving the principles and techniques of management, is taught at colleges and business schools; the entrepreneurial sense, recognizing and making use of opportunities, predicting market needs and trends, and achieving one's goals by sustained drive, skilful negotiation, and articulate advocacy, are not so easily taught, although contact with the market place in association with a successful entrepreneur will encourage an inherent ability to develop.

**Top management** includes the *chief executive (*see also* managing director) of an organization, his deputy or deputies, the board of *directors, and the managers in charge of the divisions or departments of the organization. **Middle management** consists largely of the managers to whom top management delegates the day-to-day running of the organization.

Management is usually broken down into the categories formalized in *line and staff management: the line managers organize the production of the goods or oversee the services provided by the organization, while the staff management provides such support as personnel management, transport management, service management, etc. *See also* British Institute of Management. **2.** The people involved in the running of an organization.

**management accountant** An accountant whose primary role is to advise the management of a company on the consequences of their activities. He is likely to be involved in the budgetary procedures of his organization and will be responsible for producing monthly management figures; he will also be involved in advising on new developments, pricing, cash-flow requirements, and the financial consequences of various projects. The professional organization of management accountants in the UK is the Chartered Institute of Management Accountants (*see* chartered accountant).

**management agreement** A formal agreement setting out the objectives and services provided by an investment advisor or stockbroker in managing a client's portfolio and the costs of this service.

**management audit** An independent review of the management of an organization, carried out by a firm of *management consultants specializing in this type of review. The review will cover all aspects of running the organization, including the control of production, marketing, sales, finance, personnel, warehousing, etc.

**management buyout** The buying by its managers of a company that is in trouble, the target of an unwelcome *takeover bid, or about to be floated. Usually the managers require the backing of a bank or an institutional investor to finance the purchase of sufficient shares to gain control of the company. The hope of both the managers and the backers is that the financial involvement of the managers, who know more about the business than anyone else, will make the company sufficiently successful for its shares to rise in price on the market, enabling the managers to sell at least a part of their holdings at a profit. *See also* leveraged buyout.

**management by exception** A management technique that seeks to highlight differences between actual and budgeted costs or any abnormal feature in the running of an organization in order that managerial time should be devoted to those exceptional items. Smoothly running operations and accurately budgeted costs thus receive less attention than the problem areas.

**management company** A company that manages *unit trusts. Its fees, known as **management charges**, are usually stated in the agreement setting up the trust; they are paid by the unit holders.

**management consultant** A professional adviser who specializes in giving advice to organizations on ways for improving their efficiency and hence their profitability. They come into an organization as total outsiders, uninfluenced by either internal politics or personal relationships, and analyse the way the business is run. At the end of a period, during which two or more members of the consultant firm have spent a considerable time in the organization, they provide a detailed report, giving their suggestions for improving efficiency. Their advice usually spans board-level policy-making and planning, the use of available resources (department by department), the best use of manpower, and a critical assessment of industrial relations, production, marketing, and sales. *See* Management Consultants' Association; Institute of Management Services.

**Management Consultants' Association (MCA)** An association founded in 1956 to establish and maintain the professional standards of the management consultancy profession in the UK.

**management information system (MIS)** A database held within a company, to which only management has access. It enables all the managers in the organization to have the same basic data on which to formulate their decisions.

**managerial theory of the firm** In economics, the theory that the managers of a firm make choices on behalf of their firm on the basis of their own utility rather than of *profit maximization. Factors determining the managerial utility function include salary, prestige, market share, and job security. Although this assumption may be more realistic than those based on profit maximization, it makes it much more difficult to make predictions about the behaviour of firms and markets. *See also* satisficing behaviour.

**managing director (MD)** The company director responsible for the day-to-day running of a company. Second in the hierarchy only to the *chairman, if there is one, the managing director is the company's *chief executive, a title that is becoming increasingly popular in both the USA and the UK. In the USA the *president is often the equivalent of the MD. If a company has more than one MD, they are known as **joint managing directors**.

**mandate** A written authority given by one person (the **mandator**) to another (the **mandatory**) giving the mandatory the power to act on behalf of the mandator. It comes to an end on the death, mental illness, or bankruptcy of the mandator. *See also* dividend mandate.

**man-hour** The amount of work done by one man in one hour. It is sometimes convenient in costing a job to estimate the number of man-hours it will take.

**manifest** A list of all the cargo carried by a ship or aircraft. It has to be signed by the captain (or first officer) before being handed to the customs on leaving and arriving at a port or airport.

**manpower** The total work force, both men and women, in a country, region, or industry.

**Manpower Services Commission** *See* Training Agency.

**manufacturer brand** A *brand name created by a manufacturer. Kellogs and Polaroid are examples.

**manufacturer's agent** A commission agent who usually has a *franchise to sell a particular manufacturer's products in a particular country or region for a given period.

**manufacturers' recommended price (MRP)** *See* recommended retail price.

**manufacturing account** The part of a profit-and-loss account in which the manufactured cost of goods sold is calculated by taking *direct costs, factory overheads, etc., and adjusting for changes in stocks of raw material and work in progress.

**Mareva injunction** An order of the court preventing the defendant from dealing with specified assets. Such an order will be granted in cases in which the plaintiff can show that there will be a substantial risk that any judgment he obtains against the defendant will be worthless, because the defendant will dissipate his assets to avoid paying. It is usually granted to prevent assets leaving the jurisdiction of the English courts, but may in exceptional circumstances extend to assets abroad. It is named after the 1975 case *Mareva Compania Naviera SA* v *International Bulkcarriers SA*.

**margin 1.** The percentage of the cost of goods that has to be added to the cost to arrive at the selling price. *See* gross margin. **2.** The difference between the prices at which a *market maker or commodity dealer will buy and sell. **3.** In commodity and currency dealing, the amount advanced by a speculator or investor to a broker or dealer, when buying futures. **4.** Money or securities deposited with a stockbroker to cover any possible losses a client may make. *See also* margin call.

**marginal analysis** *See* Marginalists.

**marginal cost** The additional cost of producing an additional unit of output. In conditions of perfect competition, marginal cost would be equal to the market price, which firms would be unable to influence (*see* price-taking behaviour). These circumstances lead to *Pareto optimality. However, if the marginal cost is less than the market price, firms could increase their profits by raising output. Similarly, if the marginal cost is more than the market price, firms could increase their profits by reducing output. In a *monopoly or *oligopoly or in monopolistic competition, marginal costs are not generally equal to the market price and the economy is not at a Pareto optimum.

**marginal costing** The process of costing products or activities by taking account only of the *direct costs of the product or activity. The normal procedure of marginal costing is to compare the direct costs with the selling price of the product or service in order to see what *contribution the item makes towards fixed overheads and profit. *Compare* absorption costing.

**Marginalists** Followers of a school of thought dating from the 1870s, on which modern neoclassical economics is based. Developed independently by W S Jevons (1835–82), Carl Menger (1840–1921), and M E L Walras (1834–1910), it argues that economic value is determined by the rate of exchange (or price) of the last unit of a good supplied in the market. This broke with the *Classical school's view of value, which was defined as the quantity of labour embodied in a unit of output. **Marginal analysis** was then applied to the behaviour of consumers (*see* marginal utility) and the behaviour of the firm. When applied to a particular firm or to households, this analysis yields a *partial equilibrium explanation of choice; applied to the economy as a whole it provides a general equilibrium solution, describing the efficient allocation of all resources.

**marginal productivity** The additional output that a producer will achieve by the addition of one unit of a *factor of production, such as an additional employee.

**marginal propensity to consume** The additional *consumption generated by a unit increase in income. Coined by Keynes, the phrase features in his *general theory in association with the *multiplier.

**marginal rate of substitution** The ratio of the *marginal utilities of two goods. A condition for Pareto economic efficiency (*see* Pareto optimality) is that the marginal rate of substitution of any two goods should equal their price ratio. If this were not the case, some individuals could achieve the same level of utility at a lower cost by shifting consumption between goods, which would constitute a Pareto improvement.

**marginal relief** Relief given by a tax authority if a marginal increase in a person's earnings brings him into a higher tax bracket and results in an unfair tax burden. One option for the taxpayer would be to pay tax at the lower rate, adding to the tax the amount by which his income exceeds the limit for this lower rate. For example, if the rate of tax is 25% up to £30,000 and 30% for incomes that exceed this figure, a taxpayer earning say £30,100 might opt to pay tax at 25% on the £30,000 and pay over to the tax authorities the additional £100.

**marginal revenue** The additional revenue that a producer will achieve by selling one additional unit of production.

**marginal tax rate** The additional tax paid for each unit increase in income. In the UK, for example, a taxpayer paying tax at the *higher rate of 40% (1989) will pay tax at this rate on all his additional earnings and also on all his taxable capital gains (*see* capital-gains tax). Under a *progressive tax system there is a tendency for the marginal tax rates of the poor and the rich to be very high, giving rise to the *poverty trap for the poor and, it is claimed, work disincentives for the rich (*see* Laffer curve).

**marginal utility** The additional *utility derived by an individual from the consumption of an additional unit of a good or service. One condition for the Pareto economic efficiency (*see* Pareto optimality) of an economy is that the ratios of marginal utilities of all goods to their prices should be equal. If this were not the case some individuals could switch from consumption of one good at the margin to another and retain the same utility at a lower cost; this would be a Pareto improvement.

**margin call** A call to a client from his commodity broker or stockbroker to increase his *margin, i.e. the amount of money or securities deposited with the broker as a safeguard. This usually happens if the client has an *open position in a market that is moving adversely for this position.

**marine insurance** The insurance of ships or their cargo against specified causes of loss or damage that might be encountered at sea. The definition has widened over the years to include the transit of cargo over land at each end of the voyage and the term 'vessel' now extends to include ships under construction or repair and drilling rigs.

**marked cheque** A cheque that the bank on which it is drawn has marked "good for payment". This practice has been replaced in the UK by *bank drafts, although it is still used in the USA, where such cheques are called **certified checks.**

**market 1.** The arena in which buyers and sellers meet to exchange items of *value. A physical market in which traders haggle for the best price is one of the basic concepts of trade and a considerable part of economics concerns the operation of such markets. The absence of a physical market for some goods is usually seen as one of the major sources of economic inefficiency. **2.** An organized gathering in which trading in securities (*see* stock exchange), commodities (*see* commodity market), currencies, etc., takes place. **3.** The demand for a particular product or service, often measured by sales during a specified period.

**marketable security** A security (stock, share, bond, etc.) that can be bought or sold on a *stock exchange. *See also* London Stock Exchange; alpha stocks. *Compare* non-marketable securities.

**Market and Opinion Research International (MORI)** A market research organization engaged in a wide variety of research activities. It is best known for its social and political research and its MORI opinion polls.

**market assessment** The identification and evaluation of a market for a particular good or service to ascertain its size and to estimate the price that the product or service would command. *See also* marketing research.

**market capitalization (market valuation)** The value of a company obtained by multiplying the number of its issued shares by their *market price.

**market clearing assumption** The assumption in economics that the interaction of choices made by utility-maximizing individuals ensures that the quantity of goods supplied equals the quantity demanded in any particular market. If this were not the case, a mutually advantageous trade could take place, to provide a Pareto improvement (*see* Pareto optimality). Market clearing is an important assumption in, for example, *new classical macroeconomics, which denies the possibility of an effective government *monetary policy or *fiscal policy. Keynesians, however, argue that market clearing may not occur due to inefficiences inherent in the economy (*see* efficiency wage theory; hysteresis; insider-outsider theory), which could provide a possible justification for monetary policy or fiscal policy.

**market forces** The forces of supply and demand that in a *free market determine the quantity available of a particular product or service and the price at which it is offered. In general, a rise in demand will cause both supply and price to increase, while a rise in supply will cause both a fall in price and a drop in demand, although many markets have individual features that modify this simple analysis. In practice, most markets are not free, being influenced either by restrictions on supply or by government intervention that can affect demand, supply, or price.

**market imperfection** *See* imperfect competition.

**marketing** The process of identifying, maximizing, and satisfying consumer demand for a company's products. Marketing a product involves such tasks as anticipating changes in demand (usually on the basis of *marketing research), promotion of the product (*see* sales promotion), ensuring that its quality, availability, and price meet the needs of the market, and providing after-sales service. *See also* direct marketing.

**marketing audit** A review of an organization's marketing capabilities based on a structured appraisal of a marketing department's internal strengths and weaknesses, which helps the company decide how best to respond to external opportunities and threats. Sales are analysed, the effectiveness of the *marketing mix is assessed, and, externally, the *marketing environment is monitored. When the marketing audit has been completed, a *marketing plan is drawn up based on its results.

**Marketing Board** An organization established in the UK under the provisions of the Marketing Acts by producers, with government support, in order to achieve orderly supply and marketing of products when differentiation between the products of individual producers is difficult. Such boards have been set up to market milk, eggs, potatoes, wool, etc.

**marketing environment** The combined influence of all the factors external to a company that could affect its sales. These factors include: cultural traditions, technological developments, competitors' activity, government policies, and changes in distribution channels. The marketing environment cannot be controlled by the company and, because it may change frequently, requires constant monitoring.

**marketing intelligence** Piecing together whatever information one can obtain to form a picture of one's competitors' activities, capabilities, and intentions in the marketing of a product.

**marketing logistics** *See* logistics.

**marketing mix** The factors controlled by a company that can influence consumers' buying of its products. The four components of a marketing mix (often called the **four Ps**) are: the product (quality, branding, packaging, and other features); pricing (recommended retail price, discounts for large orders, and credit terms); promotion (*see* sales promotion); and place (where to sell the product, which distributors and transport services to use, and desirable stock levels). The potential profitability of a particular marketing mix and its acceptability to its market are assessed by *marketing research.

**marketing plan** A detailed statement (usually prepared annually) of how a company's *marketing mix will be used to achieve its *market objectives. A marketing plan is usually prepared following a *marketing audit.

**marketing research** The systematic collection and analysis of data to resolve problems concerning *marketing, undertaken to reduce the risk of inappropriate marketing activity. Data is almost always collected from a *sample of the target market, by such methods as observation, interviews, and audit of shop sales. Interviews (*see* open-ended question; structured interview) are the most common technique, and can be carried out face-to-face, by telephone, or by post. When the results have been analysed (usually by computer), recommendations regarding the original problem can be made. *See also* qualitative marketing research; quantitative marketing research. **Market research** uses marketing research techniques with the restricted objective of discovering the size of the market for a particular brand or product.

**Market Intelligence Reports (MINTEL)** Market research reports published by Mintel Publications Ltd, which provide an important source of information on consumer markets in the UK. The company also reports on developments in retailing and on the leisure sector of the market.

**market leader** A company that has the largest *market share for a particular product or service in an area or country.

**market maker** A dealer in securities on the *London Stock Exchange who undertakes to buy and sell securities as a principal and is therefore obliged to announce prices at which he will buy or sell a particular security at a particular time. Before October 1986 (*see* Big Bang) this function was performed by a *stockjobber, who was then obliged to deal with the public through a *stockbroker. However since October 1986, when the rules changed, market makers attempt to make a profit by dealing in securities as principals (selling at a higher price than that at which they buy; *see* margin) as well as acting as agents, working for a commission. While this dual role may create a conflict of interest for market makers (*see* Chinese wall; front running), it avoids the restrictive trade practice of the former system and reduces the cost of dealing in the market.

**market objectives** The objectives to be fulfilled by a *marketing plan. Marketing objectives are determined by interpreting the relevant parts of a corporate plan (*see* business plan) in the light of a *marketing audit.

**market order** An order given to a stockbroker, commodity broker, dealer, etc., to buy or sell specified securities or commodities immediately at the prevailing *market price. *Compare* limit order (*see* limit).

**market orientation** *See* product orientation.

**market penetration 1.** The process of entering a market to establish a new brand or product. Market penetration may be achieved by offering the brand or product at a low initial price to familiarize the public with its name. **2.** A measure of the extent to which a market has been penetrated equal to the ratio of all the owners of that brand or product to the total number of potential owners, usually expressed as a percentage. *Compare* market share.

**market price 1.** The price of a raw material, product, service, security, etc., in an open market. In formal markets, such as a stock exchange, commodity market, foreign-exchange market, etc., there is often a *margin between the buying and selling price; there are, therefore, two market prices. In these circumstances the market prices often quoted are the average of the buying and selling price (*see* middle price). **2.** The economic concept of the price at which commodities are exchanged in a market, either for money or for each other (*see* barter).

**market rate of discount** *See* bill rate.

**market research** *See* marketing research.

**Market Research Society (MRS)** A professional association in the UK for those who use survey techniques for marketing, social, and economic research. The MRS aims to maintain professional standards, to provide its members with training programmes and information about new techniques, and to represent the interests of its members to government and commerce.

**market segmentation** The division of a *market into homogeneous groups of consumers, each of which can be expected to respond to a different *marketing mix. There are numerous ways of segmenting markets, the more traditional being by age, sex, family size, income, occupation, and social class; more recently **geodemographic segmentation**, which identifies housing areas in which people share a common lifestyle and will be more likely to buy certain types of products, has become more popular. Another frequently used method is **benefit segmentation**; for example, the toothpaste market can be segmented into those primarily concerned with dental protection and those concerned with a fresh taste. Once a segment has been identified, the marketer can then develop a unique marketing mix to reach it, for example by advertising only in the newspapers read by that market segment. Therefore, to be of practical value a market setment must be large enough to warrant the development costs.

**market share** The share of the total sales of all brands or products competing in the same market that is captured by one particular brand or product, usually expressed as a percentage. For example, if brands A, B, and C are the competing brands of a product and in a particular month they achieved sales of £48,000, £62,000, and £90,000, respectively, brand A's market share would be (48,000/(48,000 + 62,000 + 90,000)) × 100 = 24%.

**market skimming** A *marketing tactic in which a new *brand or product is launched at a high price to sell to the small market segment that is indifferent to price or is attracted by the prestige value of paying premium prices. If necessary, the price can be reduced later.

**market testing** *See* test marketing.

**market valuation** *See* market capitalization.

**market value** The value of an asset if it were to be sold on the open market at its current *market price. When land is involved it may be necessary to distinguish between the market value in its present use and that in some alternative use; for example, a factory site may have a market value as a factory site, and be so valued in the company's accounts, which may be less than its market value as building land.

**markings** The official number of bargains that have taken place during a working day on the *London Stock Exchange.

**marking up** The raising of prices by *market makers on the *London Stock Exchange in anticipation of an increased demand for a particular security.

**mark-up** *See* gross margin.

**married couple's allowance** *See* personal allowance; separate taxation of a wife's earnings.

**Marshallian (*or* uncompensated) demand function** A function that expresses the quantities of goods demanded by an individual in terms of the price of a good and the income of the individual. The sum of Marshallian demands represents the *indirect utility function. An increase in the price of a good will affect Marshallian demands in two ways: through the price change itself (the *substitution effect) and through the effective change in the value of the individual's income (the *income effect). The concept was introduced by Alfred Marshall (1842–1924).

*Compare* Hicksian (*or* compensated) demand function.

**Marxist economics** A branch of Classical economics developed by Karl Marx (1818–83), which added a strong political flavour to economic thought. Developing Adam Smith's concept of labour as the source of economic value (*see* labour theory of value), Marx argued that capitalists extract surplus value from workers by means of the production process, leaving them only a subsistence wage. In Marx's view, capitalist economies would be subject to ever deepening crises, which would eventually destroy capitalism, leaving the state open to control by the workers. Marx's predictions have largely proved inaccurate; although crises have continued to occur, living standards in capitalist countries have risen, enabling capitalism to enjoy the support of many workers. Moreover, as Marxist economic thought has developed little since Marx's writings, it has receded in importance.

**mass media** *See* media.

**mass production** The manufacture of large quantities of identical articles that may or may not be of high quality but that do not involve individual craftsmanship. Usually making use of continuous automated processes in which high-speed machines are operated by relatively small numbers of relatively unskilled employees, mass-production methods are used for an increasingly wide range of products.

**master** The captain of a merchant ship.

**matched bargain** The practice on the *London Stock Exchange of matching a sale of a particular quantity of stock with a purchase of the same quantity of the same stock. Transactions of this kind are carried out by **matching brokers**.

**material fact 1.** Any important piece of information that a person seeking insurance must disclose to the insurer to enable him to decide whether or not to accept the insurance and to calculate the premium if he wishes to do so. What is and what is not material may have to be decided by a court of law, but very often it is obvious. While a person cannot be penalized for not revealing facts that he does not know or cannot reasonably be expected to know, his insurance contract can become void if he deliberately conceals facts from the insurer that might influence whether or not he accepts the risk, the premium he charges, or the exceptions he makes. **2.** Relevant information about a company that must be made public in its *prospectus, if it is seeking a flotation on a stock exchange. **3.** Any information that could be provided by a witness in court proceedings and could influence the decision of the court.

**materiality** The state of having sufficient significance to require separate disclosure in accounting. A statement issued by the Institute of Chartered Accountants in England and Wales says "a matter is material if its non-disclosure, misstatement, or omission would be likely to distort the view given by the accounts or other statement under consideration".

**materials management** *See* logistics.

**maternity rights** The rights a woman has from her employer when she is absent from work wholly or partly because of her pregnancy or confinement. In the UK an employee is entitled to statutory maternity pay if she has been working for the same employer continuously for at least six months ending in the 15th week before the baby is due, pays National Insurance, and leaves work between the 11th and 6th week before the expected date of confinement. Women who have worked more than 16 hours a week for two years, or more than eight hours a week for five years, are entitled to a higher statutory rate (nine-tenths of their normal average weekly pay) for the first six weeks of their absence in addition to the lower rate for a further 12 weeks. Employers recover the amount of the payments by setting the amount against their National Insurance payments. The employee is also entitled to return to her former job at any time within 29 weeks after the beginning of the week of the baby's birth provided she notifies her employer in writing three weeks before she leaves work.

**mate's receipt** A document signed by the mate of a ship as proof that the goods specified in the document have been loaded onto his ship, especially if they have been delivered direct to the ship, rather than from a quayside warehouse. The mate's receipt functions as a document of title, which may be required as proof of loading in an FOB contract, pending the issue of the *bill of lading.

**matrix printer** *See* dot matrix printer.

**maturity date** The date on which a document, such as a *bond, *bill of exchange, or insurance policy, becomes due for payment. In some cases, especially for redeemable government stocks, the maturity date is known as the **redemption date**. *See also* redemption.

**maximization of utility** The assumption in neoclassical economics that individuals and firms maximize both *utility and *profits in the light of available information. Although maximization depends on the concept of *rationality, which has been frequently subject to question, the maximizing assumption has been extremely helpful in practice; in particular it has enabled economists to introduce mathematical calculus into economics, from which general theories and predictions have been developed. While many question the validity of these methods, few constructive alternatives have been presented.

**maximum fluctuation** The maximum daily price fluctuation that is permitted in some markets. *See* limit.

**maximum investment plan (MIP)** A unit-linked endowment policy marketed by a life-assurance company that is designed to produce maximum profit rather than life-assurance protection. It calls for regular premiums, usually over ten years, with options to continue. These policies normally enable a tax-free fund to be built up over ten years and, because of the regular premiums, *pound cost averaging can be used, linked to a number of markets.

**maximum slippage** The period between the date on which a new company expects to start earning income and the date up to which it can survive on its venture capital. After this date has passed, the company would be unable to raise further funds and would sink into insolvency. *See also* death-valley curve.

**MB** Abbreviation for *megabyte.

**MBA** Abbreviation for Master of Business Administration.

**MCA** Abbreviation for *Management Consultants' Association.

**MCT** Abbreviation for *mainstream corporation tax.

**MD** Abbreviation for *managing director.

**mean** An average. *See* arithmetic mean; geometric mean; median.

**mean deviation** In statistics, the *arithmetic mean of the deviations (all taken as positive numbers) of all the numbers in a set of numbers from their arithmetic mean. For example, the arithmetic mean of 5, 8, 9, and 10 is 8, and therefore the deviations from this mean are 3, 0, 1, and 2, giving a mean deviation of 1.5.

**mean price** *See* middle price.

**means test** Any assessment of the income and capital of a person or family to determine their eligibility for benefits provided by the state or a charity.

**measured daywork** A method of assessing wages in which a daily production target is set and a daily wage agreed on this basis. If the target is reached the worker receives the agreed daily wage; if it is not reached (or is exceeded) the worker is paid pro rata.

**measurement tonnage** *See* tonnage.

**media (mass media)** The means, such as television, radio, newspapers, and magazines, by which advertisers, politicians, etc., communicate with large numbers of members of the general public.

**media analysis (media research; media planning)** An investigation into the relative effectiveness and the relative costs of using the various advertising *media in an advertising campaign. Before committing an advertising budget it is necessary to carry out market research on potential customers, their reading habits, television-watching habits, how many times the advertisers wish the potential customers to see an advertisement, how great a percentage of the market they wish to reach, etc. These elements all need to be considered and balanced to plan a campaign that will effectively reach its target audience at a reasonable cost.

**median** A form of *mean in which a set of numbers is arranged in an ascending or descending scale and the middle number (if there are an odd number in the set) or the arithmetic mean of the middle two numbers (if there are an even number) is taken as the median. This can give a more representative average in some circumstances than an *arithmetic mean or a *geometric mean.

**mediation** The intervention of a neutral third party in an industrial dispute. The object is to enable the two sides to reach a compromise solution to their differences, which the mediator usually does by seeing representatives of each side separately and then together. If the mediator has power to make binding awards the process is known as *arbitration; if he can only suggest means of settling the dispute it is known as **conciliation**. *See* Advisory Conciliation and Arbitration Service.

**medical insurance** *See* private health insurance.

**medium-dated gilt** *See* gilt-edged security.

**medium of exchange** A substance or article of little intrinsic value that is used to pay for goods or services. In primitive economies various articles, such as sea shells, have been used for this purpose but *money is now used universally.

**mediums** *See* gilt-edged security.

**medium-sized company** A company, as defined by the UK Companies Act (1981), that falls below any two of the following three size criteria: (1) gross assets £2,800,000; (2) turnover £5,750,000; (3) average number of employees 250. If these companies are not *public limited companies or banking, insurance, or shipping companies, they may file abbreviated *profit and loss accounts with the Registrar of Companies (*see* modified accounts), although they must provide their own shareholders with the full statutory information. *Compare* small company.

**medium-term liabilities** *Liabilities falling due in, say, more than one but less than ten years.

**meeting of creditors** *See* bankruptcy; creditors' voluntary liquidation; liquidator.

**megabyte (MB)** A measure of computer memory capacity, equal to one thousand *kilobytes, or 1 024 000 bytes.

**member bank** A bank that belongs to a central banking or clearing system. In the UK a member bank is a *commercial bank that is a member of the *Association for Payment Clearing Services. In the USA it is a commercial bank that is a member of the *Federal Reserve System.

**member firm** A firm of brokers or *market makers that is a member of the *London Stock Exchange (International Stock Exchange). There are some 360 member firms with some 5300 individual members. Banks, insurance companies, and overseas securities houses can now become corporate members.

**member of a company** A shareholder of a company whose name is entered in the *register of members. Founder members (*see* founders' shares) are those who sign the *memorandum of association; anyone subsequently coming into possession of the company's shares becomes a member.

**members' voluntary liquidation (members' voluntary winding-up)** The winding-up of a company by a special resolution of the members in circumstances in which the company is solvent. Before making the winding-up resolution, the directors must make a *declaration of solvency. It is a criminal offence to make such a declaration without reasonable grounds for believing that it is true. When the resolution has been passed, a *liquidator is appointed; if, during the course of the winding-up, the liquidator believes that the company will not be able to pay its debts, a meeting of creditors must be called and the winding-up is treated as a members' *compulsory liquidation.

**memorandum of association** An official document setting out the details of a *company's existence. It must be signed by the first *subscribers and must contain the following information (as it applies to the company in question): the company name; a statement that the company is a public company; the address of the registered office; the objects of the company (called the **objects clause**); a statement of limited liability; the amount of the guarantee; and the amount of authorized share capital and its division.

**memorandum of satisfaction** A document stating that a mortgage or charge on property has been repaid. It has to be signed by all the parties concerned and a copy sent to the *Registrar of Companies, if the mortgage or charge was made by a company.

**memory** The storage section of a computer in which data and programs are held. A computer has two different types of memory; the *main store and the *backing store. *See also* random-access memory; read-only memory.

**menu** A list of choices displayed by a computer. When many options are available the user may be presented first with a main menu, from which more detailed menus can be selected. A well-designed menu system can make a complicated program simple to use.

**mercantile agent** *See* factor.

**mercantile law** The commercial law, which includes those aspects of a country's legal code that apply to banking, companies, contracts, copyrights, insolvency, insurance, patents, the sale of goods, shipping, trademarks, transport, and warehousing.

**mercantilism** The economic theory prevalent between 1500 and 1800, mainly in England and France. Its main characteristics were a belief that exports created wealth for a nation, while imports diminished wealth; that gold and silver bullion (specie) should be accumulated by a country in order to encourage trade internally; and that to achieve these ends governments should encourage exports of manufactured goods and imports of bullion while restricting exports of bullion and imports of manufactured goods. The implication of this theory is that trade benefits only one country; it became discredited when such economists as Adam Smith (1723–90) showed that trade can be beneficial to both sides. *See also* comparative advantage.

**merchandising** The promotion by a retailer in his shop of selected products. Commonly used techniques include displays to encourage *impulse buying, free samples and gifts, and temporary price reductions. Merchandising policy is usually designed to influence the retailer's sales pattern and is influenced by such factors as the firm's *market, the speed at which different products sell, *margins, and service considerations. Sometimes merchandising is used to attract customers into the shop rather than to promote the product itself.

**merchant** A trader who buys goods for resale, acting as a principal and usually holding stocks. Typically a merchant sells goods in smaller lots than he buys and often exports goods or is involved in *entrepôt trade.

**merchantable quality** An implied condition respecting the state of goods sold in the course of business. Such goods should be as fit for their ordinary purpose as it is reasonable to expect, taking into account any description applied to them, the price (if relevant), and all the other relevant circumstances. The condition does not apply with regard to defects specifically drawn to the buyer's attention or defects that he should have noticed if he examines the goods before the contract is made.

**merchant bank** A bank that formerly specialized in financing foreign trade, an activity that often grew out of its own merchanting business. This led them into accepting *bills of exchange and functioning as *accepting houses. More recently they have tended to diversify into the field of *hire-purchase finance, the granting of long-term loans (especially to companies), advising companies on flotations and *takeover bids, underwriting new issues, and managing investment portfolios and unit trusts. Many of them are old-established and some offer a limited banking service. Their knowledge of international trade makes them specialists in dealing with the large *multinational companies.

**merge** To combine items from two or more computer *files. Many computers, for example, have programs that merge names and addresses held on a mailing list file with a word-processing file to produce a mail shot.

**merger** The combination of two or more organizations for the benefit of all of them. The objective is invariably to increase efficiency and sometimes to avoid competition, although approval of the *Monopolies and Mergers Commission may be required and the merger must be conducted on lines sanctioned by the *City Code on Takeovers and Mergers. Mergers are normally amicably arranged by all the parties concerned, unlike some takeovers.

**merit bonus** A *bonus granted to an employee as a reward for good work.

**merit good (*or* bad)** A commodity regarded (usually by a government) as intrinsically good (or bad). There is no widely accepted criterion for establishing exactly which goods are merit goods. However, most people would probably agree that education is a merit good and cigarettes a merit bad.

**method study** *See* work study.

**metrication** *See* metric system.

**metric system** A system of measurement based on the decimal system. It was first formalized in France at the end of the 18th century and by the 1830s was being widely adopted in Europe. In the UK, bills for its compulsory adoption were defeated in 1871 and 1907 and *Imperial units remained supreme until 1963, when the yard was redefined as 0.9144 metre and the pound as 0.453 592 37 kilogram. The Metrication Board set up in 1969 failed to achieve its target of the **metrication** of British industry by 1975 and metrication now proceeds on a voluntary basis, in which it is envisaged that pints of beer, miles per hour, yards, and feet will persist until the end of the century. However, the Weights and Measures Act (1985) lists certain units that may no longer be used for trade: these include the *hundredweight, *ton, bushel, square mile, cubic yard, and cubic foot. It is hoped that before the end of the century such units as the therm and British thermal unit will have been abandoned. For all scientific purposes and many trade and industrial purposes the form of the metric system known as *SI units is now in use. In the USA metrication has been even slower than in the UK, although it is hoped to achieve at least some metrication by 1992.

**metric ton** *See* ton.

**metrology** The scientific study of weights and measures. The subject has long been closely associated with trade and commerce, which is based on the exchange of measured quantities of goods or services for money or measured quantities of other goods or services. *See* Imperial units; metric system; SI units.

**mezzanine** Denoting an intermediate stage in some financial process. **Mezzanine funding** is an intermediate stage in the funding of a new company, lying between the provision of a loan and the taking of a share in its equity.

**MICR** Abbreviation for *magnetic ink character recognition.

**microcomputer (personal computer)** The smallest and cheapest type of *computer. A microcomputer is built around a *microprocessor. Although microcomputers have limited abilities compared with the larger *mainframes or *minicomputers, they are sufficiently powerful to meet the needs of many small businesses, their uses including producing payrolls, wordprocessing, invoicing, and compiling mailing lists. Because of rapid developments in the design of microprocessors, microcomputers are constantly increasing in power. Some models now offer computing power greater than that of small minicomputers.

**microeconomics** The analysis of economic behaviour at the level of the firm or the individual (*compare* macroeconomics). For the individual or household it is concerned with the optimal allocation of a given budget, the labour supply choice, and the effects of taxation. For the firm it is largely concerned with the production process, costs, and the marketing of output, dependent on the type of competition faced (*see* imperfect competition; monopoly; oligopoly; perfect competition).

**microfoundations** A basis for a *macroeconomic model in which economic events are controlled by the rational *utility-maximizing behaviour of individuals. It was realized in the late 1960s that traditional macroeconomic models, such as the *ISLM model, had no such basis; much of macroeconomics since then has therefore focused on establishing these foundations. This has polarized the debate between monetarists and Keynesians of the 1950s and 1960s much more sharply. On the one hand, traditional monetarists emphasize the *market clearing assumption, *Pareto optimality, and rational expectations (*see* new classical macroeconomics) as a basis of analysis, while traditional Keynesians now emphasize the possibility of market failures (*see* efficiency wage theory; hysteresis; insider-outsider theory).

**microprocessor** In computer technology, an *integrated circuit that contains on one chip the arithmetic, logical, and control functions of the *central processing unit. Microprocessors are used as control systems in many devices, including photosetting machines, car electronics, and cameras, as well as in computers.

**middleman** A person or organization that makes a profit by trading in goods as an intermediary between the producer and the consumer. Middlemen include agents, brokers, dealers, merchants, factors, wholesalers, distributors, and retailers. They earn their profit by providing a variety of different services, including finance, bulk buying, holding stocks, breaking bulk, risk sharing, making a market and stabilizing prices, providing information about products (to consumers) and about markets (to producers), providing a distribution network, and introducing buyers to sellers.

**middle management** *See* management.

**middle price (mean price)** The average of the *offer price of a security, commodity, currency, etc., and the *bid price. It is the middle price that is often quoted in the financial press.

**migrant worker** A worker who has come from another country or region to work. Attracted by better wages in the country or region to which he has migrated, he may or may not have brought his family with him. In many cases he will not have done so and will therefore be sending a substantial proportion of his earnings back to his native country. This can have an adverse effect on the *balance of payments of countries that attract substantial numbers of migrant workers and conversely can be advantageous to countries from which many workers migrate. Migrant workers can also cause problems with the trade unions in the countries to which they migrate.

**milk round** An annual visit to universities by the personnel managers of large companies seeking graduates to join their organizations.

**Mincing Lane** The street in the City of London that is the centre of commodity trading. Plantation House, lying between Mincing Lane and the parallel Mark Lane, houses the offices of many commodity brokers and dealers and formerly housed the London Commodity Exchange (*see* London FOX). It is named after the *mynchens*, or nuns, of the nearby St Helen's in Bishopsgate.

**minicomputer** A *computer that is less powerful than a *mainframe but more powerful than a *microcomputer. It is unable to handle as many people or programs at the same time as a mainframe, but it is adequate for a wide range of business and scientific applications.

**minimum lending rate (MLR)** The successor, between 1971 and 1981, of the *bank rate. In this decade it was the minimum rate at which the Bank of England would lend to the *discount houses. This was a published figure; the present more informal *base rate does not have the same status. When the government suspended MLR in 1981 it reserved the right to reintroduce it at any time, which it did for one day in January 1985.

**minimum subscription** The minimum sum of money, stated in the *prospectus of a new company, that the directors consider must be raised if the company is to be viable.

**minimum wage** The minimum wage that an employer may pay an employee. In the UK there is no national statutory minimum but in certain industries *wages councils prescribe minimum rates of pay. In some other countries there are national minimum wages.

**minority interest** The interest of individual shareholders in a company more than 50% of which is owned by a holding company. For example, if 60% of the ordinary shares in a company are owned by a holding company, the remaining 40% will represent a minority interest. These shareholders will receive their full share of profits in the form of dividends although they will be unable to influence company policy as they will always be outvoted by the majority interest held by the holding company.

**minority protection** Remedies evolved to safeguard a minority of company members from the abuse of majority rule. They include just and equitable winding-up, applying for relief on the basis of *unfair prejudice, bringing a *derivative or representative action, and seeking an *inspection and investigation of the company.

**mint** A factory, owned by a government or a bank, in which coins and banknotes are manufactured. *See* Royal Mint.

**MINTEL** *See* Market Intelligence Reports.

**mint par of exchange** The rate of exchange between two currencies that were on the *gold standard. The rate was then determined by the gold content of the basic coin.

**MIP** **1.** Abbreviation for monthly investment plan. **2.** Abbreviation for *marine insurance policy. **3.** Abbreviation for *maximum investment plan.

**MIRAS** Abbreviation for mortgage interest relief at source. *See* mortgage.

**MIS** Abbreviation for *management information system.

**misfeasance** **1.** The negligent or otherwise improper performance of a lawful act. **2.** An act by an officer of a company in the nature of a breach of trust or breach of duty, particularly if it relates to the company's assets.

**misfeasance summons** An application to the court by a creditor, contributory, liquidator, or the official receiver during the course of winding up a company. The court is asked to examine the conduct of a company officer who is suspected of a breach of trust or duty and it can order him to make restitution to the company.

**misrepresentation** An untrue statement of fact, made by one party to the other in the course of negotiating a contract, that induces the other party to enter into the contract. The person making the misrepresentation is called the **representor**, and the person to whom it is made is the **representee**. A false statement of law, opinion, or intention does not constitute a misrepresentation; nor does a statement of fact known by the representee to be untrue. Moreover, unless the representee relies on the statement so that it becomes an inducement (though not necessarily the only inducement) to enter into the contract, it is not a misrepresentation. The remedies for misrepresentation vary according to the degree of culpability of the representor. If he is guilty of **fraudulent misrepresentation** (i.e. if he did not honestly believe in the truth of his statement, which is not the same as saying that he knew it to be false) the representee may, subject to certain limitations, set the contract aside and may also sue for *damages. If he is guilty of **negligent misrepresentation** (i.e. if he believed in his statement but had no reasonable grounds for doing so) the representee may also rescind (*see* rescission) the contract and sue for damages. If the representor has committed merely an **innocent misrepresentation** (one he reasonably believed to be true) the representee is restricted to rescinding the contract.

**mistake** A misunderstanding or erroneous belief about a matter of fact (**mistake of fact**) or a matter of law (**mistake of law**). In civil cases, mistake is particularly important in the law of contract. Mistakes of law have no effect on the validity of agreements, and neither do many mistakes of fact. When a mistake of fact does do so, it may render the agreement void under common-law rules (in which case it is referred to as an **operative mistake**) or it may make it voidable, i.e. liable, subject to certain limitations, to be set aside under the more lenient rules of equity.

When both parties to an agreement are under a misunderstanding, the mistake may be classified as either a **common mistake** (i.e. a single mistake shared by both) or a **mutual mistake** (i.e. each misunderstanding the other). In the case of common mistake, the mistake renders the contract void only if it robs it of all substance. The principal (and almost the only) example is when the subject matter of the contract has, unknown to both parties, ceased to exist. A common mistake about some particular attribute of the subject matter (e.g. that it is an original, not a copy) is not an operative mistake. However, a common mistake relating to any really fundamental matter will render a contract voidable. In the case of mutual mistake, the contract is valid if only one interpretation of what was agreed can be deduced from the parties' words and conduct. Otherwise, the mistake is operative and the contract void. When only one party to a contract is under a misunderstanding, his mistake may be called a **unilateral mistake** and it makes the contract void if it relates to the fundamental nature of the offer and the other party knew or ought to have known of it. Otherwise, the contract is valid so far as the law of mistake is concerned, though the circumstances may be such as to make it voidable for *misrepresentation.

A deed or other signed document (whether or not constituting a contract) that does not correctly record what both parties intended may be rectified by the courts. When one signatory to a document was fundamentally mistaken as to the character or effect of the transaction it embodies, he may (unless he was careless) plead his mistake as a defence to any action based on the document.

**mitigation of damage** Minimizing the loss incurred by the person who suffered the loss and is claiming *damages as a result of it. The injured party has a duty to take all reasonable steps to mitigate any loss and the courts will not, therefore, award damages to compensate for a loss that could have been avoided by reasonable action.

**mixed economy** An economy in which some goods and services are produced by the government and some by private enterprise. A mixed economy lies between a *command economy and a complete *laissez-faire economy. In practice, however, most economies are mixed; the significant feature is whether an economy is moving towards or away from a more laissez-faire situation. Although most western economies retreated from laissez-faire policies in the post-war era this tended to be reversed in the 1980s.

**mix variance** A *variance arising because the mix of goods actually sold or produced differs from that budgeted for.

**MLR** Abbreviation for *minimum lending rate.

**MMC** Abbreviation for *Monopolies and Mergers Commission.

**mobility of labour** The extent to which workers are willing to move from one region or country to another (**geographical mobility**) or to change from one occupation to another (**occupational mobility**). In **horizontal mobility** there is no change of status, whereas in **vertical mobility** there is. An upward change in status will increase a worker's mobility, whereas a downward change will reduce it. The more highly skilled a worker, the less his occupational mobility will be, but he will often be highly geographically mobile. An unskilled worker will often be both occupationally and geographically mobile. In the UK, many government retraining schemes aim to increase occupational mobility; at the same time considerable effort goes into encouraging new industries into areas of high unemployment to reduce the need for geographical mobility.

**mock auction** An auction during which a lot is sold to someone at a price lower than his highest bid, part of the price is repaid or credited to the bidder, the right to bid is restricted to those who have bought or agreed to buy one or more articles, or articles are given away or offered as gifts. Under the Mock Auction Act (1961) it is an offence to promote or conduct a mock auction.

**modem** Abbreviation for *mo*dulator-*dem*odulator. This device enables a computer to transmit and receive data by means of a communications link, such as a telephone line. Modems are necessary because the digital electrical signals produced by a computer are unsuitable for sending along transmission lines. The process of converting computer signals into a form suitable for transmission is called modulation. The reverse process, making the transmitted signals intelligible to the computer, is called demodulation. There are two kinds of modem: the *acoustic coupler, which uses a standard telephone handset, and the direct coupler, which does not.

**modified accounts** A form of statutory *annual accounts for *small companies and *medium-sized companies in the UK. Currently all companies have to submit full accounts to their shareholders, although consideration is being given to shortened accounts even for large companies. However, the small and medium-sized companies need file only modified accounts with the *Registrar of Companies. For small companies, modified accounts consist only of a balance sheet, certain specified notes, and a copy of the auditors' report. In the case of medium-sized companies the profit and loss account may omit some information but otherwise full information must be filed.

**monadic testing** A technique used in marketing research in which consumers are presented with a product to test on its own, rather than being asked to compare it with a competing product (*see* paired comparisons).

**monetarism** A school of thought in economics that places money at the centre of *macroeconomic policy. Based on the *quantity theory of money, and first expressed by the Scottish philosopher David Hume (1711–76), it relates the price level to the quantity of money in the economy. Latterly, monetarism has been the main opponent to *Keynesianism, claiming that monetary factors are a major influence on the economy and that, in particular, government expansion of the money supply will tend to generate inflation rather than employment. Pure monetarism has, however, become confused with more general criticisms of Keynesianism by economic theorists and the crude distinction between monetarism and Keynesianism is misleading (*see* new classical macroeconomics).

**monetary aggregate** Any of several measures of the *money supply from the narrow M0 to the broad M5 (PSL5).

**monetary assets and liabilities** Amounts receivable (assets) or payable (liabilities) that appear in a company's accounts as specific sums of money, e.g. cash and bank balances, loans, debtors and creditors. These are to be distinguished from such nonmonetary items as plant and machinery, stock in trade, or equity investments, which although they are also expressed in accounts at a value (frequently cost) are not necessarily realizable at that value.

**monetary compensatory amount** Subsidies and taxes on farm products produced within the EC that form part of the *Common Agricultural Policy. They are used to bridge the gap between the green pound (*see* green currencies), and foreign exchange rates to prevent fluctuation in these rates from altering the farm prices. The object is to enable agricultural products to cost the same in all member countries and to prevent trading between countries in these products purely to make a profit as a result of changes in the exchange rate.

**monetary economy** An economy in which *money is the *medium of exchange. An economy based on *barter is now extremely rare.

**monetary inflation** The theory that *inflation is related to the expansion of the *money supply. *See* monetary policy; quantity theory of money.

**monetary policy** A means by which governments try to affect macroeconomic conditions by increasing or decreasing the supply of money. Three main options are available: (i) printing more money (now rarely used in practice); (ii) direct controls over money held by the monetary sector; (iii) *open-market operations. The traditional Keynesian view has been that monetary policy is at best a blunt instrument (money does not matter), while monetarists have held the opposite view. The new classical macroeconomists, using the theory of *rational expectations, have argued that monetary policy is ineffective if it is anticipated. In practice, governments have tended to employ 'tight' monetary policies, in the belief that this restrains inflation. *Compare* fiscal policy.

**monetary reform** The revision of a country's currency by the introduction of a new currency unit or a substantial change to an existing system. Examples include decimalization of the UK currency (1971) and the introduction of the cruzado in Brazil (1986).

**monetary system 1.** The system used by a country to provide the public with money for internal use and to control the exchange of its own currency with those of foreign countries. It also includes the system used by a country for implementing its *monetary policy. **2.** A system used to control the exchange rate of a group of countries. *See* European Monetary System.

**monetary theory** Any theory concerned with the influence of the quantity of money in an economic system. *See* monetary policy; quantity theory of money.

**monetary unit** The standard unit of currency in a country. The monetary unit of each country is related to those of other countries by a *foreign exchange rate.

**money** A *medium of exchange that functions as a unit of account and a store of value. Originally it enhanced economic development by enabling goods to be bought and sold without the need for *barter. However, throughout history money has been beset by the problem of its debasement as a store of value as a result of *inflation. Now that the supply of money is a monopoly of the state, most governments are committed in principle to stable prices. The central debate in economics over the past 50 years has been whether *fiscal policy and *monetary policy can have any effect other than to create inflation. The word *money* is derived from the Latin *moneta*, which was one of the names of Juno, the Roman goddess whose temple was used as a mint.

**money at call and short notice** One of the assets that appears in the balance sheet of a bank. It includes funds lent to discount houses, money brokers, the stock exchange, bullion brokers, corporate customers, and increasingly to other banks. 'At call' money is repayable on demand, whereas 'short notice' money implies that notice of repayment of up to 14 days will be given. After cash, money at call and short notice are the banks' most liquid assets. They are usually interest-earning secured loans but their importance lies in providing the banks with an opportunity to use their surplus funds and to adjust their cash and liquidity requirements.

**money broker** In the UK, a broker who arranges short-term loans in the *money market, i.e. between banks, discount houses, and dealers in government securities. Money brokers do not themselves lend or borrow money; they work for a commission arranging loans on a day-to-day and overnight basis.

**money illusion** The belief that an increase in wages is of benefit to wage earners, even if prices have increased by the same percentage. Keynesian theory concerning the role of the *Phillips curve in reducing unemployment implicitly relied on the existence of the money illusion, which was one of the main reasons for the fall of *Keynesianism.

**moneylender** A person whose business it is to lend money, other than pawnbrokers, friendly or building societies, corporate bodies with special powers to lend money, banks, or insurance companies. The Consumer Credit Act (1974) replaces the earlier Moneylenders Acts and requires all moneylenders to be registered, to obtain an annual licence to lend money, and to state the true *annual percentage rate (APR) of interest at which a loan is made.

**money market 1.** The UK market for short-term loans in which *money brokers arrange for loans between the banks, the government, the *discount houses and the *accepting houses, with the Bank of England acting as *lender of last resort. The main items of exchange are *bills of exchange, Treasury bills, and trade bills. The market takes place in and around Lombard Street in the City of London. Private investors, through their banks, can place deposits in the money market at a higher rate of interest than bank deposit accounts, for sums usually in excess of £10,000. **2.** The foreign-exchange market and the bullion market in addition to the short-term loan market.

**money supply** The quantity of money issued by a country's monetary authorities (usually the Central Bank). If the *demand for money is stable, the widely accepted *quantity theory of money implies that increases in the money supply will lead directly to an increase in the price level, i.e. to inflation. Since the 1970s most western governments have attempted to reduce inflation by controlling the money supply. This raises two issues:

(i) how to measure the money supply;

(ii) how to control the money supply (*see* interest-rate policy).

In the UK various measures of the money supply have been used, from the very narrow M0 to the very broad PSL2. They are usually defined as:

M0 – notes and coins in circulation plus the banks' till money and the banks' balances with the Bank of England;
M1 – notes and coins in circulation plus private-sector current accounts and deposit accounts that can be transferred by cheque;
M2 – notes and coins in circulation plus non-interest-bearing bank deposits plus building society deposits plus National Savings accounts;
£M3 – M1 plus all other private-sector bank deposits plus certificates of deposit;
M3 – £M3 plus foreign currency bank deposits;
PSL1 – M1 plus most private-sector bank deposits plus holdings of money-market instruments (e.g. Treasury bills); PSL stands for private-sector liquidity;
PSL2 – PSL1 plus building society deposits.
The Bank of England has recently renamed £M3 as M3, M3 as M3c, PSL1 as M4, and PSL2 as M5.

**money-supply rules** A policy proposed by the new classical macroeconomists in which a government states in advance the extent to which it intends to expand the *money supply. It is based on the belief that *fiscal policy and *monetary policy cannot affect the real variables in an economic situation but that uncertainty concerning government intentions can destabilize markets. In these circumstances a stable-policy rule is the best a government can achieve.

**money wages** Wages expressed in money terms only, i.e. without taking *inflation into account. *Compare* real wages.

**Monopolies and Mergers Commission (MMC)** A commission established in 1948 as the Monopolies and Restrictive Practices Commission and reconstructed under its present title by the Fair Trading Act (1973). It investigates questions referred to it on unregistered monopolies relating to the supply of goods in the UK, the transfer of newspapers, mergers qualifying for investigation under the Fair Trading Act, and uncompetitive practices and restrictive labour practices, including public-sector monopolies as laid down in the provisions of the Competition Act (1980).

**monopoly** A market in which there is only a single seller (producer). If there is a single seller and a single buyer the situation is called **bilateral monopoly** (*see also* monopsony). A monopoly in which there are many buyers (consumers) is usually considered inefficient, since self-interest will lead the monopolist to produce less than is Pareto-optimal (*see* Pareto optimality) and at a higher price. In the process the monopolist will earn pure economic *profits, reflecting his privileged position. While governments usually try to eliminate monopolies, regulate them, or nationalize them, there is some reason to believe that natural monopolies exist (e.g. electricity distribution). Marxist economists believe that an increasing tendency towards monopoly is an inevitable consequence of capitalist competition. *See also* price discrimination.

**monopsony** A market in which there is only a single buyer. The combination of monopsony and *monopoly is a **bilateral monopoly**. If there are many sellers, the buyer may be able to exploit his market position – a situation in which governments frequently find themselves. However, a monopsonist will not be able to drive prices below those necessary to achieve normal economic *profits (in the same way that a monopolist drives prices higher), since in this case an entrepreneur would either become bankrupt or leave the market.

***Monthly Digest of Statistics*** A monthly publication of the UK *Central Statistical Office providing statistical information on industry, national income, and the UK population.

**moonlighting** Having two jobs, one a full-time daytime job, the other a part-time evening job. Often the second job is undertaken on a self-employed basis and income is not returned for tax purposes.

**moral hazard** The incentive to cheat in the absence of penalties for cheating. This term, used in the insurance world, is often associated with **adverse selection**, the incentive to conceal information about one's true nature. A typical example of a person exposed to moral hazard is the owner of an insured car, who has little or no incentive to guard against theft; an example of adverse selection occurs when a person purchasing health insurance has no incentive to reveal that he is more likely to require health treatment than the average person. The effect is to make health insurance more expensive for everyone else.

Moral hazard and adverse selection are pervasive problems in economics, particularly in recruiting and supervising the labour force, the actions of company directors, or the buying and selling of shares through stockbrokers or market makers. To combat these problems it is usually necessary to provide incentives (e.g. a reward for honesty plus a penalty for dishonesty), which generally make market activities more expensive and create inefficiency. For example, it has been suggested that moral hazard can force firms to pay high wages, reducing the demand for labour and thus possibly creating voluntary *unemployment.

**moratorium** **1.** An agreement between a creditor and a debtor to allow additional time for the settlement of a debt. **2.** A period during which one government permits a government of a foreign country to suspend repayments of a debt. **3.** A period during which all the trading debts in a particular market are suspended as a result of some exceptional crisis in the market. In these circumstances, not to call a moratorium would probably lead to more insolvencies than the market could stand. The intention of such a moratorium is, first, that firms should be given a breathing space to find out exactly what their liabilities are and, secondly, that they should be given time to make the necessary financial arrangements to settle their liabilities.

**MORI** Abbreviation for *Market and Opinion Research International.

**mortality rate (death rate)** The crude death rate, i.e. the number of deaths per 1000 of the average population in a given year. It can be subdivided into different rates for different age groups of the population and for different regions.

**mortality table (life table)** An actuarial table prepared on the basis of mortality rates for people in different occupations in different regions of a country. It provides life-assurance companies with the information they require to quote for life-assurance policies, annuities, etc.

**mortgage** An interest in property created as a security for a loan or payment of a debt and terminated on payment of the loan or debt. The borrower, who offers the security, is the **mortgagor**; the lender, who provides the money, is the **mortgagee**. *Building societies and banks are the usual mortgagees for house purchasers. In either case the mortgage is repaid by instalments over a fixed period (often 25 years), either of capital and interest (**repayment mortgage**) or of interest only, with other arrangements being made to repay the capital, for example by means of an *endowment assurance policy (this is known as an **endowment mortgage**). Business uses of the mortgage include using property to secure a loan to start a business. Virtually any property may be mortgaged (though land is the most common).

Under the Law of Property Act (1925), which governs mortgage regulations in the UK, there are two types of mortgage, legal and equitable. A **legal mortgage** confers a legal estate on the mortgagee; the only valid mortgages are (a) a lease granted for a stated number of years, which terminates on repayment of the loan at or before the end of that period; and (b) a deed expressed to be a *charge by way of legal mortgage. An **equitable mortgage** can be created if the mortgagee has only an equitable interest in the property (for example, when he is a beneficiary under a trust of the property). Provided that this is done by *deed, the rights of the parties are very similar to those under a legal mortgage. An equitable mortgage can also be created of a legal or equitable interest by an informal agreement, e.g. the mortgagor hands his title deeds to the mortgagee as security for a loan. Such a mortgagee has the remedies of possession and foreclosure only (see below). A *second mortgage or subsequent mortgage may be taken out on the same property, provided that the value of the property is greater than the amount of the previous mortgage (s). All mortgages of registered land are noted in the *register of charges on application by the mortgagee (*see* land registration), and a charge certificate is issued to him. When mortgaged land is unregistered, a first legal mortgagee keeps the title deeds. A subsequent legal mortgagee and any equitable mortgagee who does not have the title deeds should protect his interests by registration.

If the mortgaged property is the mortgagor's main residence, he is entitled to **mortgage interest relief**, an income-tax allowance on the value of the interest paid on mortgages up to a specified figure. For mortgages made on or after 1 August 1988, the limit of mortgage relief applies to the property rather than to the borrower. Thus when two or more people share a residence, the relief is allocated between them in equal shares. Previously, they were each entitled to the full relief. Under the **MIRAS** (mortgage interest relief at source) scheme, interest payments made to a bank, building society, etc., are made after deduction of an amount equivalent to the relief of due income tax at the basic rate, and therefore no other relief is necessary, unless the person paying the mortgage pays tax at a higher rate.

Under the **equity of redemption**, the mortgagor is allowed to redeem his property at any time on payment of the loan together with interest and costs; any provisions in a mortgage deed to prevent redemption (known as **clogs**) are void.

In theory, the mortgagee always has the right to take possession of mortgaged property even if there has been no default. This right is usually excluded by building-society mortgages until default, and its exclusion may be implied in any instalment mortgage. Where residential property is concerned, the court has power to delay the recovery of possession if there is a realistic possibility that the default will be remedied in a reasonable time. In case of default, the mortgagor has a statutory right to sell the property, but this will normally be exercised after obtaining possession first. Any surplus after the debt and the mortgagee's expenses have been met must be paid to the mortgagor. The mortgagee also has a statutory right to appoint a *receiver to manage mortgaged property in the event of default; this power is useful where business property is concerned. As a final resort, a mortgage may be brought to an end by *foreclosure, in which the court orders the transfer of the property to the mortgagee. This is not common in times of rising property prices, as the mortgagor would lose more than the value of the debt, so the court will not order foreclosure where a sale would be more appropriate.

**mortgage debenture** A loan made to a company by an investor, secured on the real property of the company. *See* debenture.

**mortgagee in possession** A mortgagee (*see* mortgage) who has exercised his right to take possession of the mortgaged property; this may happen at any time, even if there has been no default by the mortgagor. However, the mortgage deed may contain an agreement not to do this unless there is default and a court order will be needed to obtain possession in the case of a dwelling house. The court may adjourn the hearing to allow the mortgagor time to pay. The mortgagee will either receive the rents and profits if the property has been let or manage the property himself. He will be liable to account strictly for his actions and is not entitled to reap any personal benefit beyond repayment of the interest and the principal debt. He must carry out reasonable repairs and must not damage the property.

**most-favoured-nation clause** A clause in a trade agreement between two countries stating that each will accord to the other the same treatment as regards tariffs and quotas as they extend to the most favoured nation with which each trades. Both GATT and the EEC have used this concept.

**motion** A proposal put before a meeting for discussion. Usually members of a company who attend a meeting of the company receive a **notice of motion** to enable them to consider the motion before the meeting. If the motion is passed by a majority of the members of the company it becomes a *resolution. If the meeting decides to amend the motion, the amended motion, known as the **substantive motion**, is then discussed and voted upon.

**motivational research** A form of *marketing research in which the motivations for consumers preferring one product rather than another are studied. It may form the basis of a plan to launch a competitive product or of a campaign to boost sales of an existing product.

**motor insurance** A form of insurance covering loss or damage to motor vehicles and any legal liabilities for bodily injury or damage to other people's property. Drivers have a legal obligation to be covered against third-party claims (*see* third-party insurance), but most drivers or owners of vehicles have a **comprehensive insurance**, providing wide coverage of all the risks involved in owning a motor vehicle. An intermediate cover is known as third-party, fire, and theft. This leaves the owner of a vehicle uninsured for damage to his own vehicle in the case of an accident. *See also* Motor Insurers Bureau; no-claim bonus.

**Motor Insurers Bureau** A body, formed in 1946, to which all motor insurers must belong. It has two functions. The first is to deal with domestic claims for bodily injury or damage to third-party property caused by uninsured or untraced drivers. As all drivers must have third-party insurance cover by law, it would be unfair for victims to suffer if the driver could not be traced or was uninsured. When such a claim is submitted to the Motor Insurers Bureau, it selects, on a rota basis, an insurance company to deal with the claim as if it had insured the driver. The second function of the Bureau is to issue international motor-insurance certificates (*see* green card system) to UK drivers and to administer the international agreements that form the basis of the green-card system. It also acts as a paying bureau for claims arising in other countries caused by UK green-card holders and as a handling bureau for claims arising from accidents caused by foreign drivers visiting the UK.

**mountain** A surplus of agricultural produce. Mountains of butter, meat, etc., have accumulated in the EEC as a result of the *Common Agricultural Policy. According to this, farmers are guaranteed a minimum price for their produce, which the EEC has to buy. Unable to sell the mountain within the EEC, which would depress prices, the EEC have to dispose of mountains to charitable causes or sell them outside the EEC at very low prices.

**mouse** A computer input device that is moved across the surface of a desk, causing a *cursor to move across the computer screen. When the cursor is pointing at the item required, a button on the mouse is pressed and the computer takes appropriate action.

**moving average** A series of *arithmetic means calculated from data in a time series, which reduces the effects of temporary seasonal variations. For example, a moving average of the monthly sales figures for an organization might be calculated by averaging the 12 months from January to December for the December figure, the 12 months from February to January for the January figure, and so on.

**MRP** Abbreviation for manufacturers' recommended price. *See* recommended retail price.

**MRS** Abbreviation for *Market Research Society.

**MSC** Abbreviation for Manpower Services Commission. *See* Training Commission.

**MS-DOS** *Trademark* Abbreviation for MicroSoft Disk Operating System. This *disk operating system for *microcomputers is produced by the US corporation Microsoft. Designed originally for the IBM Personal Computer (on which it is known as **PC-DOS** or, in IBM literature, **DOS**), it is one of the most popular operating systems for larger microcomputers.

**multiaccess** *See* multiuser.

**multilateral trade agreement** A trading arrangement between a number of nations in which there is agreement to abolish quotas and tariffs or to accept that there will be a surplus or deficit on the balance of payments.

**multinational** A corporation that has production operations in more than one country for various reasons, including securing supplies of raw materials, utilizing cheap labour sources, and bypassing protectionist barriers. Multinationals may be seen as an efficient form of organization, making effective use of the world's resources and transferring technology between countries. On the other hand, some have excessive power, are beyond the control of governments, and are able to exploit host countries, especially in the third world.

**multiple** *See* price-earnings ratio.

**multiple application** The submission of more than one application form for a new issue of shares that is likely to be oversubscribed (*see* allocation). In many countries it is illegal to do so either if the applications are made in the same name or if false names are used.

**multiple exchange rate** An exchange rate quoted by a country that has more than one value, depending on the use to which the currency is put. For example, some countries have quoted a specially favourable rate for tourists or for importers of desirable goods.

**multiple shops (chain stores)** A large chain of shops (usually ten or more) owned by the same *retailer. Compared with *independent retailers, they benefit from *economies of scale and so can charge lower prices to consumers. Multiple shops have therefore become dominant in many areas of retailing, such as packaged groceries.

**multiple taxation** Taxation of the same income by more than two countries. *Compare* double taxation.

**multiplier** The feedback effect generated by a change in an economic variable. For example, an increase in total *investment will raise national income by an amount equal to its monetary value, but in addition it will have a wider positive feedback effect by stimulating other parts of the economy, thus creating new jobs and additional demand for goods. First popularized by J M Keynes in his *general theory, the multiplier when applied to government expenditure and aggregate demand has a much broader connotation, applying to any situation in which there are feedback effects. Multipliers may be negative as well as positive.

**multitasking** Describing a computer system that can run more than one program at the same time.

**multiuser (multiaccess)** A computer system that can be used simultaneously by more than one person. Multiuser systems are usually *multitasking, and are essential for all large-scale computer applications. For example, they allow large databases to be accessed and manipulated by many users at the same time.

**municipal bond** A bond issued by a local government authority, especially one in the USA.

**mutual fund** The US name for a *unit trust.

**mutual life-assurance company** A type of life-assurance company that grew out of the Friendly Societies; there are no shareholders and apart from benefits and running expenses there are no other withdrawals from the fund; thus any profits are distributed to policyholders.

**mutual mistake** *See* mistake.

# N

**NAIRU** Abbreviation for non-accelerating inflation rate of *unemployment. A modification of the *natural rate of unemployment, it represents inverse relationships between unemployment and the rate of increase in *inflation, rather than inflation itself. Historically, increases in the rate of inflation have been associated with falls in unemployment and it may be that governments can exploit this relationship to reduce unemployment. This possibility, and the reasons for its existence, remains a central issue in *macroeconomics.

**naked debenture** An unsecured *debenture.

**name** *See* Lloyd's; syndicate.

**narrow money** An informal name for M0, or sometimes M1: the part of the *money supply that can directly perform the function of a *medium of exchange. *Compare* broad money.

**narrow-range securities** *See* trustee investments.

**Nash equilibrium** An equilibrium in an economy, in which every individual is maximizing his or her utility, taking into account the actions of all other individuals. A central concept in *game theory (which permits the solution of problems in *bargaining theory), the Nash equilibrium has been known for many years but was first formally stated by John Nash in 1950.

**national banks** US commercial banks established by federal charter, which requires them to be members of the *Federal Reserve System. They were created by the National Bank

Act (1863) and formerly issued their own banknotes. *Compare* state banks.

**National certificates and diplomas** *See* BTEC.

**National Chamber of Trade (NCT)** A non-profit making organization founded in 1897, with headquarters in Henley-on-Thames, that links and represents local *chambers of trade and commerce, national trade associations, and individual businesses in the UK. It maintains lobbies scrutinizing legislation in both Westminster and Brussels and provides expert advice for affiliated chambers and members.

**National Coal Board (NCB)** *See* British Coal Corporation.

**National Consumer Council** An organization set up by the UK government in 1975 to watch over consumer interests and speak for the consumer to the government, nationalized industries, and independent industry and commerce. It deals only with issues of policy.

**national debt** The debts of a central government, both internal and overseas. Net government borrowing each year is added to the national debt. By the end of the financial year 1987, the UK national debt of £186 billion was composed of:

£6 billion (non-sterling)

£180 billion sterling of which:

- £137 billion was in gilt-edged stock
- £27 billion in National Savings
- £16 billion tax deposits, Treasury Bills, etc.

The non-sterling debt is important because interest on it adversely affects the *balance of payments. Management of the national debt, which can be an important aspect of government monetary policy, is in the hands of the **National Debt Commissioners** of the Bank of England.

**National Economic Development Office** A UK government organization whose council, the National Economic Development Council (**NEDC** *or* **Neddy**) of which the Chancellor of the Exchequer is chairman, brings together members of the government (including the secretaries of state for Education and Science, Employment, Energy, Environment, and Trade and Industry), management, and the unions to consider issues concerning employment and economic growth.

**National Enterprise Board (NEB)** *See* British Technology Group.

**National Girobank** *See* giro.

**national income** The total annual money value of the goods and services produced by a country. *See* gross national product; national income accounts; net national product.

**national income accounts** Accounts that provide figures for the main macroeconomic variables, such as *gross national product, *consumption, and *investment. Almost all countries produce national income accounts, which are widely used for evaluating national economic performances. Although the UN provides a standard system for measurement of national income accounts, many countries do not follow these and many disagreements remain as to how they should be measured. *See also* circular flow of income.

**National Institute of Economic and Social Research** An organization founded in 1938 with the aim of increasing knowledge of the social and economic conditions of contemporary society. It conducts research by its own staff and in cooperation with the universities and other academic bodies. The Institute publishes a quarterly analysis of the economic situation and prospects in the *National Institute Economic Review*.

**National Insurance** A levy in the UK for social security purposes, notionally intended to fund sickness and unemployment benefits and national retirement pensions. There are four classes of payment: Class 1 primary and secondary, paid by employees and employers respectively, based on the wages and salaries of employees; Class 2, a weekly sum paid by the self-employed; Class 3, voluntary contributions to keep up contribution requirements; Class 4, a further levy on the self-employed based on levels of profit. *See also* State Earnings-Related Pension Scheme (SERPS).

**nationalization** The process of bringing the assets of a company into the ownership of the state. Examples of nationalized industries in the UK are the National Coal Board and British Rail. Historically, nationalization has been achieved through compulsory purchase, although this need not necessarily be the case. Nationalization has often been pursued as much for political as economic ends and the economic justifications themselves are varied. One strong argument for nationalization is that if a company possesses a *natural monopoly, the pure economic *profits it earns should be shared by the whole population through state ownership. Another argument might be that particular industries are strategically important for the nation and therefore cannot be entrusted to private enterprise. In many cases, nationalization has followed the commercial collapse of a company (e.g. British Leyland in the 1970s) because the government believed that it could not allow the company to disappear. In the 1980s the Conservative governments have tended to reverse Labour's nationalization of the 1950s, 1960s, and 1970s with a series of *privatization measures.

**nationalized industries** Companies or industries that have been taken into public ownership. *See* nationalization; privatization.

**national plan** An economic plan formulated by a government as a blueprint for its economic development over a stated period, usually five or ten years.

**National Research and Development Corporation (NRDC)** *See* British Technology Group.

**National Savings** The UK Department for National Savings was established in 1969, having previously been known as the Post Office Savings Department. It is responsible for administering a wide range of schemes for personal savers, including *premium bonds, *income bonds, *deposit bonds, and *yearly savings plans. In addition the department has offered a range of **National Savings Certificates** costing either £10 (up to 1981) or £25 (from 1981), some of which have been index-linked (*see* indexation). The income they pay is income-tax free and the element of capital gain is free of capital-gains tax. *See also* National Savings Bank; National Savings Stock Register.

**National Savings Bank** A savings bank operated by the Department for *National Savings through the agency of the Post Office. It offers ordinary accounts with a minimum deposit of £1 and a maximum of £10,000, and investment accounts paying a higher rate of interest for deposits of between £5 and £100,000.

**National Savings Stock Register** An organization run by the Department for *National Savings to enable members of the public to purchase certain Treasury stock and other gilts as an alternative to the main Bank of England Register. Purchases and sales are made by post and income is taxable, but is paid before deduction of tax (unlike the Bank of England Register). Because transactions are carried out by post, this method does not provide the maximum flexibility for dealing in a moving market.

**natural increase** The difference between the number of births and the number of deaths in a particular population over a particular period. To find out whether a population is increasing or decreasing it is also necessary to know the net migration rate (the difference between emigration and immigration).

**natural justice** The minimum standard of fairness to be applied when resolving a dispute. The main rules of natural justice include: (1) the right to be heard – each party to the dispute should be given an opportunity to answer any allegations made by the other party; (2) the rule against bias – the person involved in settling the dispute should act impartially, in particular by disclosing any interest he may have in the outcome of the dispute. The rules of natural justice apply equally in judicial as well as in administrative proceedings. Alleging a breach of natural justice is the method commonly used to challenge an administrative decision before the courts.

**natural monopoly** A monopoly in which the minimum efficient scale of production is greater than or equal to the total demand, as, for example, in electricity distribution. More generally, monopolies may exist as a result of barriers to entry, government privilege, or limited information. It is important that legislators know which type of monopoly they are confronting, since enforcing competition in markets in which there are barriers to entry is likely to increase efficiency, while the same policy directed at a natural monopoly can be expected to reduce efficiency. It is usually argued that natural monopolies are best harnessed for the benefit of the public by regulation, taxation, or nationalization.

**natural rate of interest** *See* interest.

**natural rate of unemployment** The rate of unemployment consistent with the productive potential of an economy. The concept was introduced by Milton Friedman (1912– ), who argued that any attempts by governments to reduce unemployment below the natural rate would necessarily fail. In the context of the *Phillips curve, this means that there is no long-term inverse relationship between inflation and unemployment. More recently it has been argued that there is not even a short-term inverse relationship (*see* new classical macroeconomics). Econometric attempts to estimate the natural rate of unemployment have proved inconclusive, although some have claimed that increasing unemployment in the 1970s and 1980s has been due to a steady rise in the natural rate. However, it is not clear why the natural rate should vary with time. *See also* NAIRU.

**natural wastage** The method by which an organization can contract without making people redundant, relying on resignations, retirements, or deaths. If the time is available, this method of reducing a work force causes the least tension. *See also* redundancy.

**NAV** Abbreviation for *net asset value.

**NBA** Abbreviation for *Net Book Agreement.

**NBV** Abbreviation for *net book value.

**NCB** Abbreviation for National Coal Board. *See* British Coal Corporation.

**NCT** Abbreviation for *National Chamber of Trade.

**NDP** Abbreviation for *net domestic product.

**near letter-quality (NLQ)** *See* letter-quality printer.

**near money** An asset that is immediately transferable and may be used to settle some but not all debts, although it is not as liquid as banknotes and coins. *Bills of exchange are examples of near money. Near money is not included in the *money supply indicators.

**NEB** Abbreviation for National Enterprise Board. *See* British Technology Group.

**necessary good** A good for which demand increases as income rises but less than proportionately (*compare* luxury good; inferior good). More precisely, a necessary good is one for which the income *elasticity of demand is less than one but greater than zero. *See also* normal good; Engel curve.

**NEDC (Neddy)** Abbreviation for National Economic Development Council. *See* National Economic Development Office.

**negative cash flow** A *cash flow in which the outflows exceed the inflows.

**negative income tax (NIT)** A means of targeting social security benefits to those most in need. The payments would be made through the income-tax system by granting personal allowances to taxpayers so that the *standard rate of income tax on these allowances would constitute a minimum amount required for living. Those with high incomes would obtain that amount as an income-tax relief, while those with incomes lower than the allowance would have a negative income-tax liability and be paid the appropriate sums. The principal objection to the system is that to cover the needs of the disadvantaged the wealthier would obtain excessively high allowances.

**negligence** A tort in which a breach of a **duty of care** results in damage to the person to whom the duty is owed. Such a duty is owed by a manufacturer to the consumers who buy his products, by professional persons to their clients, by a director of a company to its shareholders, etc. A person who has suffered loss or injury as a result of a breach of the duty of care can claim damages in tort.

**negotiability** The ability of a document that entitles its owner to some benefit to change hands, so that legal ownership of the benefit passes by delivery or endorsement of the document. For a document to be negotiable it must also entitle the holder to bring an action in law if necessary. *See* negotiable instrument.

**negotiable instrument** A document of title that can be freely negotiated (*see* negotiability). Such documents are *cheques and *bills of exchange, in which the stated payee of the instrument can negotiate the instrument by either inserting the name of a different payee or by making the document 'open' by endorsing it (signing one's name), usually on the reverse. Holders of negotiable instruments cannot pass on a better title than they possess. Bills of exchange, including cheques, in which the payee is named or that bear a restrictive endorsement, such as "not negotiable", are **non-negotiable instruments**.

**negotiate 1.** To confer with a view to arriving at mutually acceptable terms for a contract or agreement. **2.** To transfer a *bill of exchange or cheque to another for consideration (*see* negotiability; negotiable instrument).

**nem con** (abbreviation of Latin *nemine contradicente*: no-one disagreeing) These words are often used in the minutes of a meeting when no-one has voted against a proposition.

**neoclassical school** The mainstream school of thought in economics, deriving from the work of the *Marginalists, who defined value in relation to scarcity (*see* labour theory of value) and regarded the balance of supply and demand as determining *equilibrium prices. This method was first applied in *microeconomics and used to describe the *utility and profit-maximizing behaviour of individuals and firms (*see* profit maximization). The neoclassical approach was set out by Alfred Marshall (1842–1924) in his *Principles of Economics* (1890); modern microeconomic textbooks remain remarkably similar to this work. The application of neoclassical principles to *macroeconomics has been somewhat slower, since it was not immediately accepted that economic aggregates reflect the sum of individual choices. However, the development of general *equilibrium theory has enabled **neoclassical macroeconomists** to conform to a similar pattern to that earlier established in microeconomics (*see* new classical macroeconomics).

**net assets** The assets of an organization less its *current liabilities. The resultant figure is equal to the *capital of the organization. Opinion varies as to whether long-term liabilities should be treated as part of the capital and are therefore not deductible in arriving at net assets, or whether they are part of the liabilities and therefore deductible. The latter view is probably technically preferable. *See also* net current assets.

**net asset value (NAV)** The total assets of an organization less all liabilities and all capital charges (including debentures, loan stocks, and preference shares). The **net asset value per share** is the NAV divided by the total number of ordinary shares issued.

**Net Book Agreement (NBA)** An agreement between publishers and booksellers according to which booksellers will not offer books to the public below the price marked on the cover of the book. Exceptions include school textbooks, remainders, and books offered in a national book sale. Set up in 1899, the agreement is registered under the Restrictive Trade Practices Act (1956) and the Resale Prices Act (1964) as being in the public interest, although it has been challenged by some booksellers in recent years on the grounds that it keeps the price of books artificially high. Its defenders claim that if it was abolished, the most profitable (bestselling) books would be creamed off and sold at a discount in supermarkets and elsewhere, leaving the bookshops unable to survive or at least unable to keep large stocks of slow-selling books.

**net book value (NBV)** The value at which an asset appears in the books of an organization (usually as at the date of the last balance sheet) less any depreciation that has been applied since its purchase or its last revaluation.

**net current assets** *Current assets less *current liabilities. The resultant figure is also known as *working or circulating capital, as it represents the amount of the organization's capital that is constantly being turned over in the course of its trade. *See also* net assets.

**net domestic product (NDP)** The *gross domestic product of a country less *capital consumption (i.e. depreciation).

**net income** **1.** The income of a person or organization after the deduction of the appropriate expenses incurred in earning it. **2.** *Gross income from which tax has been deducted.

**net investment** The addition to the stock of capital goods in an economy during a particular period (the **gross investment**) less *capital consumption (i.e. depreciation).

**net national product (NNP)** The *gross national product less *capital consumption (i.e. depreciation) during the period. NNP is therefore equal to the national income, i.e. the amount of money available in the economy for expenditure on goods and services. However, NNP cannot be considered a very accurate measure, as it is difficult to calculate depreciation reliably.

**net present value (NPV)** The economic value of a project calculated by summing its costs and revenues over its full life and deducting the former from the latter. If the calculation yields a positive NPV then the project should be profitable. Future costs and revenues should be discounted by the relevant interest rate (e.g. the organization's cost of capital). *See* discounted cash flow. *Compare* accounting rate of return (ARR). Calculating NPVs can be difficult and often involves highly subjective judgments (such as estimating future interest rates). Frequently, therefore, simpler calculations, such as *payback period, are used. *See also* internal rate of return.

**net price** The price a buyer pays for goods or services after all discounts have been deducted.

**net profit 1. (net profit before taxation)** The profit of an organization when all receipts and expenses have been taken into account. In trading organizations, net profit is arrived at by deducting from the *gross profit all the expenses not already taken into account in arriving at the gross profit. **2. (net profit after taxation)** The final profit of an organization, after all appropriate taxes have been deducted from the net profit before taxation. *See also* profit and loss account.

**net-profit ratio** The proportion that *net profit bears to the total sales of an organization. This ratio is used in analysing the profitability of organizations and is an indicator of the extent to which sales have been profitable.

**net realizable value (NRV)** The net value of an asset if it were to be sold, i.e. the sum received for it less the costs of the sale and of bringing it into a saleable condition.

**net receipts** The total amount of money received by a business in a specified period after deducting costs, raw materials, taxation, etc. *Compare* gross receipts.

**net relevant earnings** A person's non-pensionable earned income before personal allowances have been deducted but after deduction of expenses, capital allowances, losses, or any stock relief agreed with the Inland Revenue.

**net reproduction rate** The number of female children in a population divided by the number of female adults in the previous generation. This figure gives a good guide to population trends; if it exceeds unity the population is expanding.

**net return** The profit made on an investment after the deduction of all expenses, either before or after deduction of capital-gains tax.

**net tangible assets** The *tangible assets of an organization less its current liabilities. In analysing the affairs of an organization the net tangible assets indicate its financial strength in terms of being solvent, without having to resort to such nebulous (and less easy to value) assets as *goodwill. *See also* price–net tangible assets ratio.

**net tonnage (register tonnage)** *See* tonnage.

**net weight** *See* gross weight.

**network** A number of computers connected together. Usually the computers in a network, which are called **nodes**, are dispersed throughout an organization, a region, or even across a continent. They communicate with each other using special communications links, such as satellite links, or ordinary telephone lines, using a *modem. Networks allow the computers to share each other's facilities. Often, one node is a large central computer, which may, for example, have fast printers or hold a large database of information. Networks also allow the computers to communicate with each other, so that memos and other messages can be sent electronically. *See also* local area network.

**net worth** The value of an organization when its liabilities have been deducted from the value of its assets. Often taken to be synonymous with *net assets (i.e. the total assets as shown by the balance sheet less the current liabilities), net worth so defined can be misleading in that balance sheets rarely show the real value of assets; in order to arrive at the true net worth it would normally be necessary to assess the true market values of the assets rather than their *book values. It would also be necessary to value *goodwill, which may not even appear in the balance sheet.

**net yield** *See* gross yield.

**neutrality of money** A belief, originating from the theories of *new classical macroeconomics, that the quantity of money in the economy can only affect prices (i.e. inflation), rather than such real variables as *investment or the level of employment. As the economy will always operate at the *natural rate of unemployment and government policies will be fully anticipated by individuals with *rational expectations, choices will be adjusted to counter the effects of government policy. For example, if a government tries to raise investment, the price of investment goods will rise; this will discourage previously planned purchases, which will no longer be made, so all that will have changed will be prices, which will have increased.

**new classical macroeconomics** A school of thought that developed in the 1970s by applying the concept of *rational expectations to macroeconomic theory. Keynesians and monetarists argued that *fiscal or *monetary policy could be used to raise the level of output and employment in the economy at least in the short term; the new classicists, however, claimed that this was not true. In their view any *reflation of the economy will be fully anticipated by individuals and firms, who will adjust their behaviour in such a way that the economy will remain unchanged in *real terms. For example, if the government increased public expenditure to create jobs, taxpayers – realizing that this would have to be paid for through higher taxes in the future – would reduce their present expenditure by an equal amount in order to save money to pay future tax bills. Nevertheless, the extra money injected into the economy would raise prices and cause inflation (*see also* nominal terms).

**new entrants 1.** Those looking for work for the first time. The main group comprises school leavers but housewives are increasingly important, especially with the recent opportunities for job sharing. **2.** Firms entering an industry for the first time. They may be new organizations or established organizations entering a new field.

**new for old** The basis for household insurance policies in which payments of claims are not subject to a deduction for wear and tear. As a result, a claim for an old and worn-out table would be met by the payment of the price of a new table of a similar type.

**new issue** A share being offered on a *stock exchange for the first time. *See* flotation.

**new product development (NPD)** A *marketing procedure in which new ideas are developed into viable new products or extensions to existing products or product ranges. New ideas, which are generated either internally (e.g. by scientific research) or by feedback from consumers, are first screened for prima facie viability; the few that remain are further reduced by *concept tests and detailed analysis of their potential profitability. Any ideas that survive these obstacles are subjected to extensive product development. Prototypes are made and tested within the company and among consumers, and improvements made. This cycle is repeated until satisfactory marketing research results are obtained, when the new product will be launched (possibly at first in a restricted area; *see* test marketing).

**new time** The purchase or sale of securities on the *London Stock Exchange during the last two dealing days of an account, for settlement during the following account. When making a bargain for new time this must be clearly understood between investor, broker, and market maker.

**New York Stock Exchange** The main US stock exchange, situated on New York City's Wall Street. Founded in 1792, it is an unincorporated organization with over 1500 members. *See* Dow Jones Industrial Average; Standard and Poor's 500 Stock Index.

**niche marketing** *See* concentrated segmentation.

**NIF** Abbreviation for *note issuance facility.

**night safe** A service provided by commercial banks, enabling customers to deposit cash with the bank after banking hours. By using this service shopkeepers, etc., can avoid keeping large sums overnight. A special wallet provided by the bank is inserted into a safe in the outside wall of the bank to which customers are given the key. The following day the wallet is opened, either by the customer himself or by a bank clerk, and his account is credited accordingly.

**Nikkei Dow Index** An index of Japanese industrial blue chip shares that gives an indication of the movement of share prices at the Tokyo Stock Exchange.

**NIT** Abbreviation for *negative income tax.

**NL** Abbreviation for no liability. It appears after the name of an Australian company, being equivalent to the British abbreviation *plc (denoting a public limited company).

**NLQ** Abbreviation for near letter-quality. *See* letter-quality printer.

**NNP** Abbreviation for *net national product.

**no-claim bonus** A reward, in the form of a premium discount, given to policyholders if they complete a year or more without making a claim. The system is mostly used in motor insurance, in which discounts of 33% for one claim-free year can rise to 60% of a premium for four successive years. In every case, the bonus is allowed for remaining claim-free and is not dependent on blame for a particular accident. So, for example, a no-claim bonus is lost if a vehicle is stolen through no fault of the insured.

**node** *See* network.

**nominal account** A ledger *account that is not a *personal account in that it bears the name of a concept, e.g. light and heat, bad debts, investments, etc., rather than the name of a person. These accounts are normally grouped in the *nominal ledger. *See also* impersonal account; real account.

**nominal capital (authorized capital)** *See* share capital.

**nominal damages** *See* damages.

**nominal ledger** The *ledger containing the *nominal accounts and *real accounts necessary to prepare the accounts of an organization. This ledger is distinguished from the personal ledgers, such as the *sales and *purchases ledgers, which contain the accounts of customers and suppliers respectively.

**nominal partner** *See* partnership.

**nominal price 1.** A minimal price fixed for the sake of having some consideration for a transaction. It need bear no relation to the market value of the item. **2.** The price given to a security when it is issued, also called the **face value, nominal value,** or *par value. For example, XYZ plc 25p ordinary shares have a nominal price of 25p, although the market value may be quite different. The nominal value of a share is the maximum amount the holder can be required to contribute to the company.

**nominal terms** The value of a good expressed only in money terms, rather than in terms of the quantity of another good for which it can be exchanged. Because the overall price level tends to change from year to year, and because different countries have different currencies, it is usually necessary to convert values from nominal terms to *real terms by applying some factor (e.g. a *rate of exchange or a price index). Macroeconomists are particularly interested in separating the effects of government policy into real effects and nominal effects.

**nominal value** *See* nominal price.

**nominal yield** *See* yield.

**nomination** The person to whom the proceeds of a life-assurance policy should be paid as specified by the policyholder. *See* assignment of life policies.

**nominee** A person named by another (the **nominator**) to act on his behalf, often to conceal the identity of the nominator. *See* nominee shareholding.

**nominee shareholding** A shareholding held in the name of a bank, stockbroker, company, individual, etc., that is not the name of the beneficial owner of the shares. A shareholding may be in the name of nominees to facilitate dealing or to conceal the identity of the true owner. Although this cover was formerly used in the early stages of a takeover, to enable the bidder clandestinely to build up a substantial holding in the target company, this is now prevented by the Companies Act (1981), which makes it mandatory for anyone holding 5% or more of the shares in a public company to declare that interest to the company. The earlier Companies Act (1967) made it mandatory for directors to openly declare their holdings, and those of their families, in the companies of which they are directors.

**non-accelerating inflation rate of unemployment** *See* NAIRU.

**non-acceptance** The failure by the person on whom a *bill of exchange is drawn to accept it on presentation.

**non-assented stock** *See* assented stock.

**non-business days** *See* Bank Holidays.

**non-contributory pension** A *pension in which the full premium is paid by an employer or the state and the pensioner makes no contribution. *Compare* contributory pension.

**non-domiciled** Denoting a person who is not domiciled in his country of origin but resides permanently abroad. *See* domicile.

**non-durables** Consumer non-durables. *See* consumer goods.

**non-executive director** *See* executive director.

**non-impact printer** *See* printer.

**non-marketable securities** UK government securities that cannot be bought and sold on a stock exchange (*compare* marketable security). Non-marketable securities include savings bonds and National Savings Certificates, tax-reserve certificates, etc., all of which form part of the *national debt.

**non-monetary advantages and disadvantages** Those aspects of an employment that are not connected with its financial remuneration. They include the employee's subjective opinion of his working environment, the stimulation or boredom of the work itself, the companionship or isolation he experiences in his place of work, the distance he has to travel to reach work, etc. These aspects of an employment, together with the salary, bonuses, commission, and fringe benefits, make a person decide either to stay in his present job or seek another.

**non-negotiable instruments** *See* negotiable instrument.

**non-price competition** A form of competition in which two or more producers sell goods or services at the same price but compete to increase their share of the market by such measures as advertising, sales promotion campaigns, improving the quality of the product or service, and (for goods) improving the packaging, offering free servicing and installation, or giving free gifts of unrelated products or services.

**non-profitability sampling** A technique in *marketing research in which the researcher uses his own discretion, instead of random selection, in choosing a representative sample of people to interview. This reduces the statistical validity of the results, which are then not sufficiently reliable to use as a basis for estimating profits.

**non-profit marketing** Applying the concept of marketing to organizations that are not profit-orientated. Examples include symphony orchestras, museums, and charities.

**non-qualifying policy** A life-assurance policy that does not satisfy the qualification rules contained in Schedule 1 of the Income and Corporation Taxes Act (1970): no tax relief can be obtained on the premium payments of such a policy. *Compare* qualifying policy.

**non-resident** A British person employed overseas on a contract of service for at least a full tax year, who does not pay UK tax provided that he does not spend more than 90 days in the UK during that tax year. *See also* domicile.

**non-tariff barrier (NTB)** *See* Tokyo round.

**non-tariff office** *See* tariff office.

**non-taxable income** Income that is not taxed by statute, e.g. interest on certain National Savings bonds.

**non-voting shares** *See* A shares.

**no-par-value (NPV)** Denoting a share issued by a company that has no *par value (*see also* nominal price). Dividends on such shares are quoted as an amount of money per share rather than as a percentage of the nominal price. No-par-value shares are not allowed by UK law but they are issued by some US and Canadian companies.

**normal economic profit** The theoretical minimum profit required to keep an entrepreneur in a particular business. It must be at least as much as he could earn by investing his capital in some other business. If the entrepreneur was to earn abnormally high profits, new firms would enter the industry and the entrepreneur's profits would fall; if the entrepreneur's profit was too low he would leave the industry, allowing others to make better profits. Thus in *perfect competition only normal profits can be made. In a *monopoly, abnormally high profits can be earned as a result of *barriers to entry.

**normal good** A good for which the absolute level of demand increases as income expands. This does not necessarily mean that a consumer will spend the same proportion of his income on a normal good as his income expands, but it does mean that he will spend more in absolute terms. Another way of saying this is that the income *elasticity of demand is greater than zero. The assumption that goods are normal is usually made in economic analysis for the purposes of simplification. *See also* Engel curve; Giffen good; inferior good; luxury good; necessary good.

**normal price** The theoretical price of a good or service when there is equilibrium between supply and demand. Although the actual price will be influenced by a number of short-term factors, in the long term it will tend to move towards the normal price.

**normal retirement age** The age of an individual when he retires. This is normally 65 for a man and 60 for a woman in the UK. It is at these ages that state pensions begin. However, other policies can nominate other pre-agreed dates, which thc Inland Revenue will accept in certain cases.

**normative economics** An economic analysis that includes judgments about what ought to be done, rather than simply theorizing (*compare* positive economics). For example, Keynesian analysis contains both positive elements (the study of *involuntary unemployment) and normative elements (the recommendation that *fiscal policy should be used to reduce unemployment). Unfortunately, it is usually difficult to separate the positive from the normative in an economic theory.

**nostro account** A bank account conducted by a UK bank with a bank in another country, usually in the currency of that country. *Compare* vostro account.

**notary public** A legal practitioner, usually a solicitor, who is empowered to attest deeds and other documents and notes (*see* noting) dishonoured *bills of exchange.

**note issuance facility (NIF)** A means of enabling short-term borrowers in the *eurocurrency markets to issue euronotes, with maturities of less than one year, when the need arises rather than having to arrange a separate issue of euronotes each time they need to borrow. **Revolving underwriting facility (RUF)** achieves the same objective.

**notice in lieu of distringas** *See* stop notice.

**notice of abandonment** *See* abandonment.

**notice of motion** *See* motion.

**noting 1.** The procedure adopted if a *bill of exchange has been dishonoured by non-acceptance or by non-payment. Not later than the next business day after the day on which it was dishonoured, the holder has to hand it to a *notary public to be noted. The notary re-presents the bill; if it is still unaccepted or unpaid, he notes the circumstances in a register and also on a **notarial ticket**, which he attaches to the bill. The noting can then, if necessary, be extended to a *protest. **2.** In advertising research, noticing a particular advertisement when first looking through a newspaper or magazine in which it appears. *See* noting score.

**noting score** The average number of readers *noting a particular advertisement or editorial item in a newspaper or magazine, expressed as a percentage of the total readership.

**notional income** Income that is not received although it might be deemed to be properly chargeable to income tax. An example is the rental income one could receive from one's own home, if one were not living in it. This was formerly taxed under Schedule A. A further example might be interest foregone on the grant of an interest-free loan.

**not negotiable** Words marked on a *bill of exchange indicating that it ceases to be a *negotiable instrument, i.e. although it can still be negotiated, the holder cannot obtain a better title to it than the person from whom he obtained it, thus providing a safeguard if it is stolen. A cheque is the only form of bill that can be crossed 'not negotiable'; other forms must have it inscribed on their faces.

**NPD** *See* new product development.

**NPV 1.** Abbreviation for *net present value. **2.** Abbreviation for *no-par-value.

**NRV** Abbreviation for *net realizable value.

**NTB** Abbreviation for non-tariff barrier. *See* Tokyo round.

**nudum pactum** (Latin: nude contract) An agreement that is unenforceable in British law because no consideration is mentioned. *See* contract.

**numbered account** A bank account identified only by a number. This service, offered by some Swiss banks, encourages funds that have been obtained illegally to find their way to Switzerland.

**NV** Abbreviation for *Naamloze Vennootschap*. It appears after the name of a Dutch company, being equivalent to the British abbreviation Ltd (i.e. denoting a limited liability company).

**NVQ** Abbreviation for National Vocational Qualification. *See* vocational training.

# O

**O & M** *See* organization and methods.

**objectivity** The quality of being able to be independently verified, especially in accounting practice. It has been suggested that accounts produced on a historical cost basis (*see* historical-cost accounting) are objective, whereas those based on *inflation accounting are thought to be rather more subjective. However, such comments tend to overlook the considerable areas of subjectivity in historical cost accounts, such as choice of stock valuation methods, periods of asset life chosen for depreciation, and amounts to be set aside as provisions for bad debts.

**objects clause** *See* memorandum of association.

**obsolescence 1.** Depreciation of plant and equipment that may not have actually worn out but has become out of date because technology has advanced and more efficient plant has become available. It also applies to consumer durables (*see* consumer goods), in which a change of style may render a serviceable piece of equipment, such as a car or washing machine, out of date. **Built-in obsolescence** or **planned obsolescence** is a deliberate policy adopted by a manufacturer to limit the durability of his product in order to encourage the consumer to buy a replacement more quickly than he otherwise might have to. The morality of this technique has been frequently questioned but is usually defended on the grounds that many western economies depend on strong consumer demand; if such consumer durables as cars and washing machines were built to last for their purchaser's lifetime, consumer demand would be reduced to a level that would create enormous unemployment. **2.** The depreciation of a piece of plant or machinery as a result of its age.

**occupational hazard** A risk of accident or illness at one's place of work. Dangerous jobs usually command higher salaries than those involving no risks, the increase being known as **danger money**. *See also* Health and Safety Commission.

**occupational pension scheme** A pension scheme open to employees within a certain trade or profession or working for a particular firm. An occupational pension scheme can either be insured or self-administered. If it is insured, an insurance company pays the benefits under the scheme in return for having the premiums to invest. In a self-administered scheme, the pension-fund trustees are responsible for investing the contributions themselves. In order to run an occupational pension scheme, an organization must satisfy the Occupational Pension Board that the scheme complies with the conditions allowing employers to contract out of the *State Earnings-Related Pension Scheme. After 1988 certain regulations relating to occupational pensions schemes were introduced. *See also* personal pension scheme.

**OCR** Abbreviation for *optical character recognition.

**octal notation** A number system that uses eight digits, 0–7, to represent numbers. For example, the decimal number 26 is written as 32 in octal notation. Although less popular than *hexadecimal notation, octal notation is used in computer programming as it is easier to follow than *binary notation, yet is easy to convert to binary if required.

**ODA** Abbreviation for *Overseas Development Administration.

**odd-even pricing** The pricing of a product so that the price ends in an odd number of pence, which is not far below the next number of pounds. For example, £4.99 might be used in preference to £5.00 in order to make the product appear cheaper (at least to marketing personnel if not to consumers).

**odd lot** *See* round lot.

**OECD** Abbreviation for *Organization for Economic Cooperation and Development.

**off-balance sheet reserve** *See* hidden reserve.

**off-card rate** An advertising charging rate that is different from that shown on the *rate card, having been separately negotiated.

**offer** The price at which a seller makes it known that he is willing to sell something. If there is an *acceptance of the offer a legally binding *contract has been entered into. In law, an offer is distinguished from an **invitation to treat**, which is an invitation by one person or firm to others to make an offer. An example of an invitation to treat is to display goods in a shop window. *See also* firm offer; offer price; quotation.

**offer by prospectus** An offer to the public of a new issue of shares or debentures made directly by means of a *prospectus, a document giving a detailed account of the aims, objects, and capital structure of the company, as well as its past history. The prospectus must conform to the provisions of the Companies Act (1985). *Compare* offer for sale.

**offer document** A document sent to the shareholders of a company that is the subject of a *takeover bid. It gives details of the offer being made and usually provides shareholders with reasons for accepting the terms of the offer.

**offer for sale** An invitation to the general public to purchase the stock of a company through an intermediary, such as an *issuing house or *merchant bank (*compare* offer by prospectus); it is one of the most frequently used means of *flotation. An offer for sale can be in one of two forms: at a fixed price (the more usual), which requires some form of balloting or rationing if the demand for the shares exceeds supply; or an *issue by tender, in which case individuals offer to purchase a fixed quantity of stock at or above some minimum price and the stock is allocated to the highest bidders. *Compare* introduction; placing; public issue.

**offer price** The price at which a security is offered for sale by a *market maker and also the price at which an institution will sell units in a unit trust. *Compare* bid price.

**offer to purchase** *See* takeover bid.

**Office of Fair Trading** A government department that, under the Director General of Fair Trading, reviews commercial activities in the UK and aims to protect the consumer against unfair practices. Established in 1973, it is responsible for the administration of the Fair Trading Act (1973), the Consumer Credit Act (1974), the Restrictive Trade Practices Act (1976), the Estate Agents Act (1979), the Competition Act (1980), and the Control of Misleading Advertisements Regulations (1988). Its five main areas of activity are: consumer affairs, consumer credit, monopolies and mergers, restrictive trade practices, and anti-competitive practices.

**Office of Telecommunications (OFTEL)** A UK government body set up in 1984 to supervise the telecommunications industry. It is responsible for issuing licences, regulating competition in the industry, and protecting the interests of the consumer.

**officer of a company** A person who acts in an official capacity in a company. Company officers include the directors, managers, the company secretary, and in some circumstances the company's auditors and solicitors. The Companies Act (1985) empowers a court dealing with a *liquidation to investigate the conduct of the company's officers with a view to recovering any money they may have obtained illegally or incorrectly.

**Official List 1.** A list of all the securities traded on the *main market of the *London Stock Exchange. *See* listed security; listing requirements; Yellow Book. **2.** A list prepared daily by the London Stock Exchange, recording all the bargains that have been transacted in listed securities during the day. It also gives dividend dates, rights issues, prices, and other information.

**official rate** The rate of exchange given to a currency by a government. If the official rate differs from the market rate, the government has to be prepared to support its official rate by buying or selling in the open market to make the two rates coincide.

**official receiver** A person appointed by the Secretary of State for Trade and Industry to act as a *receiver in *bankruptcy and winding-up cases. The High Court and each county court that has jurisdiction over insolvency matters has an official receiver, who is an officer of the court. Deputy official receivers may also be appointed. The official receiver commonly acts as the *liquidator of a company being wound up by the court.

**official strike** *See* strike.

**off-line** Denoting computer equipment that is not usable, either because it is not connected to a computer or because the system has been forbidden to use it. *Compare* on-line.

**offprint** A reprint or photocopy of an article from a periodical or paper.

**offshore fund 1.** A fund that is based in a tax haven outside the UK to avoid UK taxation. Offshore funds operate in the same way as *unit trusts but are not supervised by the Department of Trade and Industry. **2.** A fund held outside the country of residence of the holder. Offshore centres provide advantageous deposit and lending rates, because of advantageous taxation, liberal exchange controls, and low reserve requirements for banks. Some countries have made a lucrative business out of offshore banking; the Cayman Islands is currently the world's largest offshore centre. The USA and more recently Japan have both established domestic offshore facilities, where business with non-residents can be conducted under more liberal regulations than domestic transactions. Their objective is to stop funds moving outside the country.

**Offshore Supplies Office** A department of the UK Department of Energy set up in Glasgow in 1973, now with offices also in Aberdeen and London. It encourages, monitors, advises, and assists the offshore oil and gas industries, providing finance and research and development.

**off-the-peg research** Marketing research that uses existing data, rather than a fresh investigation of a market. *See also* desk research.

**off-the-shelf company** A company that is registered with the *Registrar of Companies although it does not trade and has no directors. It can, however, be sold and reformed into a new company with the minimum of formality and expense. Such companies are easily purchased from specialist brokers.

**OFTEL** Abbreviation for *Office of Telecommunications.

**old-age pension** *See* pension.

**Old Lady of Threadneedle Street** An affectionate name for the *Bank of England, coined by the English politician and dramatist R B Sheridan (1751–1816). The street in which the Bank stands (since 1734 in a Renaissance building by George Sampson) probably takes its name from the thread and needle used by the Merchant Taylors, a guild whose hall is in the same street.

**oligopoly** A market in which relatively few sellers supply many buyers, each seller recognizing that he can control his prices to a certain extent and that his competitors' actions will influence his profits. This is a departure from the assumption, in *perfect competition, that there are a large number of firms in any particular market. If there are only a few firms, each one may be able to influence the market price by controlling the amount they produce, thus earning higher profits. Oligopoly is a common feature of the real world; studies have shown that in developed countries markets tend to become more oligopolistic. There is some doubt, however, as to whether prices really are higher under oligopoly. *See also* contestable markets theory.

**Ombudsman** *See* Commissioner for Local Administration; Financial Ombudsman; Parliamentary Commissioner for Administration.

**omnibus research** Marketing research surveys based on multipart questionnaires sent out regularly to a panel of respondents. Space on the questionnaire is available to companies that have specific marketing research needs, especially those having a limited number of questions to ask, which would not alone justify setting up a separate research study.

**on approval (on appro)** The practice of allowing potential buyers to take possession of goods in order to decide whether or not they wish to buy them. The potential buyer is a bailee (*see* bailment) and is obliged to return the goods in perfect condition if he decides not to purchase them. Many department stores and mail-order firms allow their customers to take goods on appro.

**on consignment** *See* consignment.

**oncosts** The costs of a product or service over and above the *direct costs.

**on demand** Denoting a *bill of exchange that is payable on presentation. An uncrossed cheque is an example of such a bill.

**on-line 1.** Denoting either a *real-time operation that is accomplished by a dialogue between a human operator and a computer system, or a system based on such operations. Examples include airline-ticket reservation systems. **2.** Denoting computer equipment that is connected to a computer and is usable by it. *Compare* off-line.

**on stream** Denoting that a specified investment or asset is bringing in the income expected of it.

**OPEC** Abbreviation for *Organization of the Petroleum Exporting Countries.

**open charter** *See* chartering.

**open cheque** *See* cheque.

**open cover (open policy)** A marine cargo insurance policy in which the insurer agrees to cover any voyage undertaken by the policyholder's vessel(s) or any cargo shipped by a particular shipper. A policy condition requires a declaration to be completed on a weekly, monthly, or quarterly basis indicating the vessels involved, the commodities carried, and the voyages undertaken. Insurers use the declaration to calculate the premium. The main advantage of a policy of this kind is that the insurer can be confident that all cargos have insurance cover without the need to notify insurers of the details before the voyage. In some cases there may be a policy limit above which the insurance ceases. On such policies it is necessary to keep a running total of the sums insured to make sure that the policy limit is not exceeded.

**open credit 1.** Unlimited credit offered by a supplier to a trusted client. **2.** An arrangement between a bank and a customer enabling the customer to cash cheques up to a specified amount at a bank or branches other than his own. This practice is less used since *credit cards and *cash cards were introduced.

**open-door policy** An import policy of a country in which all goods from all sources are imported on the same terms, usually free of import duties.

**open economy** An economy in which a significant percentage of its goods and services are traded internationally. The degree of openness of an economy usually depends on the amount of overseas trade in which the country is involved or the political policies of its government. Thus the UK economy is relatively open, as the economy is significantly dependent on foreign trade; the US economy is relatively closed as overseas trade is not very important to its economy, while the USSR has historically been closed as a result of political choice, although this policy is now being modified.

**open-ended question** A question used in marketing research that has to be answered in the respondent's own words rather than by "yes", "no", or "don't know". *Compare* structured interview.

**open-end trust** A form of unit trust in which the managers of the trust may vary the investments held without notifying the unit holders. Open-end trusts are used in the USA.

**open general licence** An import licence for goods on which there are no import restrictions.

**open indent** An order to an overseas purchasing agent to buy certain goods, without specifying the manufacturer. If the manufacturer is specified this is a **closed indent**.

**opening prices** The bid prices and offer prices made at the opening of a day's trading on any security or commodity market. The opening prices may not always be identical to the previous evening's *closing prices, especially if any significant events have taken place during the intervening period.

**open-market operations** The purchase or sale by a government of bonds (gilt-edged securities) in exchange for money. This is the main mechanism by which *monetary policy in developed economies operates. To buy (or sell) more bonds the government must raise (or lower) their price and hence reduce (or increase) interest rates. In Keynesian theory, lower interest rates will stimulate investment and so raise national output. However, in such theories as those of *new classical macroeconomics, this will only have the effect of raising the level of prices (i.e. fuelling *inflation).

**open outcry** *See* callover.

**open policy** *See* open cover.

**open position** A trading position in which a dealer has commodities, *securities, or currencies bought but unsold or unhedged (*see* hedging), or sales that he has neither covered nor hedged. In the former position he has a **bull position**; in the latter, a **bear position**. In either case he is vulnerable to market fluctuations until he closes or hedges the position. *See also* option.

**open-pricing agreement** An agreement between firms operating in an oligopolistic market in which prices and intended price changes are circulated to those taking part in the agreement in order to avoid a *price war.

**open shop** *See* closed shop.

**operating budget** A forecast of the financial requirements for the future trading of an organization, including its planned sales, production, cash flow, etc. An operating budget is normally designed for a fixed period, usually one year, and forms the plan for that period's trading activities. Any divergences from it are usually monitored and, if appropriate, changes can then be made to it as the period progresses.

**operating costs** *See* overhead costs.

**operating profit (***or* **loss)** The profit (or loss) made by a company as a result of its principal trading activity. This is arrived at by deducting its **operating expenses** from its *trading profit, or adding its operating expenses to its trading loss; in either case this is before taking into account any extraordinary items.

**operating system** The collection of programs that controls the basic operation of a *computer. Usually purchased with the computer, it controls such tasks as start-up routines, input and output, and memory allocation. It is also responsible for loading and executing programs. In small systems, the operating system loads a program and then gives it control of the machine. In larger *multitasking systems, the operating system always retains control of the machine, ensuring that the separate programs do not interfere with each other and controlling the amount of processing time each receives. *See also* CP/M; MS-DOS.

**operational research (operations research)** A scientific method of approaching industrial and commercial problems in order to arrive at the most efficient and economic method of achieving the desired objective. It basically consists of making a clear statement of the problem, designing a model to represent the possible solutions using different strategies, and applying the solutions obtained from the analysis of the model to the real problem. It makes use of *game theory, *critical-path analysis, simulation techniques, etc.

**operation job card** A card or form on which an employee writes up the details of a particular task and the length of time it took to complete. It is used in *work studies.

**opinion poll** A poll using marketing research techniques that is specifically concerned with political opinion. Public opinion polls are usually commissioned by the media (especially television and newspapers) in order to keep their readers or viewers informed. Private opinion polls are usually commissioned by the political parties in order to help develop campaign strategies.

**opportunities to see** The number of times a member of the public (or an audience) may, on average, be expected to see a particular advertisement.

**opportunity cost** The benefits lost by not employing an economic resource in the most profitable alternative activity. For example, the opportunity cost to a self-employed person is the highest salary he could earn elsewhere. Economists use the concept of opportunity cost to decide whether or not the allocation of resources is efficient. In the example above, if efficiency is judged by income alone, self-employment is efficient only if the income earned exceeds the best alternative salary, i.e. the opportunity cost. Opportunity cost is a much broader concept than accounting cost, and therefore the former is generally preferred when weighing up the costs and benefits of investment decisions (*see also* cost-benefit analysis). For example, in a situation in which investment in competing and mutually exclusive projects is being considered, the opportunity cost of selecting one project is the revenue obtainable from the next best option.

**optical character recognition (OCR)** The recognition of printed characters by light-sensitive optical scanners. The scanner recognizes the shape of a letter by scanning it with a very fine point of light. It then uses a computer to compare the pattern of reflected light with the patterns of the letters of the alphabet stored in its memory. Usually special easily recognizable characters are necessary but some scanners will read typewritten characters. Ordinary typeset text cannot yet be scanned reliably. OCR is used by banks, for example, to read the information printed on cheques.

**optimum population** The population of an economic community in which the income per head is a maximum, i.e. in which there are sufficient people to provide an adequate labour force but not so many that there is high unemployment. Countries often have long-term demographic policies, for example India to reduce its population and Australia to increase it.

**option** The right to buy or sell a fixed quantity of a commodity, currency, or *security at a particular date at a particular price (the *exercise price). Unlike futures, the purchaser of an option is not obliged to buy or sell at the exercise price and will only do so if it is profitable; he may allow the option to lapse, in which case he loses only the initial purchase price of the option (the *option money). In London, options in securities (**stock options**) are bought and sold through the *London Stock Exchange, while options in commodity futures are bought and sold on the various commodity exchanges (*see* London FOX), and options on share indexes, foreign currencies, and interest rates are dealt with through the *London International Financial Futures Exchange.

An option to buy is known as a **call option** and is usually purchased in the expectation of a rising price; an option to sell is called a **put option** and is bought in the expectation of a falling price or to protect a profit on an investment. Options, like futures, allow individuals and firms to hedge against the risk of wide fluctuations in prices; they also allow speculators to gamble for large profits with limited liability.

Professional traders in options make use of a large range of potential strategies, often purchasing combinations of options that reflect particular expectations or cover several contingencies (*see* butterfly; straddle).

**Traded options** can be bought and sold on a stock exchange, at all times, i.e. there is a trade in the options themselves. The traded option market started in London in 1978 and enables options in a limited number of companies to be bought and sold. **Traditional options**, however, once purchased, cannot be resold. In a **European option** the buyer can only exercise his right to take up the option or let it lapse on the *expiry date, whereas with an **American option** this right can be exercised at any time up to the expiry date. European options are therefore cheaper than American options. *See also* exercise notice; expiry date; hedging; intrinsic value; option to double; time value.

**option dealer** A dealer who buys and sells either traded *options or traditional options on a stock exchange, commodity exchange, or currency exchange.

**option money** The price paid for an *option. The cost of a call option is often known as the **call money** and that for a put option as the **put money**.

**option to double 1.** An *option by a seller to sell double the quantity of securities for which he has sold an option, if he wishes. In some markets this is called a **put-of-more option. 2.** An option by a buyer to buy double the quantity of securities for which he has bought an option, if he wishes. In some markets this is called a **call-of-more option.**

**option to purchase 1.** A right given to shareholders to buy shares in certain companies in certain circumstances at a reduced price. **2.** A right purchased or given to a person to buy something at a specified price on or before a specified date. Until the specified date has passed, the seller undertakes not to sell the property to anyone else and not to withdraw it from sale.

**Oracle** *See* Teletext.

**order cheque** *See* cheque.

**order of business** The sequence of the items on the *agenda of a business meeting. It is usual to adopt the following order: apologies for absence; reading and signing of the minutes of the last meeting; matters arising from these minutes; new correspondence received; reading and adoption of reports, accounts, etc.; election of officers and auditors; any special motions; any other business; date of next meeting. *See also* annual general meeting; extraordinary general meeting.

**ordinary resolution** *See* resolution.

**ordinary share** A fixed unit of the *share capital of a company. Shares in publicly owned quoted companies (*see* plc; quotation) are usually traded on *stock exchanges and represent one of the most important types of security for investors. Shares yield dividends, representing a proportion of the profits of a company (*compare* fixed-interest security; preference share). In the long term, ordinary shares, by means of *capital growth, yield higher rewards, on average, than most alternative forms of securities, which compensates for the greater element of risk they entail. *See also* convertible; growth stocks.

**organizational buying** The way in which an organization (as opposed to an individual consumer) identifies, evaluates, and chooses the products it buys. *See* decision-making unit.

**organization and methods (O & M)** A form of *work study involving the organization of procedures and controls in a business and the methods of implementing them in management terms. It is usually applied to office procedures rather than factory production.

**Organization for Economic Cooperation and Development (OECD)** An organization formed in 1961, replacing the Organization for European Economic Cooperation (OEEC), to promote cooperation among industralized member countries on economic and social policies. Its objectives are to assist member countries in formulating policies designed to achieve high economic growth while maintaining financial stability, contributing to world trade on a multilateral basis, and stimulating members' aid to developing countries. Members are Australia, Austria, Belgium, Canada, Denmark, Finland, France, West Germany, Greece, Iceland, Ireland, Italy, Japan, Luxembourg, the Netherlands, New Zealand, Norway, Portugal, Spain, Sweden, Switzerland, Turkey, UK, and USA (Yugoslavia participates with a special status).

**Organization of the Petroleum Exporting Countries (OPEC)** An organization created in 1960 to unify and coordinate the petroleum policies of member countries and to protect their interests, individually and collectively. Present members are Algeria, Ecuador, Gabon, Indonesia, Iran, Iraq, Kuwait, Libya, Nigeria, Qatar, Saudi Arabia, UAE, and Venezuela.

**organized market** A formal market in a specific place in which buyers and sellers meet to trade according to agreed rules and procedures. Stock exchanges and commodity markets are examples of organized markets.

**origin 1.** The country from which a commodity originates. **Shipment from origin** denotes goods that are shipped directly from their country of origin, rather than from stocks in some other place. *See also* certificate of origin. **2.** The country from which a person comes. A person's country of origin is not necessarily his country of *domicile or residence.

**original-entry book** *See* book of prime entry.

**original goods** Natural products that have no economic value until *factors of production are applied to them. They include virgin land, wild fruit, natural waterways, etc.

**OTC market** Abbreviation for *over-the-counter market.

**OTE** Abbreviation for on-target earnings, which represents the salary and commission a salesman should be able to earn.

**outcry** *See* callover.

**outlay** Expenditure on a specified good or service, e.g. the raw-materials outlay is the money spent during a specified period on raw materials. In accountancy this is the same as total cost, but in economics it would be necessary to add an increment for the *opportunity cost.

**outlay tax** *See* expenditure tax.

**out-of-the-money option** *See* intrinsic value.

**output device** A device used for taking data out of a computer system. The data is presented either in a form understandable by humans or in a form suitable for input to another machine. The *visual display unit (VDU) and the *printer are the most common output devices. Other output systems include speech synthesis and *computer output microfilm.

**output tax** The VAT that a trader adds to the price of the goods or services he supplies. He must account for this output tax to HM Customs and Excise, having first deducted the *input tax.

**outside broker** A stockbroker who is not a member of a stock exchange but acts as an intermediary between the public and a stockbroker who is a member.

**outside director** A member of the board of directors of a company who is not employed directly by that company, but may be employed by a holding company or associated company. Outside directors are usually non-executive directors (*see* executive director).

**outside money** Money issued from outside an economic system, usually either by a government or by other countries (i.e. foreign currency). It has been argued that a fall in the level of prices may increase the value of outside money in an economy, creating a positive *wealth effect (*compare* inside money). However, even outside money issued by a government may have no wealth effect (*see* Ricardian equivalence theorem).

**outwork** Work carried out in a person's own home or workshop rather than on a company's own premises. Outworkers are still used in the upmarket end of the garment trade.

**overbought** **1.** Having purchased more of a good than one needs or has orders for. **2.** Having purchased more securities or commodities than are covered by margins deposited with a broker or dealer. In a falling market a bull speculator can become overbought without having made a fresh purchase. **3.** Denoting a market that has risen too rapidly as a result of excessive buying. An overbought market is unstable and likely to fall if unsupported. *Compare* oversold.

**overcapitalization** A condition in which an organization has too much *capital for the needs of its business. If a business has more capital than it needs it is likely to be overburdened by interest charges or by the need to spread profits too thinly by way of dividends to shareholders. Businesses can now reduce overcapitalization by repaying long-term debts or by buying their own shares.

**overdraft** A loan made to a customer with a cheque account at a bank or building society, in which the account is allowed to go into debit, usually up to a specified limit (the **overdraft limit**). Interest is charged on the daily debit balance. This is a less costly way of borrowing than taking a *bank loan (providing the interest rates are the same) as, with an overdraft, credits are taken into account.

**over-entry certificate** A document issued by HM Customs and Excise stating that imported goods have paid excessive duty on entry into the country, which may be reclaimed. If too little duty has been paid, **post-entry** duty is claimed.

**over-full employment** A situation in which there are more jobs than people seeking work. This situation tends to be inflationary as employers increase the wages offered, to attract employees, and in addition employers have to resort to less well-qualified employees, which lowers the efficiency of industry.

**overfunding** A policy available to the UK government in which it sells more government securities than it needs to pay for public spending. The objective of the policy is to absorb surplus money and so curb *inflation.

**overhead costs (indirect costs; operating costs)** Costs of production of goods or services that are not *direct costs, i.e. costs over and above those for materials and labour employed in producing the goods or services. These are divided into **fixed costs** and **variable costs**, the former being those that do not change over broad levels of activity (e.g. factory rent, depreciation of plant and machinery) and the latter being those costs that do vary with levels of production (e.g. fuel and power).

**overhead variance** A *variance arising because overhead costs differ from those budgeted for.

**overheating** The state of an economy during a boom, with increasing *aggregate demand leading to rising prices rather than higher output. Overheating reflects the inability of some firms to increase output as fast as demand; they therefore choose to profit from the excess demand by raising prices.

**overinsurance** The practice of insuring an item for a greater amount than its value. This is pointless as insurers are only obliged to pay the full value (usually the replacement value) of an insured item and no more, even if the sum insured exceeds this value. If insurers find a policyholder has overinsured an item, the premium for the cover above the true value is returned.

**overinvestment** Excessive investment of capital, especially in the manufacturing industry towards the end of a boom as a result of over-optimistic expectations of future demand. When the boom begins to fade, the manufacturer is left with surplus capacity and he therefore makes no further capital investments, which itself creates unemployment and fuels the imminent recession.

**overnight loan** A loan made by a bank to a bill *broker to enable him to take up *bills of exchange. Initially the loan will be repayable the following day but it is usually renewable. If it is not, the broker must turn to the *lender of last resort, i.e. the Bank of England in the UK.

**overseas company** A company incorporated outside the UK that has a branch or a subsidiary company in the UK. Overseas companies with a place of business in the UK have to make a return to the Registrar of Companies, giving particulars of their memorandum of association, directors, and secretary as well as providing an annual balance sheet and profit-and-loss account.

**Overseas Development Administration (ODA)** The UK government department dealing with the administration of aid to overseas countries, including both financial and technical assistance. Previously the Ministry for Overseas Development, the ODA is now part of the Foreign and Commonwealth Office, with a minister solely responsible for ODA.

**overseas-income taxation** Taxation of income arising outside the national boundaries of the taxing authority. Most countries tax the worldwide income of their permanent residents, as well as the incomes arising in the country to outsiders. This can involve double (or multiple) taxation, for which reliefs are often provided by double-taxation treaties.

**overseas investment** Investment by the government, industry, or members of the public of a country in the industry of another country. For members of the public this is often most easily achieved by investing through foreign stock exchanges.

**overshooting** A jump in the value of an asset followed by a slow adjustment to *equilibrium, caused by a change in expectations. It is usually applied to an analysis of the real exchange rates and has been used to explain the extreme volatility of exchange rates in the 1970s and 1980s (for example, between 1980 and 1984 the US dollar rose by around 70% against foreign currencies and between 1985 and 1987 fell by 40%). Overshooting provides an argument for more government intervention in the determination of exchange rates.

**overside delivery** The unloading of a cargo over the side of a ship into *lighters.

**oversold 1.** Having sold more of a product or service than one can produce or purchase. **2.** Denoting a market that has fallen too fast as a result of excessive selling. It may therefore be expected to have an upward reaction.

**oversubscription** A situation that arises when there are more applications for a *new issue of shares than there are shares available. In these circumstances, applications have to be scaled down according to a set of rules devised by the company issuing the shares or their advisors. Alternatively some companies prefer to allocate the shares by ballot (*see* allotment). Oversubscription usually occurs because of the difficulty in arriving at an issue price that will be low enough to attract sufficient investors to take up the whole issue and yet will give the company the maximum capital. Speculative purchases by *stags also make it difficult to price a new issue so that it is neither oversubscribed or undersubscribed. In the case of **undersubscription**, which is rare, the *underwriter has to take up that part of the issue that has not been bought by the public. Undersubscription can occur if some unexpected event occurs after the announcement of the issue price but before the issue date.

**over-the-counter market (OTC market)** A market in which shares are traded outside the jurisdiction of a recognized stock exchange, such as the London or New York Stock Exchanges. Such markets tend to deal in smaller quantities of stocks (and bonds), they tend to be less liquid, and they provide less investor protection. The world's largest OTC market is NASDAQ in the USA. In the UK, OTC markets have generally been considered less respectable and may disappear altogether following the introduction of the *third market and the changes in regulations arising from the Financial Services Act (1986).

**overtime** Hours worked in excess of an agreed number per week or per day. The payment made for overtime work is usually higher than the basic rate of pay. Unions therefore have an interest in making the basic working week as short as possible, so that those working longer hours earn as much as possible. Employers, on the other hand, want to make the working week as long as possible.

**overtime ban** A form of *industrial action in which employees refuse to work *overtime, thus causing considerable dislocation to normal working but clearly less than would occur in a total *strike. Overtime bans are usually supported and often organized by unions. *See also* work-to-rule.

**overtrading** Trading by an organization beyond the resources provided by its existing capital. Overtrading tends to lead to *liquidity problems as too much stock is bought on credit and too much credit is extended to customers, so that ultimately there is not sufficient cash available to pay the debts as they arise. The solution is either to cut back on trading or to raise further permanent capital.

**own brand (own label; private brand; house brand)** A product sold under a *distributor's own name or trademark through its own outlets. These items are either made specially for the distributor or are versions of the manufacturer's equivalent *brand. Own-brand goods are promoted by the distributor rather than the manufacturer and are typically 10–20% cheaper to the distributor than an equivalent brand. Own brands are sold principally by *multiple shops.

**ownership** Rights over property, including rights of possession, exclusive enjoyment, destruction, etc. In UK common law, land cannot be owned outright, as all land belongs to the Crown and is held in tenure by the 'owner'. However, an owner of an estate in land in fee simple is to all intents and purposes an outright owner. In general, ownership can be split between different persons. For example, a trustee has the legal ownership of trust property but the beneficiary has the equitable or beneficial ownership. If goods are stolen, the owner still has ownership but not possession. Similarly, if goods are hired or pledged to someone, the owner has ownership but no immediate right to possession.

# P

**PA 1.** Abbreviation for personal account, used to denote a transaction made by a professional investment advisor for his own account rather than for the firm for which he works. **2.** Abbreviation for *personal assistant. **3.** Abbreviation for *power of attorney. **4.** Abbreviation for particular average. *See* average.

**PABX** Abbreviation for private automatic branch exchange. *See* private branch exchange.

**package** A set of computer programs designed to be sold to a number of users. They are used in computerized accounting systems for purchase and sales ledgers, stock control, payroll records, etc. Buying a software package from firms specializing in them saves users a considerable amount of time and money in developing their own.

**package deal** An agreement that encompasses several different parts, all of which must be accepted. A package deal may have involved either or both parties in making concessions on specific aspects of the package in order to arrive at a compromise arrangement.

**packaging 1.** The design of wrappers or containers for a product. **2.** The wrappers or containers themselves. **3.** An activity that combines several operations and enables the packager to deliver a finished or near-finished product to a selling organization. For example, many non-fiction books are produced by **book packagers**, who write, illustrate, typeset, and print books, which they deliver to publishers, who sell them to the public under their own imprint.

**page proof** A set of proofs of printed matter that has been broken down into separate pages. *Compare* galley.

**paid-up capital (fully paid capital)** The total amount of money that the shareholders of a company have paid to the company for their shares. *See* share capital.

**paid-up policy** An *endowment assurance policy in which the assured has decided to stop paying premiums before the end of the policy term. This results in a *surrender value, which instead of being returned in cash to the assured is used to purchase a single-premium *whole-of-life assurance. In this way the life assurance protection continues (for a reduced amount), while the policyholder is relieved of the need to pay further premiums. If the original policy was a *with-profits policy, the bonuses paid up to the time the premiums ceased would be included in the surrender value. If it is a unit-linked policy, capital units actually allocated would be allowed to appreciate to the end of the term.

**paid-up share** A share the par value of which has been paid in full. *See* share capital.

**paired comparisons** A technique used in marketing research in which consumers are presented with pairs of competing products and asked to choose the one they prefer. This technique can be used to compare, say, a number of brands of soap, giving respondents two at a time to compare. The number of times each brand is selected as a preference in a large number of such tests will reveal an order of brand preference. *Compare* monadic testing.

**pallet** A wooden frame on which certain goods are stacked in warehouses and during transport. Pallets are designed to be lifted by fork-lift trucks or pallet trucks.

**Panel on Takeovers and Mergers** *See* City Code on Takeovers and Mergers.

**paper money 1.** Legal tender in the form of banknotes. **2.** Banknotes and any form of paper that can be used as money, such as cheques, *bills of exchange, promissory notes, etc., even though they are not legal tender.

**paper profit** A profit shown by the books or accounts of an organization, which may not be a realized profit because the value of an asset has fallen below its book value, because the asset, although nominally showing a profit, has not actually been sold, or because some technicality of book-keeping might show an activity to be profitable when it is not. For example, a share that has risen in value since its purchase might show a paper profit but this would not be a real profit since the value of the share might fall again before it is sold.

**par** *See* par value.

**paradox of thrift** A paradoxical fall in the actual amount of money saved in an economy as a result an increase in the desire to save (thriftiness). The paradox, which has been associated with many economists, including the mercantilists, Marx, and J M Keynes, arises because thriftiness leads to a fall in consumption, causing goods to remain unsold; this leads firms to invest less and employ fewer people, causing incomes to fall; ultimately, although the percentage of incomes saved may have risen, incomes themselves will have fallen by so much that the actual amount saved is reduced. Neoclassical economists argue that movements in interest rates can usually pre-empt this chain of events.

**parallel processing** A method of computing in which two or more parts of a program are executed simultaneoulsy rather than sequentially. Strictly, parallel processing is only possible on computers with more than one arithmetic and logical unit in the *central processing unit, but it is often simulated on other machines by such techniques as *time sharing.

**parent company** *See* holding company.

**Pareto optimality** An efficiency condition devised by the economist and political scientist Vilfredo Pareto (1848–1923). An allocation or distribution of goods and services in an economy is said to be Pareto optimal if no alternative allocation could make at least one individual better off, without making anyone worse off. This condition is in fact very weak, since absurdly uneven distributions of income could still be Pareto efficient. Furthermore, there are a potentially infinite number of Pareto-optimal allocations in any economy, depending on the initial distribution; moreover any two Pareto-optimal allocations are Pareto non-comparable. Nonetheless, the concept has been a very useful tool for analysing the efficiency of economic systems and for formulating economic policies. *See also* general equilibrium analysis.

**pari passu** (Latin: with equal step) Ranking equally. When a new issue of shares is said to rank pari passu with existing shares, the new shares carry the same dividend rights and winding-up rights as the existing shares.

**Paris Club** *See* Group of Ten.

**parity grid** *See* European Monetary System.

**parking** Putting company shares that one owns in the name of someone else or of nominees in order to hide their real ownership. This is often illegal. *See also* warehousing.

**Parkinson's laws** The propositions made by the British writer Cyril Northcote Parkinson (1909– ) in his book *Parkinson's Law*. Facetious but true, they include such aphorisms as: work expands to fill the time available in which to do it; expenditure rises to meet income; and subordinates multiply at a fixed rate that is independent of the amount of work produced. They were devised with large organizations in mind and it is to such organizations that they apply, unless managers take steps to ensure that they do not.

**Parliamentary Commissioner for Administration** The Ombudsman responsible for investigating complaints referred to him by an MP from members of the public against maladministration by government departments and certain public bodies. The **Health Service Commissioners** are responsible for investigating complaints against the National Health Service. *See also* Commissioner for Local Administration; Financial Ombudsman.

**par of exchange** The theoretical *rate of exchange between two currencies in which there is equilibrium between the supply and demand for each currency. The par value lies between the market buying and selling rates. *See also* mint par of exchange.

**partial equilibrium analysis** The analysis of the behaviour of a particular individual, firm, household, or industry in isolation from the rest of the economy. In partial equilibrium analysis it is unrealistically assumed that the choices of the agents studied have no effects on the rest of the economy. This assumption is made to simplify the analysis and often produces some of the most interesting results in economics. However, these results must be treated with caution. *Compare* general equilibrium analysis.

**partial loss** *See* average.

**participating preference share** *See* preference share.

**participation rate** The percentage of a population that is available for employment. For example, in the UK the participation rate in 1984, between the ages of 16 and retirement, was 90% for men and 60% for women. The latter tends to vary greatly from year to year, whereas the former tends to remain constant. Supply-side economists sometimes claim that cuts in taxation would raise participation rates and therefore raise national income (*see* Laffer curve; supply-side economics). The evidence, however, it not generally accepted.

**particular average (PA)** *See* average.

**partly paid shares** Shares on which the full nominal (par) value has not been paid. Formerly, partly paid shares were issued by some banks and insurance companies to inspire confidence, i.e. because they could always call on their shareholders for further funds if necessary. Shareholders, however, did not like the liability of being called upon to pay out further sums for their shares and the practice largely died out. It has been revived for large new share issues, especially in *privatizations, in which shareholders pay an initial sum for their shares and subsequently pay one or more *calls on specified dates (*see also* share capital).

**partner** A member of a *partnership.

**partnership** An association of two or more people formed for the purpose of carrying on a business. Partnerships are governed by the Partnership Act (1890). Unlike an incorporated *company, a partnership does not have a legal personality of its own and therefore partners are liable for the debts of the firm. **General partners** are fully liable for these debts, **limited partners** only to the extent of their investment. A **limited partnership** is one consisting of both general and limited partners and is governed by the Limited Partnership Act (1907). A **partnership-at-will** is one for which no fixed term has been agreed. Any partner may end the partnership at any time provided that he gives notice of his intention to do so to all the other partners. A **nominal partner** is one who allows his name to be used for the benefit of the partnership, usually for a reward but not for a share of the profits. He is not a legal partner.

Partnerships are usually governed by a **partnership agreement** that lays down the way in which profits are to be shared, the procedure to be adopted on the death, retirement, or bankruptcy of a partner, and the rules for withdrawing capital from the partnership. Partners do not draw salaries and are not paid interest on their capital.

**partnership-at-will** *See* partnership.

**par value (face value; nominal value)** The *nominal price of a share or other security. If the market value of a security exceeds the nominal price it is said to be **above par**; if it falls below the nominal price it is **below par**. Gilt-edged securities are always repayed **at par** (usually £100), i.e. at the par value.

**Pascal** A popular *high-level language used to program computers. It is based on an earlier language, *Algol. Originally used mainly for teaching programming, it is becoming increasingly popular for applications programming, especially on microcomputers. The language was named after the French mathematician Blaise Pascal, who built the first desk calculator in 1642.

**passing a name** The disclosure by a broker of the name of the principal for whom he is acting. In some commodity trades, if a broker discloses the name of his buyer to the seller, he does not guarantee the buyer's solvency, although he may do so in some circumstances. However, if the broker does not pass his principal's name, it is usual for him to guarantee his solvency. Thus, to remain anonymous, a buyer may have to pay an additional brokerage.

**passing off** Conducting a business in a manner that misleads the public into thinking that one's goods or services are those of another business. The commonest form of passing off is marketing goods with a design, packaging, or trade name that is very similar to that of someone else's goods. It is not necessary to prove an intention to deceive; innocent passing off is actionable.

**paste-up** A layout of a page in a book, article, etc., in which *galley proofs are pasted onto pages, with illustrations in position or with spaces marked for illustrations.

**patent** The grant of an exclusive right to exploit an invention. In the UK patents are granted by the Crown through the *Patent Office, which is part of the Department of Trade and Industry. An applicant for a patent (usually the inventor or his employer) must show that the invention is new, is not obvious, and is capable of industrial application. An expert known as a **patent agent** often prepares the application, which must describe the invention in considerable detail. The Patent Office publishes these details if it grants a patent. A patent remains valid for 20 years from the date of application (the **priority date**) provided that the person to whom it has been granted (the **patentee**) continues to pay the appropriate fees. During this time, the patentee may assign his patent or grant licences to use it. Such transactions are registered in a public register at the Patent Office. If anyone infringes his monopoly, the patentee may sue for an *injunction and *damages or an *account of profits. However, a patent from the Patent Office gives exclusive rights in the UK only: the inventor must obtain a patent from the European Patent Office in Munich and patents in other foreign countries if he wishes to protect the invention elsewhere.

**patent agent** A member of the Chartered Institute of Patent Agents, who gives advice on obtaining *patents and prepares patent applications.

**Patent Office** A UK government office that administers the Patent Acts, the Registered Designs Act, and the Trade Marks Act. It also deals with questions relating to the Copyright Acts, and provides an information service about *patent specifications.

**pawnbroker** A person who lends money against the security of valuable goods used as collateral. Borrowers can reclaim their goods by repaying the loan and interest within a stated period. However, if the borrower defaults the pawnbroker is free to sell the goods. The operation of pawnbrokers is governed by the Consumer Credit Act (1974).

**payable to bearer** Describing a *bill of exchange in which neither the payee or endorsee are named. A holder, by adding his name, can make the bill *payable to order.

**payable to order** Describing a *bill of exchange in which the payee is named and on which there are no restrictions or endorsements; it can therefore be paid to the endorsee.

**pay-as-you-earn** *See* PAYE.

**payback period** A method of appraising capital projects, in which the principal criterion for acceptance is the length of time the project will take to recover the initial outlay it requires. The assumption is that any further recoveries are then pure profit. This is a relatively unsophisticated method of appraising a capital project (*compare* discounted cash flow; net present value), although it is frequently used in industry. *See also* internal rate of return.

**PAYE** Pay-as-you-earn. This means of collecting tax arose under Schedule E of the UK income-tax legislation on wages and salaries. Because it is often difficult to collect tax at the end of the year from wage and salary earners, the onus is placed on employers to collect the tax from their employees as payments are made to them. There is an elaborate system of administration to ensure that broadly the correct amount of tax is deducted week by week or month by month and that the employer remits the tax collected to the Inland Revenue very quickly. Although technically called pay-as-you-earn, the system would be better called pay-as-you-get-paid. Schedule E is on an earnings basis, while PAYE is on a payment basis.

**payee** A person or organization to be paid. In the case of a cheque payment, the payee is the person or organization to whom the cheque is made payable.

**paying banker** The bank on which a *bill of exchange (including a cheque) has been drawn and which is responsible for paying it if it is correctly drawn and correctly endorsed (if necessary).

**paying-in book** A book of slips used to pay cash, cheques, etc., into a bank account. The counterfoil of the slip is stamped by the bank if the money is paid in over the counter.

**Paymaster General** A UK political appointment. A junior minister, not usually in the cabinet, the Paymaster himself has few duties, which enables the prime minister to appoint a Paymaster General and give him responsibility for some other special task. The Paymaster General's Office, in the Treasury, was formed in 1835 to act as paying agent for certain government departments. It also pays public-service pensions.

**payment by results** A system of payment in which an employee's pay is directly linked to his performance. The majority are *premium bonus schemes. *See also* piece rate; lump system.

**payment for honour** *See* acceptance supra protest.

**payment in advance (prepayment)** Payment for goods or services before they have been received. In company accounts, this often refers to rates or rents paid for periods that carry over into the next accounting period.

**payment in due course** The payment of a *bill of exchange when it matures (becomes due).

**payment in kind** A payment that is not made in cash but in goods or services. It is often in the form of a discount or an allowance, e.g. cheap fares for British Airways employees, staff prices for food at supermarkets, etc. Now regarded as a bonus or *fringe benefit, payment in kind in the 19th century was widespread; the Truck Acts (1831) were passed to discourage employers who insisted on paying their employees in goods instead of cash (as often the cash value of goods bought in bulk by the employer was much less than their value to the employee).

**payment on account** **1.** A payment made for goods or services before the goods or services are finally billed. *See also* deposit. **2.** A payment towards meeting a liability that is not the full amount of the liability. The sum paid will be credited to the account of the payer in the books of the payee as a part-payment of the ultimate liability.

**payment supra protest** *See* acceptance supra protest.

**payment terms** The agreed way in which a buyer pays the seller for goods. The commonest are cash with order or cash on delivery; prompt cash (i.e. within 14 days of delivery); cash in 30, 60, or 90 days from date of invoice; *letter of credit; *cash against documents; *documents against acceptance; or *acceptance credit.

**payphone** A telephone provided for the use of the public requiring cash or a *Phonecard to be used to pay for the call. Some payphones will accept standard credit cards. Mobile radiopayphones are available on some mainline rail routes, ferry routes, and long-distance coach routes.

**payroll tax** A tax based on the total of an organization's payroll. The main function of such a tax would be to discourage high wages or overemployment. The current employers' Class 1 *National Insurance contributions constitute such a tax.

**PBX** Abbreviation for *private branch exchange.

**PC** Abbreviation for personal computer. *See* microcomputer.

**PDR (P/D ratio)** Abbreviations for *price–dividend ratio.

**peaceful picketing** *See* picketing.

**peak time** The period of television airtime, usually the middle part of the evening, for which the highest advertising rate is charged as during this period the highest number of people are viewing.

**pegging** **1.** **(pegging the exchange)** The fixing of the value of a country's currency on foreign exchange markets. *See* crawling peg; fixed exchange rate. **2.** **(pegging wages)** The fixing of wages at existing levels by government order to prevent them rising during a period of *inflation. The same restraint may be applied to prices **(pegging prices)** in order to control inflation.

**penalty** An arbitrary pre-arranged sum that becomes payable if one party breaches a contract or undertaking. It is usually expressly stated in a **penalty clause** of the contract. Unlike liquidated *damages, a penalty will be disregarded by the courts and treated as being void. Liquidated damages will generally be treated as a penalty if the amount payable is extravagant and unconscionable compared with the maximum loss that could result from the breach. However, use of the words 'penalty' or 'liquidated damages' is inconclusive as the legal position depends on the interpretation by the courts of the clause in which they appear.

**penetration** The extent to which a product or an advertisement has been accepted by, or has registered with, the total number of possible users. It is usually expressed as a percentage.

**penetration strategy** A marketing strategy based on low prices and extensive advertising to increase a product's market share. For penetration strategy to be effective the market will have to be large enough for the seller to be able to sustain low profit margins.

**penny shares** Securities with a very low market price (although they may not be as low as one penny) traded on a stock exchange. They are popular with small investors, who can acquire a significant holding in a company for a very low cost. Moreover, a rise of a few pence in a low-priced share can represent a high percentage profit. However, they are usually shares in companies that have fallen on hard times and may, indeed, be close to bankruptcy. The investor in this type of share is hoping for a rapid recovery or a takeover.

**pension** A specified sum paid regularly to a person who has reached a certain age or retired from employment. It is normally paid from the date of reaching the specified age or the retirement date until death. A widow may also receive a pension from the date of her husband's death. In the UK, contributory **retirement pensions** are usually paid by the state from the age of 65 for men or 60 for women, irrespective of whether or not the pensioners have retired from full-time employment. A non-working wife or widow also receives a state pension based on her husband's contributions. Self-employed people are also required to contribute towards their pensions. The state pays non-contributory **old-age pensions** to people over 80 if they are not already receiving a retirement pension.

Since 1978 state pensions have been augmented by the *State Earnings-Related Pension Scheme (SERPS) to relate pensions to inflation and to ensure that men and women are treated equally. Employers can contract out of the earnings-related part of the state pension scheme relating to retirement and widows, provided that they replace it with an *occupational pension scheme that complies with the Social Security Act (1986). Employees whose employers have not provided such an occupational pension scheme may also contract out of SERPS by starting their own approved *personal pension scheme.

The private sector of the insurance industry also provides a wide variety of pensions, *annuities, and *endowment assurances. *See also* executive pension plan; single-life pension.

**pensionable earnings** The part of an employee's salary that is used to calculate the final pension entitlement. Unless otherwise stated, overtime, commission, and bonuses are normally excluded.

**pensioneer trustee** A person authorized by the Superannuation Funds Office of the Inland Revenue to oversee the management of a *pension fund in accordance with the provisions of the Pension Trust Deed.

**pension funds** State and private pension contributions invested to give as high a return as possible to provide the funds from which pensions are paid. In the UK, pension funds managed by individual organizations work closely with insurance companies and investment trusts, being together the *institutional investors that have a dominant influence on many securities traded on the London Stock Exchange. The enormous amount of money accumulated by these pension funds, which grows by weekly and monthly contributions, needs prudent management; real estate and works of art are often purchased for investment by pension funds in addition to stock-exchange securities. *See also* pensioneer trustee; Superannuation Funds Office.

**PEP** Abbreviation for *personal equity plan.

**peppercorn rent** A nominal rent. In theory one peppercorn (or some other nominal sum) is payable as a rent to indicate that a property is leasehold and not *freehold, the peppercorn representing the *consideration. In practice it amounts to a rent-free lease.

**P/E ratio** Abbreviation for *price–earnings ratio.

**per capita income** The average income of a group, obtained by dividing the group's total income by its number of members. The **national per capita income** is the ratio of the national income to the population.

**perceptual mapping** The use of mathematical psychology to understand the structure of a market. Consumer images of different brands of a product are plotted on a graph or map; the closer two brands are on the map, the closer they are as competitors. Perceptual mapping is also used in the development of a new product; the closer a new product appears on a map to an ideal product, the more likely the new product is to succeed.

**per contra** Denoting a book-keeping entry of a particular amount on the opposite side of an *account. It is often used when two entries, a debit and a credit, are on opposite sides of the same page, as, for example, a transfer from cash to bank or vice versa, both recorded in the same cash book.

**per diem** (Latin: per day) Denoting a fee charged by a professional person who is paid a specified fee for each day that he is employed.

**perfect competition (perfect market)** A market in which there are sufficiently large numbers of buyers and sellers for no individual to be able to influence the market price. Perfect competition assumes that sellers (if they are firms) pursue *profit maximization, all agents have perfect knowledge of the market, all the factors of production are perfectly mobile, there are no government regulations interfering with the market, the products of each firm are identical, and there is free entry into or exit from the market. On these assumptions perfect competition will ensure that in the long run the market price is equal to the average and *marginal cost of production, that output is equal to the economically efficient level, and that economic welfare will be maximized. However, if any of the assumptions are unjustified the market cannot be guaranteed to be efficient. Although few markets in the real world satisfy these conditions, many come quite close to doing so, e.g. commodity markets and the stock market. Economists usually make use of the idea of a perfectly competitive market as a benchmark in discussing economic policy, rather than believing that such markets actually exist.

**perfecting the sight** *See* bill of sight.

**perfect market** A market in which there is *perfect competition.

**peril (risk)** An event that can cause a financial loss, against which an *insurance contract provides cover. *See also* risk. An **excepted peril** is one that is not normally covered by an insurance policy. Excepted perils in the carriage of goods include *acts of God, *inherent vice, *negligence, and loss resulting from any action of the **Queen's enemies**.

**period bill (term bill)** A *bill of exchange payable on a specific date rather than on demand.

**period of grace** The time, usually three days, allowed for payment of a *bill of exchange (except those payable at sight or on demand) after it matures.

**peripheral device** An input, output, or storage device connected to a computer, for example a printer, disk drive, VDU screen, or joystick.

**permanent establishment** A fixed place of business that a trader of one country has in another country, thus rendering him liable to that second country's taxation. It will include a place of management, branch, office, factory or workshop, and long-term construction projects but it may exclude some sales offices and storage depots. The concept is important in double-taxation treaties.

**permanent health insurance (PHI)** A form of health insurance that provides an income (maximum 75% of salary) up to normal retirement age (or pension age) to replace an income lost by prolonged illness or disability in which the insured is unable to perform any part of his or her normal duties. Premiums are related to age and occupation and normally are fixed; benefits, which are not paid for the first 4–13 weeks of disability, are tax free for one year and thereafter are taxed. *Compare* sickness and accident insurance.

**permanent income hypothesis** The hypothesis that individuals base their *consumption decisions on the value of their expected average income, ignoring any windfall gains or losses in any particular year. This hypothesis was proposed by Milton Friedman as an alternative to the Keynesian view, in which a temporary rise in income engineered by a government tends to raise consumption and output by means of the *multiplier process. According to the permanent income hypothesis, consumers will ignore any temporary rise in income created by government action, consumption will be unchanged, and therefore the multiplier process will not operate. *Compare* life-cycle hypothesis.

**permission to deal** Permission by the *London Stock Exchange to deal in the shares of a newly floated company. It must be sought three days after the issue of a *prospectus.

**perpetual debenture** **1.** A bond or *debenture that can never be redeemed. *See* irredeemable securities. **2.** A bond or debenture that cannot be redeemed on demand.

**perpetual inventory** A method of continuous stock control in which an account is kept for each item of stock; one side of the account records the deliveries of that type of stock and the other side records the issues from the stock. Thus, the balance of the account at any time provides a record of either the number of items in stock or their values, or both. This method is used in large organizations in which it is important to control the amount of capital tied up in the running of the business. It also provides a means of checking pilferage. Less sophisticated organizations rely on annual *stocktaking to discover how much stock they have.

**perpetual succession** The continued existence of a corporation until it is legally dissolved. A corporation, being a separate legal person, is unaffected by the death or other departure of any member but continues in existence no matter how many changes in membership occur.

**Perpetuities and Accumulations Act (1964)** An act reforming the **rule against perpetuities**. This rule exists to prevent a donor of a gift from directing its destination too far into the future and thus creating uncertainty as to the ultimate ownership of the property. Under the old law, the ownership had to become certain within the period of an existing lifetime (i.e. the lifetime of a person living at the date of the gift) plus 21 years. If there was any possibility, however remote, that this would not happen, the gift was void. For example, a gift to "the first of A's daughters to marry" fails under the old law, unless A is dead or already has a married daughter, because all of A's living daughters may die unmarried and then A might have another daughter (who was not an existing lifetime or life in being at the date of the gift), who might marry more than 21 years after A's death. This was so even if A was a woman past the age of childbearing at the date of the gift. The Perpetuities and Accumulations Act changed the old law by allowing a "wait and see" period, i.e. by allowing a gift to remain valid until it becomes clear that it will in fact offend against the rule against perpetuities. The Act also allows the donor to choose a perpetuity period of 80 years instead of the uncertain "life in being plus 21 years". The Act also allows the age at which a person is to benefit from a gift to be reduced, if this will save the gift from infringing the rule against perpetuities.

**per proc (per pro; p.p.)** Abbreviations for *per procurationem* (Latin: by procuration): denoting an act by an agent, not acting on his own authority but on that of his principal. The abbreviation is often used when signing letters on behalf of a firm or someone else, if formally authorized to do so. The firm or person giving the authority accepts responsibility for documents so signed.

**perquisite (perk)** A benefit given to an employee in addition to his normal pay. A subsidized canteen, generous travel allowances, and free products are typical perks.

**personal accident and sickness insurance** *See* sickness and accident insurance.

**personal account 1.** An *account in a *ledger that bears the name of an individual or of an organization; it records the state of indebtedness of the named person to the organization keeping the account or vice versa. Personal accounts are normally kept in the sales (or total debtors) ledger and the purchases (or total creditors) ledger. **2.** *See* PA.

**personal allowances** Sums deductible from taxable income under an income-tax system to allow for personal circumstances. The principal allowances in the UK system are a personal allowance to which all taxpayers are entitled, a married couple's allowance, a single parent's allowance, and *age relief. *See also* additional personal allowance; separate taxation of a wife's earnings.

**personal assistant (PA)** A person who is appointed to help a manager or director and who has wider responsibilities than a secretary.

**personal computer (PC)** *See* microcomputer.

**personal equity plan (PEP)** A UK government scheme to encourage individuals to invest directly in UK quoted companies, offering investors certain tax benefits. The investment is administered by an authorized plan manager. Plans are either discretionary (in which the plan manager makes the investment decisions) or non-discretionary (in which the investor makes the decisions). Investors may put in a lump sum or regular monthly amounts. Re-invested dividends are free of *income tax and *capital gains tax is not incurred, as long as the investment is retained in the plan for at least a complete calendar year. There is a limit on the amount an individual can invest in a plan in any year.

**personal identification number (PIN)** A number memorized by the holder of a cash card or a credit card and used in *electronic funds transfer at point of sale to identify the card owner. The number is given to the cardholder in secret and is memorized so that if the card is stolen it cannot be used.

**personal loan** A loan to a private person by a bank or building society for domestic purposes, buying a car, etc. There is usually no security required and consequently a high rate of interest is charged. Repayment is usually by monthly instalments over a fixed period. This is a more expensive way of borrowing from a bank than by means of an *overdraft.

**personal pension scheme (*or* plan)** A pension scheme entered into by an employee, who wishes to contract out of the *State Earnings-Related Pension Scheme (SERPS). An employee may start his own pension scheme, whether or not his employer has an *occupational pension scheme. Personal pension schemes must provide a half-rate widowers benefit. They must be approved by the Occupational Pension Board before a person can contract out of SERPS. An employee with a personal pension in place of SERPS or an occupational pension scheme pays National Insurance contributions at the full ordinary rate. The Department of Social Security pays the difference between the lower contracted-out rate and the full ordinary rate for the personal pension scheme.

**personal property (personalty)** Any property other than *real property (realty). This distinction is especially used in distinguishing property for *inheritance tax. Personal property includes money, shares, chattels, etc.

**personal representative** A person whose duty is to gather in the assets of the estate of a deceased person, to pay his liabilities, and to distribute the residue. Personal representatives of a person dying testate are known as *executors; those of a person dying intestate are known as his *administrators.

**personalty** *See* personal property.

**personnel management** The management of people in relation to their work and within an organization as a whole, both for the benefit of the organization and the employees. During the inflationary period of 1960–1980, high employment and the trend towards computerization gave rise to new needs for employers; this, coupled with an increase in legislation relating to employment, made the **personnel officer's** role of particular importance. Although all managers are to a greater or lesser extent concerned with personnel management, in this period many large organizations appointed specialist personnel managers to centralize manpower planning policies, recruitment and selection, education and training of staff, and the terms and conditions of employment, with special reference to industrial relations, health and safety at work, and the legal aspects of employers and employees. With the rise in unemployment in the 1980s, the emphasis shifted from personnel management to *personnel selection. Changing technology and the need for higher efficiency and greater flexibility to meet the needs of increased international competition meant more and more time spent on manpower planning, training, and a general restructuring of organizations to encourage growth without an expansion in personnel. The current trend is to place greater emphasis on personnel morale and ways of achieving consistent job satisfaction through variation, especially by training employees to do more than one job. *See also* Institute of Personnel Management.

**personnel selection** The method of choosing the most suitable applicant for a vacancy within an organization. Ideally, a **job description** is written, detailing the essential qualifications and experience required for the job together with any other relevant information. The job description enables a **job specification** to be created, showing how the vacancy would fit in the structure of the organization, the salary level, and any *fringe benefits. A search for the right person then begins, using advertisements in newspapers, trade papers, or employment agencies; if the vacancy is sufficiently senior a head hunter may be employed (*see* employment agency). Initial selection is then made by inviting letters of application (or using application forms) supported by a *curriculum vitae and listing the applicants that most closely match the job specification. Screening interviews may then be held, which may include practical tests so that a short list of final candidates can be drawn up. Second interviews are sometimes required and, when appropriate, *psychological tests are used. The number of interviews and the extent of testing depend on the level of the position within the organization. Interviews may be conducted by a single person or by a panel; applicants may be seen singly or in groups. Moreover the extent to which the personnel department is involved in the selection will depend on the particular organization.

**PEST** A guide to examining the environment of a business. The examination should cover the four areas: *p*olitical, *e*nvironmental, *s*ocial, and *t*echnological.

**PET** Abbreviation for *potentially exempt transfer.

**petrodollars** Reserves of US dollars deposited with banks as a result of the steep rises in the price of oil in the 1970s. The export revenues of the oil-exporting nations increased rapidly in this period, leading to large current-account surpluses, which had an important impact on the world's financial system.

**petroleum revenue tax (PRT)** A tax on the profits from oil exploration and mining occurring under the authority of licences granted in accordance with the Petroleum (Production) Act (1934) or the Petroleum (Production) Act (Northern Ireland) (1964). This tax was the principal means enabling the UK government to obtain a share in the profits made from oil in the North Sea.

**petty cash** The amount of cash that an organization keeps in notes or coins on its premises to pay small items of expense. This is to be distinguished from cash, which normally refers to amounts held at banks. Petty-cash transactions are normally recorded in a petty-cash book, the balance of which should agree with the amounts of petty cash held at any given time.

**PHI** Abbreviation for *permanent health insurance.

**Phillips curve** A curve on a graph plotting the rate of *inflation against unemployment; it is based on empirical evidence produced by the economist A W H Phillips (1914–75) for the UK between 1880 and 1950. It showed that when the rate of inflation was high, unemployment tended to be low, and vice versa. A theoretical interpretation of this result was then spuriously added to Keynesianism, suggesting that governments could achieve fine tuning of the economy by changes in fiscal and monetary policy to obtain acceptable rates of inflation and unemployment. The *stagflation of the 1970s showed that this was not, in fact, possible; this led to the view that, in the long run, there is a *natural rate of unemployment from which the economy cannot deviate, irrespective of government policy. Monetarists, such as Milton Friedman, believed that in the short run government policy could regulate inflation and unemployment, although more recently the new classical macroeconomists have argued that even this is not possible.

**Phonecard** A plastic card issued by British Telecom to enable holders to make telephone calls from those public payphones designed to accept them. Phonecards, which entitle the user to a specific number of units, are on sale at post offices, tobacconists, newsagents, etc.

**phototypesetting (filmsetting)** A method of typesetting in which photographic film of the text, generated by means of a cathode-ray tube, is used to make the printing plates. This method of setting type has largely replaced *hot-metal typesetting.

**physical capital** Items such as plant and machinery, buildings, and land that can be used to produce goods and services. It is compared to financial capital, i.e. money, and *human capital.

**physical controls** Direct measures used by a government to regulate an economy, compared to indirect controls, which influence the price mechanism. For example, the imposition of a quota on a specific import would be a physical control, whereas a surcharge on that import would be an indirect control.

**physical distribution management** *See* logistics.

**physicals** *See* actuals.

**physiocrats** A group of French economists working in the late 18th century led by F Quesnay (1694–1774), whose Tableau Economique is generally thought to be the earliest formulation of *national income accounts and the *circular flow of income. Contrary to the mercantilists (*see* mercantilism), they believed that agriculture was the only source of *wealth, as trade and manufacture were seen to be merely the exchange of goods of equal *value. While most of their doctrines, such as the proposal that taxes should be imposed on land alone, are now generally ignored, they are credited with being the earliest advocates of *free trade and with laying down the foundations of the work of Adam Smith (1723–90).

**pica** A unit of length, used in typography and printing, equal to 12 points or 0.167 inches (4.24 mm).

**picketing** A form of *industrial action in which employees gather outside a workplace in which there is a trade dispute, usually a strike. The pickets so gathered often form a **picket line**, past which they attempt to discourage other workers, delivery lorries, and customers' collection lorries from passing. The purpose is to reinforce the effects of the strike and to encourage the maximum number of employees to join it. The right to **peaceful picketing** at one's own place of work was established by the Trade Union Act (1975). It is not lawful if it has not first been authorized by a ballot of the union involved and if the reason for the action is because the employer involved is employing a non-union employee. **Secondary picketing**, picketing other people's place of work (i.e. picketing employers not otherwise involved in the dispute), is a civil offence under the Employment Acts (1980; 1982). **Flying pickets**, pickets who join a picket line although they are neither employees of the organization being picketed nor union representatives of employees, have no immunity from civil action.

**pictogram** A diagram providing quantitative information, using stylized drawings of people and things, in which the number of such individual drawings represents the actual number of people or things involved.

**piece rate** A payment scheme in which an employee is paid a specific price for each unit made. The rate is therefore directly related to output and not to time (*compare* time rate).

This method is often combined with a basic salary and takes the form of a productivity bonus, in which individual effort is rewarded. In modern factories, many tasks have been mechanized, making this method of payment less common, although it is simple to operate and popular with employees. Piece rate, to be attractive to workers, depends on speed of production, therefore the standard of quality and safety may suffer if adequate precautions are not taken. *See also* payment by results; premium bonus.

**pie chart** A diagram providing a visual representation of the proportions into which something is divided. It consists of a circle divided in sectors, the area of each sector representing the proportion of a specified component.

**Pigou effect** *See* wealth effect.

**pilot production** The small-scale production of a new product in a **pilot plant**. The object is to check and, if necessary, improve the production method. The product itself is often used in a marketing exercise to monitor its reception by the market and, in some cases, to improve it.

**pilot study** A small-scale *marketing research study conducted as a trial so that any problems can be eliminated before a full study is undertaken. For example, a pilot study might show that changes are needed in a questionnaire because questions are ambiguous, miss the point, etc.

**PIMS** Abbreviation for *profit impact of market strategy.

**PIN** Abbreviation for *personal identification number.

**pink form** An *application form in a flotation that is printed on pink paper and usually distributed to employees of the company to give them preference in the allocation of shares.

**piracy 1.** An illegal act of violence, detention, robbery, or revenge committed on a ship or aircraft. It excludes acts committed for political purposes and during a war. In marine insurance it extends to include any form of plundering at sea. **2.** Infringement of copyright. The usual remedy is for the copyright holder to obtain an injunction to end the infringement. This may not be possible in a foreign country and many British and American books are pirated in foreign countries by copying them photographically and printing them cheaply.

**pit** An area of a stock market or commodity exchange in which a particular stock or commodity is traded, especially one in which dealings in certain commodities take place by open outcry (*see* London International Financial Futures Exchange; ring trading). A member who is allowed to trade on the floor but wishes to conceal his identity may use a **pit broker** to carry out his transactions.

**placing** The sale of shares by a company to a selected group of individuals or institutions. Placings can be used either as a means of *flotation or to raise additional capital for a quoted company (*see also* pre-emption rights; rights issue). Placings are usually the cheapest way of raising capital on a *stock exchange and they also allow the directors of a company to influence the selection of shareholders. The success of a placing usually depends on the placing power of the company's stockbroker. *Compare* introduction; issue by tender; offer for sale; public issue.

**planned economy** *See* command economy.

**planned location of industry** Government intervention in the location of new industries in an area or country. As old industries in a region, e.g. mining, shipbuilding, gradually decrease, new industries are required to take their place to avoid high unemployment and skilled workers moving away. Most governments provide assistance to those wishing to set up new industries in these areas.

**planned obsolescence** *See* obsolescence.

**planned shopping centres** Groups of businesses, primarily retailers, sharing a single building or a related set of buildings; they are usually controlled by a management belonging to the developer of the centre. Planned shopping centres are common in the USA and some are opening in the UK. The Trocadero in London is one example.

**planning** *See* business plan.

**planning blight** Difficulty in selling or developing a site, building, etc., because it is affected by a government or local-authority development plan. Planning blight may be ended by a compulsory purchase by the government or local authority, but it may continue indefinitely if the government plans fail to mature or if the site itself is not required by the development but is rendered unsaleable or less valuable by its proximity to a development.

**planning permission** Permission that must be obtained from a local authority in the UK before building on or developing a site or before changing the use of an existing site or building, in accordance with the Town and Country Planning Act (1971).

**plant 1.** The area of an organization in which production or some other technological process takes place. **2.** The assets used by a business for the purpose of carrying on its activities (often referred to as plant and machinery). In tax terms plant was formerly defined (*Yarmouth* v *France*, 1887) as "whatever apparatus is used by a businessman for carrying on his business – not his stock-in-trade which he buys or makes for sale; but all goods and chattels, fixed or moveable, live or dead, which he keeps for permanent employment in his business". However, the UK courts have now endeavoured to distinguish between those assets that businesses use in carrying on their trade, which are accepted as plant and for which capital allowances can be obtained, and the setting in which the trade is carried on (for which capital allowances are normally denied).

**plant and machinery register** A record kept by an organization of the various items of plant and machinery it owns. The record will normally show dates and costs of purchase, the location of the assets, amounts provided for *depreciation, dates of disposal, selling or scrap values, etc.

**Plantation House** An office block in *Mincing Lane in the City of London that houses many firms involved in the commodity trade and formerly housed the London Commodity Exchange (*see* London FOX).

**plc** Abbreviation for *public limited company. This (or its Welsh equivalent) must appear in the name of a public limited company. *Compare* Ltd.

**pledge** An article given by a borrower (**pledgor**) to a lender (**pledgee**) as a security for a debt. It remains in the ownership of the pledgor although it is in the possession of the pledgee until the debt is repaid. *See also* pawnbroker.

**Plimsoll line** *See* load line.

**ploughed-back profits** *See* retained profits.

**PLR** Abbreviation for *public lending right.

**plug compatibility** *See* compatibility.

**plutocracy** A form of government in which power is in the hands of the rich, i.e. landowners, industrialists, and bankers.

**pluvial insurance (pluvius insurance)** An insurance policy covering loss of income or profits caused by rain or other weather conditions. The main demand comes from the organizers of outdoor summer events, who could suffer financially if rain caused the event to be curtailed or abandoned. For events in which any profits would be uncertain, e.g. a summer fete, the policy specifies that claims are to be based on the number of millimetres of rain that fall in an agreed period at an agreed weather station near to the site of the event. If the amount of lost revenue can be easily calculated from ticket sales at an event, such as a cricket test match or tennis match, a policy paying for lost profit can be arranged.

**point 1.** A unit of length used in typography and printing, equal to 1/72 inch (0.0353 mm). *Compare* pica. **2.** *See* tick.

**point of sale (POS)** The place at which a consumer makes a purchase, usually a retail shop. It may, however, also be a doorstep (in door-to-door selling), a market stall, or a mail-order house.

**poison pill** A tactic used by a company that fears an unwanted takeover by ensuring that a successful takeover bid will trigger some event that substantially reduces the value of the company. Examples of such tactics include the sale of some prized asset to a friendly company or bank or the issue of securities with a conversion option enabling the bidder's shares to be bought at a reduced price if the bid is successful. Poison pills are used all over the world but were developed in the USA. *See also* porcupine provisions; staggered directorships.

**policy** *See* insurance policy.

**policy mix** A combination of fiscal, monetary, and other policies employed by a government to achieve an economic objective.

**policy proof of interest (PPI)** An insurance policy (usually *marine insurance) in which the insurers agree that they will not insist on the usual requirement that the insured must prove an *insurable interest existed in the subject matter before a claim is paid. The possession of the policy is all that is required. These policies are a matter of trust between insurer and insured as they are not legally enforceable. They are issued for convenience in cases in which an insurable interest exists but is extremely difficult to prove or in which an insurable interest might come into existence later in the voyage.

**poll tax** A tax that is the same for each individual, i.e. a lump sum per head (from Middle Low German *polle*: head). The benefits of such taxes are that they do not distort choices, being an equal levy from everyone's resources whatever their circumstances; however, poll taxes are sometimes criticized as being regressive in that they do not take account of persons' ability to pay or the marginal utility of income. The *community charge, an alternative to *rates, is a poll tax introduced by the Conservative government in 1990 (1989 in Scotland).

**population projection** A forecast of the size of a population at some future date. For example, the population projection for the UK in 2001–2006 is 58,957,000, based on a 1985 projection.

**population pyramid** A diagram that illustrates the distribution of a population by age. The youngest and most numerous age group forms a rectangle at the base of the pyramid; the oldest and least numerous is a small rectangle at its apex.

**porcupine provisions (shark repellents)** Provisions made by a company to deter *takeover bids. They include *poison pills, *staggered directorships, etc.

**portfolio** The list of holdings in securities owned by an investor or institution. In building up an investment portfolio an institution will have its own investment analysts, while an individual may make use of the services of a *merchant bank that offers **portfolio management**. The choice of portfolio will depend on the mix of income and capital growth its owner expects, some investments providing good income prospects while others provide good prospects for capital growth.

**port mark** The markings on the packages of goods destined for export, giving the name of the overseas port to which they are to be shipped or sent by air freight.

**POS** Abbreviation for *point of sale.

**position** *See* long position; open position; short position.

**position audit** A systematic assessment of the current situation of an organization. This normally involves drawing up a report (either internally or through external consultants) of the strengths and weaknesses of the organization and the opportunities or threats that it faces (*see* SWOT). The position audit is a vital tool in designing the future strategies of a large organization.

**positive economics** An economic analysis that is free of value judgments, i.e. independent of any particular ethical position or normative judgments (*compare* normative economics). Milton Friedman (1912– ) first postulated that this is the proper field of economic analysis and it is on this postulate that the claim of economics to be a science rests. However, many have argued that a pure positive economics is not possible.

**possession** Actual physical control of goods or land. Possession has a wide variety of meanings in English law, depending on the nature of the property and the circumstances. For example, a person may still have possession of goods he has lost or mislaid, provided that he has not abandoned them and that no-one else has physical possession of them. It is sometimes used in the sense of a right to possession, e.g. the right of a person who has pledged goods to get them back. Possession may be divorced from *ownership; for example, a thief has legal possession of stolen goods but not ownership. Possession usually requires intention to possess – a person cannot possess something which he does not know that he has.

**post** To make a book-keeping entry in one account, possibly from another account but normally from a *book of prime entry.

**postcode** A code added to an address on a letter or parcel sent through the mail to facilitate sorting. The UK coding system of letters and numbers enables the mail to reach the delivery postman, whereas mail marked with numbers-only codes will only reach a particular town. *See also* zip code.

**post-date** To insert a date on a document that is later than the date on which it is signed, thus making it effective only from the later date. A **post-dated** (or **forward-dated**) **cheque** cannot be negotiated before the date written on it, irrespective of when it was signed. *Compare* ante-date.

**post-entry duty** *See* over-entry certificate.

**poster** A placard or bill exhibited in a public place as an advertisement. Although there are over 200 000 poster sites in the UK, poster advertising accounts for less than 10% of the total advertising expenditure. Illegal attachment of posters or bills to empty shop windows, company hoardings, etc., is called **flyposting**. Flyposting is often undertaken at night by a small number of companies, all of whom jealously guard their regular territories. Flyposting is particularly popular with record companies, underground magazines, and some film companies.

**poste restante** A service provided by the post offices of most countries in which letters, parcels, etc., are sent to a named post office in any of these countries for collection by the addressee (on providing some proof of identity).

**post-Keynesianism** The economic theories that attempt to re-establish the policy prescriptions of J M Keynes by the application of newer theories. Thus, former Keynesians often believe that the theories underpinning Keynes' arguments are no longer acceptable but continue to believe that government intervention to alleviate *involuntary unemployment is justified. Examples of the newer kinds of theories used by post-Keynesians are the *efficiency-wage theory and *implicit contract theory. Some post-Keynesians apply more radical approaches, including Marxism, to defend Keynesian policies. In general, post-Keynesianism is an area in which there has been much research but limited success.

**Post Office Savings Bank** *See* National Savings Bank.

**potential entrant** An organization that is poised to enter a market and would do so if there was a small price rise or a reduction in *barriers to entry. In monopoly situations, the existence of potential entrants holds potential profits in check.

**potentially exempt transfer (PET)** A gift that is not subject to inheritance tax when it is made, although it may become taxable if the donor dies within a specified period. *See* inter vivos gifts.

**pound** The standard UK currency unit. It dates back to the 8th century AD when Offa, King of Mercia, coined 240 pennyweights of silver from 1 pound of silver. When *sterling was decimalized in 1971 it was divided into 100 newly defined pence. The pound is also the currency unit of several former British territories or countries with strong trading links with the UK.

**poundage** *See* rates.

**pound cost averaging** A method of accumulating capital by investing a fixed sum of money in a particular share every month (or other period). When prices fall the fixed sum will buy correspondingly more shares and when prices rise fewer shares are bought. The result is that the average purchase price over a period is lower than the arithmetic average of the market prices at each purchase date (because more shares are bought at lower prices and fewer at higher prices).

**poverty trap** A situation in which an increase in the income of a low-earning household causes either a loss of state benefits or an increase in taxation that approximately equals the increase in earnings, i.e. the household faces a marginal tax rate of 100% (in some instances the marginal tax rate can exceed 100%). The poverty trap creates a disincentive to earning and is often demoralizing for those caught in it. Most *progressive tax systems create poverty traps and policies for removing them are difficult to find.

**power of attorney (PA)** A formal document giving one person the right to act for another. A power to execute a *deed must itself be given by a deed. An attorney may not delegate his powers unless specifically authorized to do so.

**p.p.** Abbreviation for *per procurationem*. *See* per proc.

**PPI** **1.** Abbreviation for *policy proof of interest. **2.** Abbreviation for *producer price index.

**PPP** Abbreviation for *purchasing power parity.

**PR** Abbreviation for *public relations.

**preacquisition profit** The retained profit of one company before it is taken over by another company. Preacquisition profits should not be distributed to the shareholders of the acquiring company by way of dividend, as such profits do not constitute income to the parent company but a partial repayment of its capital outlay on the acquisition of the shares.

**precautionary demand for money** The amount of money that individuals or firms will wish to hold to deal with unexpected events. An element of Maynard Keynes' theory of *liquidity preference, the precautionary motive is affected by the prevailing rate of *interest, since higher interest rates will raise the *opportunity cost of holding cash balances. The precautionary demand will also be affected by the level of confidence in the economy. *Compare* speculative demand for money; transactions demand for money.

**precious metals** The metals silver, gold, and platinum. *Compare* base metals.

**predatory pricing** The pricing of goods or services at such a low level that other firms cannot compete and are forced to leave the market. While it has long been accepted that some firms resort to predatory pricing on occasions, the application of *game theory to *strategic behaviour has shown that predatory pricing is unlikely to occur very often as it is at least as painful for the predator as for the victim. This encourages most potential predators to look for a more cooperative plan.

**pre-emption** First refusal: the right of a person to be the first to be asked if he wishes to enter into an agreement at a specified price; for example, the right to be offered a house at a price acceptable to the vendor before it is put on the open market.

**pre-emption rights** A principle, established in company law, according to which any new shares issued by a company must first be offered to the existing shareholders as the legitimate owners of the company. To satisfy this principle a company must write to every shareholder (*see* rights issue), involving an expensive and lengthy procedure. Newer methods of issuing shares, such as *vendor placings or *bought deals, are much cheaper and easier to effect, although they violate pre-emption rights. In the USA pre-emption rights have now been largely abandoned but controversy is still widespread in the UK.

**pre-empt spot** An advertising spot for a particular period of airtime on television, bought in advance at a discount for use only if another advertiser does not offer to take up that time at the full rate.

**preference share** A share in a company yielding a fixed rate of interest rather than a variable dividend. A preference share is an intermediate form of security between an *ordinary share and a *debenture. Preference shares, like ordinary shares but unlike debentures, usually confer some degree of ownership of the company. However, in the event of liquidation, they are less likely to be paid off than debentures, but more likely than ordinary shares. Preference shares may be redeemable (*see* redeemable preference share) at a fixed or variable date; alternatively they may be undated. Sometimes they are *convertible. The rights of preference shareholders vary from company to company and are set out in the *articles of association. Voting rights are normally restricted, often only being available if the interest payments are in arrears.

**Participating preference shares** carry additional rights to dividends, such as a further share in the profits of the company, after the ordinary shareholders have received a stated percentage. *See also* preferred ordinary share.

**preferential creditor** A creditor whose debt will be met in preference to those of other creditors and who thus has the best chance of being paid in full on the bankruptcy of an individual or the winding-up of a company. Preferential creditors, who are usually paid in full after *secured debts and before ordinary creditors, include: the Inland Revenue in respect of PAYE, Customs and Excise in respect of VAT and car tax, the DSS in respect of National Insurance Social Security contributions, the trustees of occupational pensions schemes, and employees in respect of any remuneration outstanding.

**preferential duty** A specially low import duty imposed on goods from a country that has a trade agreement of a certain kind with the importing country. For example, in the days of the British Empire, imports from member countries were granted **Imperial Preference** on imported materials, which later became **Commonwealth Preference**.

**preferential payment** A payment made to a *preferential creditor.

**preferred ordinary share** A share issued by some companies that ranks between a *preference share and an *ordinary share in the payment of dividends.

**preferred position** A desirable position for an advertisement in a publication, for which a premium is charged.

**preliminary expenses** Expenses involved in the formation of a company. They include the cost of producing a *prospectus, issuing shares, and advertising the flotation.

**prelims** The preliminary pages in a book or journal, giving the title, copyright details, table of contents, etc.

**premium 1.** The consideration payable for a contract of insurance or life assurance. *See also* renewal notice. **2.** An amount in excess of the nominal value of a share or other security. **3.** An amount in excess of the issue price of a share or other security. When dealings open for a new issue of shares it may be said that the market price will be at a premium over the issue price (*see* stag).

**premium bonds** UK government securities first issued in 1956 and now administered by the Department for *National Savings. No regular income or capital gain is offered but bonds enter weekly and monthly draws for tax-free prizes. The prize fund (calculated at 6.5% from 1988) is distributed in a range of prizes, winners being drawn by ERNIE (electronic random number indicating equipment). Bonds are in £1 denominations with a minimum purchase of £10 and a maximum holding of £10,000. Bonds are repaid at their face value at any time.

**premium bonus** A *payment by results method of paying workers, in which a standard time is set for the production of each unit and bonuses are paid related to the actual time taken. Depending on the needs of the organization, this scheme can be varied to achieve a balance between quality and speed of production; it also provides an incentive for employees. The standards were formerly set by experienced staff but more scientific attempts at classifying work units have resulted in the introduction of systematic *work-study techniques. These schemes can lead to friction between employee and employer as standard times are a permanent source of conflict. In the 1960s and 1970s *measured daywork became a popular alternative system. *See also* piece rate.

**premium offer** A special offer of a domestic product advertised on the package of some other product. The purchaser sends a set number of package tops or labels with a small cash payment to an address given and receives the domestic product by post. Premium offers are handled by **premium houses**, who provide all the services required to distribute the products.

**prepayment** *See* payment in advance.

**present value** *See* future value.

**president 1.** The chief executive of a US company, equivalent to the chairman of the board of a UK company. **2.** A title sometimes given to a past chairman or managing director of a UK company, usually for an honorary position that carries little responsibility for running the company.

**Prestel** A public *Viewdata service provided by British Telecom in the UK. Each user has a number and password and can use either a Prestel set or a microcomputer linked to the telephone network through a *modem. The service offers news, information on exports, investments, and companies, as well as tax guides, etc. It is a two-way system, enabling users to send messages to other users and to the information providers. Response pages enable bookings to be made and other services to be purchased. *See also* Teletext.

**pre-tax profit** The profit of a company before deduction of corporation tax.

**pretesting** The use of marketing research to predict the effectiveness of an advertisement before it is issued for wide exposure.

**price** That which must be given up in order to acquire a good or service, usually expressed in terms of money. The relationship between price and value has long been a source of dispute in economics. Classical economists attempted to use the *labour theory of value to find a stable relationship between price and value. Subsequently Marxist economists have attempted to solve what they call the transformation problem. To neoclassical economists price and value are equal at the margin, i.e. the price will reflect the value of the last unit purchased, while the value of all other units sold will exceed the price, giving rise to a consumer surplus.

**price control** Restrictions by a government on the prices of consumer goods, usually imposed on a short-term basis as a measure to control inflation. *See also* prices and income policy.

**price discrimination** The sale of the same product at different prices to different buyers. Usually practised by monopolists, it requires that a market can be subdivided to exploit different sets of consumers and that these divisions can be sustained. Pure price discrimination rarely exists, since sellers usually differentiate the product slightly (as in first-class rail travel and different types of theatre seat). Governments may use price discrimination in order to redistribute wealth but usually it is a monopolist's way of extracting a higher *pure economic profit.

**price–dividend ratio (PDR; P/D ratio)** The current market price of a company share divided by the dividend per share for the previous year. It is a measure of the investment value of the share.

**price–earnings ratio (P/E ratio)** The current market price of a company share divided by the *earnings per share (eps) of the company. The P/E ratio usually refers to the annual eps and is expressed as a number (e.g. 5 or 10), often called the **multiple** of the company. Loosely, it can be thought of as the number of years it would take the company to earn an amount equal to its market value. High multiples, usually associated with low *yields, indicate that the company is growing rapidly, while a low multiple is associated with dull no-growth stocks. The P/E ratio is one of the main indicators used by fundamental analysts to decide whether the shares in a company are expensive or cheap, relative to the market.

**price index** *See* Retail Price Index.

**price leadership** The setting of the price of a product by a dominant firm in an industry in the knowledge that competitors will follow this lead in order to avoid the high cost of a *price war. This practice is often found in *oligopolies and has the effect of a *cartel. As the price leader is as anxious as the other members of the industry to achieve a stable market, it is in his interest to set the price at a sensible level; if he fails to do so he will soon be replaced as leader by another member of the group. *Compare* price ring.

**price level** The average level of prices of goods and services in an economy, usually calculated as a figure on a price index. *See* Retail Price Index.

**price method** The US term for *piece rate.

**price–net tangible assets ratio** The current market price of a company share divided by its *net tangible assets. The higher the ratio, the more attractive the share as an investment.

**price ring** A group of firms in the same industry that have agreed amongst themselves to fix a minimum retail price for their competing products, thus forming a *cartel. Price rings are illegal in many countries unless they can be shown to be in the public interest. *See* consumer protection; price leadership.

**prices and income policy** A government policy to curb *inflation by directly imposing wage restraint (*see* wage freeze) and *price controls. Some less interventionist governments prefer indirect methods using fiscal and monetary policies.

**price support** A government policy of providing support for certain basic, usually agricultural, products to stop the price falling below an agreed level. Support prices can be administered in various ways: the government can purchase and stockpile surplus produce to support the price or it can pay producers a cash payment as a *subsidy to raise the price they obtain through normal market channels. *See also* Common Agricultural Policy.

**price-taking behaviour** The assumption made in the theory of *perfect competition that individual sellers or producers constitute too small a percentage of a market to influence the market price by changing the amount they are willing to sell. *Monopoly, *oligopoly, and monopolistic competition are all concerned with the consequences for markets when this assumption does not apply. In general, Pareto-optimal results (*see* Pareto optimality) can only be obtained when price-taking behaviour occurs (*see* competition).

**price theory** A theory that, at the *microeconomic level, defines the role of prices in consumer demand and the supply of goods by the firm. At a broader level, it explains the role of prices in markets as a whole. At the *macroeconomic level it treats wages and interest rates as the prices of particular goods, thus implying that governments must take market forces into account when formulating such policies as *incomes policy and *monetary policy.

**price variance** A *variance arising when the actual price of goods sold or purchased differs from the standard price.

**price war** Competition between two or more firms in the same industry that are seeking to increase their shares of the market by cutting the prices of their products. Although this can give short-term advantages in some circumstances to one participant, in the longer run it is a situation from which no one profits. To avoid the perils of selling products at a loss in order to outdo the competition, in most industries prices are not allowed by competitors to fall below a certain level, competition being restricted to other methods, such as advertising, improving the packaging, etc.

**prima facie** (Latin: *prima facies* first appearance) At first appearances. **Prima facie evidence** is evidence that appears to be conclusive on first appearances but is not necessarily conclusive.

**primage 1.** A percentage added to a freight charge to cover the cost of loading or unloading a ship. **2.** An extra charge for handling goods with special care when they are being loaded or unloaded from a ship, aircraft, etc.

**primary market** *See* secondary market.

**primary production** *See* production.

**prime costs** The *direct costs of the production of goods or services. Prime costs normally refer to the materials and labour immediately attributable to a particular cost unit. *Compare* oncosts; overhead costs.

**prime entry books** *See* book of prime entry.

**prime rate** The rate of interest at which US banks lend money to first-class borrowers. It is similar in operation to the *base rate in the UK.

**prime time** The time of the day when radio and TV audiences are expected to be at a peak and therefore advertising rates are highest.

**principal 1.** A person on whose behalf an agent or broker acts. **2.** A sum of money on which *interest is earned.

**principles of taxation** A set of criteria, largely determined by the economist Adam Smith (1723–90), for determining whether a given tax or system of taxation is good or bad. The main principles are that taxes should be equitable and certain. Subsidiary principles are that they should distort choices that would otherwise be made as little as possible and that the cost of collection should be as low as possible. Some economists argue that a further principle might be that the tax should be effective in redistributing income.

**printer** A computer-operated device for printing out data, programs, etc. onto paper. **Impact printers** include *line printers, which print a whole line at a time, and character printers, of which the *daisywheel printer is used for good-quality business correspondence, while the *dot matrix printer is cheaper but produces lower-quality printing. **Non-impact printers** include the *laser printer, a high-speed printer used for newsletters, reports, and correspondence to a standard approaching book-quality printing. Non-impact printers also include *thermal printers, which burn characters onto a special paper, and ink jet printers, which squirt a fine stream of ink onto the paper. The product of a computer printer is called **printout**.

**print run** The number of copies of a document, article, book, etc., made at a specific printing.

**priority date** *See* patent.

**priority percentage (prior charge)** The proportion of any profit that must be paid to holders of fixed-interest capital (*preference shares and loan stock; *see* debenture) before arriving at the sums to be distributed to *ordinary shareholders. These percentages help to assess the security of the income of ordinary shareholders and are related to the gearing of the company's capital (*see* capital gearing; leverage).

**prior-year adjustment** An adjustment to the profit and loss account of an organization necessitated by matters relating to earlier years, which give a misleading view of the position. Such an adjustment might arise as a result of under- or overvaluation of assets or liabilities or because of changes in accounting policies. For the purpose of accounting standards the term is limited to material adjustments arising from either changes in accounting policies or the correction of fundamental errors. Minor adjustments to estimates are not prior-year adjustments. Prior-year adjustments are to be made by adjusting the opening balances of reserves.

**prisoners' dilemma** A situation in which an individual's decision to maximize *utility turns out to be to his own and everyone else's detriment. A problem in *game theory, its name arose from the dilemma of two mythical prisoners (partners in crime) who accept that silence would be in their best interests but end up informing on each other, because they each expect their partner to do the same. The prisoners' dilemma has many applications in economics, in which it tends to undermine the belief in the efficiency of competitive economies (*see also* invisible hand).

**private bank 1.** A *commercial bank owned by one person or a partnership (*compare* joint-stock bank). Popular in 19th-century Britain, they have been superseded by joint-stock banks. They still exist in the USA. **2.** A bank that is not a member of a *clearing house and therefore has to use a clearing bank as an agent. **3.** A bank that is not owned by the state.

**private branch exchange (PBX)** A private telephone exchange in an organization that routes calls between extensions and the outside telephone network and between internal extensions. The automatic version, a **private automatic branch exchange (PABX)**, enables calls to be dialled direct from extensions. Incoming calls may be taken by an operator or callers can dial internal extensions direct.

**private brand** *See* own brand.

**private carrier** *See* carrier.

**private enterprise (free enterprise)** An economic system in which citizens are allowed to own capital and property and to run their own businesses with a minimum of state interference. The free-enterprise system encourages *entrepreneurs and the making of profits by private individuals and firms.

**private health insurance** A form of insurance that covers private medical treatment and all normal associated costs. This includes payment for private rooms in private or National Health hospitals, fees for surgeons, anaesthetists or physicians, and also charges for X-rays, pathology tests, and physiotherapy. Cover can be taken out on an individual basis or by a family, or a company can organize group cover with substantial discounts depending on the number of employees involved. Many large organizations offer private health insurance for senior executives and their families as a *fringe benefit.

**private limited company** Any *limited company that is not a *public limited company. Such a company is not permitted to offer its shares for sale to the public and it is free from the rules that apply to public limited companies.

**private sector** The part of an economy that is not under government control. In a *mixed economy most commercial and industrial firms are in the private sector, run by *private enterprise. *Compare* public sector.

**private-sector liquidity** *See* money supply.

**private treaty** Any contract made by personal arrangement between the buyer and seller or their agents, i.e. not by public *auction.

**privatization** The process of selling a publicly owned company (*see* nationalization) to the private sector. Privatization may be pursued for political as well as economic reasons. The economic justification for privatization is that a company will be more efficient under private ownership, although most economists would argue that privatization will only achieve this if it is accompanied by increased *competition. Recently, privatizations in the form of *share offers to the general public have been advocated as a means of increasing the participation of individuals in the capitalist system. The process can also be called **denationalization**.

**PRO** Abbreviation for public relations officer. *See* public relations.

**probability** The likelihood that an event or a particular result will occur. It can be represented on a scale by a number between 0 (zero probability of the event happening, i.e. it is certain not to) and 1 (certainty that it will occur). It is treated mathematically on this basis in statistics.

**probate** A certificate issued by the Family Division of the High Court, on the application of *executors appointed by a will, to the effect that the will is valid and that the executors are authorized to administer the deceased's estate. When there is no apparent doubt about the will's validity, probate is granted in **common form** on the executors filing an *affidavit. Probate granted in common form can be revoked by the court at any time on the application of an interested party who proves that the will is invalid. When the will is disputed, probate in **solemn form** is granted, but only if the court decides that the will is valid after hearing the evidence on the disputed issues in a **probate action**.

**probate price** The price of shares or other securities used for inheritance-tax purposes on the death of the owner. The price is taken as either one quarter of the interval between the upper and lower quotations of the day on which the owner died added to the lower figure or as half way between the highest and lowest recorded bargains of the day, whichever is the lower.

**probate value** The value of the assets at the time of a person's death, agreed with the Capital Taxes Office of the Inland Revenue for the purposes of calculating inheritance tax.

**procuration** *See* per proc.

**produce** *See* commodity.

**produce broker** *See* commodity broker.

**producer good** *See* capital good.

**producer price index (PPI)** A measure of the rate of *inflation among goods purchased and manufactured by UK industry (replacing the former **wholesale price index**). It measures the movements in prices of about 10 000 goods relative to the same base year. *Compare* Retail Price Index.

**producer surplus** The amount by which the market price exceeds the price at which producers would be willing to sell at least some goods. A concept analogous to consumer surplus, producer surplus is maximized in *perfect competition, which is another way of saying that perfect competition shows *Pareto optimality.

**producer theory** The branch of *microeconomics that analyses the methods by which goods and services are supplied to a market by firms or *entrepreneurs. The combination of producer theory and *consumer theory provides an explanation of the interaction of supply and demand in a market. This combination enables predictions to be made both about the behaviour of a particular market (*see* partial equilibrium analysis) and the whole economy (*see* general equilibrium analysis).

**product differentiation** The distinction between products that fulfil the same purpose but are made by different producers and therefore compete with each other. The distinction may be real (i.e. one product is better than others) or illusory. Producers in a competitive market use packaging, advertising, etc., to enhance the illusion that differences exist where none in fact do. In economic theory, if there is product differentiation, the products are not perfectly substitutable for each other and competition between them is therefore imperfect.

**product elimination** The process used to withdraw a product from a market in an orderly fashion so that it does not disrupt the sale of other products marketed by the same organization.

**production** The process by which inputs are converted into outputs. The production process takes many forms, from basic agriculture to large-scale manufacture. Inputs are typically combinations of labour, capital, land, and raw materials. These are known as the *factors of production; the demand for factors of production is known as a derived demand, since the demand exists only to the extent that they contribute to production. Neoclassical economists usually analyse the production process by means of a *production function, which establishes the combinations of inputs required to produce a set of outputs efficiently. Marxists, however, view the production process as the root of capitalist exploitation, the capitalist extracting labour power from workers and skimming off the surplus for himself.

**Primary production** in economic theory is concerned with reaping such natural benefits as those obtained by agriculture, forestry, hunting, fishing, mining, etc. **Secondary production** involves the manufacture of goods, buildings, roads, etc., from raw materials, while **tertiary production** concerns the services provided by transport companies, banks, insurance companies, and the professions.

**production function** A function defining the maximum possible output from different combinations and quantities of inputs; this can be represented as a curve on a graph. In *producer theory, if certain assumptions are made, a unique combination of inputs exists that will minimize the cost of a given level of output; this level of output is a point on a curve representing the production function. Production functions are widely used in economic analysis; in fact, the form of the production function can represent the technological possibilities of an economy at a particular time. *See also* cost minimization; profit maximization.

**productive expenditure** Money spent by a government on public services, schools, hospitals, etc., i.e. for benefits in the future as opposed to the immediate benefits that follow from current consumption.

**productivity** A measure of the output of an organization or economy per unit of input (labour, raw materials, capital, etc.). Economists are usually interested in changes in productivity, particularly with respect to labour. Increasing labour productivity is one means by which an economy can enhance its *competitiveness. *See also* unit labour costs.

**productivity agreement** An agreement between an employer and a union in which an increase in wages is given for a measured increase in *productivity. To arrive at such an agreement, **productivity bargaining** is needed to reach a compromise between the increase in wages demanded by the unions and the increase in productivity demanded by the employers.

**product life cycle** A *marketing model that describes the changes of a product's level of sales over a specified period. The model describes four phases, each of which represents a different opportunity for the marketer. A new product starts in the **introduction phase**, characterized by low sales: buyers are unsure about the product and it is not stocked by all distributors. Sales can be increased by introductory price offers and advertising support. If purchasers are satisfied with the product, its reputation will spread and it will enter the **growth phase**: the product will become more widely available and sales will increase. Competitors' versions will then appear and eventually the **maturity phase** will be reached, in which supply and demand are matched and sales stabilize. The maturity phase is the longest period, characterized by intense competition, but eventually a better way of satisfying consumers' needs will almost certainly be provided by another new product and the **decline phase** will be entered. The duration of each phase is unique to each product. Due to the poor profitability of the introduction phase, it is common for marketing expenditure to be high to encourage a quick transition to the growth phase; however, this expenditure has to be weighed against the need to recoup product development costs in the event that the product is a failure.

**product line** A group of closely related products marketed by one company. *See also* family brand.

**product-market strategy (Ansoff matrix)** A *marketing planning model. Companies can either sell existing or new products; and they can sell them either in markets familiar to them (existing markets) or in new markets. The resulting two-by-two matrix gives four alternative strategies for increasing sales. One is to concentrate on selling more existing products into existing markets (i.e. increasing *market penetration), by such means as price reductions and increased advertising; this is regarded as a low-risk strategy. Alternatively, an organization can modify or improve its existing products and sell these to current customers (**product development**); for example, a lawn-mower manufacturer might install a more powerful motor in its products. The third option is to sell existing products to a new market (**market development**) – for example, to export the current range of lawn mowers to America. These two strategies are medium-risk. The highest-risk strategy is to develop new products for new markets – for example, for the lawn-mower manufacturer to develop printing machinery for the publishing industry.

**product orientation** The attitude of a company that believes the product comes first and persuading customers to buy it follows. **Market orientation** typifies the attitude of a company that will only produce what it believes it can sell.

**products-guarantee insurance** An insurance policy covering financial loss as a consequence of a fault occurring in a company's product. A product-guarantee claim would be made, for example, to pay for the cost of recalling and repairing cars that are found to have a defective component. This type of policy would not pay compensation to customers or members of the public injured as a result of the defect. Such claims would be met by a *products-liability insurance.

**products liability** The liability of manufacturers and other persons for defective products. Under the Consumer Protection Act (1987), passed to conform with the requirements of the *European Community (EC) law, the producer of a defective product that causes death or personal injury or damage to property is strictly liable for the damage. A claim may only be made for damage to property if the property was for private use or consumption and the value of the damage caused exceeds £275. The persons liable for a defective product are the producer (i.e. the manufacturer, including producers of component parts and raw materials), a person who holds himself out to be producer by putting his name or trade mark on the product, a person who imports the product into the EC, and a supplier who fails, when reasonably requested to do so by the person injured, to identify the producer or importer of the product.
The purchaser of a defective product may sue the seller for *breach of contract in failing to supply a product that conforms to the contract. Under the same Act suppliers of consumer goods must ensure that the goods comply with the general safety requirement. Otherwise they commit a criminal offence. *See also* consumer protection.

**products-liability insurance** An insurance policy that pays any compensation the insured is legally liable to pay to customers who are killed, injured, or have property damaged as a result of a defect in a product that they have manufactured or supplied. Costs incurred as a consequence of the defect that are not legal *damages would not be covered by a policy of this kind. A *products-guarantee insurance is intended to cover these costs.

**product testing** Anonymous testing and evaluation of a product, often used in the early stages of new product development to assess its marketability.

**professional-indemnity insurance** A form of *third-party insurance that covers a professional man, such as a solicitor, surveyor, accountant, businessman, etc., against compensation he has to pay if he is successfully sued for professional negligence. This can include the giving of defective advice if the businessman advertises himself as an expert in his given field. *See also* public-liability insurance.

**professional valuation** An assessment of the value of an asset (share, property, stock, etc.) in the balance sheet or prospectus of a company, by a person professionally qualified to give such a valuation. The professional qualification necessary will depend on the asset; for example, a qualified surveyor may be needed to value property, whereas unquoted shares might best be valued by a qualified accountant.

**profit** **1.** For a single transaction, the excess of the selling price of the article or service being sold over the costs of providing it. **2.** For a period of trading, the surplus of net assets at the end of a period over the net assets at the start of that period, adjusted where relevant for amounts of capital injected or withdrawn by the proprietors. As profit is notoriously hard to define, it is not always possible to derive one single figure or profit for an organization from an accepted set of data. **3.** In economics, the return on capital as a *factor of production (*see* normal economic profit; pure economic profit; super profit).

**profitability** The capacity or potential of a project or an organization to make a *profit. Measures of profitability include return on capital employed, positive net cash flows, and the ratio of net profit to sales.

**profit and loss account 1.** An *account in the books of an organization showing the profits (or losses) made on its business activities with the deduction of the appropriate expenses. **2.** A statement of the profit (or loss) of an organization derived from the account in the books. It is one of the statutory accounts that, for most limited companies, has to be filed annually with the UK Registrar of Companies. The profit and loss account usually consists of three parts. The first is a trading account, showing the total sales income less the costs of production, etc., and any changes in the value of stock or work in progress from the last accounting period. This gives the *gross profit (or loss). The second part gives any other income (apart from trading) and lists administrative and other costs to arrive at a *net profit (or loss). From this net profit before taxation the appropriate corporation tax is deducted to give the net profit after taxation. In the third part, the net profit after tax is appropriated to dividends or to reserves. The UK Companies Act (1985) gives a choice of four formats, one of which must be used to file a profit and loss account for a registered company.

**profit à prendre** A right to take something off another person's land; it is a form of incorporeal *hereditament. The thing to be taken must consist either of part of the land (e.g. crops) or wild animals existing on it. A right to take water is therefore not a profit à prendre.

**profit centre** Any unit within an organization that is required to show a profit.

**profiteer** A person who makes excessive profits by charging inflated prices for a commodity that is in short supply, especially during a war or national disaster. In a free-enterprise society that encourages entrepreneurs to make profits, the distinction between profit-making and profiteering may not always be clear cut.

**profit forecast** A forecast by the directors of a public company of the profits to be expected in a stated period. If a new flotation is involved, the profit forecast must be reported on by the reporting accountants and the sponsor to the share issue. An existing company is not required to make a profit forecast with its *accounts, but if it does it must be reported on by the company's auditors.

**profit function** A function relating the amount a firm chooses to produce, the costs of its inputs, and its level of profits; this can be represented as a curve on a graph. In *producer theory, making certain assumptions, there will be a unique combination of inputs for a given level of output that maximizes profits; this is a point on the curve representing the profit function. However, a firm may choose to maximize some objective other than profits (*see* managerial theory of the firm; satisficing behaviour), which will lead it to pick some other point on the curve of its profit function.

**profit impact of market strategy (PIMS)** A study of the major factors that influence profits in a wide variety of industries.

**profit maximization** The assumption that firms or *entrepreneurs will choose outputs and inputs to gain the highest level of profit that is feasible (*see* profit motive). Although this is one of the most frequently made assumptions in economics (because it yields simple results), there are several alternative assumptions (*see* satisficing behaviour; managerial theory of the firm). As with all the other assumptions made in *microeconomics, profit maximization, in a perfectly competitive economy, will yield

Pareto-optimal results (*see* Pareto optimality).

**profit motive** The assumption in economics that *entrepreneurs seek to maximize *profits. The assumption is little more than an observation that most human beings aspire to be wealthy. *See also* profit maximization.

**profit sharing** The distribution of part of the profits of a company to its employees, in the form of either cash or shares in the company. There are many workers' participation schemes that use profit sharing as a means of increasing the motivation of employees; some schemes relate the share of the profit to salary or wages, others to length of service, and yet others give equal shares to all who have been employed by the company for a minimum period.

**profits tax** Any tax on the profits of a company. *Corporation tax in the UK is a form of profits tax.

**profit taking** Selling commodities or securities at a profit, either after a market rise or because they show a profit at current levels but will not do so if an expected fall in prices occurs.

**proforma invoice** An invoice sent in certain circumstances to a buyer, usually before some of the invoice details are known. For example, in commodity trading a proforma invoice may be sent to the buyer at the time of shipment, based on a notional weight, although the contract specifies that the buyer will only pay for the weight ascertained on landing the goods at the port of destination. When the missing facts are known, in this case the landed weight, a *final invoice is sent.

**program** *See* computer programming; programming language.

**programming language** A language used to give instructions (a **program**) to a *computer. There are two main types of programming languages: *low-level languages and *high-level languages. High-level languages are the easier to use as they are designed to reflect the needs of the programmer rather than the capabilities of the computer. Programs written in these languages have to be translated into the *machine code that the computer understands before they can be run. Low-level languages, which are close to machine code, produce programs that run more swiftly and make more efficient use of computer resources. Other types of programming languages are used for special purposes (*see* query language; report program generator).

**program trading** Trading on international stock exchanges using a computer program to exploit differences between stock index futures and actual share prices on world equity markets. It is said to account for some 10% of the daily turnover (1989–90) on the New York Stock Exchange and has been partly blamed for the market crash in October 1987.

**progress chaser** A person who is responsible for following the progress of work being done in a factory or office and for seeing that it is completed on time.

**progressive tax** A tax in which the rate of tax increases with increases in the tax base. The most common of these is income tax but progressive rates are also applied to National Insurance contributions, inheritance tax, and to a limited extent corporation tax. Such taxes are generally linked to the ability-to-pay principle (*see* ability-to-pay taxation). *Compare* proportional tax.

**progress payment** An instalment of a total payment made to a contractor, when a specified stage of the operation has been completed.

**promissory note** A document that is a negotiable instrument and contains a promise to pay a certain sum of money to a named person, to his order, or to the bearer. It must be unconditional, signed by the maker, and delivered to the payee or bearer. They are widely used in the USA but not in common use in the UK. A promissory note cannot be reissued, unless the promise is made by a banker and is payable to the bearer, i.e. unless it is a *banknote.

**promoter** A person involved in setting up and funding a new company, including preparing its *articles and *memorandum of association, registering the company, finding directors, and raising subscriptions. The promoter is in a position of trust with regard to the new company and may not make an undisclosed profit or benefit at its expense. A promoter may be personally liable for the fulfilment of a contract entered into by, or on behalf of, the new company before it has been formed.

**promotion** *See* sales promotion.

**prompt cash** Payment terms for goods or services in which payment is due within a few days (usually not more than 14) of delivery of the goods or the rendering of the service.

**prompt day (prompt date) 1.** The day (date) on which payment is due for the purchase of goods. In some commodity spot markets it is the day payment is due and delivery of the goods may be effected. **2.** The date on which a contract on a commodity exchange, such as the *London Metal Exchange, matures.

**proof** A printed version of text, produced for checking to enable errors to be found (by **proofreading**) and corrected. *See* galley; page proof.

**property** Something capable of being owned (*see* personal property; real property). It may be tangible, such as a building or work of art, or intangible, such as a right of way or a right under a contract.

**property bond** A bond issued by a life-assurance company, the premiums for which are invested in a fund that owns property.

**property insurance** Insurance covering loss, damage, or destruction of any form of item from personal jewellery to industrial plant and machinery. Property-insurance policies are a form of *indemnity in which the insurer undertakes to make good the loss suffered by the insured. The policy may state the specific compensation payable in the event of loss or damage; if it does not, the policy will normally pay the intrinsic value of the insured object, taking into account any appreciation or depreciation on the original cost. Such policies usually have a maximum sum for which the insurers are liable.

**property register** *See* land registration.

**property tax** A tax based on the value of property owned by the taxpayer. The most common property taxes are the *rates raised by local authorities.

**proportional spacing** A form of typesetting in which each character occupies a space proportional to its size. In standard typewriters the characters have a uniform width, but in some electronic typewriters, and in typesetting machines, it is usual to have proportional spacing.

**proportional tax** A tax in which the amount of tax paid is proportional to the size of the tax base, i.e. a tax with a single rate. *Compare* progressive tax.

**proprietary company** *See* Pty.

**proprietorship register** *See* land registration.

**pro rata** (Latin: in proportion) Denoting a method of dividing something between a number of participants in proportion to some factor. For example, some of the profits of a company are shared, pro rata, among the shareholders, i.e. in proportion to the number of shares each shareholder owns.

**prospective damages** *See* damages.

**prospectus** A document that gives details about a new issue of shares and invites the public to buy shares or debentures in the company. A copy must be filed with the Registrar of Companies. The prospectus must conform to the provisions of the Companies Act (1985), describe the aims, capital structure, and any past history of the venture, and may contain future *profit forecasts. There are heavy penalties for knowingly making false statements in a prospectus.

**protectionism** The policy of using tariffs, import quotas, and other restrictions to shield domestic manufacturers from foreign competition. A number of conflicting issues have to be resolved before such a policy is implemented. Retaliation is an obvious reason for not trying to protect domestic industries in this way. On the other hand an industry of national importance in wartime could be eliminated by peacetime foreign competition if some protection is not given. Also, the general good of the community must be taken into account; protecting a particular industry may increase profits and maintain employment in that industry but it may do so at the expense of the consumers, who have to pay more for the products of that industry than they would if they were allowed to buy untariffed imports.

**protective duty** A tariff imposed on an import to protect domestic manufacturers from foreign competition. *See* protectionism.

**protest** A certificate signed by a *notary public at the request of the holder of a *bill of exchange that has been refused payment or acceptance. It is a legal requirement after *noting the bill (*see also* acceptance supra protest). The same procedure can also be used for a *promissory note that has been dishonoured.

**protocol** The rules governing the communications between different computers or computer peripherals. The protocol is a formal statement of how data is to be set out when messages are exchanged, the order and priority of various messages, and so on.

**provision** An amount set aside out of profits in the accounts of an organization for a known liability (even though the specific amount might not be known) or for the diminution in value of an asset. Common provisions are for bad debts and for depreciation and also for accrued liabilities. According to the UK Companies Act (1981) notes must be given to explain every material provision in the accounts of a limited company.

**provisional liquidator** A person appointed by a court after the presentation of a winding-up petition on a company (*see* compulsory liquidation). His job is to protect the interests of all the parties involved until the winding-up order is made, and his powers are limited. The *official receiver is normally appointed to this post. *See also* liquidator.

**provision for bad debts** A sum credited to an *account in the books of an organization to allow for the fact that some of the debtors might not pay their debts in full. This amount is normally deducted from the total debtors in the balance sheet.

**provision for depreciation** A sum credited to an *account in the books of an organization to allow for the *depreciation of a fixed asset. The aggregate amounts set aside from year to year are deducted from the value of the asset in the balance sheet to give its *net book value.

**proximate cause** The dominant and effective cause of an event or chain of events that results in a claim on an insurance policy. The loss must be caused directly, or as a result of a chain of events initiated, by an insured peril. For example, a policy covering storm damage would also pay for items in a freezer that deteriorate because of a power cut caused by the storm, which is the proximate cause of the loss of the frozen food.

**proximo** *See* instant.

**proxy** A person who acts in the place of a member of a company at a company meeting at which one or more votes are taken. The proxy need not be a member of the company but it is quite common for directors to offer themselves as proxies for shareholders who cannot attend a meeting. Notices calling meetings must state that a member may appoint a proxy and the appointment of a proxy is usually done on a form provided by the company with the notice of the meeting; it must be returned to the company not less than 48 hours before the meeting. A **two-way proxy form** is printed so that the member can state whether he wants the proxy to vote for or against a particular resolution. A **special proxy** is empowered to act at one specified meeting; a **general proxy** is authorized to vote at any meeting.

**PRT** Abbreviation for *petroleum revenue tax.

**prudence concept** A principle of accounting designed to ensure that unrealized profits are not distributed to shareholders by way of dividend. According to this principle, unrealized profits are not taken account of until they are realized; on the other hand foreseeable losses are taken account of as soon as they can be foreseen. Some accountants find it difficult to reconcile this concept with the 'true and fair view' required of published accounts; the prudence concept must give a view with a pessimistic bias. However, most accountants accept that taking a 'true and fair view' means doing so in accordance with the prudence concept.

**PSBR** Abbreviation for *Public Sector Borrowing Requirement.

**PSL** Abbreviation for private-sector liquidity. PSL1 and PSL2 are used as measures of the *money supply.

**psychological tests** Tests designed to assess the personalities and abilities of individuals to determine their suitability for a particular job and to make best use of their talents. With the increasing use of computers to analyse information, the tests have become more and more complex in *personnel selection; choosing the wrong person for a senior job can be costly and have far-reaching effects in a competitive market.

**PTN** Abbreviation for *public telephone network.

**Pty** Abbreviation for proprietary company, the name given to a *private limited company in Australia and the Republic of South Africa. The abbreviation Pty is used after the name of the company as Ltd is used in the UK.

**public company** A company whose shares are available to the public through a stock exchange. *See* public limited company.

**public corporation** A state-owned organization set up either to provide a national service (such as the British Broadcasting Corporation) or to run a nationalized industry (such as the *British Coal Corporation, formerly the National Coal Board). The chairman and members of the board of a public corporation are usually appointed by the appropriate government minister, who retains overall control and accountability to parliament. The public corporation attempts to reconcile public accountability for the use of public finance, freedom of commercial operation on a day-to-day basis, and maximum benefits for the community.

**public debts** The debts of the *public sector of the economy, including the *national debt.

**public deposits** The balances to the credit of government departments held at the Bank of England.

**public examination** *See* bankruptcy.

**public finance** **1.** The financing of the goods and services provided by national and local government through taxation or other means. **2.** The economic study of the issues involved in raising and spending money for the public benefit.

**public finance accountant** A member of the Chartered Institute of Public Finance and Accountancy. The principal function of the members of this body is to prepare the financial accounts and act as management accountants for government agencies, local authorities, nationalized industries, and such bodies as publicly owned health and water authorities. As many of these bodies are non-profit making and are governed by special statutes, the skills required of public sector accountants differ from those required in the private sector.

**public good** A good that is simultaneously available to all the individuals in an economy. The consumption of a public good by one individual does not reduce the quantity available for consumption by another individual. While many goods have a public aspect, it is hard to find an example of a pure public good; perhaps the closest example is national defence. It is usually argued that public goods will be undersupplied by the market since a calculation of the ratio of the cost to the benefit to an individual will usually ignore the public aspect. This is usually considered the minimum justification for a government to intervene in the operation of a free market.

**public issue** A method of making a *new issue of shares, loan stock, etc., in which the public are invited, through advertisements in the national press, to apply for shares at a price fixed by the company. *Compare* introduction; issue by tender; offer for sale; placing.

**public lending right (PLR)** An attempt to compensate authors for loss of *royalties on books that are borrowed from libraries rather than bought. The scheme, which was introduced in the UK in 1983, pays authors a fee proportional to the number of times their books were borrowed in the previous year from a sample of 16 public libraries situated throughout the UK. The sum to be divided in this way is fixed by parliament. Individual authors may not earn more than £5000 from the PLR.

**public-liability insurance** An insurance policy that pays compensation to a member of the public and court costs in the event of the policyholder being successfully sued for causing death, injury, or damage to property by failing to take reasonable care in his actions or those of his employees. A business whose work brings it into contact with the public must have a public-liability policy. *See also* employers'-liability insurance; professional-indemnity insurance; third-party insurance.

**public limited company (plc)** A company registered under the Companies Act (1980) as a public company. Its name must end with the initials 'plc'. It must have an authorized share capital of at least £50,000, of which at least £12,500 must be paid up. The company's memorandum must comply with the format in Table F of the Companies Regulations (1985). It may offer shares and securities to the public. The regulation of such companies is stricter than that of private companies. Most public companies are converted from private companies, under the re-registration procedure in the Companies Act.

**public policy** The interests of the community. If a contract is (on common-law principles) contrary to public policy, this will normally make it an *illegal contract. In a few cases, however, such a contract is void but not illegal, and is treated slightly more leniently (for example, by *severance). Contracts that are illegal because they contravene public policy include any contract to commit a crime or a tort or to defraud the revenue, any contract that prejudices national safety or the administration of justice, and any immoral contract. Contracts that are merely void include contracts in *restraint of trade.

**public relations (PR)** Influencing the public so that they regard an individual, firm, charity, etc., in a favourable light in comparison to their competitors. Some media personalities, large companies, and national charities employ their own **public relations officers (PROs)** to deal with the media, provide information in the form of handouts, and to represent their principals at press conferences, etc. Others use **public relations agencies** to fulfil these functions. PR does not involve paid advertising, which is a quite separate activity. While an advertising agent will plan an advertising campaign, charging a percentage of the money spent, PR agencies, for a flat fee (plus expenses), will seek to promote their principals by persuading newspapers to feature them in articles, TV and radio personalities to interview them or otherwise feature them in their programmes, etc.

**public sector** The part of an economy in a *mixed economy that covers the activities of the government and local authorities. This includes education, the National Health Service, the social services, public transport, the services, the police, local public services, etc., as well as state-owned industries and *public corporations. *Compare* private sector.

**Public Sector Borrowing Requirement (PSBR)** The amount by which UK government expenditure exceeds its income (i.e. the **public sector deficit**); this must be financed by borrowing (e.g. by selling gilt-edged securities) or by printing money. As an indicator of government fiscal policy the PSBR has acquired increased status since the late 1970s. By that time many economists had come to accept that a high PSBR is inflationary or leads to the crowding out of private expenditure; this remains a widely held view. While printing money simply causes prices to rise (*see* quantity theory of money), selling gilts has the effect of raising interest rates, reducing private investment, and curbing private expenditure. *See also* Central Government Borrowing Requirement.

**public sector deficit** *See* Public Sector Borrowing Requirement.

**public telephone network (PTN)** The telephone network provided in the UK by British Telecom. Telephone equipment in offices, factories, homes, etc., connected to this network may be bought or rented from British Telecom or from outside suppliers. However, by government order, all such equipment must be clearly marked with a green circle to show that it has been approved for connection to the network. Equipment marked with a red triangle is not approved for connection.

**public trustee** A state official in charge of the Public Trust Office, a trust corporation set up for certain statutory purposes. Being a *corporation sole, the office exists irrespective of the person performing it. He may act as *administrator of small estates, as *trustee for English trusts where required, and as *receiver when directed to do so by a court. He may hold funds of registered *Friendly Societies and trade unions.

**public warehouse** *See* warehouse.

**public works** Government-sponsored construction work, especially that undertaken during a *recession or a *depression on such activities as house building or road building. Aimed at increasing the level of employment and *aggregate demand, it is an activity generally associated with *Keynesianism. *See also* pump priming; infrastructure; reflation; effective demand.

**published accounts** Accounts of organizations published according to UK law. The most common, according to the UK Companies Act (1981), are the accounts of *limited companies, which must be provided for their shareholders and filed with the Registrar of Companies at Companies House, Cardiff. The accounts comprise the *balance sheet, the *profit and loss account, the statement of *sources and application of funds, the *directors' report, and the *auditors' report. In the case of *groups of companies, consolidated accounts are also required. *Small companies and *medium-sized companies, as defined by the Act, need not file some of these documents.

**puisne mortgage** A legal *mortgage of unregistered land that is not protected by the deposit of title deeds. It should instead be protected by registration.

**pump priming** An addition to *aggregate demand generated by a government in order to set off the *multiplier process. A policy advocated by Keynesians, it usually involves allowing government expenditure to exceed receipts, thus creating a *budget deficit. Pump priming was widely employed by governments in the post-war era in order to maintain *full employment; however, it became discredited in the 1970s when it failed to halt rising unemployment and was even held to be responsible for *inflation. *See also* demand management; stagflation.

**punter** A speculator on a stock exchange or commodity market, especially one who hopes to make quick profits.

**purchase day book (purchase journal)** The *book of prime entry in which the invoices of an organization's suppliers are recorded. These may only include invoices for goods supplied for resale but in most organizations they include invoices for all goods and services supplied; this often requires a number of columns to analyse the invoices into such account categories as motor expenses, light and heat, etc. Postings are made from the purchase day book to the personal accounts of the suppliers in the *purchase ledger, while the totals of the analysis columns are posted to the *nominal ledger.

**purchased life annuity** An *annuity in which a single premium purchases an income to be paid from a specified future date for the rest of the policyholder's life.

**purchase ledger** The *ledger in which the personal accounts of an organization's suppliers are recorded. The total of the balances in this ledger represents the organization's *trade creditors.

**purchase ledger control account** An account in the *nominal ledger to which the totals of the entries in the *purchase day book are posted at regular intervals. With this procedure the balance on the purchase ledger control account should, at any time, equal the aggregate of the balances on all the individual accounts in the ledger. The balance also represents the total of *trade creditors.

**purchasing officer** An employee of a manufacturer, who is responsible for purchasing the raw materials used in the manufacturing process. If these include raw materials that fluctuate in price, the purchasing officer has to show that his average purchase price for the year is not excessive. He is also responsible for maintaining adequate stocks to tide the company over any break in the continuity of supply, without tying up excessive amounts of capital.

**purchasing power parity (PPP)** Parity between two currencies at a *rate of exchange that will give each currency exactly the same purchasing power in its own economy. The belief that exchange rates adjust to reflect PPP dates back at least to the 17th century, but in the short term, at least, is demonstrably false (*see* overshooting). It may well hold in the long term, and is often used as a benchmark to indicate the levels that exchange rates should achieve (although measurement is difficult and controversial).

**pure economic profit** The surplus or *rent extracted by a producer or seller in exploiting a privileged position in a market. This position may be a result of *monopoly, *oligopoly, limited information, or even government regulations. It represents a return in excess of that required to bring an entrepreneur to the market and is therefore generally disliked. This is not the same as saying that monopoly is not Pareto optimal (*see* Pareto optimality), although the two statements are often confused. *See also* profit; superprofit.

**pure endowment assurance** An assurance policy that promises to pay an agreed amount if the policyholder is alive on a specified future date. If the policyholder dies before the specified date no payment is made and the premium payments cease. The use of the word 'assurance' for this type of contract is questionable as there is no element of life-assurance cover.

**push money** A cash inducement given to a retailer by a manufacturer or wholesaler to be used to reward sales personnel who are particularly successful in selling specified products.

**put-of-more option** *See* option to double.

**put option** *See* option.

**put through** Two deals made simultaneously by a *market maker on the London Stock Exchange, in which a large quantity of shares is sold by one client and bought by another, the market maker taking a very small turn.

**pyramid selling** A method of selling to the public using a hierarchy of part-time workers. Usually a central instigator (at the apex of the pyramid) sells a franchise to regional organizers for a certain product, together with an agreed quantity of the goods. These organizers recruit district distributors, who each take some of the stock and, in turn, recruit door-to-door salesmen, who take smaller proportions of the stock, which they attempt to sell. In some cases the last stage consists of people who sell the goods to their friends. As the central instigator can sell more goods by this means than are likely to be bought at the base of the pyramid, and someone on the way down is liable to be caught with unsaleable stock, the system is illegal in the UK.

# Q

**QL** Abbreviation for *query language.

**qualified acceptance** An *acceptance of a *bill of exchange that varies the effect of the bill as drawn. The holder may refuse to take a qualified acceptance; if he does he should notify the drawer and any endorsers or they will no longer be liable. If the holder takes a qualified acceptance, he releases all previous signatories who did not assent from liability.

**qualified report** An *auditors' report in which the auditors, for one reason or another, have been unable to satisfy themselves that the accounts give a true and fair view of a company's affairs. For smaller companies, in which *internal control cannot be perfect, qualified reports are not uncommon. However, it is normally regarded as a serious matter for a public limited company to receive a qualified report.

**qualifying policy** A life-assurance policy that the UK Inland Revenue has agreed is eligible for tax relief on the premiums and/or the payment of the sum assured. Tax relief on life-assurance premiums was abolished on 13 March 1984, therefore policies eligible for life-assurance premium relief have now disappeared. *Compare* non-qualifying policy.

**qualitative marketing research** *Marketing research techniques that use small *samples of respondents to gain an impression of their beliefs, motivations, perceptions, and opinions. Such unstructured methods of data collection as *depth interviews and *group discussions are used to explore topics in considerable detail. Qualitative marketing research is frequently used to test an advertisement's effectiveness or to explore new products. *See also* concept test. In general, it is used to show why people buy a particular product, whereas *quantitative marketing research reveals how many people buy it.

**quality control** A systematic inspection of products, or a sample of the products on a production line, at various stages of production. The purpose is to ensure that requisite standards are being maintained and that acceptable tolerances are not being exceeded. In mass production, the statistical analysis of parameters measured on a random sample of the end product is most important. The larger the sample tested, the higher will be the manufacturer's reputation for producing goods of a high standard. **Quality control charts** are frequently used in mass production; the horizontal axis of the chart (graph) is calibrated in units of time, while the vertical axis represents the values of such variables as the percentage of defective products. The cause of any sustained rise in the variable must be examined immediately.

**quango** Acronym for quasi-autonomous government organization. Such bodies, some members of which are likely to be civil servants and some not, are appointed by a minister to perform some public function at the public expense. Examples are the *Advisory Conciliation and Arbitration Service and the *Health and Safety Commission. While not actually government agencies, they are not independent and are usually answerable to a government minister.

**quant** A computer specialist with a background in the quantitative sciences (hence the name), employed at a high salary by a city institution (e.g. portfolio management company, bond research house, merchant bank, etc.) to develop systems that map past movements in financial markets with a view to predicting future equity, commodity, and currency values.

**quantitative marketing research** *Marketing research techniques that use large *samples of respondents to quantify behaviour and reactions to marketing activities. Typically, a structured questionnaire is used to obtain data that quantifies the numbers and proportions of respondents falling into each predetermined category; for example, a study might show how many people per thousand of the population buy a particular product. *Compare* qualitative marketing research.

**quantity theory of money** A theory, first proposed by the philosopher David Hume (1711–76), stating that the *price level is proportional to the quantity of money in the economy. Formally, it is usually stated in the equation: $MV = PT$, where $M$ is the quantity of money, $V$ is its velocity of circulation, $P$ is the price level, and $T$ the number of transactions in the period. Milton Friedman (1912– ) made this equation the central pivot of *monetarism, with the additional assumption that $V$ is more or less constant; thus, for a given number of transactions, the relationship between $M$ and $P$ is direct. This implies that any increase in the *money supply will lead to an increase in the price level, i.e. to inflation. *See* demand for money; monetary policy.

**quantity variance** A *variance arising when the quantities sold, purchased, or produced differ from the standard.

**quantum meruit** (Latin: as much as has been earned) **1.** Denoting a payment for goods or services supplied in partial fulfilment of a contract, after the contract has been breached. **2.** Denoting a payment for goods or services supplied and accepted, although no price has been agreed between buyer and seller.

**quarter days** Four days traditionally taken as the beginning or end of the four quarters of the year, often for purposes of charging rent. In England, Wales, and Northern Ireland they are Lady Day (25 March), Midsummer Day (24 June), Michaelmas (29 September), and Christmas Day (25 December). In Scotland they are Candlemas (2 February), Whitsuntide (15 May), Lammas (1 August), and Martinmas (11 November).

**quarter up** The means of arriving at the *probate price of a share or other security.

**quasi-contract** A legally binding obligation that one party has to another, as determined by a court, although no formal contract exists between them.

**Queen's Awards** Two separate awards instituted by royal warrant in 1976 to replace the Queen's Award to Industry, which was instituted in 1965. The **Queen's Award for Export Achievement** is given for a sustained increase in export earnings to an outstanding level for the products or services concerned and for the size of the applicants' organizations. The **Queen's Award for Technological Achievement** is given for a significant advance in technology to a production or development process in British industry. The awards are announced on the Queen's actual birthday (21 April). Awards are held for five years and entitle the holders to fly a special flag and display on their packages, stationery, etc., the emblem of the award.

**Queen's enemies** *See* peril.

**query language (QL)** A special-purpose computer *programming language, used to extract information from a *database. In general, each database management system has its own query language.

**quick assets** *See* liquid assets.

**quid pro quo** (Latin: something for something) Something given as compensation for something received. *Contracts require a quid pro quo; without a *consideration they would become unilateral agreements.

**quorum** The smallest number of persons required to attend a meeting in order that its proceedings may be regarded as valid. For a company, the quorum for a meeting is laid down in the *articles of association.

**quota 1.** A limit on the import or export of a particular product imposed by a government. Import quotas may be imposed for protectionist reasons (*see* protectionism); export quotas may be imposed in countries that depend on the export of a particular raw material, as a means of stabilizing prices. Quotas are usually controlled by the issue of licences. **2.** An agreed quantity of goods that a member of a cartel is permitted, by the terms of the cartel, to produce in a given period.

**quotation 1.** The representation of a security on a recognized *stock exchange (*see* listed company). A quotation allows the shares of a company to be traded on the stock exchange and enables the company to raise new capital if it needs to do so (*see* flotation; listing requirements; rights issue). **2.** An indication of the price at which a seller might be willing to offer goods for sale. A quotation does not, however, have the status of a firm *offer. **3.** A *quoted price.

**quoted company** *See* listed company.

**quoted price** The official price of a security or commodity. On the London Stock Exchange, quoted prices are given daily in the *Official List. Quoted prices of commodities are given by the relevant markets and recorded in the financial press.

# R

**rack rent** *See* rent.

**radiopaging** A method of contacting members of an organization when they are not in their offices. A high-frequency signal is received by an outside radiopaging unit carried by a user when away from his desk, causing it to bleep and alerting him to phone in to his headquarters. Several messages can be stored on a pocket receiver and messages can be received from more than one source by using different bleeps for different callers.

**raider** An individual or organization that specializes in exploiting companies with undervalued assets by initiating hostile *takeover bids.

**rally** A rise in prices in a market, such as a stock exchange or commodity market, after a fall. This is usually brought about by a change of sentiment. However, if the change has occurred because there are more buyers than sellers, it is known as a **technical rally**. For example, unfavourable sentiment might cause a market to fall, in turn causing sellers to withdraw at the lower prices. The market will then be sensitive to the presence of very few buyers, who, if they show their hand, may bring about a technical rally.

**RAM** Abbreviation for *random-access memory.

**R & D** Abbreviation for *research and development.

**random-access memory (RAM)** The part of computer *main store memory that can be both read from and written to (*compare* read-only memory). The amount of RAM determines the amount of memory the user has in which to run programs. Most types of RAM lose their contents when the computer is turned off.

Strictly, a random-access memory is any type of memory, such as *magnetic disks, in which an arbitrary location can be accessed immediately, without passing over all preceding locations (*see* sequential access). However, the term is rarely (and the abbreviation RAM never) used in this sense.

**random sample** A group selected from a population of people or things in such a way that every individual has an equal chance of being selected. A **stratified random sample** is obtained by dividing the total population into subgroups according to some desired criterion before selecting the random sample. For example, in selecting a random sample of people to interview one might first stratify the population by age or income.

**random-walk theory** The theory that share prices move, for whatever reason, without any memory of past movements and that the movements therefore follow no pattern. This theory is used to refute the predictions of chartists, who do rely on past patterns of movements to predict present and future prices.

**ratchet effect** An irreversible change to an economic variable, such as prices, wages, exchange rates, etc. For example, once a price or wage has been forced up by some temporary economic pressure, it is unlikely to fall back when the pressure is reduced. This rise may be reflected in parallel sympathetic rises throughout the economy, thus fuelling *inflation.

**rateable value** *See* rates.

**rate capping** *See* rates.

**rate card** A list of prices charged for advertising space, TV or radio time, etc. *See also* off-card rate.

**rate of exchange (exchange rate)** The price of one currency in terms of another. It is usually expressed in terms of how many units of the home country's currency are needed to buy one unit of the foreign currency. However, in some cases, notably in the UK, it is expressed as the number of units of foreign currency that one unit of the home currency will buy. Two rates are usually given, the buying and selling rate; the difference is the profit or commission charged by the organization carrying out the exchange.

**rate of interest** *See* interest.

**rate of return** The annual amount of income from an investment, expressed as a percentage of the original investment. This rate is very important in assessing the relative merits of different investments. It is therefore important to note whether a quoted rate is before or after tax, since with most investments, the after-tax rate of return is most relevant. Also, because some rates are payable more frequently than annually, it may be important, in order to make true comparisons, to consider the annual percentage rate (APR), which most investment institutions are required to state by law.

**rate of time preference** *See* time preference.

**rate of turnover** The frequency, expressed in annual terms, with which some part of the assets of an organization is turned over. The total sales revenue is often referred to as 'turnover' and in order to see how frequently stock is turned over, the sales revenue (or if a more accurate estimate is needed the cost of goods sold) is divided by the average value of the stock to give the number of times the stock is turned over. This provides a reasonable measure in terms of stock. However, some accountants divide the sales figure by the value of the fixed assets to arrive at turnover of fixed assets. This is less realistic, although it does express the relationship of sales to the fixed assets of the organization, which in some organizations could be significant.

**rates** A local-authority tax calculated as a poundage on the rateable value of property in the area of the rating authority. The **poundage** is fixed annually by the rating authority as the number of pence that must be paid for each pound of rateable value. Under the Rates Act (1984), the Secretary of State for the Environment is empowered to limit the rates a local council can charge ratepayers (a process known as **rate capping**). The **rateable value** (or **net annual rentable value**) is determined by the Board of Inland Revenue for each property by deducting, from the rack *rent that the property would earn, certain allowable costs. Rates are being replaced in the UK by the *community charge (*see also* poll tax).

**rational expectations** Expectations about the economy that are held by rational individuals and that utilize all the available *information without systematic error. First proposed by J F Muth (1961), the theory of rational expectations has caused a revolution in *macroeconomics. Keynesian theory, prior to Muth, did not involve rational expectations, implying that individuals used information inefficiently. In the current theory, however, when combined with the *market clearing assumption, rational expectations imply that *fiscal policy and *monetary policy cannot influence real variables, provided that everyone has equal access to information. One direction taken by *post-Keynesian theory has been to investigate the consequences of the unequal distribution of information and the economic inefficiencies that this creates.

**rationality** The assumption made in neoclassical economics that individuals will compare all possible combinations of goods when making their choices and will always prefer more of a good than less. Since, in practice, rationality could involve a large amount of calculation and information gathering, the assumption has always been considered controversial. Without it, however, most of the results of *consumer theory and *producer theory are not valid and alternative assumptions do not appear to yield useful predictions (*see also* bounded rationality).

**rationalization** A reorganization of a firm, group, or industry to increase its efficiency and profitability. This may include closing some manufacturing units and expanding others (horizontal *integration), merging different stages of the production process (vertical integration), merging support units, closing units that are duplicating effort of others, etc.

**rationing by price** Making use of the price mechanism to ration goods in short supply, i.e. by increasing their price so that only the rich can afford to buy them. The term is usually used pejoratively, i.e. although the supply of Rolls Royce cars is rationed by price in a sense, it is more usual to speak of an essential commodity, such as food, being rationed by price during a famine in a poor country.

**reach** The proportion of a total market that an advertiser hopes to reach at least once within a given period in an advertising campaign.

**reaction** A reversal in a market trend as a result of overselling on a falling market (when some buyers are attracted by the low prices) or overbuying on a rising market (when some buyers are willing to take profits).

**read-only memory (ROM)** A type of computer *main store memory whose contents may be read but not altered and are not lost when the computer is switched off. ROM generally stores the essential programs required by the computer. For example, ROM usually contains a short program to load the *operating system when the machine is switched on, and sometimes the whole operating system itself.

**real account** A ledger *account for some types of property (e.g. land and buildings, plant, investments, stock) to distinguish it from a *nominal account, which would be for revenue or expense items (e.g. sales, motor expenses, discount received, etc.). This distinction is now largely obsolete and both sets of accounts are maintained in the same ledger, usually referred to as the *nominal ledger.

**real effect** *See* real terms.

**real estate** The US name for *real property.

**real investment** Investment in capital equipment, such as a factory, plant and machinery, etc., or valuable social assets, such as a school, a dam, etc., rather than in such paper assets as securities, debentures, etc.

**realizable asset** *See* liquid assets.

**realization account** An *account used to record the disposal of an asset or assets and to determine the profit or loss on the disposal. The principle of realization accounts are that they are debited with the book value of the asset and credited with the sale price of the asset. Any balance therefore represents the profit or loss on disposal.

**realized profit (*or* loss)** A profit (or loss) that has arisen from a completed transaction (usually the sale of goods or services or other assets). In accounting terms, a profit is normally regarded as having been realized when an asset has been legally disposed of and not when the cash is received, since if an asset is sold on credit the asset being disposed of is exchanged for another asset, a debtor. The debt may or may not prove good but that is regarded as a separate transaction.

**real property (realty)** Any property consisting of land or buildings as distinct from *personal property (personalty).

**real terms** A representation of the value of a good or service in terms of money, taking into account fluctuations in the *price level. Economists are usually interested in the relationship between the prices of goods in real terms, i.e. by adjusting prices according to a price index or some other measure of inflation. In neoclassical economics, any adjustment in the relative prices of goods leads to changes in the quantities supplied and demanded – this is referred to as a **real effect**. When assessing a government policy, economists are interested in real effects.

**real-time operation** A computer operation in which time is significant. Either it is essential that the computer coordinates its activities with external events (e.g. in the control of industrial processes) or any delay in response should be minimal (e.g. in such point of sale terminals as airline reservation systems).

**realtor** The US name for an estate agent or land agent.

**realty** *See* real property.

**real value** A monetary value expressed in *real terms.

**real wages** Wages expressed in *real terms, i.e. the quantity of goods and services a money wage will buy at any particlar time. If money wages increase faster than the *price level in an economy the wagearners are better off; otherwise they are not.

**rebate 1.** A discount offered on the price of a good or service, often one that is paid back to the payer, e.g. a tax rebate is a refund to the taxpayer. **2.** A discount allowed on a *bill of exchange that is paid before it matures.

**recall test** A test used in marketing research to ascertain how much consumers can remember about an advertisement. **Spontaneous tests** reveal which advertisements a respondent can remember without guidance or assistance. **Prompted** (*or* **aided) tests** indicate which advertisements respondents can remember from a series they have seen in a campaign.

**receipt** A document acknowledging that a specified payment has been made. For payments made by cheque, the cheque itself functions as the receipt. *See also* dock receipt; mate's receipt.

**received bill** A *bill of lading that has been stamped "Goods received for shipment". This does not imply that they have been loaded but that they have been received for loading (*compare* shipped bill).

**receiver** A person exercising any form of *receivership. In bankruptcy, the *official receiver becomes receiver and manager of the bankrupt's estate (*see* bankruptcy). Where there is a *floating charge over the whole of a company's property and a crystallizing event has occurred, an **administrative receiver** may be appointed to manage the whole of the company's business. He will have wide powers under the Insolvency Act to carry on the business of the company, take possession of its property, commence *liquidation, etc. A receiver appointed in respect of a *fixed charge can deal with the property covered by the charge only, and has no power to manage the company's business.

**receivership** A situation in which a lender holds a mortgage or charge (especially a *floating charge) over a company's property and, in consequence of a default by the company, a receiver is appointed to realize the assets charged in order to repay the debt.

**receiving order** Formerly, an order made during the course of bankruptcy or insolvency. It is now called a bankruptcy order (*see* bankruptcy).

**recession** A slowdown or fall in the rate of growth of *GNP. A severe recession is called a *depression. Economic growth usually follows a cycle from boom to recession and back again (*see* business cycle). Recession is associated with falling levels of *investment, rising *unemployment, and sometimes falling prices. The conditions of the major economies in the 1980s has been called **growth recession**, since despite very high levels of unemployment, economic growth has continued at a reasonable pace. Interventionist economists, whether Keynesian or monetarist, advocate government intervention through fiscal or monetary stimulus during recession, whilst many economists now argue that government intervention is ineffective (*see* new classical macroeconomics).

**reciprocity** A form of negotiation in which one party agrees to make a concession in return for a reciprocal, or broadly equivalent, action by the other. Most international economic negotiations take this form. For example, agreements to lower protectionist barriers over markets are nearly always done on a reciprocal basis. Economists often argue that this is inefficient, since unilateral action is frequently beneficial by itself.

**recommended retail price (RRP)** The price that a manufacturer recommends as the retail price for his product. The manufacturer has no legal power to enforce this recommendation when an agreement exists under the Restrictive Trade Practices Act (1976). For example, the RRP for books is printed on the jacket because according to the *Net Book Agreement, which is so registered, booksellers must not sell below this price. The RRP is also called the **manufacturers' recommended price** (MRP). *See also* resale price maintenance.

**record** A collection of related information treated as a unit within a computer *file or *database file. For example, a company may keep a file of payroll data, with one record per employee. Records are usually divided into *fields, one field for each item of information – name, salary, tax code, and so on.

**recorded delivery** A postal service offered by the UK Post Office that provides a record of posting and delivery of inland letters for an extra fee; advice of delivery is also offered for a further fee. *Compare* registered post.

**recourse** *See* without recourse.

**recourse agreement** An agreement between a hire-purchase company and a retailer, in which the retailer undertakes to repossess the goods if the buyer fails to pay his regular instalments.

**recovery stock** A share that has fallen in price but is believed to have the potential of climbing back to its original level.

**rectification of register 1.** An alteration to any of the land registers (*see* land registration), where it is thought necessary by the registrar or the court. Examples include a mistake that has been made as to the ownership of land or an entry obtained by fraud. Anyone suffering loss because of rectification or anyone who has suffered loss as a result of a refusal to rectify a mistake in the register may be entitled to state compensation. **2.** An alteration of the register of the members of a company by the court if any members have been omitted, a person's name has been wrongly entered, or delay has occurred in removing the name of a former member. The court may award damages as well as rectifying the register.

**recto** The right-hand page, normally odd numbered, of an open book, pamphlet, etc. *Compare* verso.

**redeemable gilts** *See* gilt-edged security.

**redeemable preference share** A *preference share that a company reserves the right to redeem, either out of profits or out of the proceeds of a further issue of shares. It may or may not have a fixed redemption date.

**redemption** The repayment at maturity of a *bond (*see also* gilt-edged security) or other document certifying a loan by the borrower to the lender (or whoever owns the bond at that date). Thus the **redemption date** specifies when repayment takes place and is usually printed on the bond certificate itself.

**redemption date** *See* redemption.

**redemption yield** *See* yield.

**rediscounting** The discounting of a *bill of exchange or *promissory note that has already been discounted by someone else, usually by a *discount house.

**redistribution of income** *See* income redistribution.

**reducing-balance depreciation** *See* depreciation.

**redundancy 1.** The loss of a job by an employee because his job has ceased to exist or because there is no longer work for him. It involves dismissal by the employer, with or without notice, for any reason other than a breach of the contract of employment by the employee, provided that no reasonable alternative employment has been offered by the same employer. In these circumstances a **redundancy payment** must be made by the employer, the amount of which will depend on the employee's age, length of service, and his rate of pay. The employer can claim back part of the redundancy payment from the government. **2.** *See* compensation for loss of office.

**re-exports** Goods that have been imported and are then exported without having undergone any material change while in the exporting country. Countries with a major re-export or *entrepôt trade distinguish re-exports from domestic exports in their balance of payments accounts. Any import duty paid on goods imported for re-export can be reclaimed (*see* drawback).

**referee 1.** A person who will, if asked, give an applicant for a job a **reference**. An approach to a referee can be in writing or by telephone and the referee should be given a brief job description of the job the applicant is trying to secure. **2.** A person appointed by two arbitrators, who cannot agree on the award to be made in a dispute to be settled by arbitration. The procedure for appointing a referee in these circumstances is usually laid down in the terms of arbitration. **3.** A person or organization named on some *bills of exchange as a **referee** (*or* **reference**) **in case of need**. If the bill is dishonoured its holder may take it to the referee for payment.

**reference** *See* referee.

**reference group** A group upon which consumers model their behaviour or their buying habits. A reference group might consist of friends, neighbours, colleagues, or a more remote group of people that is admired or to which the consumer aspires. Advertisers, recognizing the influence of reference groups, often associate a product with an appropriate reference group in order to enhance its appeal.

**refer to drawer** Words written on a cheque that is being dishonoured by a bank, usually because the person who drew it has insufficient funds in his account to cover it and the manager of the bank is unwilling to allow the account to be overdrawn or further overdrawn. Other reasons for referring to the drawer are that the drawer has been made bankrupt, that there is a garnishee order against him, that the drawer has stopped it, or that something in the cheque itself is incorrect (e.g. it is wrongly dated, words and figures don't agree, etc.).

**refinance credit** A credit facility enabling a foreign buyer to obtain credit for a purchase when the exporter does not wish to provide it. The buyer opens a credit at a branch or agent of his bank in the exporting country, the exporter being paid by sight draft on the buyer's credit. The bank in the exporting country accepts a *bill of exchange drawn on the buyer, which is discounted and the proceeds sent to the bank issuing the credit. The buyer only has to pay when the bill on the bank in the exporting country matures.

**refinancing** The process of repaying some or all of the loan capital of a firm by obtaining fresh loans, usually at a lower rate of interest.

**reflation** A policy aimed at expanding the level of output of the economy by government stimulus, either *fiscal or *monetary policy. This could involve increasing the money supply and government expenditure on investment, public works, subsidies, etc., or reducing taxation and interest rates. Usually advocated in times of increasing unemployment, this policy is favoured by Keynesian economists (*see* Keynesianism). Monetarists and new classical macroeconomists have argued that reflation can only lead to *inflation.

**refugee capital** *Hot money belonging to a foreign government, company, or individual that is invested in the country offering the highest interest rate, usually on a short-term basis.

**registered capital** *See* share capital.

**registered company** *See* company.

**registered land** Land the title to which has been registered under the Land Registration Act (1925). *See* land registration; rectification of register.

**registered land certificate** The document that has replaced title deeds for *registered land. It provides proof of ownership of the land and will be in the possession of the land owner unless the land is subject to a mortgage, in which case it will be retained by the *Land Registry.

**registered name** The name in which a UK company is registered. The name, without which a company cannot be incorporated, will be stated in the *memorandum of association. Some names are prohibited by law and will not be registered; these include names already registered and names that in the opinion of the Secretary of State for Trade and Industry are offensive. The name may be changed by special resolution of the company and the Secretary of State may order a company to change a misleading name. The name must be displayed at each place of business, on stationery, and on bills of exchange, etc., or the company and its officers will be liable to a fine.

**registered office** The official address of a UK company, to which all correspondence can be sent. Any change must be notified to the Registrar of Companies within 14 days and published in the *London Gazette*. Statutory registers are kept at the registered office, the address of which must be disclosed on stationery and in the company's annual return.

**registered post** A postal service offered by the UK Post Office in which letters can be registered (a certificate of posting given and a receipt signed on delivery); the fee is related to the compensation given if the letter is lost. *Compare* recorded delivery.

**registered stock** *See* inscribed stock.

**register of charges** **1.** The register maintained by the Registrar of Companies on which certain charges must be registered by companies. A charge is created when a company gives a creditor the right to recover his debt from specific assets. The types of charge that must be registered in this way, and the details that must be given, are set out in the Companies Act (1985). Failure to register the charge within 21 days of its creation renders it void, so that it cannot be enforced against a liquidator or creditor of the company. The underlying debt remains valid, however, but ranks only as an unsecured debt. **2.** A list of charges that a company must maintain at its registered address or principal place of business. Failure to do so may render the directors and company officers liable to a fine. This register must be available for inspection by other persons during normal business hours. **3.** *See* Land Charges Register. **4.** The charges register. *See* land registration.

**register of companies** *See* company; Companies House; Registrar of Companies.

**register of debenture-holders** A list of the holders of *debentures in a UK company. There is no legal requirement for such a register to be kept but if one exists it must be kept at the company's registered office or at a place notified to the Registrar of Companies. It must be available for inspection, to debenture-holders and shareholders free of charge and to the public for a small fee.

**register of directors and secretaries** A register listing the directors and the secretary of a UK company, which must be kept at its *registered office. It must state the full names of the directors and the company secretary, their residential addresses, the nationality of directors, particulars of other directorships held, the occupation of directors, and in the case of a public company, their dates of birth. If the function of director or secretary is performed by another company, the name and registered office of that company must be given. The register must be available for inspection by members of the company free of charge and it may be inspected by the public for a small fee.

**register of members** A list of the *members of a company, which all UK companies must keep at their *registered office or where the register is made up, provided that this address is notified to the Registrar of Companies. It contains the names and addresses of the members, the dates on which they were registered as members and the dates on which any ceased to be members. If the company has a share capital, the register must state the number and class of the shares held by each member and the amount paid for the shares. As legal, rather than beneficial, ownership is registered, it is not always possible to discover from the register who controls the shares. The register must be available for inspection by members free of charge for at least two normal office hours per working day. Others may inspect it on payment of a small fee. The register may be rectified by the court if it is incorrect.

**Registrar of Companies** An official charged with the duty of registering all the companies in the UK. There is one registrar for England and Wales and one for Scotland. The registrar is responsible for carrying out a wide variety of administrative duties connected with registered companies, including maintaining the **register of companies** and the *register of charges, issuing certificates of incorporation, and receiving annual returns.

**registration fee** A small fee charged by a company whose shares are quoted on a stock exchange when it is requested to register the name of a new owner of shares.

**regressive tax** A tax in which the rate of tax decreases as income increases. Indirect taxes fall into this category. Regressive taxes are said to fall more heavily on the poor than on the rich; for example, the poor spend a higher proportion of their incomes on VAT than the rich. A *proportional tax may also be regressive; because of the theory of the marginal utility of money (i.e. the next pound is of more value to the poor than to the rich), it is thought that a flat rate of tax bears unfairly on the poor.

**regulated company** *See* company.

**regulated tenancy** A tenancy on a dwelling with full rent protection under the Rent Act (1977). The maximum rent payable under such a tenancy is that agreed between landlord and tenant or a registered fair rent set by a local Rent Officer on the application of either party, with appeal to the Rent Assessment Committee. Under the Housing Act (1988), new tenancies are known as **assured tenancies**, and rents are fixed by Rent Assessment Committees if the landlord and tenant cannot agree. Previous regulated tenancies remain protected.

**regulation** The imposition by a government of controls over the decisions of individuals or firms. Usually it refers to the control of industries in which there is *monopoly or *oligopoly, in order to prevent firms from exploiting their market power to extract *pure economic profits. Regulation may be seen as an alternative to *nationalization. For example, the types of industries that have formerly been nationalized in the UK have usually been regulated in the USA. The main vehicle for regulation in the UK has been the *Monopolies and Mergers Commission. To an economist, the most important reason for regulation by a government is to encourage *competition. *See also* deregulation.

**reinstatement of the sum insured** The payment of an additional premium to return the sum insured to its full level, after a claim has reduced it. Insurance policies are, in effect, a promise to pay money if a particular event occurs. If a claim is paid, the insurance is reduced by the claim amount (or if a total loss is paid, the policy is exhausted). If the policyholder wishes to return the cover to its full value, a premium representing the amount of cover used by the claim must be paid. In the case of a total loss the whole premium must be paid again.

**reinsurance** The passing of all or part of an insurance risk that has been covered by an insurer to another insurer in return for a premium. The policyholder is usually not aware that reinsurance has been arranged as no mention is made of it on the policy. Reinsurance is a similar process to the bookmaker's practice of laying off bets with other bookmakers when too much money is placed on a particular horse. Very often an insurer will only accept a risk with a high payout if a total loss occurs (e.g. for a jumbo jet) if he knows he can reduce his potential loss by reinsurance. *See also* facultative reinsurance. *Compare* coinsurance.

**reintermediation** *See* disintermediation.

**relaunch** To reintroduce an existing product or brand into the market after changes have been made to it.

**relocation** The removal of an office, factory, or other place of work from one place to another, often as a result of a merger, takeover, etc. Sometimes employees are offered a **relocation allowance**, as an incentive for them to remain with the organization.

**remoteness of damage** Loss or injury that has resulted from unforeseen or unusual circumstances. In the law of negligence, a man is presumed to intend the natural consequences of his acts. If he is negligent he will be liable for all the direct and immediate consequences of his negligence. If, however, the consequences of his acts could not be reasonably foreseen or anticipated, the resulting damage is said to be too remote. Similarly, in the law of contract, damage resulting from breach of contract will be deemed too remote unless it arises directly from the breach complained of. No damages will be awarded for damage the courts deem to be too remote.

**remuneration** **1.** A sum of money paid for a service given. *See* auditor (for **auditors' remuneration**). **2.** A salary.

**rendu** *See* franco.

**renewal notice** An invitation from an insurer to continue an insurance policy that is about to expire by paying the **renewal premium**. The renewal premium is shown on the notice; it may differ from the previous premium, either because insurance rates have changed or because the insured value has changed. Many insurers increase the insured value of certain objects automatically, in line with inflation.

**renouncable documents** Documents that provide evidence of ownership, for a limited period, of unregistered shares. An allotment letter (*see* allotment) sent to shareholders when a new issue is floated is an example. Ownership of the shares can be passed to someone else by *renunciation at this stage.

**rent** **1.** A payment for the use of land, usually under a *lease. The most usual kinds are *ground rent and **rack rent**. Rack rent is paid when no capital payment (premium) has been made for the lease and the rent therefore represents the full value of the land and buildings. The amount of rent may be regulated by statute in certain cases (*see* regulated tenancy). *See also* peppercorn rent. **2. (economic rent)** The extra amount earned by a *factor of production in order to keep that factor from being used elsewhere. The factor may be land, labour, or capital. The use of the term arose because land was originally the factor of production that earned rent, but it gradually widened in 19th-century economic theory to include any factor of production.

**rent control** *See* regulated tenancy.

**rentes** Non-redeemable government bonds issued by several European governments, notably France. The interest, which is paid annually, for ever, is called **rente**.

**rentier** A person who lives on income from *rentes or on receiving rent from land. The meaning is sometimes extended to include anyone who lives on the income derived from his assets rather than a wage or salary.

**renting back** *See* lease-back.

**rent regulation** *See* regulated tenancy.

**renunciation 1.** The surrender to someone else of rights to shares in a rights issue. The person to whom the shares are allotted by allotment letter (*see* allotment) fills in the **renunciation form** (usually attached) in favour of the person to whom he wishes to renounce his rights. **2.** The disposal of a unit-trust holding by completing the renunciation form on the reverse of the certificate and sending it to the trust managers.

**repairing lease** *See* lease.

**repatriation 1.** The return of a person to his country of origin. **2.** The return of capital from a foreign investment to investment in the country from which it originally came.

**replacement cost** The price at which the assets of an organization could be replaced, broadly in their existing state. The significance of the replacement cost is in *current-cost accounting; instead of valuing assets at their historical cost, less depreciation where appropriate, assets are valued at their current cost, which in most cases is taken to be the replacement cost.

**replacement ratio** The percentage of a redundant worker's former income that is replaced by unemployment benefit or other state benefits. Supply-side economists (*see* supply-side economics) have argued that replacement ratios have risen in the postwar period, encouraging workers to scrounge off the dole rather than find new jobs. This assertion, however, is not generally accepted.

**reporting accountants** A firm of accountants who report on the financial information provided in a *prospectus. They may or may not be the company's own auditors. It is usual for reporting accountants to have had previous experience of new issues and the preparation of prospectuses.

**report of the auditor(s)** *See* auditors' report.

**report of the directors** *See* directors' report.

**report program generator (RPG)** A simple business-oriented computer programming language. RPG allows users to generate reports easily from information in *files and simple RPG programs can perform complicated business tasks.

**repositioning** An attempt to make a product appeal to a different segment of a market, especially a product that has not performed as profitably as expected. Repositioning usually involves changing some of the product's features, its packaging, or its price.

**representations** Information given by a person wishing to arrange insurance about the nature of the risk that is to be covered. If it contains a *material fact it must be true to the best knowledge and belief of the person wishing to be insured. *See* utmost good faith.

**repudiation 1.** The refusal of one party to pay a debt or honour a contract. It often refers to the refusal by the government of a country to settle a national debt incurred by a previous government. **2.** *See* breach of contract.

**repurchase 1.** The purchase from an investor of a unit-trust holding by the trust managers. **2.** **(repurchase agreement)** An agreement in which a security is sold and later bought back at an agreed price. The seller is paid in full and makes the agreement to raise ready money without losing his holding. The buyer negotiates a suitable repurchase price to give him a profit equivalent to interest on his money.

**reputed owner** A person who acts as if he owns certain goods, with the consent of the true owner. If that person becomes bankrupt, the goods are divided among his creditors and the true owner may be estopped (*see* estoppel) from claiming them.

**requisitioning** The taking possession of property by a government, usually during a national emergency, with or without the consent of the owner. When the emergency has ended, the property may be **derequisitioned**, i.e. returned to its owner. A requisition differs from a *compulsory purchase in that ownership does not pass in a requisition as it does in a compulsory purchase.

**resale price maintenance (RPM)** An agreement between a manufacturer, wholesaler, or retailer not to sell a specified product below a fixed price, usually one given by the manufacturer. The purpose is to ensure that all parties are allowed to make a reasonable profit without cut-throat competition. Such an agreement, however, is not in the public interest as it stifles competition; RPM contracts are now illegal in the UK unless it can be shown that they are in the public interest. *See* recommended retail price.

**rescission** The right of a party to a contract to have it set aside and to be restored to the position he was in before the contract was made. This is an equitable remedy, available at the discretion of the court. The usual grounds for rescission are mistake, misrepresentation, undue influence, and unconscionable bargains (i.e. those in which the terms are very unfair). No rescission will be allowed where the party seeking it has taken a benefit under the contract (affirmation), or where it is not possible to restore the parties to their former position (restitutio in integrum), or where third parties have already acquired rights under the contract.

**research and development (R & D)** The search for new and improved products and industrial processes by governments and industrial firms. Research may be pure (seeking clarification of scientific principles with no end product in view) or applied (using the techniques of pure research to some specified commercial end); development usually refers to the improvement of a product or process by scientists in conjunction with engineers. R & D is the lifeblood of many industries and, indeed, provides one of the main justifications for free enterprise and the making of profits. Vast sums are spent by industrial conglomerates in seeking new products and the means of producing them cheaply, efficiently, and safely.

**research brief** A document that defines the objectives of a *marketing research study before the research is undertaken.

**reservation price** The minimum price at which a supplier will sell goods or services in the market. The term is most frequently applied to *labour; the **reservation wage** is usually taken to be the level of unemployment benefit received.

**reserve** Part of the *capital of a company, other than the share capital, largely arising from retained profit or from the issue of share capital at more than its nominal value. Reserves are distinguished from *provisions in that for the latter there is a known diminution in value of an asset or a known liability, whereas reserves are surpluses not yet distributed and, in some cases (e.g. share premium account or capital redemption reserve), not distributable. The directors of a company may choose to earmark part of these funds for a special purpose (e.g. a **reserve for**

**obsolescence** of plant). However, reserves should not be seen as specific sums of money put aside for special purposes as they are represented by the general net assets of the company. Reserves are subdivided into *retained earnings (revenue reserves), which are available to be distributed to the shareholders by way of dividends, and *capital reserves, which for various reasons are not distributable as dividends, although they may be converted into permanent share capital by way of a bonus issue.

**reserve capital (uncalled capital)** That part of the issued *capital of a company that is held in reserve and is intended only to be called up if the company is wound up.

**reserve currency** A foreign currency that is held by a government because it has confidence in its stability and intends to use it to settle international debts. The US dollar and the pound sterling fulfilled this role for many years but the Japanese yen and the Deutschmark are now preferred by most governments.

**reserve for obsolescence** *See* reserve.

**reserve price** The lowest price a seller is willing to accept for an article offered for sale by public *auction. If the reserve price is not reached by the bidding, the auctioneer is instructed to withdraw the article from sale.

**reserve tranche** The 25% of its quota to which a member of the *International Monetary Fund has unconditional access, for which it pays no charges, and for which there is no obligation to repay the funds. The reserve tranche corresponds to the 25% of quota that was paid not in the member's domestic currency but in *Special Drawing Rights (SDRs) or currencies of other IMF members. It counts as part of the member's foreign reserves. As before 1978 it was paid in gold, it was known as the **gold tranche**. Like all IMF facilities it is available only to avert balance of payments problems, although for the reserve tranche the IMF has no power either to challenge the member's assessment of need or to impose a corrective policy. Further funds are available through credit tranches but these are subject to certain conditions (*see* conditionality). Use of the reserve tranche was heaviest in the 1970s, when members needed it to gain access to the IMF's Oil Facilities (1974–76) and the USA made a record level of reserve-tranche purchases (1978).

**residual unemployment** The number of people in an economy that remain unemployed, even when there is full employment. They consist largely of people who are so physically or mentally handicapped that they are virtually unemployable; they therefore do not usually feature in unemployment statistics.

**residuary legatee** A person to whom a testator's estate is left, after specific bequests have been satisfied. If a will leaves £3000 to A, £10,000 to B, and the rest of the estate to C, then C is the residuary legatee.

**resolution** A binding decision made by the members of a company. If a *motion is put before the members of a company at a general meeting and the required majority vote in favour of it, the motion is passed and becomes a resolution. A resolution may also be passed by unanimous informal consent of the members. An **ordinary resolution** may be passed by a bare majority of the members. The Companies Act prescribes this type of resolution for certain actions, such as the removal of a director. Normally, no particular length of notice is required for an ordinary resolution to be proposed, beyond the notice needed to call the meeting. However, ordinary resolutions of the company require **special notice** if a director or an auditor is to be removed or if a director who is over the statutory retirement age is to be appointed or permitted to remain in office. In these circumstances 28 days' notice must be given to the company and the company must give 21 days' notice to the members. An **extraordinary resolution** is one for which 14 days' notice is required. The notice should state that it is an extraordinary resolution and for such a resolution 75% of those voting must approve it if it is to be passed. A **special resolution** requires 21 days' notice to the shareholders and a 75% majority to be effective. The type of resolution required to make a particular decision may be prescribed by the Companies Acts or by the company's articles. For example, an extraordinary resolution is required to wind up a company voluntarily, while a special resolution is required to change the company's articles of association.

**respondentia bond** *See* hypothecation.

**restitution** The legal principle that a person who has been unjustly enriched at the expense of another should make restitution, e.g. by returning property or money. This principle is not yet fully developed in English law but has a predominant place in the law of Scotland and the USA.

**restraint of trade** A term in a contract that restricts a person's right to carry on his trade or profession. An example, in a contract covering the sale of a business, would be a clause that attempts to restrict the seller's freedom to set up in competition to the buyer. Such clauses are illegal, unless they can be shown not to be contrary to the public interest.

**restrictive covenant 1.** A clause in a contract that restricts the freedom of one of the parties in some way. Employment contracts, for example, sometimes include a clause in which an employee agrees not to compete with the employer for a specified period after leaving the employment. Such clauses may not be enforceable in law. *See* restraint of trade; restrictive trade practices. **2.** A clause in a contract affecting the use of land. *See* covenant.

**restrictive endorsement** An endorsement on a *bill of exchange that restricts the freedom of the endorsee to negotiate it.

**restrictive trade practices** Agreements between traders that are not considered to be in the public interest. Under the Restrictive Trade Practices Acts (1956; 1968; 1976) and the Fair Trading Act (1973) any agreement between two or more suppliers of goods or services restricting prices, conditions of sale, quantities offered, processes, areas and persons to be supplied, etc., must be registered with the Director General of Fair Trading and, where appropriate, investigated by the **Restrictive Practices Court**. Such agreements are presumed to be contrary to the public interest unless the parties are able to prove to the court that the agreement is beneficial. An example of an agreement permitted by the court is the *Net Book Agreement. *See also* Office of Fair Trading.

In 1976 the service industries were included in the Restrictive Trade Practices Act. This led to an examination by the Office of Fair Trading of the *London Stock Exchange and its subsequent (1986) change or rules (*see* Big Bang).

**retailer** A *distributor that sells goods or services to consumers (*compare* wholesaler). There are, broadly, three categories of retailer: *multiple shops, *cooperative retailers, and *independent retailers.

**Retail Price Index (RPI)** An index of the prices of goods and services in retail shops purchased by average households, expressed in percentage terms relative to a base year, which is taken as 100. For example, if 1975 is taken as the base year for the UK (i.e. average prices in 1975 = 100), then in 1948 the RPI stood at 23.1, and in 1988 at 286.3. The RPI is compiled by the Department of Employment on a monthly basis and includes the prices of some 130 000 different commodities. The RPI is one of the standard measures of the rate of *inflation. In the USA and some other countries the RPI is known as the **consumer price index**. *See also* gross national product deflator; producer price index; tax and price index.

**retained earnings (retentions)** Profits earned to date but not yet distributed to shareholders by way of dividends. Such earnings are also called revenue reserves (*see* reserve) and form an important part of the *capital of most companies.

**retained profits (ploughed-back profits)** The part of the annual profits of an organization that are not distributed to its owners but are invested in the assets of the organization. Retained profits are normally the means that enable organizations to grow; it is usually easier to retain profits than to raise new capital.

**retention 1.** An amount held back from payment by the customer of a contractor until an agreed period has elapsed as a safeguard against any faults being found in the work done. **2.** *See* retained earnings.

**retirement pension** *See* pension.

**retirement relief** A relief from capital-gains tax given to persons disposing of business assets at or over 60 years of age, or earlier if retiring due to ill-health.

**retiring a bill** The act of withdrawing a *bill of exchange from circulation when it has been paid on or before its due date.

**return** The income from an investment, frequently expressed as a percentage of its cost. *See also* return on capital.

**returned cheque** A cheque on which payment has been refused and which has been returned to the bank on which it was drawn. If the reason is lack of funds, the bank will mark it *refer to drawer; if the bank wishes to give the customer an opportunity to pay in sufficient funds to cover it, they will also mark it "please represent". The collecting bank will then present it for payment again, after letting his customer know that the cheque has been dishonoured.

**return on capital** The profits of an organization expressed as a percentage of the *capital employed. This is an important indicator of the efficiency with which the assets of the organization are used; it provides a useful comparison of companies in the same industry or, for the investor, a comparison between various industrial sectors. However, it is important to know what is meant by capital. Frequently, return on capital is calculated by comparing the profits with the *book value of the net assets. This tends to undervalue the assets, so that a capital figure based on market value would be more helpful, although not as readily obtainable.

**returns inwards** Goods returned to an organization by customers, usually because they are unsatisfactory.

**returns outwards** Goods returned by an organization to its suppliers, usually because they are unsatisfactory.

**returns to scale** The relationship between the quantities of input required to produce different levels of output. At very low levels of production, doubling the amount of input will usually more than double the amount of output; this is called **increasing returns to scale**. At intermediate levels of production it is usually assumed that double the amounts of input will simply double the amount of output, called **constant returns to scale**. At very high levels of output it is usually thought that doubling input leads to less than a doubling of output (for example, because it becomes increasingly difficult to manage large-scale organizations). This last stage is called **decreasing** (*or* **diminishing**) **returns to scale**. Assumptions concerning returns to scale are often important in microeconomic analysis.

**revalorization of currency** The replacement of one currency unit by another. A government often takes this step if a nation's currency has been devalued frequently or by a large amount. *Compare* revaluation of currency.

**revaluation of assets** A revaluation of the assets of a company, either because they have increased in value since they were acquired or because *inflation has made the balance-sheet values unrealistic. The Companies Act (1985) makes it obligatory for the directors of a company to state in the directors' report if they believe the value of land differs materially from the value in the balance sheet. The Companies Act (1980) lays down the procedures to adopt when fixed assets are revalued.

**revaluation of currency** An increase in the value of a currency in terms of gold or other currencies. It is usually made by a government that has a persistent balance of payments surplus. It has the effect of making imports cheaper but exports become dearer and therefore less competitive in a country; revaluation is therefore unpopular with governments. *Compare* devaluation; revalorization of currency.

**revenue account 1.** An *account recording the income from trading operations or the expenses incurred in these operations. **2.** A budgeted amount that can be spent for day-to-day operational expenses, especially in public-sector budgeting. *Compare* capital account.

**revenue reserve** *See* reserve; retained earnings.

**reverse takeover 1.** The purchasing of a public company by a private company. This may be the cheapest way that a private company can obtain a listing on a stock exchange, as it avoids the expenses of a *flotation and it may be that the assets of the public company can be purchased at a discount. However, on the *London Stock Exchange there are regulations stipulating that the nature of the target company's business must be compatible with that of the private company. Usually the name of the public company is changed to that of the private company, who takes over the listing. **2.** The buying of a larger company by a smaller company.

**reverse yield gap** *See* yield gap.

**reversible lay days** *See* lay days.

**reversionary annuity** *See* contingent annuity.

**reversionary bonus** A sum added to the amount payable on death or maturity of a *with-profits life-assurance policy. The bonus is added if the life-assurance company has a surplus or has a profit on the investment of its life funds. Once a reversionary bonus has been made it cannot be withdrawn if the policy runs to maturity or to the death of the insured. However, if the policy is cashed, the bonus is usually reduced by an amount that depends on the length of time the policy has to run.

**revocable letter of credit** *See* letter of credit.

**revolving credit** A bank credit that is automatically renewed until cancelled. It may provide for an unlimited amount to be drawn at any one time, a limited amount at any one time but an unlimited total amount, or a limited total that is automatically renewed as debts are paid off.

**revolving underwriting facility** *See* note issuance facility.

**Ricardian equivalence theorem** The theory, formulated by David Ricardo (1772–1823), that a tax cut financed by government borrowing will have no effect on an individual's wealth and therefore will not increase either consumption or *investment. The reason is that additional government borrowing will need to be repaid in the future through taxation, thus offsetting the original tax cut. This theory has recently been popularized by the US economist Robert Barro, providing a basis for the belief that government *fiscal policy cannot alter such real variables as employment or output. The version of the theory popularized by Barro is sometimes known as **debt neutrality**.

**rigging a market** An attempt to make a profit on a market, usually a security or commodity market, by overriding the normal market forces. This often involves taking a *long position or a *short position in the market that is sufficiently substantial to influence price levels, and then supporting or depressing the market by further purchases or sales.

**right of resale** The right that the seller in a contract of sale has to resell the goods if the buyer does not pay the price as agreed. If the goods are perishable or the seller gives notice to the buyer of his intention to resell and the buyer still does not pay within a reasonable time, the seller may resell them and recover from the first buyer damages for any loss.

**rights issue** A method by which quoted companies on a stock exchange raise new capital, in exchange for new shares. The name arises from the principle of *pre-emption rights, according to which existing shareholders must be offered the new shares in proportion to their holding of old shares. For example in a 1 for 4 rights issue, shareholders would be asked to buy one new share for every four they already hold. As rights are usually issued at a discount to the market price of existing shares, those not wishing to take up their rights can sell them in the market (*see* rights letter; renunciation). *Compare* bought deal; vendor placing; scrip issue.

**rights letter** A document sent to an existing shareholder of a company offering him shares in a *rights issue on advantageous terms. If the recipient does not wish to take advantage of the offer himself he can sell the letter and the attendant rights on the stock exchange (*see* renunciation).

**ring 1.** A number of manufacturers, dealers, or traders who agree among themselves to control the price or conditions of sale of a product for their own benefit. Such agreements are illegal in most countries, unless they can be shown to be in the public interest. *See* restrictive trade practices. **2.** An association of dealers in an auction sale, especially a sale of antiques or paintings, who agree not to bid against each other but to allow one of their number to buy an article being auctioned at an artificially low price, on the understanding that he will auction it again exclusively to members of the ring. The difference between the purchase price and the final price paid is shared among the members of the ring. This is an illegal practice.

**ring trading** *See* callover.

**riot and civil commotion** A risk not usually covered in householders' insurance policies or commercial policies, unless a special clause has been added to cover loss or damage due to riot or civil commotion.

**risk** The possibility of suffering some form of loss or damage (*see also* peril). If it is one that can be described sufficiently accurately for a

calculation to be made of the probability of it happening, on the basis of past records, it is called an **insurable risk**. Fire, theft, accident, etc., are all insurable risks because underwriters can assess the probability of having to pay out a claim and can therefore calculate a reasonable *premium. If the risk is met so infrequently that no way of calculating the probability of the event exists, no underwriter will insure against it and it is therefore an **uninsurable risk** (*see also* actuary).
In the sale of goods, the risk of loss or damage to the goods passes with title, i.e. when the title passes from seller to buyer, the insurance risk also passes. In marine insurance, the seller must give the buyer sufficient information to enable him to cover the insurance; if he does not do so the risk remains with the seller.

**risk capital (venture capital)** Capital invested in a project in which there is a substantial element of risk, especially money invested in a new venture or an expanding business in exchange for shares in the business. Risk capital is normally invested in the equity of the company; it is not a loan.

**road haulage** The transport of goods by road vehicles. With the advent of motorways in Europe, substantial quantities of goods that formerly were transported by rail are now sent by road in vehicles carrying up to 38 tonnes.

**rolling launch** The process of gradually introducing a new product into the market.

**roll-on roll-off ferry (RORO)** A ferry that loads motor vehicles by allowing them to drive on and drive off. Such ferries are widely used for transporting road haulage vehicles across the English Channel.

**roll-over relief** A relief from capital-gains tax allowing tax on the disposal of an asset to be postponed, by deducting the capital gain from the base cost of another asset acquired by the person making the original disposal. The effect is that the gain on the disposal of the second asset will be increased because the consideration for the purchase of the asset is treated as artificially reduced by the gain on the earlier asset. This type of relief is available in a variety of transactions, in which the person disposing of the asset will not be in funds to pay the tax, e.g. the replacement of business assets, gifts of assets from one person to another, and the formation of companies in exchange for shares. *See also* employee share-ownership plan.

**ROM** Abbreviation for *read-only memory.

**RORO** Abbreviation for roll-on roll-off. *See* roll-on roll-off ferry.

**rotation of directors** Under the articles of association of most UK companies, one-third of the directors are obliged to retire each year (normally at the *annual general meeting), so that each director retires by rotation every three years. Retiring directors may be re-elected.

**round lot** A round number of shares or a round amount of stock for which market makers will sometimes offer better prices than for **odd lots**.

**roundtripping** A transaction that enables a company to borrow money from one source and lend it at a profit to another, by taking advantage of a short-term rise in interest rates.

**Royal Institution of Chartered Surveyors** An institution, founded in 1868 and given a Royal Charter in 1881, that aims to provide education and training for surveyors and to define and control their professional conduct. Qualification as a Chartered Surveyor is part academic (passing the Institution's own professional examinations or a recognized degree or diploma) and part practical (assessed professional experience over a two-year period). A candidate is then able to apply for election as an Associate, using the initials ARICS after his name. After 12 years' further experience an Associate may apply for a Fellowship (FRICS).

**Royal Mint** The organization in the UK that has had the sole right to manufacture English coins since the 16th century. Controlled by the Chancellor of the Exchequer, it was formerly situated in the City of London but moved to Llantrisant in Wales in 1968. It also makes bank notes, UK medals, and some foreign coins.

**royalty** A payment made for the right to use the property of another person for gain. This may be an *intellectual property, such as a book (*see* copyright) or an invention (*see* patent). It may also be paid to a landowner, who has granted mineral rights to someone else, on sales of minerals that have been extracted from his land. A royalty is regarded as a wasting asset as copyrights, patents, and mines have limited lives.

**RPG** Abbreviation for *report program generator.

**RPI** Abbreviation for *Retail Price Index.

**RPM** Abbreviation for *resale price maintenance.

**RRP** Abbreviation for *recommended retail price.

**RUF** Abbreviation for revolving underwriting facility. *See* note issuance facility.

**rule against perpetuities** *See* Perpetuities and Accumulations Act (1964).

**rummage** A search of a ship or aircraft by customs officers looking for contraband, illegal drugs, etc.

**running broker** A bill broker who does not himself discount *bills of exchange but acts between bill owners and discount houses or banks on a commission basis.

**running days** Consecutive days including Saturdays, Sundays, and Bank Holidays. In chartering a vessel the period of the charter is given in running days, rather than working days.

**running-down clause** A clause in a *hull insurance policy covering a shipowner against liability for compensation as a result of a collision with another vessel.

**running yield** *See* yield.

**run of paper** Denoting an advertisement in a paper, magazine, etc., in which the publisher decides where the advertisement shall be placed. This is offered at a lower rate than an advertisement set in a specified position in the publication.

# S

**SA** **1.** Abbreviation for *société anonyme*. It appears after the name of a French, Belgian, or Luxembourg company, being equivalent to the British abbreviation plc (i.e. denoting a public limited company). **2.** Abbreviation for *sociedad anónima.* It appears after the name of a Spanish public company. **3.** Abbreviation for *sociedade anónima.* It appears after the name of a Portuguese public company. *Compare* Sarl.

**sacrifice** Loss of welfare as a result of paying a tax. One of the principles of taxation is that a tax should be formulated so that it involves **equality of sacrifice** on the part of the taxpayers. It is assumed that taxpayers with equal incomes will sacrifice equal amounts of welfare by paying identical taxes.

**s.a.e.** Abbreviation for stamped addressed envelope.

**safe custody** A service offered by most UK commercial banks, in which the bank holds valuable items belonging to its customers in its strong room. These items are usually documents, such as house deeds and bearer bonds, but they may also include jewellery, etc. The bank is a bailee for these items and its liability will depend on whether or not it has charged the customer for the service and the terms of the customer's own insurance (in the case of jewellery, etc.). In a **safe deposit** the customer hires a small lockable box from the bank or safe-deposit company, to which he has a key and to which he has access during normal banking hours.

**salary** A regular, usually monthly, payment to an employee. It is the remuneration paid to executive, managerial, and administrative employees, usually those who work in an office. Unlike a wage, it is unlikely to be related to hours worked or to the quantity of goods produced and it is usually specified in annual rather than weekly terms (e.g. a manager may earn £25,000 per annum, while a plumber may earn £480 per week).

**sale and lease-back** *See* lease-back.

**sale as seen** A sale made on the basis that the buyer has inspected the goods and is buying them as a result of this inspection and not on any guarantee of quality or condition by the seller. The seller does not guarantee or imply that the goods are suitable for any particular purpose. *Compare* sale by description; sale by sample.

**sale by description** A sale made on the basis that the quality of the goods sold will correspond to the description given of them in the contract of sale. *Compare* sale as seen; sale by sample.

**sale by instalments** *See* hire purchase.

**sale by sample** A sale made on the basis of *sample, i.e. that the quality and condition of the bulk of the goods will be at least as good as that of the selling sample. In contracts in which goods are sold by sample, the contract may stipulate how and by whom the bulk should be sampled and what tests should be used to compare it to the selling sample. The contract will also usually lay down the procedure to be adopted if the bulk sample is not as good as the selling sample. *Compare* sale as seen; sale by description.

**sale or return** Terms of trade in which the seller agrees to take back from the buyer any goods that he has failed to sell, usually in a specified period. Some retail shops buy certain of their goods on sale or return.

**sales day book** The *book of prime entry in which an organization records the invoices issued to its customers for goods or services supplied in the course of its trade. Postings are made from this book to the personal accounts of the customers, while the totals of the invoices are posted to the sales account in the *nominal ledger.

**sales forecast** An estimate of the likely sales for a product, given certain criteria and assumptions.

**sales ledger** The *ledger that records the personal accounts of an organization's customers. The total of the balances in this ledger represents the organization's trade debtors.

**sales management** The organization, direction, recruitment, training, and motivation of the sales force within a planned marketing strategy. In very large markets, such as the USA, there might be a hierarchy of sales managers and district managers, reporting to regional managers, who in turn report to a national sales manager.

**sales promotion** An activity designed to boost the sales of a product or service. It may include an advertising campaign, increased PR activity, a free-sample campaign, offering free gifts or trading stamps, arranging demonstrations or exhibitions, setting up competitions with attractive prizes, temporary price reductions, door-to-door calling, telephone selling, personal letters, etc.

**sales quota** A target set by *sales management for a specific area or *sales representative. Since the earnings of sales representatives are often directly related to the sales achieved, the setting of realistic quotas is imperative.

**sales representative** A person who sells the goods and services of an organization, either as an employee of that organization or as part of a contractual sales force. Additional responsibilities might include generating new sales leads (*see* cold calling), publicizing changes in products or prices, gathering information about competitors' activity, and *merchandising products. Sales representatives often receive a commission on what they sell.

**sales response function** The relationship between the likely sales of a particular product during a specified period and the level of marketing support it receives. The more advertising, sales promotion, or public-relations support for a product, the higher the sales are likely to be.

**sales tax** A tax based on the selling price of goods. Such taxes are not now generally favoured, since they have a cascade effect, i.e. if goods are sold on from one trader to another the amount of sales tax borne by the ultimate buyer becomes too great. *VAT was largely designed to meet this objection.

**sales territory** The unit of organization of a sales force within a company. Each *sales representative is usually assigned a geographical area, market segment, or product group in which to develop sales, and accordingly accepts the credit for the success (or otherwise) of the product within that area.

**salvage 1.** Goods, property, etc., saved from a shipwreck or from a fire. If a cargo is treated as a total loss for insurance purposes, there still may be salvagable items that have a **salvage value**; these may be sold by the insurers or allowed for in the settlement of their claim. **2.** A reward paid to persons who help to save a ship, goods, etc., voluntarily and at some danger to themselves. **Salvage money** cannot be paid to anyone responsible for the care of the ship. The reward is paid by the owners of the ship or the cargo and the **salvors** have a lien on the property rescued for the salvage money.

**sample 1.** A small quantity of a commodity, etc., selected to represent the bulk of a quantity of goods. *See* sale by sample. **2.** A small quantity of a product, given to potential buyers to enable them to test its suitability for their purposes. **3.** A small group of items selected from a larger group to represent the characteristics of the larger group. Samples are often used in *marketing research because it is not feasible to interview every member of a particular market; however, conclusions about a market drawn from a sample always contain a sampling error and must be used with caution. The larger the sample, in general, the more accurate will be the conclusions drawn from it. In **quota sampling** the composition of the sample reflects the known structure of the market. Thus, if it is known that 60% of purchasers of household DIY products are men, any sample would reflect this. An alternative sampling procedure is **random sampling**, which ensures that everyone in a particular market has an equal chance of selection (*see* random sample). Although more accurate than quota sampling, it is also more expensive.

**Samurai bond** A *bond issued in Japan by a foreign institution. It is denominated in yen and can be bought by non-residents of Japan.

**sanctions** *See* economic sanctions.

**sandbag** A stalling tactic used by an unwilling target company in a takeover bid. The management of the target company agrees to have talks with the unwelcome bidder, which it protracts for as long as possible in the hope that a *white knight will appear and make a more acceptable bid.

**Sandilands Committee** A committee chaired by Sir Francis Sandilands, set up in 1975 by the UK government to consider the most appropriate way to account for the effects of inflation in the published accounts of companies. It recommended *current-cost accounting in preference to the *current-purchasing-power accounting favoured by the accountancy bodies.

**sandwich course** A course of study in the UK for a diploma or degree in which periods of academic study in an institution of higher education are sandwiched between periods of supervised work experience in a factory or industrial organization. The course usually lasts for five years, with six-monthly periods in both college and industry.

**sans recours** *See* without recourse.

**sans serif** A form of typeface in which there are no serifs (the tails attached to the tops and bottoms of letters).

**Sarl. 1.** Abbreviation for *società a responsabilità*. It appears after the name of an Italian company, being equivalent to the British abbreviation Ltd (i.e. denoting a private limited liability company). **2.** Abbreviation for *société à responsibilité limitée*. It appears after the name of a French private limited company. *Compare* SA.

**satisficing behaviour** The behavioural assumption in economics that firms, unable to maximize profits because of *bounded rationality, pursue such goals as satisfactory profits and satisfactory growth. Although this assumption is probably more realistic than *profit maximization, it has not yet produced many useful predictions about the behaviour of firms or markets. *See also* managerial theory of the firm.

**save-as-you-earn (SAYE)** A method of making regular savings (not necessarily linked to earnings), which carries certain tax privileges. This method has been used to encourage tax-free savings in building societies or National Savings and also to encourage employees to acquire shares in their own organizations.

**savings** The residue remaining from income after expenditure on *consumption (*see also* savings ratio). Savings are an important concept in *macroeconomics as the level of savings in an economy determines the resources available for *investment. If, for example, firms plan to invest more than households save, in an *open economy resources will have to be borrowed from overseas. Governments may also force individuals in the economy to save by means of taxation, which is subsequently used for investment.

**savings account** A bank account or a building-society account into which personal savings are paid. Interest paid on a savings account is usually higher than that paid on a deposit account and withdrawals are usually restricted. Banks and building societies now offer a wide range of accounts with interest depending on the sums invested and the withdrawal terms.

**savings and loan association** The US equivalent of a UK building society. It usually offers loans with a fixed rate of interest and has greater investment flexibility than a UK building society.

**savings certificate** *See* National Savings.

**savings ratio** The ratio of *savings by individuals and households to *disposable income. Savings are estimated in the *national income accounts by deducting consumers' expenditure (*see* consumption) from disposable income. Variations in the savings ratio reflect the changing preferences of individuals between present and future consumption. Countries, such as Japan, with very high savings ratios have tended to experience faster growth in GDP than countries, such as the USA, with low savings ratios.

**SAYE** Abbreviation for *save-as-you-earn.

**Say's law** The law, attributed to the French economist J-B Say (1767–1832), that supply creates its own demand. This suggests that there is never any barrier to achieving a perfectly competitive Pareto-optimal general equilibrium (*see* Pareto optimality). The implication that in these circumstances *involuntary unemployment is impossible was denied by J M Keynes. Say's law is generally considered too strong, since *Walras' law is sufficient to demonstrate the possibility of a full employment equilibrium.

**SBU** Abbreviation for *strategic business unit.

**scale effect** *See* economies of scale.

**scarcity** In economic theory, the condition that exists if more of a good or service is demanded than can be produced. As human needs are virtually unlimited and resources are finite, most goods and services are, in the economic sense, scarce. In a free economy, allocation of scarce resources is controlled by the price mechanism; in a controlled economy, central government has to decide how resources are to be allocated.

**schedule 1.** The part of legislation that is placed at the end of an Act of Parliament and contains subsidiary matter to the main sections of the act. **2.** One of several schedules of income tax forming part of the original income-tax legislation, now used to classify various sources of income for tax purposes. Some of the schedules are further subdivided into cases. The broad classification is: Schedule A, rents from property in the UK; Schedule B, income from commercial woodlands; Schedule C, interest paid by public bodies; Schedule D, Case 1, profits from trade; Case II, profits from professions or vocations; Case III, interest not otherwise taxed; Case IV, income from securities outside the UK; Case V, income from possessions outside the UK; Case VI, other annual profits and gains; Schedule E, Cases I, II, and III, emoluments of offices or employments (the cases depending on the residential status of the taxpayer); Schedule F, dividends paid by UK companies. **3.** Working papers submitted with tax returns or tax computations. **4.** Any scale of rates.

**scheme of arrangement** *See* arrangement.

**scorched earth policy** An extreme form of *poison pill in which a company that believes it is to be the target of a *takeover bid makes its balance sheet or profitability less attractive than it really is by a reversible manouevre, such as borrowing money at an exorbitant rate of interest.

**Scottish Development Department** A department of the UK government **Scottish Office** that, with the **Industry Department for Scotland**, another department of the same office, is responsible for the development and expansion of industry in Scotland. Both departments have offices in London and Edinburgh.

**SCOUT** Abbreviation for Shared Currency Option Under Tender. A currency *option especially designed for companies who are tendering for the same overseas job in a foreign currency. To save each of them having to hedge the currency risk separately, SCOUT enables them to share the cost of a single option.

**scrip** The certificates that demonstrate ownership of *stocks, *shares, and *bonds (capital raised by sub*scrip*tion), especially the certificates relating to a *scrip issue.

**scrip issue (bonus issue; capitalization issue; free issue)** The issue of new share certificates to existing shareholders to reflect the accumulation of profits in the reserves of a company's balance sheet. The shareholders do not pay for the new shares and appear to be no better off. However, in a 1 for 3 scrip issue, say, the shareholders receive 1 new share for every 3 old shares they own. This automatically reduces the price of the shares by 25%, catering to the preference of shareholders to hold cheaper shares rather than expensive ones; it also encourages them to hope that the price will gradually climb to its former value, which will, of course, make them 25% better off. In the USA this is known as a **stock split**.

**scroll** To move the text shown on a computer screen vertically upwards. Lines appearing at the bottom edge of the screen move up the screen and disappear off the top – as if the user was reading a scroll.

**SDR** Abbreviation for *Special Drawing Rights.

**SEAF** Abbreviation for Stock Exchange Automatic Exchange Facility. This computerized system is used on the *London Stock Exchange to enable a broker to execute a transaction in a security through a SEAF terminal, which automatically completes the bargain at the best price with a *market maker, whose position is automatically adjusted. The price of the transaction is then automatically recorded on a trading report and also passes into the settlement system. The system has greatly reduced the administrative burden on brokers and market makers but has been criticized for eliminating the personal element between brokers and market makers on the floor of the exchange.

**SEAQ** Abbreviation for Stock Exchange Automated Quotations System. This computerized system is used on the *London Stock Exchange to record the prices at which transactions in securities have been struck, thus establishing the market prices for these securities; these prices are made available to brokers through *TOPIC. When a bargain is concluded, the details must be notified to the central system within certain set periods during the day.

**search and replace** A wordprocessing operation in which each occurrence of a particular word in a document, or a section of a document, is located and the word is replaced by a different word. *See also* global search.

**seasonal rate** A rate of charging that varies according to the time of the year. This is quite common in *advertising, for example the *Radio Times* charges a higher page rate for its Christmas issue than for any other issue. **Seasonal discounts** are also common, especially in advertising, when sellers wish to encourage business in a slack period.

**seasonal unemployment** *Unemployment that occurs as a result of the seasonal nature of some jobs. The building trade, for example, employs more people in the summer months than the winter months. Much of the seaside holiday trade is seasonal and a considerable amount of agricultural work is seasonal. It is for this reason that UK unemployment figures are usually **seasonally adjusted**, i.e.

adjusted to flatten the peaks and troughs of unemployment in certain seasons.

**SEC** Abbreviation for *Securities and Exchange Commission.

**secondary bank 1.** A name sometimes given to *finance houses. **2.** Any organization that offers some banking services, such as making loans, offering secondary mortgages, etc., but that does not offer the usual commercial-bank services of cheque accounts, etc.

**secondary data** Marketing research information available to researchers that they have not gathered themselves. Examples include data provided by a government department or extracted from company records.

**secondary market** A market in which existing securities are traded, as opposed to a **primary market**, in which securities are sold for the first time. In most cases a *stock exchange largely fulfils the role of a secondary market, with the flotation of *new issues representing only a small proportion of its total business. However, it is the existence of a flourishing secondary market, providing *liquidity and the spreading of *risks, that creates the conditions for a healthy primary market.

**secondary picketing** *See* picketing.

**secondary production** *See* production.

**secondary storage** *See* backing store.

**second mortgage** A *mortgage taken out on a property that is already mortgaged. Second and subsequent mortgages can often be used to raise money from a *finance house (but not a commercial bank or building society) if the value of the property has increased considerably since the first mortgage was taken out. As the deeds of the property are usually held by the first mortgagee, the second mortgagee undertakes a greater risk and therefore usually demands a higher rate of interest. The second mortgagee usually registers the second mortgage to protect himself against subsequent mortgages. In the UK, the Land Registry will issue a certificate of second charge in the case of registered land.

**second of exchange** *See* bills in a set.

**secret reserve** *See* hidden reserve.

**secured 1.** Denoting a loan (**secured loan**) in which the lender has an asset to which he can have recourse if the borrower defaults on his loan repayments. **2.** Denoting a creditor (**secured creditor**) who has a charge on the property of the debtor. *See* fixed charge; floating charge.

**secured debenture** A *debenture secured by a charge over the property of a company, such as mortgage debentures (secured on land belonging to the company). Usually a trust deed sets out the powers of the debenture holders to enforce their security in the event of the company defaulting in payment of the principal or the interest. It is usual to appoint a *receiver to realize the security.

**secured liability** A debt against which the borrower has provided sufficient assets as security to safeguard the lender in case of non-repayment.

**Securities and Exchange Commission (SEC)** A US government agency that closely monitors the activities of stockbrokers and traders in securities. It also monitors takeovers in the USA. If a person or organization acquires 5% or more of the equity of another company it must notify the SEC of the purchase within 10 days. *See also* creeping takeover.

**Securities and Investment Board (SIB)** A regulatory body set up by the Financial Services Act (1986) to oversee London's financial markets (e.g. the Stock Exchange, life assurance, unit trusts). Each market has its own *Self-Regulatory Organization (SRO), which reports to the SIB. The prime function of the SIB is to protect investors from fraud and to ensure that the rules of conduct established by the government and the SROs are followed. However, as the structure of City institutions and their regulation is not fixed for all time, the role of the SIB has to be capable of adapting to changing practices. Moreover, some City activities are outside its control; for example, takeovers remain under the supervision of the Takeover Panel (*see* City Code on Takeovers and Mergers). Members of the SIB are appointed jointly by the Secretary of State for Trade and Industry and the Governor of the Bank of England from leading City institutions; while this understandably leads to suggestions of partisanship, it is doubtful whether outsiders would have sufficient understanding of City practices to be effective as regulators. The SIB is authorized to grant recognition to investment institutions and is financed by the fees paid to achieve this recognition.

**Securities Association Ltd** *See* London Stock Exchange; Self-Regulatory Organization.

**securitization** The process that enables borrowing and lending by banks to be replaced by the issue of such securities as *eurobonds. A bank borrows money from savers (investors) and lends to borrowers, charging fees to both for its service as well as making a turn on the interest payments. If a borrower can borrow directly from an investor, by issuing them with a bond (or equity), the costs to both borrower and lender can be reduced. Securitization has occurred increasingly in the 1980s as technology has improved and investors have become more sophisticated.

**security** **1.** An asset or assets to which a lender can have recourse if the borrower defaults on his loan repayments. In the case of loans by banks and other moneylenders the security is sometimes referred to as *collateral. **2.** A financial asset, including shares, government stocks, debentures, bonds, unit trusts, and rights to money lent or deposited. It does not, however, include insurance policies. *See also* bearer security; dated security; fixed-interest security; gilt-edged security; listed security; marketable security.

**seed capital** The small amount of initial capital required to fund the research and development necessary before a new company is set up. The seed capital should enable a persuasive and accurate *business plan to be drawn up.

**self-employed taxpayers** Persons who are not employees and who trade on their own account. They are taxed on the profits of their trades rather than by *PAYE and their national-insurance contributions differ from those of employees.

**self-liquidating** **1.** Denoting an asset that earns back its original cost out of income over a fixed period. **2.** Denoting a loan in which the money is used to finance a project that will provide sufficient yield to repay the loan and its interest and leave a profit. **3.** Denoting a *sales-promotion offer that pays for itself. For example, if a seller of tea bags offers a free tea mug in exchange for a specified number of vouchers, each taken from a box of tea bags, the seller expects the extra number of tea-bag boxes sold during the promotion to pay for the costs of buying and despatching the mugs.

**Self-Regulatory Organization (SRO)** One of several organizations set up in the UK under the Financial Services Act (1986) to regulate the activities of investment businesses and to draw up and enforce specific codes of conduct. Five SROs are recognized by the *Securities and Investment Board, to whom they report. They are: **The Securities Association Ltd (TSA)**, which regulates the *London Stock Exchange; the **Association of Futures Brokers and Dealers (AFBD)**, which regulates the *London International Financial Futures Exchange (LIFFE) and the commodity exchanges; the **Financial Intermediaries, Managers and Brokers Regulatory Association Ltd (FIMBRA)**, which regulates organizations marketing and managing securities, unit trusts, and unit-linked life assurance policies; the **Life Assurance and Unit Trust Regulatory Organization (LAUTRO)**, which regulates institutions offering life assurance and unit trusts as principals; and the **Investment Management Regulatory Organization (IMRO)**, which regulates any institution that offers investment management.

**self-service** A method of retailing goods in which the customer serves himself and pays at a cash point before leaving. Widely used in food supermarkets, petrol stations, and cafés, self-service greatly reduces labour costs but offers increased opportunities for shoplifters.

**self-sufficiency** The economic condition of a country that can survive on its own, independently of the rest of the world. The USA and USSR are probably capable of self-sufficiency as they both have the necessary resources; however, it is not always economically sound to use home products if they can be purchased more cheaply elsewhere, especially if that country wishes to build up an export trade.

**self-tender** A *tender offer in which a company approaches its shareholders in order to buy back some or all of its shares. There are two circumstances in which this operation can be of use. One is in the case of a hostile bid: the directors may wish to buy back shares in their company in order to reduce the chances of the bidder being able to buy a controlling interest in the company. The other circumstance is that the board may wish to show increased earnings per share; if they are unable to increase their profits it may be appropriate for them to reduce the number of shares in the company.

**sellers' market** A market in which the demand exceeds the supply, so that sellers can increase prices. At some point, however, buyers will cease to follow the price rises and the sellers will be forced to drop prices in order to make sales, i.e. the supply will have exceeded the demand. *See* buyers' market.

**sellers over** A market in commodities, securities, etc., in which buyers have been satisfied but some sellers remain. This is clearly a weak market, with a tendency for prices to fall. *Compare* buyers over.

**selling costs** The costs incurred in selling a product or service. These include advertising, display, after-sales service, promotions, and the salaries and commissions of the sales force.

**selling out** The selling of securities, commodities, etc., by a broker because the original buyer is unable to pay for them. This invariably happens after a fall in market price (the buyer would take them up and sell them at a profit if the market had risen). The broker sells them at the best price available and the original buyer is responsible for any difference between the price realized and the original selling price, plus the costs of selling out. *Compare* buying in.

**selling short** *See* short selling.

**semisolus** An advertisement that appears on a page in a publication containing another advertisement but is not placed adjacent to it. *Compare* solus position.

**semivariable costs (stepped costs)** Costs that vary with levels of production but in discrete steps rather than continuously. An example might be the rent of workshop floor-space; new space will only be required when production reaches a given level and then not again until the new capacity has been exhausted.

**sensitive market** A market in commodities, securities, etc., that is sensitive to outside influences because it is basically unstable. For example, a poor crop in a commodity market may make it sensitive, with buyers anxious to cover their requirements but unwilling to show their hand and risk forcing prices up. News of a hurricane in the growing area, say, could cause a sharp price rise in such a sensitive market.

**separate assessment** Before April 1990, an election made by one party to a marriage in the UK enabling each party to pay his or her own tax. Unlike an election for *separate taxation of a wife's earnings, this saved no tax. It merely allocated the total tax payable by the married couple as a single unit to the two parties for payment.

**separate taxation of a wife's earnings** Before April 1990, an election made by both parties to a marriage in the UK enabling the earned income of the wife to be treated separately from that of her husband. The normal principle of income tax was that a married couple were treated as one and the total income of both was deemed to be that of the husband. Thus a married woman's income was taxed at her husband's marginal rate. By separate taxation of a wife's earnings, both parties could be treated as single persons (thus foregoing the married man's allowance) and the wife's earned income was treated as hers, although her *unearned income continued to be treated as that of her husband. For this election to be effective, the combined incomes of the two parties had to be such that more was saved by not combining the incomes than was lost by converting the married man's allowance to a single person's allowance. However, from April 1990 independent taxation of husband and wife was introduced in the UK. All taxpayers, married or single, have the same *personal allowance, in addition to a married couple's allowance. Wives pay their own tax on their own income and make a separate tax return. At the same time other anomalies favouring unmarried cohabitation were removed from the UK tax system.

**SEPON Ltd** Abbreviation for *Stock Exchange Pool Nominees Ltd.

**sequential access** A method of locating items of data stored in a computer memory that involves searching through a sequence of other irrelevant items. Magnetic-tape memories use this method of **sequential storage** since data is stored item after item; to locate a particular item all preceding items have to be passed over. Thus the *access time of sequential storage is longer than with a *random-access memory, which goes straight to the item required.

**sequestration** The confiscation of the property of a person or organization who has not complied with a court order. This is a very drastic remedy, and will be ordered in serious cases only. It is most commonly used against companies, trade unions, and in matrimonial cases.

**SERPS** Abbreviation for *State Earnings-Related Pension Scheme.

**service** An economic *good consisting of human worth in the form of labour, advice, managerial skill, etc., rather than a *commodity. **Services to trade** include banking, insurance, transport, etc. **Professional services** encompass the advice and skill of accountants, lawyers, architects, business consultants, doctors, etc. **Con-**

**sumer services** include those given by caterers, cleaners, mechanics, plumbers, etc. Industry may be divided into extractive, manufacturing, and service sectors. The **service industries** make up an increasing proportion of the national income.

**service contract (service agreement)** An employment contract between an employer and employee, usually a senior employee, such as a director, executive manager, etc. Service contracts must be kept at the registered office of a company and be open to inspection by members of the company. The Companies Acts (1980; 1985) prohibit service contracts that give an employee guaranteed employment for more than five years, without the company having an opportunity to break the employment as and when it needs to. This measure prevents directors with long service agreements from suing companies for loss of office in the event of a takeover or reorganization. *See also* compensation for loss of office; golden parachute.

**service industry** *See* service.

**set-off** An agreement between the parties involved to set off one debt against another or one loss against a gain. A banker is empowered to set off a credit balance on one account against a debit balance on another. It is usual, in these circumstances for the bank to issue a **letter of set-off**, which the customer countersigns to indicate his agreement.

**settlement** **1.** The payment of an outstanding account, invoice, charge, etc. **2.** The payment of outstanding dues on the *London Stock Exchange at the end of an *account. *See also* account day (settlement day). **3.** A disposition of land, or other property, made by deed or will under which a *trust is set up by the settlor. The settlement names the beneficiaries and the terms under which they are to acquire the property. **4.** The document in which such a disposition is made. **5.** The voluntary conclusion of civil litigation or an industrial dispute, as a result of agreement between the parties.

**settlement day** *See* account day.

**settlor** A person declaring or creating a *settlement or *trust. For tax purposes any person providing money or property for a settlement will be regarded as the settlor.

**severance** The separation of the good parts of a contract from the bad; it applies to contracts that are not illegal but are void by statute or common law. Courts will, if possible, save contracts from complete invalidity by severance of the parts that are void.

**severance payment** *See* compensation for loss of office, redundancy payment.

**SFO** Abbreviation for Superannuation Funds Office. This department of the UK Inland Revenue deals with all forms of pension funds and benefits from pensions.

**shadow director** A person in accordance with whose instructions the directors of a company are accustomed to act although that person has not been appointed as a director. A shadow director influences the running of the company and some provisions of the Companies Acts, including wrongful trading and the regulation of loans to directors, relating to directors also extend to shadow directors.

**shadow price** The price attributed to a good or service by an economist in the absence of an explicit market price. To calculate this price economists use the *opportunity cost. In a perfectly competitive economy the shadow price is equal to the market price, but in general the two may differ. It is common, therefore, to attempt to estimate shadow prices when carrying out a *cost-benefit analysis for government projects, since few believe that real-world economies are perfectly competitive. However, there is no generally accepted method for calculating shadow prices and they are usually considered unreliable.

**share** One of a number of titles of ownership in a company. Most companies are limited by shares, thus an investor can limit his liability if the company fails to the amount paid for (or owing on) the shares. A share is a *chose in action, conferring on its owner a legal right to the part of the company's profits (usually by payment of a *dividend) and to any voting rights attaching to that share (*see* voting shares; A shares). Companies are obliged to keep a public record of the rights attaching to each class of share. The common classes of shares are: *ordinary shares, which have no guaranteed amount of dividend but carry voting rights; and *preference shares, which receive dividends (and/or repayment of capital on winding-up) before ordinary shares, but which have no voting rights. Shares in public companies may be bought and sold in an open market, such as a stock exchange. Shares in a private company are generally subject to restrictions on sale, such as that they must be offered to existing shareholders first or that the directors' approval must be sought before they are sold elsewhere. *See also* cumulative preference share; deferred ordinary share; founders' shares; partly paid shares; preferred ordinary share; redeemable preference share; subscription shares; term shares.

**share account** A building society deposit account with no fixed investment period; it usually demands one month's notice for withdrawals but will often pay up to £250 without notice.

**share capital** That part of the *capital of a company that arises from the issue of *shares. Every company must commence with some share capital (a minimum of two shares). The **authorized share capital** (**registered capital** or **nominal capital**) of a company is the total amount of capital it is authorized to raise according to its *articles of association. The **issued share capital** or **subscribed share capital** is the amount of the authorized capital that shareholders have subscribed. If the shareholders have subscribed the full *par value of the share, this will constitute the **fully paid share capital**. If they have subscribed only a proportion of the issued share capital, this is called the **called-up capital**. Some capital may be subscribed on application or on allotment or as separate calls. The shares do not become fully paid until the last call has been made. *See also* reserve capital.

**share certificate** A document that provides evidence of ownership of shares in a company. It states the number and class of shares owned by the shareholder and the serial number of the shares. It is stamped by the common seal of the company and usually signed by at least one director and the company secretary. It is not a negotiable instrument. *See* bearer security.

**share exchange** A service offered by most unit-trust managements and life assurance companies, in which the trust or company takes over a client's shareholding and invests the proceeds in unit-trust funds, etc., of the client's choice. The client is thereby saved the trouble and expense of disposing of his shares and if the shares are absorbed into the trust's or company's own portfolio he may receive a better price than he would on the market (i.e. he may receive the offer price rather than the bid price).

**shareholder** An owner of shares in a limited company or limited partnership. A shareholder is a member of the company.

**share index** *See* Chambre Agent General Index; Commerzbank Index; Dow Jones Industrial Average; Financial Times Share Indexes; Nikkei Dow Index; Hang Seng Index.

**share option 1.** A benefit sometimes offered to employees, especially new employees, in which they are given an option to buy shares in the company for which they work at a favourable fixed price or at a stated discount to the market price. **2.** *See* option.

**share premium** The amount payable for shares in a company and issued by the company itself in excess of their nominal value (*see* nominal price). Share premiums received by a company must be credited to a **share premium account**, which cannot be used for paying dividends to the shareholders, although it may be used to make bonus or scrip issues.

**share premium account** *See* share premium.

**share register** The register kept by a limited company in which ownership of shares in that company is kept, together with the full names, addresses, extent of holding, and class of shares for each shareholder. Entry in the register constitutes evidence of ownership. Thus, a shareholder who loses his share certificate can obtain a replacement from the company provided that he supplies proof of his identity and that his holding is recorded in the register.

**share splitting** The division of the share capital of a company into smaller units. The effect of a share split is the same as a *scrip issue although the technicalities differ. Share splits are usually carried out when the existing shares reach such a high price that trading in them becomes difficult.

**share transfer** *See* transfer deed.

**share warehousing** The practice of building up a holding of the shares of a company that is to be the target for a takeover. The shares are bought in the name of nominees in relatively small lots and 'warehoused' until the purchaser has built up a significant interest in the company.

**shark repellents** *See* porcupine provisions.

**shark watcher** A business consultant who specializes in helping companies to identify raiders and to provide early warning of *share warehousing and other manoeuvres used as preliminaries to takeovers.

**shell company 1.** A non-trading company, with or without a stock-exchange listing, used as a vehicle for various company manoeuvres or kept dormant for future use in some other capacity. **2.** A company that has ceased to trade and is sold to new owners for a small fee to minimize the cost and trouble of setting up a new company. Some business brokers register such companies with the sole object of selling them to people setting up new businesses. The name and objects of such a company can be changed for a small charge.

**shipbroker** A broker who specializes in arranging charters, cargo space, and sometimes passenger bookings. He receives a brokerage on the business obtained. *See also* Baltic Exchange.

**shipowner's lien** The lien a shipowner has on cargo he is carrying, if the freight charges are not paid.

**shipped bill** A *bill of lading confirming that specified goods have been loaded onto a specified ship.

**shipper 1.** An organization that exports goods, which it usually owns, to a foreign country by sea or by air. **2.** In some trades, the importer who sells on to merchants or users, especially if he buys FOB port of shipment and pays the freight himself.

**shipping and forwarding agent** An organization that specializes in the handling of goods sent by sea, air, rail, or road, especially if the goods are being exported or imported. They advise on the best method of transport, book freight, cover insurance, and arrange custom clearance at both ends, if necessary. They usually do all the required paperwork and documentation. They also arrange inland transport in the exporting or importing country, arranging factory-to-factory transport, if it is required.

**shipping bill** A form used by HM Customs and Excise before goods can be exported from the UK. It is also required when removing goods from a bonded warehouse before *drawback can be claimed.

**shipping conference** An association of shipowners whose liners ply the same routes. They combine to fix freight rates, passenger rates, and terms of contracts. Not all liners belong to conferences; those that do are called **conference line ships**. Non-conference line ships are often owned by members of the eastern bloc.

**shipping documents** The documents that an exporter of goods delivers to his bank in the exporting country in order to obtain payment for the goods. The bank sends them to its branch or agents in the importing country, who only release them to the importer against payment. Once the importer has these documents he can claim the goods at the port of destination. The documents usually consist of a commercial invoice, *bill of lading, insurance policy or certificate, weight note, quality certificate, and, if required, a certificate of origin, consular invoice, and export licence.

**shipping ton** *See* ton.

**ship's certificate of registry** A document that the master of a ship must always carry. It records the ship's country of registration, home port, owner's name, registered tonnage, permitted number of passengers, master's name, etc.

**ship's papers** The documents that the master of a ship must always have available for inspection if required. They include the *ship's certificate of registry, bill of health, log book, the ship's articles (muster roll), the charterparty if the ship is on charter (*see* chartering), cargo *manifest and bills of lading, and passenger manifest if it is carrying passengers.

**ship's report** A report that the master of a ship must make to the port authorities on arrival at a port. It lists details of the ship, crew, passengers, and cargo.

**shop steward** An employee of a firm, who works in a factory or office and is the representative of a trade union. He is responsible for negotiating with the employers on behalf of the other members of the same union in the factory or office.

**short bill** A *bill of exchange that is payable at sight, on demand, or within ten days.

**short covering** The purchasing of goods that have been sold short (*see* short position) so that the open position is closed. The dealer in commodities, securities, or foreign exchanges, hopes to cover his shorts at below the price at which he sold them in order to make a profit. Dealers cover their shorts when they expect the market to turn or when it has already started to move upwards.

**short-dated gilt** *See* gilt-edged security.

**short delivery** A delivery of goods that has fewer items than invoiced or a smaller total weight than invoiced. This may be due to accidental loss, which could give rise to an insurance claim, or it may be due to some normal process, such as drying out during shipment, in which case the weight on arrival will be used in a *final invoice. It may also be an attempt by the seller to make an extra profit, in which case the buyer would be well advised to make a claim for short delivery.

**shorting** Establishing a *short position.

**short interest** In marine insurance, the difference between the insured value of goods and their market value, when the insured value exceeds the market value. In these circumstances any excess premium paid can be reclaimed.

**short position** A position held by a dealer in securities (*see* market maker), commodities, currencies, etc., in which his sales exceed his holdings because he expects prices to fall, enabling him to cover the shorts at a profit. *Compare* long position.

**short run** A period in economic theory that is sufficiently short for at least one *factor of production to remain unchanged. *Compare* long run.

**shorts** **1.** *See* gilt-edged security. **2.** Securities, commodities, currencies, etc., of which a dealer is short, i.e. has a *short position.

**short selling** Selling commodities, securities, currencies, etc., that one does not have. A short seller expects prices to fall so that he can cover his short sale at a profit before he has to deliver. A short seller is a *bear.

**short-term capital** Capital raised for a short period to cover an exceptional demand for funds over a short period. A bank loan, rather than a debenture, is an example of short-term capital.

**short-term interest rates** The rates of interest on **short-term loans**, i.e. loans that are made for a short period. Banks will usually pay higher rates for short-term loans, in which no withdrawal is permitted until the money is withdrawn on an agreed date, usually within three months. However, when banks are asked to make loans for a short term, their interest rate charged may be lower than for a long-term loan, which will involve a higher risk.

**short ton** *See* ton.

**show stopper** A legal action taken by the target firm in an unwelcome takeover bid that seeks a permanent injunction to prevent the bidder from persisting in his takeover activities, on the grounds that his bid is legally defective in some way.

**shrinkage** The goods that disappear from a retail outlet without being registered as sales. This includes goods that are broken, damaged, stolen, or – in the case of foodstuffs – past their sell-by date and therefore thrown away or given away.

**shut-down cost** The costs to be incurred in closing down some part of an organization's activities.

**SIB** Abbreviation for *Securities and Investment Board.

**sickness and accident insurance** A form of health insurance in which the benefits are paid after eight days for a total of 140 weeks after the onset of an illness or accident that prevents the insured from working. Premiums can increase each year and renewal can be refused if a claim has been made. *Compare* permanent health insurance.

**side deal** A private deal between two people, usually for the personal benefit of one of them, as a subsidiary to a transaction between the officials of a company, government, etc. For example, the chairman of a public company may agree to encourage his board to welcome a takeover bid, provided that he makes a personal profit in some side deal with the bidder. Side deals are rigorously investigated by the Panel on Takeovers and Mergers (*see* City Code on Takeovers and Mergers).

**sight bill** *See* at sight.

**sight draft** Any *bill of exchange that is payable on sight, i.e. on presentation, irrespective of when it was drawn.

**silicon chip** *See* integrated circuit.

**simple interest** *See* interest.

**simulation** Acting out a marketing situation for test purposes. In **computer simulation** all the available data is fed into a computer, which enables a range of possible strategies to be compared. In **laboratory simulation** marketing situations are realistically recreated in order to assess possible results or responses.

**sine die** (Latin: without a day) Denoting an adjournment of an action, arbitration, etc., indefinitely.

**single-capacity system** *See* dual-capacity system.

**single column centimetre** A unit used in selling advertising space in printed publications.

**single-life pension** A pension or *annuity that is paid for the lifetime of the beneficiary only, rather than for the lifetime of a surviving spouse. *Compare* joint-life and last-survivor annuities.

**single-premium assurance** A life-assurance policy in which the insured pays only one capital sum rather than regular premiums. *See also* investment bond.

**single-tax system** A system of taxation in which there would be only one major tax, usually a *comprehensive income tax, instead of several taxes, such as income tax, capital-gains tax, and national insurance, as in the UK. Arguments in favour of such taxes are that they should be less avoidable and should simplify administration. On the other hand the present variety of taxes is designed for a variety of purposes and flexibility may be lost in a single-tax system.

**sinking fund** A fund set up to replace a *wasting asset at the end of its useful life. Usually a regular annual sum is set aside to enable the fund, taking into account interest at the expected rate, to replace the exhausted asset at a specified date. Some have argued that amounts set aside for *depreciation of an asset should be equal to the annual amounts needed to be placed in a notional sinking fund.

**sister-ship clause** A clause used in marine insurance policies, enabling the insurer to make a claim as a result of a collision between two ships both owned by the same insurer. Without this clause, the insurer may have no claim as it is not possible to sue oneself.

**sit-down strike** A strike in which workers come to their place of work, but refuse to either work or to go home. In a **work-in**, the workers refuse to leave the place of work and continue working, usually in spite of management instructions not to do so.

**SI units** Système International d'Unités: an international metric system of units used in science and increasingly for commercial purposes. It is based on the seven basic units: metre, kilogram, second, ampere, kelvin, candela, and mole. Derived units with special names include the hertz (frequency), joule (energy), volt (potential), and watt (power). They are used with a standard set of multiples and sub-multiples, including hecto- (× 100), kilo- (× 1000), centi- (× 1/100), and milli- (× 1/1000).

**sleeping partner** A person who has capital in a *partnership but takes no part in its commercial activities. He has all the legal benefits and obligations of ownership and shares in the profits of the partnership in accordance with the provisions laid down in the partnership agreement.

**sliding peg** *See* crawling peg.

**slump** *See* depression.

**slush fund** Money set aside by an organization for discreet payments to influential people for preferential treatment, advance information, or other services for the benefit of the organization. Slush funds are normally used for purposes less blatant than bribes, but sometimes not much less.

**small claims** A civil action involving a relatively small sum, usually £500 in the UK. In the county courts, such claims are dealt with by arbitration before a registrar, with a simplified procedure without incurring the expense of legal representation.

**small company** A company, as defined by the UK Companies Act (1981), that falls below any two of the following three size criteria: 1) gross assets £700,000; 2) turnover £1,400,000; 3) average number of employees 50. These companies, if not *public limited companies or banking, insurance, or shipping companies, may file very abbreviated accounts (*see* modified accounts) with the Registrar of Companies (excluding even a profit and loss account), although they must provide their own shareholders with the full statutory information. *Compare* medium-sized company.

**small print** Printed matter on a document, such as a life-assurance policy or hire-purchase agreement, in which the seller sets out the conditions of the sale and the mutual liabilities of buyer and seller. The use of a very small type size and unintelligible jargon is often intended to obscure from the buyer his legal rights and safeguards. This unfair practice has largely been remedied by the various Acts that provide *consumer protection. *See also* cooling-off period.

**smoke-stack industries** An informal name for traditional British industries, especially heavy engineering manufacturers, as opposed to modern industries, such as electronics, etc.

**SMP** Abbreviation for *statutory maternity pay.

**snake** A European monetary system in which the exchange rate between the participants was restricted by a fluctuation limit of $2\frac{1}{4}$%. The participants were the Belgian franc, Danish kroner, French franc, Irish punt, Dutch gilder, German Deutschmark, and Italian lire. It was originally called a snake-in-the-tunnel to indicate that the rates of exchange could 'snake' about within the limits imposed by the tunnel. Sterling did not participate in this arrangement, which was replaced in 1979 by the *European Monetary System, which is itself regarded as a preliminary step to a common European currency and monetary union.

**social grades** A system for classifying the population according to social status. The system is based on the occupation of the head of the household and groups are designated by a letter. Developed by Research Surveys Ltd for the *Institute of Practitioners in Advertising in 1962, it is used for the national readership survey. The groups are:

A – upper middle class (managerial, professional; 3% of population)

B – middle class (administrative, professional; 14%)

C1 – lower middle class (supervisory, clerical; 22%)

C2 – skilled working class (skilled manual workers; 29%)

D – working class (semi-skilled and unskilled workers; 18%)

E – lowest levels (state pensioners, casual workers; 14%)

**social overhead capital** *See* infrastructure.

**social security** A government system for paying allowances to the sick and the unemployed, as well as maternity benefits and retirement pensions. Since 1988 in the UK this has been the responsibility of the separate Department of Social Security.

**societal marketing concept 1.** A form of marketing that takes into consideration the long-term welfare of society. For many years the individual needs of consumers dominated the practice of marketing. For example, two cars might be compared in terms of their acceleration. It is now recognized, however, that both cars may have jeopardized the long-term welfare of society as a result of the pollution they cause. Adapting both cars to run on lead-free petrol is an example of the way in which societal marketing concepts are being used. **2.** Marketing techniques applied to a social cause, such as an anti-apartheid campaign or an anti-smoking campaign.

***société anonyme*** *See* SA.

**soft commodities (softs)** Commodities, other than metals, that are traded in London futures markets (*see* futures contract). They include cocoa, coffee, grains, potatoes, and sugar.

**soft currency** The currency of a country that has a weak *balance of payments and for which there is relatively little demand. *Compare* hard currency.

**soft loan** A loan with an artificially low rate of interest. Soft loans are sometimes made to developing nations by industrialized nations for political reasons.

**soft sell** The use of unobtrusive methods of selling as opposed to those employed in a *hard sell. Implication, rather than forceful statement and relentless repetition, form the basis of a soft sell.

**software** The programs used with a computer, together with their documentation. Program listings, program libraries, and user and programming manuals are all software as they are less permanent than the physical parts of the computer system, called the *hardware. *See also* applications software; system software.

**sold note** *See* contract note.

**sole agency** An *agency that gives its holder exclusive rights to sell a product or service in a particular territory for a specified period.

**solicitor's letter** A letter written by a solicitor to a debtor who has failed to settle his debt. The letter usually threatens to take the matter to court, unless payment is received by a specified date.

**solo (sola)** A single *bill of exchange of which no other copies are in circulation.

**solus position** An isolated position for a poster or press advertisement, so that it is separated from any competitive announcement. *Compare* semi-solus.

**solus site** A retail outlet, such as a petrol station, that carries the products of only one company.

**source and application (disposition) of funds** A statement showing how an organization has raised finance for a specified period and how that finance has been applied. Sources of funds are typically trading profits, issues of shares or loan stock, sales of fixed assets, and borrowings. Applications are typically trading losses, purchases of fixed assets, dividends paid, and repayment of borrowings. Any balancing figure represents an increase or decrease in *working capital.

**special clearing** The clearing of a cheque through the UK banking system in less than the normal three days, for a small additional charge. A cheque for which a special clearing has been arranged can usually be passed through the system in one day.

**Special Commissioners** A body of civil servants who are specialized tax lawyers appointed by the Lord Chancellor to hear appeals against assessments to income tax, corporation tax, and capital-gains tax. A taxpayer may choose to appeal to the Special Commissioners, rather than the *General Commissioners, particularly in cases in which legal matters rather than questions of fact are at issue.

**special crossing** A crossing on a *cheque in which the name of a bank is written between the crossing lines. A cheque so crossed can only be paid into the named bank.

**special deposits** Deposits that the UK government may instruct the clearing banks to make at the Bank of England, as a means of restricting credit in the economy. The less money the clearing banks have at their disposal, the less they are able to lend to businesses. A similar system has been used by the Federal Reserve System in the USA.

**Special Drawing Rights (SDRs)** The standard unit of account used by the *International Monetary Fund (IMF). In 1970 members of the IMF were allocated SDRs in proportion to the quotas of currency that they had subscribed to the fund on its formation. There have since been further allocations. SDRs can be used to settle international trade balances and to repay debts to the IMF itself. On the instructions of the IMF a member country must supply its own currency to another member, in exchange for SDRs, unless it already holds more than three times its original allocation. The value of SDRs was originally expressed in terms of gold, but since 1974 it has been valued in terms of its members' currencies. SDRs provide a credit facility for IMF members in addition to their existing credit facilities (hence the name); unlike these existing facilities they do not have to be repaid, thus forming a permanent addition to members' reserves and functioning as an international reserve currency. *Compare* European Currency Unit.

**special manager** A person appointed by the court in the liquidation of a company or bankruptcy of an individual to assist the *liquidator or *official receiver to manage the business of the company or individual. He has whatever powers the court gives him.

**special notice** *See* resolution.

**special resolution** *See* resolution.

**special sort** A character in printing that is not part of a normal *fount.

**specie** Money in the form of coins, rather than bank notes or bullion.

**specific charge** *See* charge; fixed charge.

**specific damages** *See* damages.

**specific performance** An equitable legal remedy requiring the parties to carry out the contract they have entered into. This will be granted by a court only if damages are an inadequate remedy and if it is possible to perform the contract. It is the usual remedy in cases in which there is a breach of contract for the sale of land.

**speculation** The purchase or sale of something for the sole purpose of making a capital gain. For professional speculators the security, commodity, and foreign exchange markets are natural venues as they cater for speculation as well as investment and trading. Indeed, speculators help to make a viable market and thus smooth out price fluctuations. This is particularly true of commodity futures markets (*see* futures contract). Left to producers and users alone, these markets would be much more volatile than they are with speculators taking part.

**speculative demand for money** The demand for money based on the conviction of speculators that *interest rates are about to rise, causing *bond prices to fall; it will therefore be prudent to hold money, awaiting these falls before investing. A crucial part of J M Keynes' theory of *liquidity preference, it suggests that if speculative demand is great enough there will be a shortage of funds for *investment, causing recession. The theory has been severely criticized on the grounds that speculators (investors) in this case would not exhibit *rational expectations, since their expectations are based on past movements in interest rates or bond prices, not the present state of supply and demand. *Compare* precautionary demand for money; transactions demand for money.

**spelling check** A computer program used to check the spelling of the words in a document. The spelling-check program includes a dictionary giving the correct spellings of important words. When the program is run it checks each word in the document against the dictionary. If it cannot find a word in the dictionary, either because the word is misspelt or because the word is not in the dictionary, the word is highlighted on the screen. The user can then correct the word if necessary. Spelling-check programs are usually part of a larger piece of software, such as a *word-processor.

**split** *See* share splitting.

**split screen** A technique used in computing in which the output screen is divided into two or more separate areas. Wordprocessing programs, for example, often have one area devoted to the text being processed and a second area showing such data as the number of words or lines.

**split trust** An investment trust divided into holdings expected to provide growth and holdings providing a good income. All income from the combined holdings is paid into one trust and all capital gains into the other, in order to cater for the preferences of investors.

**sponsor** The *issuing house that handles a new issue for a company. It will supervise the preparation of the *prospectus and make sure that the company is aware of the benefits and obligations of being a public company.

**spot currency market** A market in which currencies are traded for delivery within two days, as opposed to the *forward dealing exchange market in which deliveries are arranged for named months in the future.

**spot goods** Commodities that are available for immediate delivery, as opposed to futures (*see* futures contract) in which deliveries are arranged for named months in the future. The price of spot goods, the **spot price**, is usually higher than the forward price, unless there is a glut of that particular commodity but an expected shortage in the future.

**spread 1.** The difference between the buying and selling price made by a *market maker on the stock exchange. **2.** The diversity of the investments in a portfolio. The greater the spread of a portfolio the less volatile it will be. **3.** The simultaneous purchase and sale of commodity futures (*see* futures contract) in the hope that movement in their relative prices will enable a profit to be made. This may include a purchase and sale of the same commodity for the same delivery, but on different commodity exchanges (*see* straddle), or a purchase and sale of the same commodity for different deliveries.

**spreadsheet** A computer program used for numerical tabular operations, such as financial forecasting and planning. It displays on the computer screen a large table of columns and rows. Numbers are entered by the user to show, for example, financial results or items of income and expenditure. If instructed, the spreadsheet can automatically calculate those numbers that are derived from figures already entered. The program can also update the figures shown in all columns when a single figure is changed by the user.

**squeeze 1.** Controls imposed by a government to restrict inflation. An **income (pay) squeeze** limits increases in wage and salaries, a **credit squeeze** limits the amounts that banks and other moneylenders can lend, a **dividend (profits) squeeze** restricts increases in dividends. **2.** Any action on a market that forces buyers to come into the market and prices to rise. In a **bear squeeze**, bears are forced to cover in order to deliver. It may be restricted to a particular commodity or security or a particular delivery month may be squeezed, pushing its price up against the rest of the market.

**SRO** Abbreviation for *Self-Regulatory Organization.

**SSP** Abbreviation for *statutory sick pay.

**stabilizers** Economic measures used in a free economy to restrict swings in prices, production, employment, etc. Such measures include progressive income tax, control of interest rates, government spending, unemployment benefits, and government retraining schemes.

**staff management** *See* line and staff management.

**stag** A speculator who applies for a new issue of shares with the sole purpose of selling any holding he is allocated as soon as dealing in them begins. Stags hope for the shares to open at a *premium on the issue price, thus providing an instant profit.

**stagflation** A combination of inflation and recession describing the economic conditions of the 1970s. At that time governments generally pursued Keynesian policies, giving credence to monetarist views that government fiscal and monetary policies can only influence the price level.

**staggered directorships** A measure used in the defence against unwanted takeover bids. If the company concerned resolves that the terms of office served by its directors are to be staggered and that no director can be removed from office without due cause, a bidder cannot gain control of the board for some years, even if he has a controlling interest of the shares. *See* poison pill.

**stagnation theory** The theory that business cycles arise in industrial economies because savings increase as people become more affluent but the opportunities for investing them decline. The supporters of this theory believe that in the course of the boom period of the cycle large and exciting investment opportunities, such as clearance of slums, etc., should be provided by governments. The theory, proposed during World War II to explain the Depression, is not widely accepted.

**STAGS** Abbreviation for Sterling Transferable Accruing Government Securities. These European sterling bonds are backed by a holding of Treasury stock. They are *deep-discount bonds paying no interest.

**stale bull** A dealer or speculator who has a *long position in something, usually a commodity, which is showing a paper profit but which the dealer cannot realize as there are no buyers at the higher levels. He, moreover, may be fully committed financially and unable to increase his bull position even if he wanted to. He is, therefore, unable to trade.

**stale cheque** A cheque that, in the UK, has not been presented for payment within six months of being written. The bank will not honour it, returning it marked 'out of date'.

**stamp duty** A tax on specific transactions collected by stamping the legal documents giving rise to the transactions. The most common in the UK are the stamp duties on the transfer of land and of securities. The stamp duty on the transfer of securities is abolished as from a date to be fixed in the fiscal year 1991–92.

**stamp duty reserve tax** A tax introduced in 1986 on agreements to transfer dutiable securities from one party to another when the transfer is not completed and stamped within a specified period.

**stand-alone** Denoting any device or system that can operate by itself. Small self-contained computers are stand-alone devices, unlike terminals that need to be connected to a central computer.

**Standard and Poor's 500 Stock Index** A US index of 425 shares in US industrial companies and 75 stocks in railway and public-utility corporations.

**standard costing** A method of budgetary control in which standard costs are fixed in advance for such items as raw material prices, labour rates, machine costs, fixed overheads, etc. Entries in the costing records are made at the agreed standard costs so that the accounts will show variances when the actual costs differ from the standard costs. This enables management to take up the matter of the differences with those responsible.

**standard rate** The first rate of tax in the schedule of rates of UK income tax. Subsequent rates are called higher rates. If deduction of tax at source is authorized, the rate applicable is normally the standard rate.

**standby agreement** An agreement between the *International Monetary Fund and a member state, enabling the member to arrange for immediate drawing rights in addition to its normal drawing rights in such cases of emergency as a temporary balance of payments crisis.

**standing order** An instruction by a customer to a bank (**banker's order**) or building society to pay a specified amount of money on a speficied date or dates to a specified payee. Standing orders are widely used for such regular payments as insurance premiums, subscriptions, etc. *See also* credit transfer.

**standstill agreement** **1.** An agreement between two countries in which a debt owed by one to the other is held in abeyance until a specified date in the future. **2.** An agreement between an unwelcome bidder for a company and the company, in which the bidder agrees to buy no more of the company's shares for a specified period.

**state banks** Commercial banks in the USA that were established by state charter rather than federal charter (*compare* national banks). The rules governing their trading are controlled by state laws and there are therefore differences in practices from state to state. State banks are not compelled to join the *Federal Reserve System, although national banks are.

**State Earnings-Related Pension Scheme (SERPS)** A scheme, started in 1978, run by the UK government to provide a pension for every employed person in addition to the basic state flat-rate pension. The contributions are paid from part of the national-insurance payments made by employees and employers. Payment of the pension starts at the state retirement age (65 for men, 60 for women) and the amount of pension received is calculated using a formula based on a percentage of the person's earnings. Persons who wish to contract out of SERPS may subscribe to an *occupational pension scheme or a *personal pension scheme.

**statement of account** A document recording the transactions of an organization with its customer for a specified period and normally showing the indebtedness of one to the other. Many firms issue statements to their customers every month to draw attention to any unpaid invoices.

**statement of affairs** A statement showing the assets and liabilities of a person who is bankrupt or of a company in liquidation.

**statements of standard accounting practice** Statements issued by the Combined Accountancy Bodies to guide accountants on how to deal with such matters as depreciation, stock, mergers, leases, etc., in accounts in order to minimize disparity of treatment of the same transactions in different companies. These statements do not have the force of law but they represent the best accountancy practice; auditors are expected to follow them as presenting a 'true and fair view', unless it would be misleading to do so.

**state ownership** *See* public sector.

**state planning** *See* command economy; mixed economy.

**Stationery Office (HMSO; Her Majesty's Stationery Office)** A central government organization, founded in 1786, that supplies printing, binding, office supplies, and office machinery for the home and overseas public services. It also publishes and sells government publications, such as White Papers, Acts of Parliament, Central Statistical Office information, Hansard, etc., at its own bookshops in London, Edinburgh, Manchester, Bristol, Birmingham, and Belfast.

**statistics 1.** The branch of mathematics concerned with the collection, classification, and presentation of information in numerical form. It is based on the assumption that if a group is sufficiently large, it will, unlike an individual, behave in a regular and reproducible manner. For groups that are not large enough for this assumption to be true, a measure of the likelihood of an event happening is its *probability. Much of statistics is concerned with calculating and interpreting probabilities. **2.** The numbers by which information is expressed. *See also* Central Statistical Office.

**statute-barred debt** A debt that has not been collected within the period allowed by law. *See* limitation of actions.

**statutory books** The books of account that the Companies Act (1985) requires a company to keep. They must show and explain the company's transactions, disclose with reasonable accuracy the company's financial position at any time, and enable the directors to ensure that any accounts prepared therefrom comply with the provisions of the Act. They must also include entries from day to day of all money received and paid out together with a record of all assets and liabilities and statements of stockholding (where appropriate).

**statutory company** *See* company.

**statutory damages** *See* damages.

**statutory maternity pay (SMP)** Pay for pregnant women in the UK who have been employed for at least six months and earned at least the lower earnings limit for national insurance contributions. It is paid by the employer at a standard rate for 18 weeks for those who have been employed for between six months and two years. For those employed for over two years SMP is related to earnings for the first six weeks, followed by twelve weeks at the standard rate.

**statutory meeting** A meeting held in accordance with the Companies Act (1985). This normally refers to the annual general meeting of the shareholders, although it could refer to any other meeting required to be held by statute.

**statutory report** A report required to be made by statute. This normally refers to the annual report and accounts required to be laid before the members of a company by the Companies Act (1985).

**statutory sick pay (SSP)** Pay for employees in the UK absent from work due to sickness or injury. It is paid by the employer for a maximum of 28 weeks in any year and can be recovered from National Insurance contributions.

**STD** Abbreviation for *subscriber trunk dialling.

**stepped costs** *See* semivariable costs.

**sterilization** The process of offsetting the inflationary-deflationary effects that result when a government intervenes in foreign-exchange markets. If a nation's currency is depreciating and its government wishes to intervene to stabilize the exchange rate, it can sell its reserves of foreign currency and buy its own currency. However, buying its own currency will take money out of circulation, which could cause interest rates to rise, followed by recession. Sterilization is the process of expanding the money supply in these circumstances to prevent increases in interest rates and any consequent recession. Conversely, if a government intervenes to prevent a currency from appreciating, sterilization would involve reducing the money supply.

**sterling** The UK *pound, as distinguished from the pounds of other countries. The name derives from the small star *steorra* (Old English) that appeared on early Norman pennies.

**Sterling Transferable Accruing Government Securities** *See* STAGS.

**stet** A mark used in correcting proofs. If a correction has been made but subsequently the author wishes this correction to be ignored, he writes 'stet' beside the passage and puts a row of dots below the text he wishes to restore.

**stock 1.** In the UK, a fixed-interest security (*see* gilt-edged security) issued by the government, local authority, or a company in fixed units, often of £100 each. They usually have a *redemption date on which the *par value of the unit price is repaid in full. They are dealt in on stock exchanges at prices that fluctuate, but depend on such factors as their *yield and the time they have to run before redemption. *See also* tap stock. **2.** The US name for an *ordinary share. **3.** *See* stock-in-trade. **4.** Any collection of assets, e.g. the stock of plant and machinery owned by a company.

**stock appreciation** (*or* **depreciation)** The amount by which the value of the *stock-in-trade of an organization has increased (or decreased) over any given period. This may refer to an increase in value due to inflation, which will be recorded in the accounts as a profit, although it is not a real profit as when the goods are sold they need to be replaced at an inflated price. To counteract this situation, adjustments have been made in the tax system, through *stock relief, and in the accounting system, through *current-cost accounting. Stock-in-trade may also appreciate (or depreciate) genuinely as a result of market fluctuations, especially of raw materials.

**stockbroker** An agent who buys and sells securities on a stock exchange on behalf of clients and receives his remuneration for this service in the form of a *commission. Before October 1986 (*see* Big Bang), stockbrokers on the *London Stock Exchange were not permitted by the rules of the Stock Exchange to act as *principals (*compare* stockjobbers) and they worked for a fixed commission laid down by the Stock Exchange. Since October 1986, however, many London stockbrokers have taken advantage of the new rules, which allow them to buy and sell as principals, in which capacity they are now known as *market makers. This change has been accompanied by the formal abolition of fixed commissions, enabling stockbrokers to vary their commission in competition with each other. Stockbrokers have traditionally offered investment advice, especially for their *institutional investors, but as many stockbroking firms are now making less income from commission their expensive research and market analysis departments have had to be cut down, and there is consequently a tendency for this service to be reduced.

**stock building** The process of accumulating stocks of goods for future sale. Most firms retain a certain level of stocks of finished products as a precaution against sudden increases in demand or loss of production because of strikes, etc. (*see* stock control). Stock building represents an *investment by the firm in its own goods, although unlike other forms of investment stock levels tend to increase in a *recession and to fall during a boom. Stockbuilding can have an important effect on the behaviour of the macroeconomy; in the UK, stocks are equivalent to about 30% of the value of the *gross domestic product.

**stock control** Regulation of the *stock-in-trade of a company so that all components or items are available without delay but without tying up unnecessarily large sums of money. Stock control is a discipline well suited to computerization. Many large manufacturers, retailers, etc., have a computerized stock-control system with automatic reordering when the stock of an item reaches a predetermined low level. In retail supermarkets, for example, the act of registering a sale at the check-out reduces the quantity of that item on the computer-held stock record; new deliveries are entered as they arrive. Thus, the stock at any instant can be read from the computer. A periodic check of the actual stock reveals the extent of pilfering.

**stock cover** The time for which a company's stock of raw materials would last, without replenishment at the current rate of sale or use.

**stock depreciation** *See* stock appreciation (or depreciation).

**stock exchange (stock market)** A *market in which *securities are bought and sold, prices being controlled by supply and demand. Stock markets have developed hand in hand with *capitalism since the 17th century, constantly growing in importance and complexity. The basic function of a stock exchange is to enable public companies, governments, and local authorities to raise *capital by selling securities to investors. The secondary market function of a stock exchange is to enable these investors to sell their securities to others, providing liquidity and reducing the *risks attached to investment. Most non-communist countries (and now even some communist ones) have stock markets; the major international stock exchanges are based in the USA, Japan, and the UK (*see* London Stock Exchange; New York Stock Exchange).

**Stock Exchange Automated Quotations System** *See* SEAQ.

**Stock Exchange Automatic Exchange Facility** *See* SEAF.

**Stock Exchange Daily Official List** A record of all the bargains made on the *London Stock Exchange. It also provides details of dividend dates, rights issues, prices, etc., of all *listed companies. It is known as the Official List.

**Stock Exchange Pool Nominees Ltd (SEPON Ltd)** The official nominee company that holds all stocks and shares sold during the course of settlement on the *London Stock Exchange, to facilitate the allocation of holdings. *See* TALISMAN.

**stock-in-trade 1.** Trading stock. The goods that an organization has in hand for the purposes of its trade. This will include raw materials, components, work-in-progress, and finished products. In the USA this is referred to as a company's **inventory**. *See also* stock appreciation; stock building; stock control; stock cover. **2.** The goods or services that an organization normally offers for sale.

**stockjobber (jobber)** A *market maker on the *London Stock Exchange prior to the Big Bang (October, 1986). Stockjobbers were only permitted to deal with the general public through the intermediary of a *stockbroker. This single capacity system was replaced by the *dual capacity system of market makers after the Big Bang. Formerly, jobbers, whose liability was traditionally unlimited, earned their living by the **jobbers' turn**, the difference between the prices at which they were prepared to buy and sell. After the Big Bang most of the London firms of stockjobbers were absorbed into larger financial institutions, usually banks.

**stock market 1.** *See* stock exchange. **2.** A market in which livestock are bought and sold.

**stock of money** *See* money supply.

**stock option** *See* option.

**stockpile 1.** An unusually large stock of a raw material held by an organization in anticipation of a shortage, transport strike, planned production increase, etc. **2.** A large stock of strategic materials, food, etc., built up by a government in anticipation of a war, a break in trading relations with a supplier nation, etc.

**stock policy** An insurance policy covering the goods stocked and sold by a commercial company for specified risks or for all risks (*see* all-risks policy). Policies of this kind are usually based on the price paid for the stock, not the price at which it might be sold, i.e. the insurance does not cover loss of profit or mark-up on the stock.

**stock provision** A book-keeping entry made in the books of account to charge the profit and loss account of an organization with an amount to compensate for the loss in value that some items of stock may have suffered (through price falls, obsolescence, etc.). The corresponding amount is deducted from the value of the *stock-in-trade in the balance sheet.

**stock relief** A method of giving relief to traders through the tax system to allow for any overstatement of their profits as a result of *stock appreciation caused by inflation. Originally, the relief was open to abuse since it could be obtained by increasing stock quantities. A more refined version dealt only with the inflationary aspects; however, the relief was removed in 1984, when inflation began to decline.

**stock split** *See* scrip issue.

**stocktaking** The process of counting and evaluating stock-in-trade, usually at an organization's year end in order to value the total stock for preparation of the accounts. In more sophisticated organizations, in which permanent stock records are maintained, stock is counted on a random basis throughout the year to compare quantities counted with the quantities that appear in the, usually, computerized records.

**stock turnover** The number of times in a year that the *stock-in-trade of an organization is deemed to have been sold. This is best found by dividing the total cost of the goods sold in a year by an average value of the stock-in-trade. The faster the stock is turned over, the more opportunities there are to make profits on it and therefore the lower the margins that are required; for example a supermarket, in which the stock is turned over frequently, makes lower margins than a jeweller's shop, in which high-priced goods are sold infrequently.

**stock watering** The creation of more new shares in a company than is justified by its tangible assets, even though the company may be making considerable profits. The consequences of this could be that the dividend may not be maintained at the old rate on the new capital and that if the company were to be liquidated its shareholders may not be paid out in full.

**stop-go** Vacillation in the economic policies of a government between those intended to reflate the economy in times of high unemployment and those intended to deflate it when it has overheated (*see* overheating).

**stop-loss order** An order placed with a broker in a security or commodity market to close an *open position at a specified price in order to limit a loss. It may be used in a volatile market, especially by a speculator, if the market looks as if it might move strongly against his position.

**stop notice** A court procedure available to protect those who have an interest in shares but have not been registered as company members. The notice prevents the company from registering a transfer of the shares or paying a dividend upon them without informing the server of the notice. It was formerly known as **notice in lieu of distringas**.

**stoppage *in transitu*** A remedy available to an unpaid seller of goods when the buyer has become insolvent and the goods are still in course of transit. If the seller gives notice of stoppage to the carrier or other bailee of the goods, he is entitled to have them redelivered to him and may then retain possession of them until the price is paid. If the right is not exercised, the goods will fall into the insolvent buyer's estate and go towards satisfying his creditors.

**storage device** A device used with a computer to store information and act as a *memory.

**store audit** A measurement of what is being purchased in a shop. Such audits are regularly conducted by retail stores for their own purposes but they are also particularly useful to marketing, advertising, and brand managers in assessing the performance of their goods and services.

**store of value** A *function of money, enabling the acquisition of goods to be delayed. In a barter economy, goods are exchanged more or less simultaneously but in a monetary economy the selection of goods can be delayed, the value of the goods parted with being stored in the money received. To be a satisfactory store of value money must be used in a stable economy; in an inflationary economy its stored value declines.

**storyboard** A sequence of sketches and cartoons to show the main elements in a television or cinema advertisement.

**stowage plan** A plan of a ship showing where all the cargo on a particular voyage was stowed.

**straddle** A strategy used by dealers in traded *options or *futures. In the traded option market it involves simultaneously purchasing put and call options; it is most profitable when the price of the underlying security is very volatile. *Compare* butterfly. In commodity and currency futures a straddle may involve buying and selling options or both buying and selling the same commodity or currency for delivery in the future, often on different markets. Undoing half the straddle is known as **breaking a leg**.

**straight** *See* eurobond.

**straight-line method of depreciation** *See* depreciation.

**stranding** The running aground of a ship as a consequence of an unusual event. For the purposes of *marine insurance, stranding does not include running aground and being refloated; it might involve being driven onto rocks when taking avoiding action or a similar event.

**strap** A triple *option on a share or commodity market, consisting of one put option and two call options at the same price and for the same period. *Compare* strip.

**strategic behaviour** The behaviour of firms or individuals that is aimed at influencing the structure of a *market. In traditional economics, such situations as *monopoly or *oligopoly were seen as the outcome of technological conditions and the state of demand. More recently, it has been observed that a particular firm or individual can influence its competitors in the market in various ways, for example by threatening a *price war if other firms attempt to enter the market; this will clearly influence the structure of the market. *See also* bargaining theory; game theory; predatory pricing.

**strategic business unit (SBU)** An autonomous division within a company responsible for planning the marketing of a particular range of products.

**stratified random sample** *See* random sample.

**street-name stocks** The US name for nominee stocks (*see* nominee shareholding).

**strike** An organized refusal on the part of a group of employees to work, in an attempt to force their employers to concede to their demands for higher pay, shorter hours, better working conditions, etc. An **official strike** is one that takes place on the instructions of a trade union, whereas an **unofficial strike** is one that takes place without union backing. A **wildcat strike** is called at short notice in contravention of an agreement not to strike and it does not have union backing. A **sympathetic strike** is one in which one union calls its members out on strike to support another union, although it is not in dispute with its own employers. A **token strike** is a short withdrawal of labour, e.g. for an hour or two or even for one day, to threaten employers of more serious action to come if the employees' demands are not met. In a **general strike** most of the trade unions in a country call out their members, virtually bringing the country to a standstill.
During some official strikes unions pay **strike pay** to their members, while funds for doing so last.

**strike pay** *See* strike.

**striking price 1.** The price fixed by the sellers of a security after receiving bids in a *tender offer (for example, in the sale of gilt-edged securities or a new stock market issue). Usually, those who bid below the striking price receive nothing, while those who bid at or above it receive some proportion of the amount they have bid for. **2.** *See* exercise price.

**strip** A triple *option on a share or commodity market, consisting of one call option and two put options at the same price and for the same period. *Compare* strap.

**stripped bond** A bond or stock that has been subjected to *dividend stripping.

**structural unemployment** *Unemployment caused by changes in technology or tastes. For example, increasing automation in manufacturing industry, encouraged by recent developments in computer technology, has made many skills obsolete. Thus whole communities may become unemployed, until new skills have been acquired, which is often a long process. Governments aim to relieve structural unemployment through regional policies. They may also stimulate *aggregate demand if it is believed that the loss of employment has had *multiplier effects.

**structured interview** An interview used in *marketing research in which the interviewer asks the questions exactly as they appear on the questionnaire, adding and explaining nothing to the respondent. The questions can only be answered "yes", "no", or "don't know". Although this type of market research produces easily tabulated data quickly, it places a heavy burden on the designer of the questionnaire to ensure that the data is not misleading.

**sub-agent** A person or firm that is employed to buy or sell goods as an *agent of an agent. A firm may hold the agency for certain goods in a particular country, for example, and employ sub-agents to represent it in outlying districts of that country.

**sub-lease** *See* head lease.

**subordinated debt** A debt that can only be claimed by an unsecured creditor, in the event of a liquidation, after the claims of secured creditors have been met. In **subordinated unsecured loan stocks** loans are issued by such institutions as banks, in which the rights of the holders of the stock are subordinate to the interests of the depositors.

**subpoena** (Latin: under penalty) An order made by a court instructing a person to appear in court on a specified date to give evidence, or to produce specified documents. The party calling for the witness must pay his reasonable expenses. Failure to comply with a subpoena is *contempt of court.

**subrogation** The principle that, having paid a claim, an insurer has the right to take over any other methods the policyholder may have for obtaining compensation for the same event. For example, if a neighbour is responsible for breaking a person's window and an insurance claim is paid for the repair, the insurers may, if they wish, take over the policyholder's legal right to claim the cost of repair from the neighbour.

**subscribed share capital** *See* share capital.

**subscriber 1.** A person who signs the *memorandum of association of a new company and who joins with other members in the company in paying for a specified quantity of shares in the company, signing the *articles of association, and appointing the first directors of the company. **2.** A person or organization that rents a telephone line, and sometimes also telephone equipment, from British Telecom.

**subscriber trunk dialling (STD)** A telephone system, now used in most countries, that enables subscribers to dial any number at home or abroad, without reference to the operator.

**subscription shares 1.** Shares in a building society that are paid for by instalments; they often pay the highest interest rates. **2.** The shares bought by the initial *subscribers to a company.

**subsidiary company** *See* group of companies.

**subsidy** A payment by a government to producers of certain goods to enable them to sell the goods to the public at a low price, to compete with foreign competition, to avoid making redundancies and creating unemployment, etc. In general, subsidies distort international trade and are unpopular but they are sometimes used by governments to help to establish a new industry in a country. *See also* Common Agricultural Policy.

**subsistence crop** A crop grown by a farmer for consumption by himself and his family, rather than for sale (a **cash crop**). *See also* catch crop.

**subsistence wage** The lowest wage upon which a worker and his family can survive. 19th-century economists, such as Malthus and Marx, believed that the long-term wages would not rise above this level, as any better standard of living would promote a population increase, which would bring the level of wages down again. This view has been shown to be fallacious in industrial countries, where workers share in national prosperity, attaining a high standard of living comparable to that of members of professions and of entrepreneurs.

**substantive motion** *See* motion.

**substitute** A good for which the demand changes in the opposite direction to the demand for some other good whose price has changed. For example, if the price of bread increases, leading to a fall in the demand for bread, and at the same time the demand for pasta increases, pasta is described as a substitute for bread. Another way of saying this is that the price *elasticity of the demand for pasta with respect to bread is greater than zero. *Compare* complement.

**substitution effect** The change in the quantity of a good demanded when the price falls (*compare* income effect). Usually, a change in the price of a good is expected to have a negative substitution effect on the good itself, i.e. a fall in price will lead to an increase in demand. The effect of a change in price on other goods will depend on whether they are *complements (in which case, the effect is negative) or *substitutes (the effect is positive).

**sue and labour clause** A clause in a *marine insurance policy extending the insurance to cover costs incurred by the policyholder in preventing a loss from occurring or minimizing one that could not be avoided. Without this clause the policy would only cover damage that had actually occurred.

**sum insured** The maximum amount the insurers will pay in the event of a claim.

**sunk capital** The amount of an organization's funds that has been spent and is therefore no longer available to the organization, frequently because it has been spent on either unrealizable or valueless assets.

**sunk costs** Past expenditure, which is often thought to be irrelevant to future decisions as the best decisions tend to maximize future cash flow.

**sunspot theory** The theory that expectations in economics can influence an economy. The economist W S Jevons (1835–82) put forward a theory, said to be based on statistical evidence, that trade cycles seemed to coincide with changes in sunspot activity. This somewhat eccentric hypothesis is in itself of no importance but has been used as an example of the economics of expectations. For example, if sufficient people believe that sunspots influence share prices, the supporters of this theory believe that they will do so. J M Keynes suggested that self-fulfilling expectations of this kind could provide one explanation for economic depressions. The literature on *rational expectations was thought to have refuted this view, but recent articles on sunspot theory have attempted to show that this is not the case.

**superannuation** An occupational pension scheme. Contributions are deducted from an employee's salary by the employer and passed to an insurance company or the trustees of a pension fund. After retirement, the employee receives a pension payment from the scheme.

**Superannuation Funds Office** *See* SFO.

**supermajority provisions** Provisions in the byelaws of a company that call for more than a simple majority of its members when voting on certain *motions, such as the approval of a merger or whether or not to agree to a takeover. In these circumstances the provisions may call for a supermajority of between 70% and 80% of the votes cast.

**superprofit** Any profit in excess of a *normal economic profit. *See also* pure economic profit.

**supplementary costs** *See* overhead costs.

**supply and demand** *See* market forces; price theory.

**supply curve** A curve on a graph relating the quantity of a good supplied to its price. Economists usually expect a supply curve to slope upwards, i.e. an increase in price will be associated with an increase in the amount supplied. The intersection of the market supply curve and the market *demand curve gives the market price and the total amount sold. *See also* supply theory.

**supply-side economics** An approach to *macroeconomics that emphasizes the importance of the conditions under which goods and services (including labour) are supplied to the market in determining the level of employment and output in an economy. The growth of output and employment, according to this approach, can be enhanced by measures to reduce government involvement in the economy and by allowing the free market to operate. This includes such controversial measures as reducing unemployment benefits, restricting the power of trade unions, and cutting both taxes and government expenditure. Supply-side economists, who gained in popularity in the 1970s and 1980s, generally reject as ineffectual Keynesian policies (*see* Keynesianism) aimed at increasing aggregate demand (*see* monetarism; new classical macroeconomics).

**supply theory** The theory, largely based on *producer theory, analysing the way in which goods and services are brought to the market by firms or *entrepreneurs. For every seller in a market there will be a *supply curve relating the amount he is willing to sell at each possible price. Adding these curves together will enable a market supply curve to be constructed; the point at which this curve intersects the market *demand curve will determine the market price and the total amount sold. Adding together all the market supply curves for an economy will give an aggregate supply curve, a concept used in *macroeconomics.

**supra protest** *See* acceptance supra protest.

**surety 1.** A guarantor for the actions of someone. **2.** A sum of money held as a *guarantee or as evidence of good faith.

**surplus value** The excess of value produced by the labour of workers over the wages they are paid. A key concept in *Marxist economics, it is the basis for the Marxist view that a 'just' economic system will only be achieved when workers control the process of production and so receive their own surplus value. Like most of Marxist economics, however, the theory of surplus value has been more successful as a political doctrine than as a description of an economic process.

**surrender value** The sum of money given by an insurance company to the insured on a life policy that is cancelled before it has run its full term. The amount is calculated approximately by deducting from the total value of the premiums paid any costs, administration expenses, and a charge for the life-assurance cover up to the cancellation date. There is little or no surrender value to a life policy in its early years.

**surtax** Formerly, a separate UK tax raised progressively on the higher incomes. This tax has now been incorporated into income tax by means of the *higher rates.

**suspense account** A temporary account in the books of an organization to record balances to correct mistakes or balances that have not yet been finalized (e.g. because a particular deal has not been concluded).

**swap** The means by which a borrower can exchange the type of funds he can most easily raise for the type of funds he wants, usually through the intermediary of a bank. For example, a UK company may find it easy to raise a sterling loan when they really want to borrow Deutsche Marks; a German company may have exactly the opposite problem. A swap will enable them to exchange the currency they possess for the currency they need. The other common type of swap is an interest-rate swap, in which borrowers exchange fixed- for floating-interest rates. Swaps are most common in the *eurocurrency markets.

**switching 1.** Using the cash from the sale of one investment to purchase another. This may, or may not, involve a liability for capital-gains tax, depending on the circumstances. **2.** Closing an *open position in a commodity market and opening a similar position in the same commodity but for a different delivery period. For example, a trader may switch his holding of sugar for October delivery FOB to an equal quantity of sugar for March delivery FOB for the next year.

**SWOT** Abbreviation for strengths, weaknesses, opportunities, and threats. During the planning of the marketing of a new product a company needs to embark on a SWOT analysis to assess its strengths and weaknesses (internally) and the opportunities and threats facing it (externally). Internal strengths could be a good distribution system and adequate cash flow. Weaknesses might be identified as an already extended product line or poor servicing facilities. Opportunities could be consumer demand for a particular product or the vulnerability of a competitor, while threats might be forthcoming government legislation or diversification by a competitor. A SWOT analysis is also the main feature of an organization's managerial *position audit.

**symbol retailer (voluntary retailer)** A voluntary group of independent retailers formed to buy in large quantities from *wholesalers at lower prices than they could achieve independently. Originally a defensive move by independents to counter the increasing dominance and expansion of the multiple stores, supermarkets, hypermarkets, etc., these groups can also provide each other with such services as promotional support and management advice. The members of a group often use a common name or symbol to identify themselves.

**sympathetic strike** *See* strike.

**syndicate** A number of *Lloyd's underwriters who accept insurance risks as a group; each syndicate is run by a syndicate manager or agent. The **names** in the syndicate accept an agreed share of each risk in return for the same proportion of the premium. The names do not take part in organizing the underwriting business. Although a syndicate underwrites as a group, each member is financially responsible for only his or her own share.

**syndicated loan** A medium-term loan made by a group of international banks, usually at a rate based on the *London Inter-Bank Offered Rate.

**syndicated research** A large-scale marketing research project undertaken by market-research companies and subsequently offered for sale to interested parties. It is not therefore undertaken on behalf of a client.

**systems analysis** The examination or analysis of the objectives and problems of a system for the purpose of developing and improving the system by the use of computers. Systems analysis entails producing a precise statement of what is to be accomplished, determining the methods of achieving the objectives that are most cost-effective, and preparing a feasibility study. If the study is accepted by the customer, the **systems analysts** will advise on the hardware and software needed, provide programmers with the details they need to write the programs, produce documentation describing the system, plan staff training on it, and monitor the system once it is installed.

**systems selling** The selling of a total system rather than an individual product. For example, some computer manufacturers sell VDUs, keyboards, printers, and the necessary software for their computers, while other companies might only supply the software for these computers.

**system software** Computer programs that control the functioning of the computer itself, rather than directly meeting the user's computing needs (*compare* applications software). Examples are *operating systems and utility programs, which perform such tasks as copying *files, checking the integrity of *magnetic disks, etc. These programs are usually supplied with the computer and may even be built into it (*see* firmware).

Programs, such as *compilers and *interpreters, which contribute to the production of applications software, are also regarded as systems software.

# T

**tachistoscope** A projection device used in advertising and marketing research to measure the extent to which the features of an advertisement or brand are registered by consumers. The tachistoscope shows a package or advertisement for a brief period (which can be varied), enabling researchers to test the effectiveness of the design, colour, lay-out, or name. Tachistoscope tests are often undertaken before the product or advertisement is released.

**takeover bid (offer to purchase)** An offer made to the shareholders of a company by an individual or organization to buy their shares at a specified price in order to gain control of that company. In a welcome takeover bid the directors of the company will advise shareholders to accept the terms of the bid. This is usually known as a *merger. If the bid is unwelcome, or the terms are unacceptable, the board will advise against acceptance. In the ensuing **takeover battle**, the bidder may improve his terms and will usually write to shareholders outlining the advantages that will, in his view, follow from the successful takeover. In the meantime bids from other sources may be made (*see* grey knight; white knight) or the original bidder may withdraw as a result of measures taken by the board of the *target company (*see* poison pill; porcupine provisions). In an **unconditional bid**, the bidder will pay the offered price irrespective of the number of shares acquired, while the bidder of a **conditional bid** will only pay the price offered if he receives sufficient shares to give him a controlling interest. Takeovers in the UK are subject to the rules and disciplines of the *City Code on Takeovers and Mergers.

**Takeover Panel** *See* City Code on Takeovers and Mergers.

**TALISMAN** Abbreviation for Transfer Accounting Lodgement for Investors and Stock Management. This is the *London Stock Exchange computerized transfer system, which covers most UK securities; it also covers claims for dividends on shares being transferred. It is operated by a special company set up for the purpose, the Stock Exchange Pool Nominees Ltd, known as SEPON Ltd.

**tallyman (tallyclerk)** A worker at a port or airport, who checks that the goods unloaded from a ship or aircraft tally with the documents covering them and the ship's or aircraft's manifest.

**talon** A printed form attached to a *bearer bond that enables the holder to apply for a new sheet of *coupons when the existing coupons have been used up.

**tangible assets** Assets that can be touched, i.e. physical objects. However, tangible assets usually also include leases and company shares. They are therefore the fixed assets of an organization as opposed to such assets as goodwill, patents, and trademarks, which are even more intangible than leases and shares.

**tanker** A ship designed to carry liquids in bulk in large storage tanks. Most carry oil but some carry wine, sugar products, and liquified gases.

**tap issue** The issue of government securities or bills to government departments without going through the stock exchange or the money market.

**tap stock** A gilt-edged security from an issue that has not been fully subscribed and is released onto the market slowly when its market price reaches predetermined levels. **Short taps** are short-dated stocks and **long taps** are long-dated taps.

**tare** *See* gross weight.

**target company** A company that is subject to a *takeover bid.

**Target Group Index (TGI)** An annual report compiled from the returned questionnaires of a random sample of 24 000 respondents on their use of hundreds of products and thousands of brands. The questionnaires are completed with no interviewer present. The TGI is an example of *syndicated research.

**target marketing** The selection of one or more segments of a market at which companies direct their marketing thrust. Since it has become almost impossible to market products that will satisfy everyone, target marketing enables companies to aim the products at specific groups of consumers.

**target price** *See* Common Agricultural Policy.

**targets** The goals of an economic policy pursued by a government. These usually include *full employment, stable prices, and a high rate of growth of gross domestic product. Debate in *macroeconomics centres on the targets it is feasible for a government to pursue, the *instruments that are suitable to implement them, and the *indicators that are appropriate to assess the success or failure of the policy. For example, a *new classical macroeconomist might argue that full employment is not a feasible target while stable prices is; he might further argue that the *money supply is the correct instrument and the growth of a particular *monetary aggregate is the proper indicator.

**tariff 1.** Any list of charges for goods or services. **2.** A list of taxes or customs duties payable on imports or exports. The Customs and Excise issue tariffs stating which goods attract duty and what the rate of duty is. **3.** A list of charges in which the charging rate changes after a fixed amount has been purchased or in which a flat fee is imposed in addition to quantity-related charge, as in two-part tariffs for gas or telephone services.

**tariff office** An insurance company that bases its premiums on a tariff arranged with other insurance companies. A **non-tariff office** is free to quote its own premiums.

**tatonnement** The process by which prices reach equilibrium is described by this French word (meaning *groping*) in the work of M E L Walras (1834–1910). Walras' *general equilibrium model, together with tatonnement, has been one of the most influential ideas in neoclassical economics. The process of tatonnement may be thought of as an auction, in which buyers and sellers bid until equilibrium is reached at a unique price. In a free market, the process of tatonnement will ensure that an economy will reach a perfectly competitive Pareto-optimal equilibrium (*see* Pareto optimality).

**TAURUS** Abbreviation for Transfer and Automated Registration of Uncertified Stock. This is the *London Stock Exchange's computerized system designed to enable stocks and shares to be transferred without the use of contract notes or share certificates. It is to come into use in the fiscal year 1991–92, when *stamp duty will be abolished.

**taxable income** Income liable to taxation, usually by reason of the statutes passed by Parliament. It is calculated by deducting *income-tax allowances from the taxpayer's gross income.

**tax and price index (TPI)** A measure of the increase in taxable income needed to compensate taxpayers for any increase in retail prices. As an index of the rate of *inflation, it is similar to the *Retail Price Index, but in addition to measuring price changes it includes changes in average tax liability; for example, a cut in income tax will cause the TPI to rise by less than the RPI.

**taxation** A levy on individuals or corporate bodies by central or local government in order to finance the expenditure of that government and also as a means of implementing its fiscal policy. Payments for specific services rendered to or for the payer are not regarded as taxation. In the UK, an individual's income is taxed by means of an *income tax (*see also* PAYE), while corporations pay a *corporation tax. Increases in individual wealth are taxed by means of *capital-gains tax and by *inheritance tax. *See also* community charge; direct taxation; indirect taxation; VAT.

**taxation brackets** Figures between which taxable income or wealth is taxed at a specified rate. For example, incomes between, say, £3,000 and £19,300 attract tax at 25%; the slice above £19,300 attracts tax at 40%. *See also* bracket indexation.

**tax avoidance** Minimizing tax liabilities legally and by means of full disclosure to the tax authorities. *Compare* tax evasion.

**tax base** The specified domain on which a tax is levied, e.g. an individual's income for *income tax, the estate of a deceased person for *inheritance tax, the profits of a company for *corporation tax.

**tax burden** The amount of tax suffered by an individual or organization. This may not be the same as the tax actually paid because of the possibility of shifting tax or the normal *incidence of taxation. As an example of the latter case, *inheritance tax is paid by the personal representatives of the deceased but the tax burden falls on the heirs, since their inheritance is reduced.

**tax clearance** An assurance, obtained from the UK Inland Revenue, that a proposed transaction, for example the reorganization of a company's share capital, will, if executed, not attract tax.

**tax credit** An amount deductible from an amount of tax payable, usually because tax has already been paid on an element of the *tax base. The most common tax credit in the UK is for income tax that has already been deducted at source.

**tax credit system** A taxation system in which individuals are given tax allowances dependent on their needs; if tax on their income is less than the total of their tax credits, they can be paid the excess credits. *See also* negative income tax.

**tax deposit certificate** A certificate issued by the UK Inland Revenue to a company that has deposited a sum of money with the collector of taxes, usually as an advance against future corporation tax or against corporation tax that has not been finally agreed. The deposit earns interest until it is used to settle a tax liability.

**tax evasion** Minimizing tax liabilities illegally, usually by not disclosing that one is liable to tax or by giving false information to the authorities. Evasion is liable to severe penalties. *Compare* tax avoidance.

**Tax Exempt Special Savings Account (Tessa)** A UK savings scheme introduced in January 1991, enabling savers to invest up to £150 per month (£1800 p.a.) in a bank or building society with no tax to pay on their interest, provided that the capital remains in the account for five years. In the first year, up to £3000 can additionally be paid into such an account, with an overall limit of £9000 for the account. Interest up to the net-of-tax level can be withdrawn at any time; the capital can also be withdrawn at any time if the tax relief is sacrificed.

**tax exile** A person with a high income or considerable wealth who chooses to live in a *tax haven in order to avoid the high taxation of his native country. For such people the cost of living in their own country may be extremely high.

**tax haven** A country or independent area that has a low rate of tax and therefore offers advantages to retired wealthy individuals or to companies that can arrange their affairs so that their tax liability falls at least partly in the low-tax haven. In the case of individuals, the cost of the tax saving is usually residence in the tax haven for a majority of the year (*see* tax exile). For multinational companies, an office in the tax haven, with some real or contrived business passing through it, is required. Monaco, Liechtenstein, the Bahamas, and the Cayman Islands are examples of tax havens.

**tax holiday** A period during which a company, in certain countries, is excused from paying corporation tax or profits tax (or pays them on only part of its profits) as an export incentive or an incentive to start up a new industry.

**tax invoice** An invoice for goods or services rendered by a trader who is registered for VAT, showing the selling price of the goods or services and the VAT charged on them. This invoice is an authority for the purchaser (if he is himself registered for VAT) to reclaim the VAT as input tax. Unregistered persons are prohibited by law from issuing tax invoices.

**tax loss** An amount deemed to be a loss for tax purposes, which can be set against an amount of profit, thus reducing the amount of the profit on which tax is payable. The loss may arise from trading at a loss or it may arise even when trade is profitable because such tax reliefs as capital allowances exceed the profits; this creates a loss for tax purposes only. Tax losses can normally be set against profits of a similar and sometimes of a different type according to stringent rules.

**tax relief** A deduction from a taxable amount, usually given by statute for such purposes as personal allowances for *income tax, retirement and private residence reliefs for *capital-gains tax, and business property and agricultural property reliefs for *inheritance tax.

**tax return** A form upon which a taxpayer makes an annual statement of income and personal circumstances enabling claims to be made for personal allowances. Statements of any capital gains, inheritances, etc., are also required in the return. The return is used by the tax authorities to assess the taxpayer's liability for tax.

**tax schedules** *See* schedules.

**tax shelter** Any financial arrangement made in order to avoid or minimize taxes.

**tax wedge** In economic theory, the difference between the marginal cost to the seller or producer of a good and the price paid by the purchaser. This difference represents the amount of tax paid.

**tax year** *See* fiscal year.

**tea auction** Auctions of Indian and Sri Lankan tea, held in Plantation House, Mincing Lane, in the City of London. The tea is shipped to the UK and handled by selling brokers, who provide sampling facilities for the buyers prior to the auction.

**TEC** Abbreviation for Training and Enterprise Council.

**technical rally** *See* rally.

**technological change** An increase in the level of output resulting from automation and computerized methods of production. Apart from increasing output, technological change can affect the ratio of capital to labour used in a factory. If it involves reducing the labour force it can lead to **technological unemployment** in an area or industry.

**technological unemployment** *See* technological change.

**telegraphic address** A single word used to identify a company in a specified city. It is used in telegrams and cables to reduce the cost of transmitting a complete name and address. Telegraphic addresses are registered by post offices throughout the world but are now much less used than formerly as a result of instant communication by *Telex and *Fax.

**telegraphic transfer (TT)** A method of transmitting money overseas by means of a cabled transfer between banks. The transfer is usually made in the currency of the payee and may be credited to his account at a specified bank or paid in cash to the payee on application and identification.

**telemarketing** *See* telephone selling.

**telephone research** A marketing research technique in which **telephone interviews** are conducted. It is cheaper than personal interviewing and, perhaps because it avoids face-to-face confrontations, seems to be more popular with respondents.

**telephone selling (telemarketing)** A method of *direct marketing in which the telephone is used to contact potential customers in order to reduce the time spent in making personal visits. Certain products, such as double glazing and central heating, are frequently marketed using this technique. *See also* cold calling.

**Teletex** An automatic international telecommunications service using as terminals electronic typewriters with Teletex adapters, electronic telephone exchanges, or computers. Simple messages can be sent around the world at 30 times the speed of *Telex.

**Teletext** A means of providing information to the home or office by means of adapted television sets (*see* Viewdata). The information is given under various headings, with a *menu call-up system, and includes latest stock exchange prices. **Ceefax** is the Teletext service provided by the BBC and **Oracle** that provided by ITV. *See also* Prestel.

**Teletext Output Price Information Computer** *See* TOPIC.

**television rating (TVR)** The measurement of the popularity of a television programme based on survey research. Equipment attached to sets in selected homes records which channel the set is tuned to, while *diary panels are used to determine how many people are watching the set. In this way ratings can be calculated for each programme and expressed as a percentage of the total households that can receive TV. Survey research is undertaken in the UK by the Broadcasters Audience Research Board (BARB).

**Telex** A method of communicating written messages over the international telephone network, using a Telex (teleprinter) machine. Telex calls are made by direct dialling, a hard copy of the message being provided for both sender and recipient. Telex messages may be received by unattended machines and a tape of a message may be prepared for sending to several different recipients at any time required. The main advantage of Telex over *Fax is that the appearance of the recipient's call-back signal on the sender's copy of the message transmitted is proof that the message has reached the recipient's Telex. *Compare* Teletex.

**teller** A bank or building society cashier.

**temporary assurance** *See* term assurance.

**tenancy agreement** An agreement to let land. *See* lease.

**tenancy in common** Two or more persons having an interest in the same piece of land. Each person can leave his interest as he wishes in his will, although the land itself has not been physically divided. Trustees must hold the legal estate and the tenancy in common exists in equity. *See also* joint tenants.

**tender** A means of auctioning an item of value to the highest bidder. Tenders are used in many circumstances, e.g. for allocating valuable construction contracts, for selling shares on a stock market (*see* offer for sale; issue by tender), or for the sale of government securities (*see* gilt-edged security).

**tenor** The time that must elapse before a *bill of exchange or *promissory note becomes due for payment, as stated on the bill or note.

**term assurance (temporary assurance)** A life-assurance policy that operates for a specified period of time. No benefit is paid if the insured person dies outside the period. This form of insurance is often used to cover the period of a loan, mortgage, etc. If the insured person dies before the loan has been repaid, it is settled by the insurance company. *See also* decreasing term assurance.

**term bill** *See* period bill.

**terminal** A computer input and output device. It commonly consists of a keyboard and display screen, connected to the computer. Terminals that are distant from the computer and connected to it by a communications link are called remote terminals. Terminals with some computing power, often incorporating a *microprocessor, are called *intelligent terminals. Specialized terminals have been developed for use in banking and retailing as well as in home information systems, such as *Viewdata.

**terminal bonus** An additional amount of money added to payments made on the maturity of an insurance policy or on the death of an insured person, because the investments of the insurer have produced a profit or surplus. Bonuses of this kind are paid at the discretion of the life office and usually take the form of a percentage of the sum assured.

**terminal loss relief** A tax loss made in the closing year of a business's life, which may be carried back and set against profits of earlier years.

**terminal market** A commodity market in a trading centre, such as London or New York, rather than a market in a producing centre, such as Calcutta or Singapore. The trade in terminal markets is predominantly in *futures, but *spot goods may also be bought and sold.

**term shares** Shares that cannot be sold for a given period (term). They are usually shares in a building society that cannot be cashed on demand and consequently carry a higher rate of interest.

**terms of trade** A measure of the trading prospects of a country in terms of the prices of its imports in relation to the prices of its exports. It may be expressed as the ratio of an index of export prices to an index of import prices, or as a change of this ratio relative to a base year. Thus, the terms of trade of a country improve if its export prices rise faster than the price of its imports. A fall in the *rate of exchange, other things being equal, will lead to a deterioration in the terms of trade.

**term structure of interest rates** The relationship between the *yields on fixed-interest securities (such as government bonds) and their maturity dates. One important factor affecting the term structure relates to expectations of changes in interest rates. For example, if both short-term and long-term bonds have the same yield initially and subsequently investors expect interest rates to fall, long-term bonds will appear relatively attractive, because, being fixed-interest securities, they will continue to pay their stated rates of interest for a long period, after market interest rates have declined. *Arbitrage will then ensure that their prices will rise and their yields will fall; thus yields on short-term bonds will exceed those on long-term bonds. A graph showing the relationship between the number of years to maturity and the yield is called the *yield curve. Other factors that may also affect the term structure include *liquidity preference and *hedging pressure.

**territorial waters** The part of the sea around a country's coastline over which it has jurisdiction according to international law. Historically, it was the sea within three miles of the coast, a practical distance determined by the range of a cannon ball. Since disputes have arisen over fishing rights, off-shore oil mining, and mineral rights below the ocean, the universal three-mile rule has been abrogated and various nations have made different claims, some, but not all, of which have been accepted internationally. Ten to fifteen miles is now quite common in definitions of territorial waters. The Territorial Sea Act (1987) fixes the UK territorial waters at 12 nautical miles. A UN convention proposes a new 200-mile **exclusive economic zone**, which would give a country sovereign rights over all the resources of the sea, seabed, and subsoil within this zone. Fishing rights extending to 200 miles are also claimed by many nations, including the UK.

**tertiary production** *See* production.

**Tessa** Abbreviation for *Tax Exempt Special Savings Account.

**testacy** The state of a person who has died leaving a valid will. Such a person is said to have died **testate**.

**testator** A person who makes a will. The feminine form is **testatrix**.

**testimonial advertisement** An advertisement that makes use of the implied or explicit patronage of a product by a well-known person or organization.

**test marketing (market testing)** A procedure for launching a new product in a restricted geographical area to test consumers' reactions. If the product is unsuccessful, the company will have minimized its costs and can make the necessary changes before a wider launch. Test marketing is typically undertaken in large towns or in areas served by a particular commercial television company. Test marketing has the disadvantage that competitors learn about the new product before its full launch. *See also* new product development.

**TGI** Abbreviation for *Target Group Index.

**theory of games** *See* game theory.

**thermal printer** A type of non-impact computer printer that uses special heat-sensitive paper. The printing head contains heated needles that are pressed against the paper to form dot matrix characters. The paper darkens where it is heated by the needles. Thermal printers are cheap and quiet in operation.

**The Securities Association Ltd (TSA)** *See* London Stock Exchange; Self-Regulatory Organization.

**thin capitalization** A form of company capitalization in which the capital of a company consists of too few shares and too much loan stock in the view of the tax authority. Some countries reserve the right in such cases to treat some of the interest on the loan stock as if it were a dividend, thus denying the right to a tax deduction on the interest payment.

**third-line forcing** Forcing a buyer to take a supply of something he does not want as a condition of supplying him with a product he wants. The practice is deprecated by the Restrictive Practices Court (*see* restrictive trade practices).

**third market** A market established by the *London Stock Exchange in January 1987 for the trading of shares unsuited to either the *unlisted securities market or the *main market. By specifying much less stringent *listing requirements the Stock Exchange was attempting to attract companies that were formerly traded on the unregulated *over-the-counter market, which is outside the jurisdiction of the Stock Exchange. However, because much less information about companies is provided on this market, investment in the third market is generally much more risky than investment in the senior markets.

**third-party insurance** Insurance of a person's legal liabilities to others. The insurer and the policyholder are the two parties to the insurance contract, any other person to whom there is a legal obligation is therefore a third party. Some forms of third-party insurance are compulsory, e.g. motorists must have third-party cover for any loss or damage suffered by members of the public as a result of their driving and employers must cover against any results of injury to employees at work. Professional indemnity insurance is a form of third-party insurance that is optional.

**third world** *See* developing countries.

**threshold agreement** An agreement between an employer and employees (or their union) that pay will increase by a specified amount if the rate of inflation exceeds a specified figure in a specified period. Threshold agreements are seen by some as providing an inflationary pressure of their own but others claim that they reduce inflation by making it unnecessary for unions to press for excessive settlements to protect their members against inflation.

**threshold effect** The point at which advertising begins to show signs of increasing sales. As advertising is expensive, the threshold effect is important because it sets the minimum level for an advertising budget.

**threshold price** *See* Common Agricultural Policy.

**tick (point)** The smallest increment of price fluctuation in a commodity market.

**tied loan** A loan made by one nation to another on condition that the money loaned is spent buying goods or services in the lending nation. It thus helps the lending nation, by providing employment, as well as the borrowing nation.

**tied outlet** A retail outlet that is obliged to sell only the products of one producer (and possibly other noncompeting products). The outlet is usually owned by the producer; in other cases the outlet agrees to become tied in return for financial concessions.

**Tigers** *See* TIGR.

**tight money** *See* dear money.

**TIGR** Abbreviation for Treasury Investment Growth Receipts. These *zero-coupon bonds are linked to US Treasury bonds. Denominated in dollars, their semi-annual compounding yield is taxed in the UK as income in the year of encashment or redemption. They are often known as **Tigers**.

**time and motion study** A method of finding the best way of performing a complex task by breaking the task into small steps and measuring the time taken to perform each step. This enables standards of performance to be set. These standards can then be used to plan and control production, estimate prices and delivery times, and devise incentive schemes.

**time bargain** A contract in which securities have to be delivered at some date in the future.

**time card** A card used to record the number of hours an employee spends in his place of work. When the employee clocks in, he inserts his time card into the time clock, which marks the time on the card. When he leaves work, he clocks out by a similar process. Wages are paid on the basis of hours spent at work as shown on the time card.

**time charter** The hire of a ship or aircraft for a specified period of time rather than for a specified number of voyages (*compare* voyage charter).

**time deposit** A deposit of money in an interest-bearing account for a specified period. In the USA, a time deposit requires at least 30 days' notice of withdrawal.

**time preference** The tendency of individuals to prefer current consumption rather than consumption in the future. The economic theory proposed by von Böhm-Bawerk (1851–1914) defines time preference as the sum required by an individual to compensate him for postponing consumption, expressed as a percentage of his income. The **rate of time preference** can be expressed as the interest rate required.

**time rate** A rate of pay expressed as a sum of money paid to an individual for the time worked, rather than for a specified output (*compare* piece rate).

**time sharing** A method of operating a computer in which several programs apparently run at the same time. Although the computer actually divides its time between the programs, it is fast enough to allow each program to operate at an acceptable speed. Most *multitasking and *multiuser systems are time sharing systems.

**time value** The market value of an *option over and above its *intrinsic value.

**title deed** A document proving the ownership of land. The document must go back at least 15 years to give a good root of title. If the land is registered, the *land certificate now takes the place of the title deed. *See* abstract of title.

**Tobin's Q** A theory stating that the level of *investment is determined by the ratio of the stock-market valuation of the physical capital of firms and the current *replacement cost of those items. Investment will rise if the former is greater than the latter and vice versa. First suggested by James Tobin (1918– ), the theory provides an improvement on both the *accelerator theory (which is theoretically weak) and the *user-cost-of-capital theory (which is weak in practice).

**token strike** *See* strike.

**tokkin** The Japanese equivalent of a unit-linked pension fund.

**Tokyo round** The sixth bargaining session held between contracting parties to *General Agreement on Tariffs and Trade, held in Tokyo. The round lasted from 1973 to 1979 and sought to build on the successes in multilateral tariff reduction in the previous Kennedy round. It continued to make progress on tariff reduction while also tackling the problems of **non-tariff barriers** (**NTB**s), which were a common way of replacing tariffs. Various codes of conduct on NTBs were agreed, including: government procurement, import valuation, and technical standards. The Tokyo round was followed by the Uruguay round, which began in 1986.

**ton** The formerly widely used **long ton** of 2240 pounds has now been largely replaced by the **metric ton** or **tonne** of 1000 kilograms. In the USA the **short ton** of 2000 lbs is still in use (consisting of 20 short hundredweights of 100 lbs each). The **shipping ton** is used for light cargoes and consists of 100 cubic feet of space; for heavy cargoes the shipping ton is taken either as the long ton or the tonne. *See also* tonnage.

**tone** The sentiment of a market. If a stock exchange or commodity market has a firm or strong tone, prices are tending upwards, whereas if the tone is weak, nervous, unsettled, etc., prices are tending to fall.

**tonnage** A measure of the volume or cargo-carrying capacity of a ship, measured in various ways. The **gross register tonnage (GRT)** is the volume of a ship below the upper deck measured in shipping *tons of 100 cubic feet. The **net register tonnage (NRT)** is the GRT less the space occupied by engines, fuel, stores, and accommodation for crew and passengers. Harbour and port dues are usually based on the NRT and dry dock dues are usually based on the GRT. The **deadweight tonnage** is the number of long tons or metric tons of cargo, stores, and fuel that a ship can carry. The **lightweight tonnage** is the weight of the ship in long tons or metric tons, without any cargo, fuel, or stores. The **displacement tonnage** is the weight in tons of the water displaced by a ship, taking one displacement ton as the weight of 35 cubic feet of water.

**tonne** *See* ton.

**tontine** *See* last-survivor policy.

**top-hat scheme** A pension plan for a senior executive of a company. *See* executive pension plan.

**TOPIC** Abbreviation for Teletext Output Price Information Computer. This computerized communication system provides brokers and market makers on the *London Stock Exchange with information about share price movements and bargains as they are transacted. Input is from *SEAQ.

**top management** *See* management.

**top slicing** A method of assessing the taxable gain on a life-assurance policy. The proceeds of the policy plus all capital withdrawals, less the premiums paid, are divided by the number of years for which the policy has been in force. This amount is added to any other income for the year in which the chargeable event occurred; if this places the taxpayer in a higher tax band, the whole gain is charged at the appropriate marginal rate, i.e. the tax rate in that band less the basic rate of tax. If the sum does not exceed the basic rate tax, no further tax is due.

**top up** To increase the benefits due under an existing insurance scheme, especially to increase the provision for a pension when a salary increase enables the insured to pay increased premiums.

**tort** A civil wrong other than one relating to a contract. The law of tort is concerned with providing *damages for personal injury and damage to property resulting from negligence. It is also concerned with protecting against defamation of character and preserving personal freedom, enjoyment of property, and commercial interests. A person who commits a tort is called a **tortfeasor**.

**total-absorption costing** A method of arriving at the cost of producing goods and services that allocates to them not only such *direct costs as labour and materials but also the other costs of the organization, such as general overheads and head-office costs. This method ensures, if the goods can be sold at the resulting price, that all costs will be covered. However, opportunities may be lost to make some contribution to overheads if they are not fully covered. *Compare* marginal costing.

**total income** The income of a taxpayer from all sources. This is often referred to as **statutory total income**, which consists of income from sources based on the income of the current year and income from other sources based on income of the preceding year. This artificial concept is used to calculate a person's income tax for a given year.

**total profits** The income of a company from all sources, including capital gains. This figure, after deduction of annual charges, is used to calculate the corporation tax payable by the company. Dividends from UK companies are not included in the figure.

**town clearing** *See* Association for Payment Clearing Services.

**TPI** Abbreviation for *tax and price index.

**trade** The activity of selling goods or services in order to make a profit. Profits from trade are taxed under income tax or corporation tax on income, rather than under capital-gains tax or corporation tax on capital gains. The concept of trade is difficult to define for the purposes of taxation. The Taxes Act merely states that it includes every trade, manufacture, or adventure or concern in the nature of trade. The Royal Commission on the Taxation of Profits and Income in 1955 identified six 'badges of trade': the subject matter of the transaction, length of ownership of the asset traded, frequency of similar transactions, work done on the asset to make it marketable, the circumstances responsible for its realization, and the motive to make a profit. However, none of these individually establishes a transaction as trading. Thus, a person who sells his car every year in order to buy a new one is not treated for taxation as a trader in cars.

**trade advertising** Advertising that is aimed at members of the distribution channel of a product or service rather than at the consumer. It is sometimes advantageous to draw the attention of the trade to a product either in addition to *consumer advertising or instead of it (usually because it is much cheaper). For example, book publishers rarely go to the great expense of consumer advertising (say, on TV) but usually advertise a new book in the trade journals seen by booksellers, so that the staff in bookshops are kept advised of forthcoming productions.

**trade agreement** A commercial treaty between two (**bilateral trade agreement**) or more (**multilateral trade agreement**) nations.

**trade association** An association of companies in the same trade, formed to represent them in negotiations with governments, unions, other trade associations, etc., and to keep members informed of new developments affecting the trade. Trade associations also frequently draw up contracts for their members to use and provide arbitration procedures to settle disputes between members.

**trade barrier** Any action by a government that restricts free trading between organizations within that country and the world outside. Tariffs, quotas, embargoes, sanctions, and restrictive regulations all present barriers to free trade.

**trade bill** A *bill of exchange used to pay for goods. They are usually either held until they mature, as they do not command a favourable discount rate compared to bank bills, or they are discounted by banks.

**trade bloc** A group of nations united by trade agreements between themselves. The EC and communist countries form trade blocs.

**trade creditor** One who is owed money by an organization for having provided goods or services to that organization.

**trade cycle** *See* business cycle.

**trade description** Any direct or indirect indication of certain characteristics of goods or of any part of them, such as their quantity, size, fitness for their purpose, time or place of origin, method of manufacture or processing, and price. Under the Trade Descriptions Act (1968), it is a criminal offence to apply a false trade description to goods or to supply or offer any goods to which a false description is applied.

**trade discount (trade terms)** A reduction on the recommended retail price of a product or service that is offered to distributors because they buy regularly in bulk. The difference between the retail price and the discounted price provides the retailer with his overheads and profit. The amount of the trade discount will often depend on the size of an order. Thus, a supermarket chain buying goods in very large quantities expects a higher trade discount than a corner shop buying in small quantities. It is also usual for members of the same trade to offer each other trade terms as a matter of course; for example, publishers are able to purchase each other's books at trade prices, i.e. the retail price of the book less the trade discount that would otherwise be given to the bookseller.

**trade dispute (industrial dispute)** A dispute between an employer and employees (or their trade union), usually about wages or conditions of working. Under the Trade Union and Labour Relations Act (1974), a person cannot be sued in *tort for an act that is committed in furtherance of a trade dispute on the grounds that it induces or threatens any breach of interference with the performance of a contract. Generally, such immunity extends only to the acts of employees against their own employer. Secondary industrial action (*see* picketing) may be unlawful when it is directed against an employer who is neither a party to the dispute nor the customer or supplier of the employer in dispute. Moreover, there is no immunity in respect of action taken to enforce a *closed-shop agreement.

The Act, as amended by the Employment Acts (1982 and 1988) and the

Trade Union Act (1984), gives similar immunity to trade unions for their acts committed in contemplation or furtherance of a trade dispute provided the act concerned is authorized by a majority vote in favour of the action in a secret ballot of the union's members. Under the Employment Act (1988) a trade union member can obtain a court order preventing industrial action being taken if it has not been authorized by a ballot.

Under the Employment Act (1982), when a trade union's immunity does not apply and it is ordered to pay damages (other than for causing personal injury or for breach of duty concerning the ownership, control, or use of property, or for *products liability under the Consumer Protection Act (1987)), the amount awarded may not exceed specified limits. These range from £10,000 for a union with under 5000 members to £250,000 for a union with 100 000 or more members.

**traded option** *See* option.

**trade gap** A deficit in a nation's *balance of payments.

**trade investment** Shares in or loans made to another company with a view to facilitating trade with the other company.

**trademark** A distinctive symbol that identifies particular products of a trader to the general public. The symbol may consist of a device, words, or a combination of these. A trader may register his trademark at the Register of Trade Marks, which is at the Patent Office (*see* patent). He then enjoys the exclusive right to use the trademark in connection with the goods for which it was registered. Any manufacturer, dealer, importer, or retailer may register a trademark. Registration is initially for seven years and is then renewable. The right to remain on the register may be lost if the trademark is not used or is misused. The owner of a trademark may assign it or, subject to the Registrar's approval, allow others to use it. If anyone uses a registered trademark without the owner's permission, or uses a mark that is likely to be confused with a registered trademark, the owner can sue for an *injunction and *damages or an *account of profits.

The owner of a trademark that is not registered in the Register of Trade Marks but is identified with particular goods through established use may bring an action for *passing off in the case of infringement.

**trade price** The price paid for goods to a wholesaler or manufacturer by a retailer. It is usually the recommended retail price less the *trade discount.

**trade reference** A reference concerning the creditworthiness of a trader given by another member of the same trade, usually to a supplier. If a firm wishes to purchase goods on credit from a supplier, the supplier will usually ask for a trade reference from another member of the same trade, in addition to a *banker's reference.

**trade secret** Knowledge of some process or product belonging to a business, disclosure of which would harm the business's interests. In the UK, the courts will generally grant injunctions to prohibit any threatened disclosure of trade secrets by employees, former employees, and others to whom the secrets have been disclosed in confidence. *See also* industrial espionage.

**trade terms** *See* trade discount.

**Trades Union Congress (TUC)** The central UK organization to which most *trade unions belong. It represents the trade-union movement in negotiations with the government and such bodies as the *Confederation of British Industry. It also acts as an arbitrator in disputes between member unions.

**trade union** An organization whose principal purposes include the regulation of relations between employees and employers or employers' associations. A **house** (*or* **company**) **union** is an association of employees all belonging to the same company, which will probably have no connection with the trade-union movement. An **industrial** (*or* **general**) **union** has members who work in the same industry; it is usually very large. A **craft union** is a small union of skilled workers. Unions' affairs are regulated by the Trade Union Act (1984) and the Employment Act (1988). These provide that secret ballots must be held for election of unions' executive committees and before any industrial action backed by the union is taken. The Employment Act (1988) gives a right to trade-union members not to be unjustifiably disciplined by their union (for example for failing to take industrial action). A member can apply to an *industrial tribunal for a declaration that he has been unjustifiably disciplined.

**trade-union official** An officer of a trade union (or of a branch or section of it) or a person elected or appointed in accordance with the union's rules to represent a group of its members. An employee who is a trade-union official and whose normal working week in his employment is of 16 hours or more is entitled to time off work, paid at his normal rate, for certain purposes. These must be official union duties concerning industrial relations between his employer and any associated employer and their employees or training approved by the *Trade Union Congress and relevant to his union duties.

**trading account** The part of a *profit and loss account in which the cost of goods sold is compared with the money raised by their sale in order to arrive at the gross profit.

**trading estate (industrial estate)** An area of land that has been granted planning permission for the development of factories, warehouses, offices, etc. First built in the 19th century (e.g. at Trafford Park, Manchester in 1896), they offer land away from residential areas that is well served by public transport and services; they also enable premises to be acquired at low rentals. Some are developed by private companies, which buy the land cheaply from the government in areas in which the government wishes to encourage development; others are run by the government itself.

**trading profit** The profit of an organization before deductions for such items as interest, directors' fees, auditors' remuneration, etc.

**trading stamps** Stamps bought from a trading-stamp company by a retailer and given to his customers when they purchase goods. The customer obtains stamps free in proportion to the goods purchased; when he has collected enough stamps he can exchange them for goods or for cash from the trading-stamp company. The issue of trading stamps is regulated by the Trading Stamp Act (1964). Each stamp must be clearly marked with a monetary value and the name of the issuing company. The retailer aims to cover the cost of the stamps by profits from the increased custom he hopes they will attract. Although trading stamps were widely used in the 1960s, their popularity has since dwindled.

**trading stock** *See* stock-in-trade.

**traditional option** *See* option.

**Training Agency** An organization that came into existence in 1989 to replace the Training Commission, which itself replaced the **Manpower Services Commission**. It is separate from the government but is responsible to the Secretary of State for Employment and, with government support, provides opportunities for training and retraining and for unemployed 16–17-year-olds to join its **Youth Training Scheme (YTS)**. This gives a two-year vocational training, 20 weeks of which is spent in a college or training centre, the rest being devoted to planned work experience; it usually leads to a vocational qualification. Entrants receive a tax-free weekly allowance. The Training Agency is also responsible for skills training and adult training. *See also* Training and Enterprise Council.

**Training and Enterprise Council (TEC)** One of a network of some 80 employer-led councils set up in 1989 under contracts with the *Training Agency to foster economic growth in England and Wales. Their specific objectives are promoting effective training by employers to meet local requirements, providing Youth Training Schemes (*see* Training Agency), developing employment training to enable unemployed people to train for local jobs, providing support for new and expanding businesses, and stimulating enterprise and economic growth by means of the *Enterprise Allowance Scheme. The office staff of an area's Training Agency is available to TECs. A separate framework for training and enterprise is available in Scotland.

**tramp ship** A cargo-carrying ship that does not have a planned itinerary (*compare* liner) but carries its cargo to any ports for which its charterers can find cargo. Tramps are available for *voyage charter or *time charter (the latter is now more common) and many are tankers. Some chartering of tramps is arranged through London's Baltic Exchange but much is now done direct through shipping agents and brokers.

**tranche** (French: slice) A part or instalment of a large sum of money. In the International Monetary Fund the first 25% of a loan is known as the **reserve** (formerly **gold**) **tranche**. In **tranche funding**, successive sums of money become available on a prearranged basis to a new company, often linked to the progress of the company and its ability to reach the targets set in its *business plan.

**transactions demand for money** The demand for money to finance current expenditure. A component of Maynard Keynes' theory of *liquidity preference, it is essentially similar to the *quantity theory of money. Keynes argued that transactions demand would be related positively to the level of activity in an economy but negatively to the rate of *interest. *Compare* precautionary demand for money; speculative demand for money.

**transactions velocity of circulation** *See* velocity of circulation.

**Transfer Accounting Lodgement for Investors** *See* TALISMAN.

**Transfer and Automated Registration of Uncertified Stock** *See* TAURUS.

**transfer deed** A deed that is used to transfer property from one person to another. On the *London Stock Exchange a **stock transfer form** has to be signed by the seller of registered securities to legalize the transaction. This will be dealt with by the computerized *TALISMAN system.

**transfer form** *See* transfer deed.

**transfer of value** The reduction in the value of a person's estate by a gratuitous transfer. Such transfers are subject to *inheritance tax in most cases if made in a given period before the date of the donor's death.

**transfer payment (transfer income)** A payment made or income received in which no goods or services are being paid for. Pensions, unemployment benefits, subsidies to farmers, etc., are transfer payments; they are excluded in calculating *gross national product.

**transferred charge call** A telephone call in which the caller asks the operator to ask the party called to pay for the call. In addition to the charge for the call, a transfer charge has also to be paid.

**transfer stamp** An impressed stamp on documents relating to the transfer of securities or land, as an acknowledgement that the *stamp duty has been paid.

**transhipment** The shipment of goods from one port to another with a change of ship at an intermediate port. Transhipments are usually made because there is no direct service between the ports of shipment and destination, or because it is necessary for some political reason, e.g. wishing to conceal the port of shipment from the buyers, whose government may have embargoed goods from the country of that port.

**transire** A two-part document in which the cargo loaded onto a coaster is detailed. It is supplied by the Customs at the port of shipment; one part has to be handed to the Customs at the port of destination to prove that the cargo comes from a home port rather than an overseas port.

**traveller's cheque** A cheque issued by a bank, building society, travel agency, credit-card company, etc., to a traveller to enable him to obtain cash in a foreign currency when he is abroad. They may be cashed at banks, exchange bureaus, restaurants, hotels, some shops, etc., abroad on proof of identity. The traveller has to sign the cheque twice, once in the presence of the issuer and again in the presence of the paying bank, agent, etc. Most traveller's cheques are covered against loss.

**treasurer** A person employed by a large organization, such as a company, charity, club, etc., to look after the funds of the organization and to make appropriate investments of surplus funds.

**treasure trove** Gold or silver found on land that has been hidden deliberately and that has no known owner. Treasure trove belongs to the Crown but the finder is recompensed for its value.

**Treasury** The UK government department responsible for the country's financial policies and management of the economy. The First Lord of the Treasury is the Prime Minister, but the Treasury is run by the Chancellor of the Exchequer.

**Treasury bill** A *bill of exchange issued by the Bank of England on the authority of the UK government that is repayable in three months. They are issued by tender each week to the *discount houses in units of £5000 to £100,000. They bear no interest, the yield being the difference between the purchase price and the redemption value. The US Treasury also issues Treasury bills.

**Treasury Investment Growth Receipts** *See* TIGR.

**Treasury stocks** *See* gilt-edged security.

**treaty 1.** Any formal agreement between nations. A **commercial treaty** relates to trade between the signatories. **2.** A transaction in which a sale is negotiated between the parties involved **(by private treaty)** rather than by auction. **3.** An agreement, usually in reinsurance, in which a reinsurer agrees automatically to accept risks from an insurer, either when a certain sum insured is exceeded or on the basis of a percentage of every risk accepted. With such a treaty an insurer has the confidence and capacity to accept larger risks than would otherwise be possible, as the necessary reinsurance is already arranged.

**trial balance** A listing of the balances on all the *accounts of an organization with debit balances in one column and credit balances in the other. If the processes of double-entry book-keeping have been accurate, the totals of each column should be the same. If they are not the same, checks must be carried out to find the discrepancy. The figures in the trial balance after some adjustments, e.g. for closing stocks, prepayments and accruals, depreciation, etc., are used to prepare the final accounts (profit and loss account and balance sheet).

**Trinity House** A UK corporation granted its first charter in 1514. It is responsible for lighthouses, aids to navigation maintained by local harbours, dealing with wrecks that present a hazard to navigation, and, until 1988, for pilotage, which was transferred to harbour authorities by the Pilotage Act (1987).

**true and fair view** Auditors of the published accounts of companies both in the UK and internationally are required by law to form an opinion as to whether the accounts they audit show a 'true and fair view' of the organization's affairs. 'True' implies that the accounts contain no false statements, 'fair' implies that the aggregate of facts shown by the 'true' statements is not misleading because, for instance, of omissions.

**trunking** Delivering goods by road over long distances, usually by means of a service that has local distribution depots, from which collections and deliveries are made.

**trust** An arrangement enabling property to be held by a person or persons (the *trustees) for the benefit of some other person or persons (the beneficiaries). The trustee is the legal owner of the property but the beneficiary has an equitable interest in it. A trust may be intentionally created or it may be imposed by law (e.g. if a trustee gives away trust property, the recipient will hold that property as constructive trustee for the beneficiary). Trusts are commonly used to provide for families and in commercial situations (e.g. pensions trusts).

**trust deed** The document creating and setting out the terms of a *trust. It will usually contain the names of the trustees, the identity of the beneficiaries, and the nature of the trust property, as well as the powers and duties of the trustees. Trusts of land must be declared in writing; trusts of other property need not be although there is often a trust deed to avoid uncertainty.

**trustee** A person who holds the legal title to property but who is not its beneficial owner. Usually there are two or more trustees of a *trust and for some trusts of land this is necessary. The trustee may not profit from his position; he must act for the benefit of the beneficiary, who may be regarded as the real owner of the property. Either an individual or a company may act as trustee. It is usual to provide for the remuneration of trustees in the trust deed, otherwise there is no right to payment. Trustees may be personally liable to beneficiaries for loss of trust property.

**trustee in bankruptcy** A person who administers a bankrupt's estate and realizes it for the benefit of the creditors (*see* bankruptcy).

**trustee investments** Investments in which trustees are authorized to invest trust property. In the UK the Trustees Investment Act (1961) regulates the investments of trust property that may be made by trustees. The Act applies unless excluded by a trust deed executed after the Act was passed. Half of the trust fund must be invested in **narrow-range securities**, largely specified in fixed-interest investments. The other half may be invested in **wider-range securities**, most importantly ordinary shares in companies quoted on the London Stock Exchange. In some cases, trustees must take advice before investing.

The Act considerably enlarged the range of trustee investments.

**Trustee Savings Bank** *See* T.S.B. Group plc.

**trust fund** A fund consisting of the assets belonging to a *trust, including money and property, that is held by the *trustees for the beneficiaries.

**TSA** Abbreviation for The Securities Association Ltd (*see* London Stock Exchange; Self-Regulatory Organization).

**T.S.B. Group plc** A *commercial bank and public limited company consisting of T.S.B. England & Wales plc, T.S.B. Northern Ireland plc, and T.S.B. Scotland plc, formed in 1986 after a change in the law allowed the individual Trustee Savings Banks of England & Wales, Scotland, Northern Ireland, and the Channel Islands to be merged and floated on the *London Stock Exchange. These four banks were themselves the product of many local savings banks, managed by boards of trustees, that grew up to assist the small saver in the 19th century.

**TT** Abbreviation for *telegraphic transfer.

**TUC** Abbreviation for *Trades Union Congress.

**turn** The difference between the price at which a *market maker will buy a security (*see* bid price) and the price at which he will sell it (*see* offer price), i.e. the market maker's profit.

**turnkey system** A computer system that is ready to start work on its assigned task as soon as it is installed. All the necessary programs and pieces of equipment are supplied with the system.

**turnover 1.** The total sales figure of an organization for a stated period. **2.** More generally, the rate at which at some asset is turned over, e.g. stock turnover is obtained by dividing the total sales figure by the value of the particular asset.

**turnover tax** A tax on the sales made by a business, i.e. on its turnover. *See* sales tax.

**TVR** Abbreviation for *television rating.

**two-tier tender offer** A tender offer in a takeover in which shareholders are offered a high initial offer for sufficient shares to give the bidder a controlling interest in the company, followed by an offer to acquire the remaining shares at a lower price. Bidders use this technique in order to provide an incentive to shareholders to accept the initial offer quickly. Such offers are usually for a combination of cash and shares in the bidder's own company.

**typeface** The style or design of a set of characters used in printing.

# U

**uberrima fides** (Latin: utmost good faith) The basis of all insurance contracts. *See* utmost good faith.

***UK Balance of Payments*** An annual publication of the Central Statistical Office. It is often known as the *Pink Book*.

***UK National Accounts*** A monthly publication of the Central Statistical Office. It is often known as the *Blue Book*. It provides figures for the *gross domestic product and separate accounts of production, income, and expenditure.

**ullage 1.** The amount by which the full capacity of a barrel or similar container exceeds the volume of the contents, as a result of evaporation or leakage. **2.** As defined by HM Customs and Excise, the actual contents of a barrel or similar container at the time of importation into the UK. The difference between this volume and the full capacity is called the **vacuity**.

**ultimo** *See* instant.

**ultra vires** (Latin: beyond the powers) Denoting an act of an official or corporation for which there is no authority. The powers of officials exercising administrative duties and of companies are limited by the instrument from which their powers are derived. If they act outside these powers, their action may be challenged in the courts. A company's powers are limited by the objects clause in its *memorandum of association. It it enters into an agreement outside these objects, the agreement may be unenforceable, although a third party may have a remedy under the Companies Act (1985) if it was dealing with the company in good faith (or there may be other equitable remedies).

**umbrella fund** An *offshore fund consisting of a *fund of funds that invests in other offshore funds.

**umpire** *See* arbitration.

**unabsorbed cost** The part of the overhead costs of a production process that are not covered by its revenue when the output falls below a specified level. Overheads are sometimes added to the *direct costs of a process and divided by the output to give a *unit cost. The output for this calculation is often set at such a level that the whole of the overheads are absorbed by the unit cost. If, in fact, the output falls below this level, the whole of the overheads will not be absorbed, the deficit being the unabsorbed cost.

**unappropriated profit** The part of an organization's profit that is neither allocated to a specific purpose nor paid out in *dividends.

**unbundling** The takeover of a large conglomerate with a view to retaining the core business as a going concern and selling off some or all of the subsidiary businesses to help pay for the takeover.

**uncalled capital** *See* reserve capital.

**uncertificated units** A small number of units purchased for an investor in a *unit trust by reinvestment of dividends. If the number of units is too small to warrant the issue of a certificate they are held on account for the investor and added to his total when he sells his holding.

**uncompensated demand function** *See* Marshallian demand function.

**unconditional bid** *See* takeover bid.

**unconfirmed letter of credit** *See* letter of credit.

**unconscionable bargain** *See* catching bargain.

**UNCTAD** Abbreviation for *United Nations Conference on Trade and Development.

**undated security** A *fixed-interest security that has no *redemption date. *See* Consols.

**undersubscription** *See* oversubscription.

**underwriter 1.** A person who examines a risk, decides whether or not it can be insured, and, if it can, works out the premium to be charged, usually on the basis of the frequency of past claims for similar risks. Underwriters are either employed by insurance companies or are members of *Lloyd's (*see also* syndicate). The name arises from the early days of marine insurance, when a merchant would, as a sideline, *write* his name *under* the amount and details of the risk he had agreed to cover on a slip of paper. *See also* Institute of London Underwriters. **2.** A financial institution, usually an *issuing house or *merchant bank, that guarantees to buy a proportion of any unsold shares when a *new issue is offered to the public. Underwriters usually work for a commission (usually 2%), and a number may combine together to buy all the unsold shares, provided that the minimum subscription stated in the *prospectus has been sold to the public.

**undifferentiated marketing** The marketing of a product aimed at the widest possible market. An early example was the Model T Ford, which was mass-produced, aimed at a very wide market, and only available in black. Ford now produce a wide range of models, with different engine capacities, colours, and prices aimed at attracting a wide variety of tastes and requirements.

**undischarged bankrupt** A person whose *bankruptcy has not been discharged. Such a person must not obtain credit (above £250) without first informing his creditor that he is an undischarged bankrupt, become a director of a company, or trade under another name. He may not hold office as a JP, MP, mayor, or councillor. If he is a peer, he may not sit in the House of Lords.

**undisclosed factoring** A form of *factoring in which the seller of goods does not wish to disclose the use he is making of a factor. In these circumstances, the factor buys the goods that have been sold (rather than the debt their sale incurred) and, as an *undisclosed principal, appoints the original seller to act as agent on his behalf to recover the debt. The factor assumes responsibility in the case of non-payment so that from the point of view of the seller, the factor offers the same service as in normal factoring.

**undisclosed principal** A person who buys or sells through an agent or broker and remains anonymous. The agent or broker must disclose when he is dealing on behalf of an undisclosed principal; if he does not do so he, himself, may be treated in law as the principal and in some circumstances the contract may be made void.

**undistributable reserves** *See* capital reserves.

**undistributed profit** Profit earned by an organization but not distributed to its shareholders by way of dividends. Such sums are available for later distribution but are frequently used by companies to finance their trade.

**undue influence** Unfair pressure exerted on a person so that he signs a contract that is not a true expression of his aims or requirements at the time, but is to the advantage of another party (either a party to the contract or a third party). Such a contract may be set aside by a court.

**unearned income** Income not derived from trades, professions or vocations, or from the emoluments of office. In the UK, until 1984, it was thought that as investment income was more permanent than earned income and did not depend on the labours of the taxpayer, it should be taxed more heavily than earned income. This was achieved by an investment-income surcharge, which was an extra 15% over the normal rate of income tax. In the UK both earned and unearned income are now taxed at the same rates.

**unemployment** An inability to find work although it is actively sought. Theoretically the unemployed are those members of the active labour force who are out of work, although in most cases statistics are based on the number of people claiming unemployment benefit. Unemployment has been a persistent problem in industrial economies during *recession over the last 200 years. J M Keynes' solution of government intervention, proposed in the 1930s, seemed successful in the post-war era; however, since then Keynesianism has lost ground both theoretically and in practice, largely as a result of the extent to which governments have seen the control of *inflation as their first priority and have begun to doubt their ability to control long-term unemployment (*see* Phillips curve). In the

1980s unemployment on a large scale has again become a problem in most western economies, with economists more divided than ever as to possible solutions. *See also* involuntary unemployment; NAIRU; natural rate of unemployment; seasonal unemployment.

**unfair dismissal** The dismissal of an employee that the employer cannot show to be fair. Under the Employment Protection (Consolidation) Act (1978) employees have the right not to be unfairly dismissed, provided they have served the required period of continuous employment (52 weeks for full-time employment) and are not over 65 or the normal retirement age for an employee in that particular job. An employee who considers he has been unfairly dismissed can apply within three months after the effective date of termination of his employment contract to an industrial tribunal for reinstatement, re-engagement, or compensation. The tribunal will make an award unless the employer shows that the principal reason for the dismissal was the employee's incapability, lack of qualifications, or conduct; redundancy; the fact that it would be illegal to continue employing him; or some other substantial reason. The employer must also show that he acted reasonably in dismissing the employee. The statutory protection does not apply to members of the police and armed forces and certain other employees.

**unfair prejudice** Conduct in the running of a company's affairs that may entitle members to a remedy under the Companies Act (1985). In small companies, the unfairness may consist of a failure by some members to honour a common understanding reached when the company was formed, e.g. that a member be allowed to assist in the management and be paid accordingly. The usual remedy sought is for the purchase of the member's shares at a fair price. *See also* minority protection.

**unfair trading** *See* consumer protection.

**unfavourable balance** A *balance of trade or *balance of payments deficit.

**unfranked income** Any type of income from investments that is not *franked investment income.

**UNIDO** Abbreviation for *United Nations Industrial Development Organization.

**uninsurable risk** *See* risk.

**unique selling proposition (USP)** A product benefit that can be regarded as unique and therefore can be used in advertising to differentiate it from the competition. The concept is not now as popular as it was as not every product can have a USP.

**unit cost** The cost of a unit of production. Examples include the cost of one refrigerator made by a refrigerator manufacturer or, in the case of a manufacturer who makes components, the cost of one component. In the case of a railway, it might be the cost of one passenger mile.

**United Nations Common Fund for Commodities (CFC)** *See* United Nations Conference on Trade and Development.

**United Nations Conference on Trade and Development (UNCTAD)** A permanent organization set up by the UN in 1964 to encourage international trade, particularly to help developing countries to finance their exports. Under its auspices the **UN Common Fund for Commodities (CFC)** was established in 1989 to provide finance for international commodity organizations to enable *buffer stocks to be maintained and to carry out research into the development of *commodity markets.

**United Nations Industrial Development Organization (UNIDO)**

An international organization that became a specialized agency of the UN in 1986, having developed from the Centre for Industrial Development, set up in 1961. With headquarters in Vienna, its aim is to promote the industrialization of developing countries, with special emphasis on the manufacturing sector. It provides planning policies and technical advice and assistance for third world countries.

**unitization** The process of changing an *investment trust into a *unit trust.

**unit labour costs** The total expenditure on labour per unit of output. A comparison between changes in *productivity and unit labour costs in different countries enables economists to make predictions about changes in their *competitiveness. For example, if unit labour costs and productivity both rise by the same amount in the UK, UK producers do not have to change the prices of their goods; if, however, Japanese unit labour costs rise by less than productivity, Japanese prices could fall, making them increasingly competitive.

**unit-linked policy** A life-assurance policy in which the benefits depend on the performance of a portfolio of shares. Each premium paid by the insured person is split: one part is used to provide life-assurance cover, while the balance (after the deduction of costs, expenses, etc.) is used to buy units in a *unit trust. In this way a small investor can benefit from investment in a *managed fund without making a large financial commitment. As they are linked to the value of shares, unit-linked policies can go up or down in value. Policyholders can surrender the policy at any time and the *surrender value is the selling price of the units purchased by the date of cancellation (less expenses). Until the 1970s the profits on *with-profits life-assurance policies were determined by the size of the bonus given at the discretion of the life-assurance company; this depended on their investment success and their profits in any particular year. Unit-linked policies, involving unitized portfolios, have produced higher benefits for policyholders than those that rely on with-profits bonuses. There has therefore been a widespread increase in **unit-linked savings plans**, both with and without life-assurance cover.

**unit of account 1.** A *function of money enabling its users to calculate the value of their transactions and to keep accounts (*compare* store of value). **2.** The standard unit of currency of a country. **3.** An artificial currency used only for accounting purposes; the EC employs such a unit for fixing its farm prices (*see* green currencies).

**unit pricing** The practice of showing the price of a single unit of a product in order that shoppers can compare the price against the cost of a multipack. A single bag of crisps in a supermarket might cost 16p. A larger bag containing six single bags could be offered at 90p, i.e. a unit price of 15p each. The shopper can therefore see at a glance that the larger purchase is the more economical.

**unit profit (*or* loss)** The profit (or loss) attributable to one unit of production. *See* unit cost.

**unit trust** A *trust formed to manage a portfolio of stock-exchange securities, in which small investors can buy units. This gives the small investor access to a diversified portfolio of securities, chosen and managed by professional fund managers, who seek either high capital gains or high yields, within the parameters of reasonable security. The trustees, usually a commercial bank, are the legal owners of the securities and responsible for ensuring that the managers keep to the terms laid down in the trust deed.

Prices of unit trusts are quoted daily, the difference between the bid and offer prices providing a margin for the management costs and the costs of buying and selling on the *London Stock Exchange. Basic-rate tax is deducted from the dividends paid by unit trusts and capital gains on the sale of a holding are subject to capital-gains tax, although transactions involved in creating the portfolio are free of capital-gains tax. UK unit trusts are authorized and controlled by the Department of Trade and Industry and most belong to the *Unit Trust Association. Many trusts are now available, specializing in various sectors of the market, both at home and abroad; there is also a wide spectrum of trusts catering for both those seeking growth and those seeking income. *See also* unit-linked policy.

**Unit Trust Association (UTA)** An association formed to agree standards of practice for the managers of unit trusts for the protection of unit-trust holders and to act as the representative body of **Managers of Authorized Unit Trusts** in dealings with the government and other authorities. It also coordinates control of commissions and charges and has a register of approved agents for selling unit trusts as intermediaries.

**Unit Trusts Ombudsman Scheme** *See* Financial Ombudsman.

**unlimited company** *See* company.

**unliquidated damages** *See* damages.

**unlisted securities** Securities (usually *equities) in companies that are not on an official stock-exchange list. They are therefore not required to satisfy the standards set for listing (*see* listed security). Unlisted securities are usually issued in relatively small companies and their shares usually carry a high degree of risk. In London, some unlisted securities are traded on the *unlisted securities market (USM) and the *third market, ensuring at least some degree of regulation. For full listing a company has to have a capital of at least £500,000 and 25% of the equity has to be available to the public. For unlisted securities the capital may be below this figure and only 10% of the equity need by available for purchase on the USM.

**unlisted securities market (USM)** A market established by the *London Stock Exchange in 1980 to trade in shares of small companies, not suitable for the *main market (although most USM companies ultimately aim for promotion). The USM is cheaper to join, and the *listing requirements are less stringent. There are currently under 400 companies traded on the USM, compared to over 2000 on the main market. *Compare* over-the-counter market; third market.

**unofficial strike** *See* strike.

**unquoted securities** Securities that are not dealt in on any stock exchange. On the London Stock Exchange there are rules for occasional unofficial trading in unquoted securities of small companies or new ventures. Unquoted securities are usually shown in balance sheets valued at cost or market value (if this can be ascertained).

**unsecured creditor** A person who is owed money by an organization but who has not arranged that in the event of non-payment specific assets would be available as a fund out of which he could be paid in priority to other creditors.

**unsecured debenture** A *debenture or loan stock in which no specific assets have been set aside as a fund out of which the debenture holders could be paid in priority to other creditors in the event of non-payment.

**unsocial hours** Working hours, other than normal office or factory hours, that prevent an employee from enjoying family life and the usual social activities. Shift workers and workers who have to work unsocial hours are often paid at higher rates than 9–5 workers.

**unvalued policy** An insurance policy for property that has a sum insured shown for each item although the insurers do not acknowledge that this figure is its actual value. As a result, if a claim is made the insured must provide proof of the value of the item lost, damaged, or stolen before a payment will be made.

**upper case** The capital letters of the alphabet, as compared to the small letters (*lower case).

**upward compatibility** *See* compatibility.

**Uruguay round** *See* General Agreement on Tariffs and Trade.

**usance 1.** The time allowed for the payment of short-term foreign *bills of exchange. It varies from country to country but is often 60 days. **2.** Formerly, the rate of interest on a loan.

**US Customary units** The units of measurement in use in the USA. They are based on *Imperial units but the US gallon is equal to 0.8327 Imperial gallons and the hundredweight is taken as 100 lbs, making the *ton 2000 lbs (instead of the Imperial 2240 lbs). The USA has fallen behind Europe in converting to metric units but intends to do so.

**user-cost-of-capital theory** The theory that the level of *investment is determined by the economic costs of the investment balanced against the potential returns. The costs are represented by the rate of interest that must be paid on borrowings and the price of investment goods. The returns are the flow of income from production. The calculation of costs and returns yields a *net present value. While more theoretically acceptable than such models as the *accelerator theory, this theory is difficult to test in practice or to use as a basis for predicting the level of investment in an economy. *See also* Tobin's Q.

**user-friendly** Denoting a computer system that is intended to be easy to use by people who are not computer specialists. The term was coined to distinguish such systems from older systems, in which the priority was efficient utilization of machine resources with few concessions to the convenience of the user. Commonly, a user-friendly system can be started up with a minimum of trouble and provides on-screen guidance to the user, usually in the form of *menus. If the user becomes completely confused, a help menu lists the action to be taken to correct all common mistakes. Some input devices, such as the *mouse, are considered easier to use than the normal keyboard, and user-friendly machines therefore make use of them.

**USM** Abbreviation for *unlisted securities market.

**USP** Abbreviation for *unique selling proposition.

**usury** An excessively high rate of interest.

**UTA** Abbreviation for *Unit Trust Association.

**utility** The subjective benefit derived by an individual from consumption of a good or service. In the *Classical school of economics, utility referred to the usefulness of an item, which was contrasted with its *value (*see* labour theory of value). It was the Marginalists who laid the basis for the *neoclassical school of economics, by observing that the value of a good is related to the subjective utility derived from that good, taking into account its scarcity. Thus, water has utility but little value, since it is generally not scarce. Both *general and *partial equilibrium analysis are based on utility and the assumption that consumers maximize utility. However, because of its subjectivity, it is difficult to measure utility and normally economists simply have to assume that consumers are rational and therefore do maximize utility. *See also* consumer preferences; consumer theory.

**utility function** A function that relates the different goods and services in an economy to the preferences of individuals (*see* consumer preference). By making certain assumptions, the utility function can be used to analyse the behaviour of *consumers and *markets. *See also* direct utility function; indirect utility function.

**utmost good faith (*uberrima fides*)** The fundamental principle of insurance practice, requiring that a person wishing to take out an insurance cover must provide all the information the insurer needs to calculate the correct premium for the risk involved. Nothing must be withheld from the insurers, even if they do not actually ask for the information on an application form. The principle is essential because an insurer usually has no knowledge of the facts involved in the risk they are being asked to cover; the only source of information is the person requiring the insurance. If an insured person is found to have withheld information or given false information, the insurer can treat the policy as void and the courts will support his refusal to pay claims.

# V

**vacuity** *See* ullage.

**valorization** The raising or stabilization of the value of a commodity or currency by artificial means, usually by a government. For example, if a government wishes to increase the price of a commodity that it exports it may attempt to decrease the supply of that commodity by encouraging producers to produce less, by stockpiling the commodity itself, or, in extreme cases, by destroying part of the production.

**value** The worth attached by someone to something. The determination of value is one of the main problems in economics; there are two principal approaches. The *Classical school and the Marxist school (*see* Marxist economics) regard value as an objective reality, which can be quantified as the input of *labour into goods or services. The *neoclassical school treats value as a subjective quality depending on the scarcity of something that is desired at a particular time. Most economists now subscribe to the neoclassical view.

**value added** The value added to goods or services by a step in the chain of original purchase, manufacture or other enhancement, and retail. For example, if a manufacturer acquires a partly made component, the value added will be the combination of labour and profit that increase the value of the part before it is sold. *See* VAT.

**value-added tax** *See* VAT.

**value analysis** An examination of every feature of a product to ensure that its cost is no greater than is necessary to carry out its functions. Value analysis can be applied to a new product idea at the design stage and also to existing products.

**valued policy** An insurance policy in which the value of the subject matter is agreed when the cover starts. As a result, the amount to be paid in the event of a total-loss claim is already decided and does not need to be negotiated.

**value for money audit** An audit of a government department, charity, or other non-profit-making organization to assess whether or not it is functioning efficiently and giving value for the money it spends.

**value received** Words that appear on a *bill of exchange to indicate that the bill is a means of paying for goods or services to the value of the bill. However, these words need not appear on a UK bill as everyone who has signed a UK bill is deemed to have been a party to it for value.

**variable costs** *See* overhead costs.

**variance** The amount by which an actual cost differs from a standard cost in the system of *standard costing. Variances may be analysed into their causes, e.g. price variances, quantity variances, efficiency variances, sales mix variances, etc.

**variation margins** The gains or losses on open contracts in future markets, calculated on the basis of the closing price at the end of each day. They are credited by the *clearing house to its members' accounts and by its members to their customers' accounts.

**VAT** Abbreviation for value-added tax. Although this is theoretically a tax on *value added, in practice it resembles a sales tax in that each trader adds the tax to his sale invoices and accounts for the tax he collects to Customs and Excise. However, he is permitted to deduct the amount of tax he himself has paid on the invoices issued to him for goods and services (but not for wages and salaries). Thus the tax is an *indirect tax, its burden being borne not by traders but by the ultimate consumers of their goods and services. The system is designed to avoid the cascade in which tax is paid on tax, as goods and services pass through long chains of activity; in VAT, all goods and services ultimately bear only the rate of tax applicable to the final sale to the consumer. VAT was introduced in the UK in 1973 to comply with its partners in the EEC. In the UK all goods and services bear VAT unless they are *zero-rated or exempt.

**VDU** Abbreviation for *visual display unit.

**velocity of circulation** The average number of times that a unit of money is used in a specified period, approximately equal to the total amount of money spent in that period divided by the total amount of money in circulation. The **income velocity of circulation** is the number of times that a particular unit of currency forms part of a person's income in a specified period. It is given by the ratio of the *gross national product to the amount of money in circulation. The **transactions velocity of circulation** is the number of times that a particular unit of currency is spent in a money transaction in a specified period, i.e. the ratio of the total amount of money spent in sales of goods or services to the amount of money in circulation.

**vendor** A person who sells goods or services.

**vendor placing** A type of *placing used as a means of acquiring another company or business. For example, if company X wishes to buy a business from company Y, it issues company X shares to company Y as payment with the prearranged agreement that these shares are then placed with investors in exchange for cash. Vendor placings have been popular with companies in recent years as a cheaper alternative to a *rights issue. *See also* bought deal.

**venture capital** *See* risk capital.

**verso** The left-hand page, normally even-numbered, of an open book, pamphlet, etc. *Compare* recto.

**vertical integration** *See* integration.

**vertical mobility** *See* mobility of labour.

**vested interest 1.** In law, an interest in property that is certain to come about rather than one dependent upon some event that may not happen. For example, a gift to "A for life and then to B" means that A's interest is **vested in possession**, because he has the property now. B's gift is also vested (but not in possession) because A will certainly die sometime and then B (or his estate if he is dead) will inherit the property. A gift to C "if he reaches the age of 30" is not vested, because he may die before reaching that age. An interest that is not vested is known as a **contingent interest. 2.** An involvement in the outcome of some business, scheme, transaction, etc., usually in anticipation of a personal gain.

**videoconferencing** A system available in the UK and internationally enabling groups of people at various locations to see and speak to each other, making it possible to hold meetings, seminars, etc., without the expense and time wasted in travel. Special studios are not required; information is available from British Telecom. In **Confravision,** public videoconferencing facilities, with studios in several major cities, are offered by British Telecom.

**Videotex** An information system in which information from a distant computer is displayed on a television screen. There are two types: in *Viewdata the information is sent to the television set using telephone lines or cables; in *Teletext, the information is broadcast as a part of the normal signal from a television station.

**Viewdata** A *Videotex system that sends information from a distant computer along telephone lines or cables. The information is displayed on a television set in the receiving home or office. Viewdata provides a vast amount of general information, such as weather reports, news, stock-exchange prices, and travel information, as well as access to specialist databases holding medical data for doctors, legal information for lawyers, and details of holidays for travel agents. The system also allows users to shop, book theatre tickets, and conduct their banking from home. *See* Prestel.

**vigilantibus non dormientibus jura subveniunt** (Latin: the law will not help those who sleep on their rights) *See* laches.

**visibles** Earnings from exports and payments for imports of goods, as opposed to services (such as banking and insurance). The *balance of trade is made up of visibles and is sometimes called the **visible balance.**

**visual display unit (VDU)** A computer output device consisting of a television-like cathode-ray screen. In home computers it is common to use a television set but most computers use special high-quality cathode-ray tubes. Often the input keyboard is attached to the screen. Specialized VDUs are used for such applications as air-traffic control.

**vital statistics** Statistics relating to a nation's or region's population, including its birth rate, death rate, marriage rate, etc.

**vocational training** Training in a trade or occupation. In the UK, the Business and Technician Education Council (BTEC) was set up in 1983 to develop a national system of vocational training. In 1991 a National Vocational Qualification (NVQ), consisting of four levels, will be introduced.

**volume 1.** The space occupied by something. **2.** A measure of the amount of trade that has taken place, usually in a specified period. On the *London Stock Exchange, for example, the number of shares traded in a day is called the volume and the value of these shares is called the **turnover**. In commodity markets, the daily volume is usually the number of lots traded in a day.

**voluntary arrangement** A procedure provided for by the Insolvency Act (1986), in which a company may come to an arrangement with its creditors to pay off its debts and to manage its affairs so that it resolves its financial difficulties. This arrangement may be proposed by the directors, an administrator acting under an *administration order, or a *liquidator. A qualified insolvency practitioner must be appointed to supervise the arrangement. He may be the administrator or liquidator, in which case he must call a meeting of the company and its creditors to consider the arrangement. The proposals may be modified or approved at this meeting but, once approved, they bind all those who had notice of the meeting. The court may make the necessary orders to bring the arrangement into effect. The arrangement may be challenged in court in the case of any irregularity. The aim of this legislation is to assist the company to solve its financial problems without the need for a winding-up (*see* liquidation).

**voluntary liquidation (voluntary winding-up)** *See* creditors' voluntary liquidation; members' voluntary liquidation.

**voluntary retailer** *See* symbol retailer.

**vostro account** A bank account held by a foreign bank with a UK bank, usually in sterling. *Compare* nostro account.

**voting shares** Shares in a company that entitle their owner to vote at the annual general meeting and any extraordinary meetings of the company. Shares that carry **voting rights** are usually *ordinary shares, rather than *A shares or *debentures. The company's articles of association will state which shares carry voting rights.

**voucher** A receipt for money or any document that supports an entry in a book of account.

**voyage charter** A charter of a ship or of cargo space for a fixed number of voyages rather than for a fixed period (*compare* time charter). The shipowner usually warrants that the ship will arrive at an agreed port on an agreed day, will be in seaworthy condition, and will be properly equipped. The charterer agrees to load and unload in the agreed number of *lay days.

# W

**WA** Abbreviation for with average. *See* average.

**wafer seal** A modern form of seal used on such documents as deeds. It is usually a small red disc stuck onto the document to represent the seal. Almost any form of seal agreeable to all the parties concerned is now acceptable in place of sealing wax.

**wage differential** The difference in earnings between workers with similar skills in different industries or between workers with different skills in the same industry. There may also be differentials between urban and non-urban wages or between regions (wages in the UK are highest in the south-east, where the cost of living is highest). The Equal Pay Act (1970) makes it illegal for a differential to exist between the pay of men and women doing like work in the same employment.

**wage drift** The difference between earnings and wage rates as a result of overtime pay and bonuses. In a period of *wage freeze, employers can circumvent the effects of the freeze by means of bonuses and overtime, thus defeating the anti-inflationary purpose of the freeze.

**wage freeze** An attempt by a government to counter inflation by fixing wages at their existing level for a specified period. For various reasons, including the *wage drift, this is no longer considered an effective policy. **Wage restraint** is a less rigidly applied attempt to control prices by obtaining union agreement to make only modest and essential claims for wage increases.

**wage restraint** *See* wage freeze.

**wagering contract** *See* gaming contract.

**wages council** A UK statutory body empowered by the Wages Council Act (1954) and the Wages Act (1986) to prescribe minimum rates of pay in a particular industry. Each council consists of representatives of employers and employees in the industry concerned and independent members. Councils have usually been established for industries in which employees' collective bargaining power is comparatively weak. The Wages Act (1986) abolished the power to create new wages councils.

**waiter** An attendant at the *London Stock Exchange and at *Lloyd's, who carries messages and papers, etc. The name goes back to the 17th-century London coffee houses from which these institutions grew, in which the waiters also performed these functions.

**waiver** The setting aside or non-enforcement of a right. This may be done deliberately or it may happen by the operation of law. For example, if a tenant is in breach of a covenant in his lease and his landlord demands rent in spite of knowing of the breach, the landlord may be held to have waived his right to terminate the lease.

**Wall St 1.** The street in New York in which the New York Stock Exchange is situated. **2.** The New York Stock Exchange itself. **3.** The financial institutions, collectively, of New York, including the stock exchange, banks, money markets, commodity markets, etc.

**Walras' Law** The law, formulated by Léon Walras (1834–1910), that the value of goods demanded in an economy is equal to the value of goods sold. This follows from the simple idea that for each individual, income equals expenditure (including borrowing and saving). This is an important step in proving the existence of a perfectly competitive *equilibrium, which is Pareto efficient (*see* Pareto optimality) in *general equilibrium theory.

**warehouse** Any building in which goods are stored; a **public warehouse** is one at or near a port in which goods are stored after being unloaded from a ship or before being loaded onto a ship (*see also* bonded warehouse). When goods are taken into a public warehouse a warehouse *warrant is issued, which must be produced before the goods can be removed.

**warehouse keeper** The person in charge of a warehouse. A *bonded warehouse also has a **warehouse officer**, employed by HM Customs and Excise, to inspect goods entering and leaving the warehouse.

**warehousing 1.** The storage of goods in a warehouse. **2.** Building up a holding of shares in a company prior to making a *takeover bid, by buying small lots of the shares and 'warehousing' them in the name of nominees. The purpose is for the bidder to remain anonymous and to avoid having to make the statutory declaration of interest. This practice is contrary to the *City Code on Takeovers and Mergers.

**war loan** A government stock issued during wartime; it has no redemption date, pays only 3½% interest, and stands at less than half its face value.

**warrant 1.** A security that offers the owner the right to subscribe for the *ordinary shares of a company at a fixed date, usually at a fixed price. Warrants are themselves bought and sold on *stock exchanges and are equivalent to stock options. Subscription prices usually exceed the market price, as the purchase of a warrant is a gamble that a company will prosper. They have proved increasingly popular in recent years as a company can issue them without including them in the balance sheet. **2.** A document that serves as proof that goods have been deposited in a public *warehouse. The document identifies specific goods and can be transferred by endorsement. Warrants are frequently used as security against a bank loan. Warehouse warrants for warehouses attached to a *wharf are known as **dock warrants** or **wharfinger's warrants.**

**warranty 1.** A statement made clearly in a contract (**express warranty**) or, if not stated clearly, understood between the parties to the contract (**implied warranty**). An unfulfilled warranty does not invalidate the contract (as it would in the case of an unfulfilled *condition) but could lead to the payment of damages. *See also* floating warranty. **2.** A condition in an insurance policy that confirms that something will or will not be done or that a certain situation exists or does not exist. If a warranty is breached, the insurer is entitled to refuse to pay claims, even if they are unconnected with the breach. For example, if a policy insuring the contents of a house has a warranty that certain locks are to be used on the doors and windows and these are found not to have been used, the insurers could decline to settle a claim for a burst pipe. In practice, however, this does not happen as insurers have agreed that they will only refuse to pay claims if the breach of warranty has affected the circumstances of the claim. **3.** A manufacturer's written promise to repair or replace a faulty product, usually free of charge, during a specified period subsequent to the date of purchase. This is often called a **guarantee.**

**wasting asset** An asset that has a finite life; for example, a lease may lose value throughout its life and become valueless when it terminates. It is also applied to such assets as plant and machinery, which wear out during their life and therefore lose value.

**watered stock** *See* stock watering.

**waybill 1.** *See* consignment note. **2.** *See* air waybill.

**wealth** The *value of the assets owned by an individual or group of individuals. Economics began as the study of wealth (e.g. Adam Smith's *The Wealth of Nations*) and how it changes during a given period. Keynesian theory tended to place a greater emphasis on *income as the object of study in *macroeconomics but it has since been accepted that income only tends to affect the behaviour of individuals as it affects their wealth.

**wealth effect (Pigou effect)** A change in the value of the assets held by an individual as a result of a change in the price level (inflation or deflation). J M Keynes argued that falling prices would reduce the level of *aggregate demand and therefore create *unemployment. However, if there is a wealth effect, falling prices will raise the value of money held and enhance aggregate demand, implying that the effect of money on the economy is neutral. *See* inside money; neutrality of money; outside money.

**wear and tear** Normal damage that occurs to assets in the course of their useful lives, causing them to depreciate in value (unless steps are taken to make good the damage).

**weather insurance** *See* pluvial insurance.

**weather working days** *Working days on which the weather permits work to be carried out. This often refers to the days allowed in a charter for loading or unloading a ship. *See* lay days.

**weighted average** An arithmetic average that takes into account the importance of the items making up the average. For example, if a person buys a commodity on three occasions, 100 tonnes at £70 per tonne, 300 tonnes at £80 per tonne, and 50 tonnes at £95 per tonne, his purchases total 450 tonnes; the simple average price would be $(70 + 80 + 95)/3 =$ £81.7. The weighted average, taking into account the amount purchased on each occasion, would be $[(100 \times 70) + (300 \times 80) + (50 \times 95)]/450$ = £79.4 per tonne.

**weighted ballot** A *ballot held if a new issue of shares has been oversubscribed, in which the allocation of shares is based on the number of shares applied for and biased towards either the smaller investor or the larger investor.

**weight note** A document produced by a seller of goods in which the *gross weights of the packages are listed. It also gives the marks and numbers of the packages and an indication of the average tare.

**weight or measurement** *See* freight.

**welfare economics** The branch of economics that is concerned with normative questions (*see* normative economics) relating to the well-being of society as a whole; in particular, it seeks to establish a ranking in terms of welfare between different allocations of resources. Unfortunately, the only widely agreed criterion for judging between alternatives is *Pareto optimality, which is very weak. Thus, although welfare economics concerns itself with such important questions as taxation, distribution, regulation of externalities, government policy, etc., it has been of limited practical use.

**wharf** The quay at which ships dock in a port, together with the adjacent warehouses in which goods are stored before loading or after unloading. The **wharfinger** is the firm or individual responsible for running the wharf.

**wharfage** The charge made by a wharfinger for the use of a wharf for loading or unloading cargo onto or from a ship.

**wharfinger's receipt** *See* dock receipt.

**wharfinger's warrant** *See* warrant.

**white-collar worker** Any non-manual office worker, including clerical, administrative, managerial, and professional personnel. *Compare* blue-collar worker.

**white goods** A class of consumer durables that includes washing machines, dishwashers, refrigerators, tumble-dryers, deep-freezers, and cookers; they are so named because they are usually finished in white enamel paint, although browns and other colours are now being used to make them fit less starkly into modern kitchens. *Compare* brown goods.

**white knight** A person or firm that makes a welcome *takeover bid for a company on improved terms to replace an unacceptable and unwelcome bid from a *black knight. If a company is the target for a takeover bid from a source of which it does not approve or on terms that it does not find attractive, it will often seek a white knight, whom it sees as a more suitable owner for the company, in the hope that a more attractive bid will be made. *Compare* grey knight.

**whole (of) life policy** A life-assurance policy that pays a specified amount on the death of the life assured. Benefits are not made for any other reason and the cover continues until the death of the life assured, provided the premiums continue to be paid, either for life or until a specified date. They may be *with-profits or *unit-linked policies.

**wholesale price index** *See* producer price index.

**wholesaler** A *distributor that sells goods in large quantities, usually to other distributors. Typically, a wholesaler buys and stores large quantities of several producers' goods and breaks into the bulk deliveries to supply *retailers with smaller amounts assembled and sorted to order.

**wider-range securities** *See* trustee investments.

**wife's earned-income relief** A personal allowance given before April 1990 to the husbands of married women who work. The allowance is equal to the amount of the wife's earned income or the amount of the single person's allowance, whichever is the lesser. This arrangement ceased with the *separate taxation of a wife's earnings. *See also* unearned income.

**wife's earnings** *See* separate taxation of a wife's earnings.

**wildcat strike** *See* strike.

**will** A document giving directions as to the disposal of a person's property after death. It has no effect until death and may be altered as many times as the person (the *testator) wishes. To be binding, it must be executed in accordance with statutory formalities. It must be in writing, signed by the testator or at his direction and in his presence. It must appear that the signature was intended to give effect to the will (usually it is signed at the end, close to the last words dealing with the property). The will must be witnessed by two persons, who must also sign the will. The witnesses must not be beneficiaries.

**Winchester disk** A type of *hard disk for computers in which the disks are permanently sealed, together with the read/write heads, in an airtight container to keep dust out. Introduced by IBM in 1973, this design is now used by many manufacturers, especially for microcomputers.

**windbill** *See* accommodation bill.

**winding-up** *See* liquidation.

**windmill** *See* accommodation bill.

**window** A rectangular area of a computer output screen that acts as a small screen within a screen. Typically, two or more windows are displayed at once. For example, a word-processor may show two different parts of a document in separate windows; or one window may be set aside for a special purpose, such as displaying calculations, while the rest of the screen is devoted to the main task. Some modern *operating systems, especially on microcomputers, use windows to facilitate *multitasking.

**window dressing** Any practice that attempts to make a situation look better than it really is. It has been used by accountants to improve the look of balance sheets, although the practice is now deprecated. For example, banks used to call in their short-term loans and delay making payments at the end of their financial years, in order to show spuriously high cash balances.

**WIP** Abbreviation for *work in progress.

**with average (WA)** *See* average.

**withholding tax** Tax deducted at source from *dividends or other income paid to non-residents of a country. If there is a *double-taxation agreement between the country in which the income is paid and the country in which the recipient is resident, the tax can be reclaimed.

**without prejudice** Words used as a heading to a document or letter to indicate that what follows cannot be used in any way to harm an existing right or claim, cannot be taken as the signatory's last word, cannot bind the signatory in any way, and cannot be used as evidence in a court of law. For example, a solicitor may use these words when making an offer in a letter to settle a claim, implying that his client may decide to withdraw the offer. It may also be used to indicate that although agreement may be reached on the terms set out in the document on this occasion, the signatory is not bound to settle similar disputes on the same terms.

**without recourse (sans recours)** Words that appear on a *bill of exchange to indicate that the holder has no recourse to the person from whom he bought it, if it is not paid. It may be written on the face of the bill or as an endorsement. If these words do not appear on the bill, the holder does have recourse to the drawer or endorser if the bill is dishonoured at maturity.

**with-profits policy** A life-assurance policy that has additional amounts added to the sum assured, or paid separately as cash bonuses, as a result of a surplus or profit made on the investment of the fund or funds of the life-assurance office. *Compare* unit-linked policy.

**word** In computing, a group of *binary digits (bits) considered as a unit and stored together in the computer memory. Usually, a word is 16 or 32 bits (2 or 4 *bytes) long.

**word-of-mouth advertising** The process in which the purchaser of a product or service tells his friends, neighbours, and associates about its virtues, especially when this happens in advance of media advertising.

**wordprocessor (WP)** A computerized text-processing system consisting of a computer unit with a typewriter keyboard, display screen, printer, and disk-memory unit. Words typed on the keyboard are displayed on the screen. Any mistakes can be corrected immediately and, when finalized, a document can be output on the printer. The document can also be stored on the disk unit for future reuse. There are normally extensive facilities for editing text: words or blocks of text can be moved or deleted; margins and line spacings can be adjusted; and many wordprocessors have a *spelling-check routine. Some can automatically incorporate names and addresses from a separate file into a letter – a technique known as **mailmerge**. Wordprocessors may be specialized systems; however, they are increasingly general-purpose *microcomputers using wordprocessing software. *See also* desktop publishing.

**wordwrap** A facility available on most *wordprocessors that causes a word at the end of a typed line to move automatically to the next line if it cannot fit within the maximum line length.

**workers' participation** *See* industrial democracy.

**work-in** *See* sit-down strike.

**working assets** *See* working capital.

**working capital** The part of the capital of a company that is employed in its day-to-day trading operations. It consists of *current assets (mainly trading stock, debtors, and cash) less current liabilities (mainly trade creditors). In the normal trade cycle – the supply of goods by suppliers, the sale of stock to debtors, payments of debts in cash, and the use of cash to pay suppliers – the working capital is the aggregate of the net assets involved, sometimes called the **working assets**.

**working days** In commerce, the days of the week excluding Sunday and, in some cases, Saturday (depending on the custom of the trade). Public holidays (*see* Bank Holidays) are also excluded from working days. In shipping and some other trades it is usual to distinguish *weather working days. *See also* lay days.

**working director** *See* executive director.

**work in progress (WIP)** Partly manufactured goods or partly completed contracts. For accounting purposes, work in progress is normally valued at its cost by keeping records of the cost of the materials and labour put into it, together with some estimate of allocated overheads. In the case of long-term contracts, the figure might also include an element of profit. In the USA the usual term is **work in process**.

**work measurement** *See* work study.

**work study** A study of the ways of doing a particular job more efficiently and cheaply. It consists of a **method study** to compare existing methods with proposed methods and **work measurement** to establish the time required by a qualified worker to do the job to a specified standard.

**work-to-rule** A form of *industrial action in which employees pay a slavish obedience to employers' regulations in order to dislocate working procedures, without resorting to a *strike. A work-to-rule usually involves an *overtime ban.

**World Bank** The name by which the *International Bank for Reconstruction and Development combined with its affiliates, the *International Development Association and the *International Finance Corporation, is known.

**WP** Abbreviation for *wordprocessor.

**writ** An order issued by a court. A **writ of summons** is an order by which an action in the High Court is started. It commands the defendant to appear before the court to answer the claim made in the writ by the plaintiff. It is used in actions in *tort, claims alleging *fraud, and claims for *damages in respect of personal injuries, death, or infringement of *patent. A **writ of execution** is used to enforce a judgment; it is addressed to a court officer instructing him to carry out an act, such as collecting money or seizing property. A **writ of delivery** is a writ of execution directing a sheriff to seize goods and deliver them to the plaintiff or to obtain their value in money, according to an agreed assessment. If the defendant has no option to pay the assessed value, the writ is a **writ of specific delivery**.

**write** To cover an insurance risk, accepting liability, under an insurance contract as an *underwriter.

**write off 1.** To reduce the value of an asset to zero in a balance sheet. An expired lease, obsolete machinery, or an unfortunate investment would be written off. **2.** To reduce to zero a debt that cannot be collected (*see* bad debt). Such a loss will be shown in the *profit and loss account of an organization.

**written-down value** The value of an asset for accounting purposes after deducting amounts for depreciation or, in the case of tax computations, for capital allowances.

**wrongful dismissal** The termination of an employee's contract of employment in a manner that is not in accordance with that contract. Thus when an employee is dismissed without the notice to which he is entitled (in circumstances that do not justify summary dismissal) or when the employer prematurely terminates the employee's fixed-term contract, the employee is entitled to claim *damages in the courts for wrongful dismissal. The court's jurisdiction concerns only the parties' contractual rights and not their statutory rights under the employment protection legislation (*compare* unfair dismissal). The Employment Protection (Consolidation) Act (1978) empowers the Secretary of State for Employment to give industrial tribunals jurisdiction to hear claims of wrongful dismissal.

**wrongful trading** Trading during a period in which a company had no reasonable prospect of avoiding insolvent *liquidation. The liquidator of a company may petition the court for an order instructing a director of a company that has gone into insolvent liquidation to make a contribution to the company's assets. The court may order any contribution to be made that it thinks proper if the director knew, or ought to have known, of the company's situation. A director would be judged liable if a reasonably diligent person carrying out the same function in the company would have realized the situation: no intention to defraud need be shown. *See* fraudulent trading.

# X

**xd** Abbreviation for *ex dividend.

**xerography** A dry electrostatic process for making photocopies and printing computer output with a *laser printer. In a **plain-paper photocopier** an electrostatic image is formed on a selenium-coated plate or cylinder. This is dusted with a resinous toner, which sticks selectively to the charged areas, the image so formed being transferred to a sheet of paper, on which it is fixed by heating.

**X-inefficiency** The amount by which the *efficiency of a firm falls below the maximum possible. For example, in 1981 two completely identical Ford plants were compared, one in Germany and one in the UK: the German plant was found to produce 50% more cars with 22% fewer workers. The X-inefficiency of the British plant could have been due to any of several causes including: inertia, incomplete *information, insulation from free competition, poor labour relations, or managerial incompetence.

# Y

**Yankee bond** A *bond issued in the USA by a foreign borrower.

**yard 1.** An Imperial unit of length, now defined in terms of the metre; 1 yard = 0.9144 metre exactly. **2.** An informal name for one billion.

**yearling bond** A UK local authority bond that is redeemable one year after issue.

**yearly savings plans** The National Savings Yearly Plans were introduced by the Department for *National Savings in 1984. Savers make 12 monthly payments of between £20 and £200; this entitles them to a **Yearly Plan Certificate**, which earns maximum tax-free interest if held for four years, annual interest being compounded.

**year of assessment** *See* fiscal year.

**Yellow Book** The colloquial name for *Admission of Securities to Listing*, a book issued by the Council of the *London Stock Exchange that sets out the regulations for admission to the *Official List and the obligations of companies with *listed securities.

**Yellow Pages** *See* Classified Directories.

**yield 1.** The income from an investment expressed in various ways. The **nominal yield** of a fixed-interest security is the interest it pays, expressed as a percentage of its *par value. For example, a £100 stock quoted as paying 8% interest will yield £8 per annum for every £100 of stock held. However, the **current yield** (also called **running yield**, **earnings yield**, or **flat yield**) will depend on the market price of the stock. If the 8% £100 stock mentioned above was standing at a market price of £90, the current yield would be 100/90 × 8 = 8.9%. As interest rates rise, so the market value of fixed-interest stocks (not close to redemption) fall in order that they should give a competitive current yield. The capital gain (or loss) on redemption of a stock, which is normally redeemable at £100, can also be taken into account. This is called the **yield to redemption** (or the **redemption yield**). The redemption yield consists approximately of the current yield plus the capital gain (or loss) divided by the number of years to redemption. Thus, if the above stock had 9 years to run to redemption, its redemption yield would be about 8.9 + 10/9 = 10%. The yields of the various stocks on offer are usually listed in commercial papers as both current yields and redemption yields, based on the current market price. However, for an investor who actually owns stock, the yield will be calculated not on the market price but the price the investor paid for it. The annual yield on a fixed-interest stock can be stated exactly once it has been purchased. This is not the case with *equities, however, where neither the dividend yield (*see* dividend) nor the capital gain (or loss) can be forecast, reflecting the greater degree of risk attaching to investments in equities. Yields on fixed-interest securities and equities are normally quoted gross, i.e. before deduction of tax. **2.** The income obtained from a tax.

**yield curve** A curve on a graph in which the *yield of fixed-interest securities is plotted against the length of time they have to run to maturity. The yield curve usually slopes upwards, indicating that investors expect to receive a premium for holding securities that have a long time to run. However, when there are expectations of changes in interest rate, the slope of the yield curve may change (*see* term-structure of interest rates).

**yield gap** The difference between the average annual dividend yield on *equities and the average annual yield on long-dated *gilt-edged securities. Before the 1960s equity yields usually exceeded gilt yields, reflecting the greater degree of risk involved in an investment in equities. In the 1960s rising equity prices led to falling dividend yields causing a **reverse yield gap**. This was accepted as equities were seen to provide a better hedge against *inflation than fixed-interest securities; thus their greater risk element is compensated by the possibility of higher capital gains.

**yield to redemption** *See* yield.

**York–Antwerp rules** A set of rules drawn up in 1877 to facilitate the task of working out shares in general *average losses in *marine insurance.

**Youth Training Scheme (YTS)** *See* Training Agency.

**YTS** Abbreviation for Youth Training Scheme. *See* Training Agency.

**yuppie** A successful and ambitious young person, especially one from the world of business. The word is formed from *y*oung *u*pwardly-mobile *p*rofessional.

# Z

**Zebra** A discounted *zero-coupon bond, in which the accrued income is taxed annually rather than on redemption.

**zero-coupon bond** A type of *bond that offers no interest payments. In effect, the interest is paid at maturity in the redemption value of the bond. Investors sometimes prefer zero-coupon bonds as they may confer more favourable tax treatment. *See also* deep-discount bond; TIGR; Zebra.

**zero-rated** Denoting goods or services that are taxable for VAT, but with a tax rate of zero. A distinction is made between exempt goods and services (such as postage stamps) and zero-rated goods or services (such as food, books, and children's clothes) as exempt goods are outside the VAT system and there is no opportunity to offset input tax, whereas zero-rated goods and services are within the system, providing an opportunity for offsetting input tax.

**zero-sum game** A game or contest between two or more individuals in which the winnings of one equal the losses of the other, i.e. gains minus losses equal zero. An cxamplc of a zero-sum game is an argument over who pays a taxi fare – one person's gain is another's loss. The *prisoners' dilemma is a nonzero-sum game that demonstrates the benefits of cooperation. Both zero- and nonzero-sum games have a wide range of applications in economics.

**zip code** The US name for a *postcode. It consists of a five- or nine-digit code, the first five digits indicating the state and postal zone or post office, the last four digits the rural route, building, or other delivery location.